Proceedings of Previous Easter Schools in Agricultural Science, published by Butterworths, London

SOIL ZOOLOGY Edited by D.K. McL, Kevan (1955)
THE GROWTH OF LEAVES Edited by F.L. Milthorpe (1956)
CONTROL OF PLANT ENVIRONMENT Edited by J.P. Hudson (1957)
NUTRITION OF THE LEGUMES Edited by E.G. Hallsworth (1958)
THE MEASUREMENT OF GRASSLAND PRODUCTIVITY Edited by J.D. Ivins (1959)
DIGESTIVE PHYSIOLOGY AND NUTIRION OF THE RUMINANT Edited by D. Lewis (1960)
NUTRITION OF PIGS AND POULTRY Edited by J.T. Morgan and D. Lewis (1961)
ANTIBIOTICS IN AGRICULTURE Edited by M. Woodbine (1962)
THE GROWTH OF THE POTATO Edited by J.D. Ivins and F.L. Milthorpe (1963)
EXPERIMENTAL PEDOLOGY Edited by E.G. Hallsworth and D.V. Crawford (1964)
THE GROWTH OF CEREALS AND GRASSES Edited by F.L. Milthorpe and J.D. Ivins (1965)
REPRODUCTION IN THE FEMALE ANIMAL Edited by G.E. Lamming and E.C. Amoroso (1967)
GROWTH AND DEVELOPMENT OF MAMMALS Edited by G.A. Lodge and G.E. Lamming (1968)
ROOT GROWTH Edited by W.J. Whittington (1968)
PROTEINS AS HUMAN FOOD Edited by R.A. Lawrie (1970)
LACTATION Edited by I.R. Falconer (1971)
PIG PRODUCTION Edited by D.J.A. Cole (1972)
SEED ECOLOGY Edited by W. Heydecker (1973)
HEAT LOSS FROM ANIMALS AND MAN: ASSESSMENT AND CONTROL Edited by J.L. Monteith and L.E. Mount (1974)
MEAT Edited by D.J.A. Cole and R.A. Lawrie (1975)
PRINCIPLES OF CATTLE PRODUCTION Edited by Henry Swan and W.H. Broster (1976)
LIGHT AND PLANT DEVELOPMENT Edited by H. Smith (1976)
PLANT PROTEINS Edited by G. Norton (1977)
ANTIBIOTICS AND ANTIBIOSIS AGRICULTURE Edited by M. Woodbine (1977)
CONTROL OF OVULATION Edited by D.B. Crighton, N.B. Haynes, G.R. Foxcroft and G.E. Lamming (1978)
POLYSACCHARIDES IN FOOD Edited by J.M.V. Blanshard and J.R. Mitchell (1979)
SEED PRODUCTION Edited by P.D. Hebblethwaite (1980)
PROTEIN DEPOSITION IN ANIMALS Edited by P.J. Buttery and D.B. Lindsay (1981)
PHYSIOLOGICAL PROCESSESS LIMITING PLANT PRODUCTIVITY Edited by C. Johnson (1981)
ENVIRONMENTAL ASPECTS OF HOUSING FOR ANIMAL PRODUCTION Edited by J.A. Clark (1981)
EFFECTS OF GASEOUS AIR POLLUTION IN AGRICULTURE AND HORTICULTURE Edited by M.H. Unsworth and D.P. Ormrod (1982)
CHEMICAL MANIPULATION OF CROP GROWTH AND DEVELOPMENT Edited by J.S. McLaren (1982)
CONTROL OF PIG REPRODUCTION Edited by D.J.A. Cole and G.R. Foxcroft (1982)
SHEEP PRODUCTION Edited by W. Haresign (1983)
UPGRADING WASTE FOR FEEDS AND FOOD Edited by D.A. Ledward, A.J. Taylor and R.A. Lawrie (1983)
FATS IN ANIMAL NUTRITION Edited by J. Wiseman (1984)
IMMUNOLOGICAL ASPECTS OF REPRODUCTION IN MAMMALS Edited by D.B Crighton (1984)
ETHYLENE AND PLANT DEVELOPMENT Edited by J.A. Roberts and G.A. Tucker (1985)
THE PEA CROP Edited by P.D. Hebblethwaite, M.C. Heath and T.C.K. Dawkins (1985)
PLANT TISSUE CULTURE AND ITS AGRICULTURAL APPLICATIONS Edited by Lindsey A. Withers and P.G. Alderson (1986)
CONTROL AND MANIPULATION OF ANIMAL GROWTH Edited by P.J. Buttery, N.B. Haynes and D.B. Lindsay (1986)
COMPUTER APPLICATIONS IN AGRICULTURAL ENVIRONMENTS Edited by J.A. Clark, K. Gregson and R.A. Saffell (1986)
MANIPULATION OF FLOWERING Edited by J.G. Atherton (1987)
NUTRITION AND LACTATION IN THE DAIRY COW Edited by P.C. Garnsworthy (1988)
MANIPULATION OF FRUITING Edited by C.J. Wright (1989)
APPLICATIONS OF REMOTE SENSING IN AGRICULTURE Edited by M.D. Steven and J.A. Clark (1990)
GENETIC ENGINEERING OF CROP PLANTS Edited by G.W. Lycett and D. Grierson (1990)
FEEDSTUFF EVALUATION Edited by J. Wiseman and D.J.A. Cole (1990)
THE CONTROL OF FAT AND LEAN DEPOSITION Edited by K.N. Boorman, P.J. Buttery and D.B. Lindsay (1992)

Proceedings of Previous Easter Schools in Agricultural Science, published by Nottingham University Press

THE GLASSY STATE IN FOODS Edited by J.M.V. Blanshard and P.J. Lillford (1993)

Resource Capture by Crops

J.L. MONTEITH
Institute of Terrestrial Ecology, Edinburgh

R.K. SCOTT and M.H. UNSWORTH
University of Nottingham

NOTTINGHAM
University Press

Nottingham University Press
Sutton Bonington Campus,
Loughborough, Leicestershire LE12 5RD

NOTTINGHAM

First published 1994

British Library Cataloguing in Publication Data
A catalogue record for this book is available from the British Library

ISBN 1–897676–21–2

Typeset by The Midlands Book Typesetting Company, Loughborough, Leicestershire
Printed and bound by Redwood Books, Trowbridge, Wiltshire

PREFACE

Both in agricultural science and ecology, resource capture is proving to be a very useful integrating concept. Analyses based on the examination of the efficiency of radiant energy conversion and water use have enabled crop scientists the world over to direct their research so that genotypes and cultural practices can be dovetailed with particular environments in a quantitative way. The Green Revolution produced a massive increase in crop yield over the last thirty years through the introduction of new varieties with a commensurate increase in the use of water, fertilisers and pesticides. As we look to the future, the quest must be to maintain increases in productivity, but with the emphasis switching to more efficient use of resources, and the research to deliver recommendations will build on an analysis of resource capture. The approach is pivotal to the activities of widespread groups concerned with estimating how national food production will be influenced by changes in the global climate. While it cannot yet be claimed that much progress has been made in investigating how pests and diseases restrict the size and efficiency of capture systems, nor that breeders have yet exploited the principles developed by physiologists and ecologists, a common language is in place. This Easter School demonstrated how physicists, chemists and biologists from a range of backgrounds can now effectively converse.

The 52nd University of Nottingham Easter School was attended by 170 participants from 20 countries. It was pleasing that so many of the group were in the early stages of their careers and gratifying also that the conference was characterised by such a lively and friendly atmosphere both in the formal and informal sessions.

Caroline Reffin and Mavis Secker did a spendid job in organising the conference as a whole. Thanks are also due to Sayed Azam-Ali, John Corrie, Mike Gubbins, David Hodson and Simon McWilliam for their capable management of the poster sessions and visual aids.

The organisers wish to acknowledge generous financial support from the Royal Society, Unilever plc, BASF plc, ICI Agrochemicals and PP Systems.

CONTENTS

1

PRINCIPLES OF RESOURCE CAPTURE BY CROP STANDS

J.L. MONTEITH
Institute of Terrestrial Ecology, Bush Estate, Penicuik, Midlothian, UK

Definitions and distinctions

That plants possess a set of structures and mechanisms for 'capturing' resources from the environment has been recognized since botany began, but until recently this terminology appeared more often in popular descriptions of the genus *Triffid* than in respectable textbooks. Indeed one of my Sutton Bonington colleagues recently reminded me that it was only a few years since I had scribbled a derogatory comment on the draft of a manuscript in which he referred to roots 'capturing' water! At this Easter School, it is hoped that we shall be able to demonstrate that both 'resource' and 'capture' are useful terms because they are capable of precise definition and measurement and because the analysis of capture for individual resources provides a sound basis for exploring how plants integrate and perhaps optimize this type of activity. Our goal should be to attain a clearer understanding of both individual and integrated processes of capture as a basis for the more efficient management of resources in sustainable systems of production.

Within the context of this meeting, a 'resource' is a form of energy or of matter that plants need in order to grow and to reproduce. 'Capture' can be used in two senses. The term is usually applied to the process by which any organ, above or below ground, removes a resource from its environment to maintain metabolic processes. It can also describe the subsequent uptake of metabolites by organs as they grow.

An important characteristic of any environmental resource is the way in which it is replenished to make good, partly or completely, what plants remove. Carbon dioxide absorbed from the surface boundary layer during photosythesis is continually replenished by atmospheric turbulence, usually fast enough to maintain an almost constant concentration during the day. Water removed from soil by roots is replenished slowly by diffusion and more rapidly by infiltration following rain. Soil nutrients, too, are replaced by diffusion as well as by mineralization. Light is attenuated within a canopy of leaves but the supply of this resource is independent of the rate at which it is absorbed by foliage.

Capture of these four major resources – carbon dioxide, light, water and nutrients – is coupled in several ways. Clearly, the assimilation of carbon dioxide (in C_3 and C_4 species) cannot take place without the simultaneous absorption of light; and the uptake of nutrients depends on the existence of a water continuum between soil pores and plant tissue carrying ions by mass flow and by diffusion. More subtly, the *optimal* growth of a plant in a given environment depends on its ability to maintain a 'functional balance' between the growth of shoots and roots, as postulated by Brouwer (1962) and others and this point will be explored later.

Because the diffusion of water and nutrients in soil is a relatively slow process, roots must grow *towards* these resources in order to increase rates of uptake during the establishment of a crop stand and to maintain a constant rate during the grand period of growth. In contrast, a crop with a fixed leaf area could, in principle, maintain a constant rate of photosynthesis until the onset of senescence. In many environments, the ability of crop plants to grow fast and yield well must depend more on the architecture and dynamics of root systems than on the corresponding attributes of foliage. Crop physiologists have put (and continue to put) a disproportionate amount of effort into the measurement and modelling of plant organs above ground because they are so much easier to harvest, measure and describe than the components of a root system.

Evolution of analyses for capture and growth

TRADITIONAL GROWTH ANALYSIS

Quantitative studies of plant growth have been based on relations between resource capture and dry matter production for more than 70 years. The first of these relations was the 'compound interest' law derived by V.H. Blackman (1919) whose analysis contained two implicit assumptions: that the rate at which leaves captured light energy was proportional to total biomass (W); and that the rate of biomass accumulation ($\mathrm{d}W/\mathrm{d}t$) was proportional to the rate of capture. With R as the constant of proportionality,

$$\mathrm{d}W/\mathrm{d}t = R\,W \tag{1}$$

which yields on integration:

$$W(t) = W(0)\exp(Rt) \tag{2}$$

where biomass is $W(0)$ when $t = 0$.

Blackman regarded R as an 'efficiency index' for dry matter production and was perhaps the first of many analysts to believe that he had found a growth parameter useful for characterizing and comparing different species or individuals because it changed only slowly with time. However, subsequent measurements in the field showed that R changed rapidly with time as well as depending on species. (Evans (1972) gives a succinct account of this and later developments in growth analysis.) Briggs, Kidd and West (1920) therefore suggested that $(\mathrm{d}W/\mathrm{d}t)/W$ should be treated as a variable and referred to as a 'relative growth rate' (RGR). The term specific growth rate is now preferred by many workers for consistency with the convention in the physical sciences that a quantity referred to unit mass is 'specific'.

Briggs *et al.* also explored the relation between increase in biomass and leaf area, both specified as mean values per plant. They referred to the rate of increase of biomass per unit leaf area as a 'unit leaf rate' but the term 'net assimilation rate' (NAR) later introduced by Gregory (1926) has been more widely used. Neither label is logical but in 60 years of crop ecology, nothing better has emerged (or if it has, it has not survived!).

Attempts to analyse crop growth in terms of RGR and NAR focused attention on three other parameters specifying related properties of the light capturing system: the leaf area per unit plant weight or 'leaf area ratio' (LAR) (not dimensionless and therefore not a ratio); the leaf area per unit leaf weight or 'specific leaf area' (SLA); and the leaf weight per unit plant weight or 'leaf weight ratio' (LWR).

These quantities are related by the identities RGR = NAR × LAR and LAR = SLA × LWR, which are valid for all values of the three plant variables they incorporate: total standing dry weight and its rate of increase, leaf dry weight and leaf area. They have been used mainly to analyse ontogenetic trends in RGR and NAR in terms of SLA and LWR and to compare adaptation to a range of environments or success in competition. Evans (1972) remains the best source of information on how this type of growth analysis is applied to field measurements.

D.J. Watson was first to recognize that the analysis of agronomic trials needed a form of growth analysis based on resources captured per unit field area rather than per plant. He introduced two new quantities (Watson, 1947) that remain the cornerstone of attempts to relate crop productivity to environmental factors. The first was a 'leaf area index' (L), the plan area of leaves per unit area of ground beneath them. The second was the rate of increase of standing biomass W per unit of ground area, a quantity 'hitherto without a name' which he called 'crop growth rate' (C). Given these, the net assimilation rate for a stand could be found from C/L and the relative growth rate from C/W. Watson would have been astonished and delighted to find that leaf area index is now a major variable in computer models of crop growth (Chapter 8); in ecological models that predict how biomes may respond to climatic change (Woodward, 1987); and in the exploitation of satellite images to estimate the productivity of both crops and indigenous vegetation (Kanemasu *et al.*, 1990).

Watson also introduced the concept of 'leaf area duration', the value of L for a stand integrated with respect to time over the growing season. He demonstrated that this quantity was well correlated with biomass accumulated over the growing season by a range of species at the same site, despite the fact that NAR declined with time. The basis of this behaviour emerged later when Sibma (1968) showed that most arable crops grown under good management in the Netherlands achieved and maintained growth rates of 15–20 g m^{-2} d^{-1} during the grand period of growth. Similar consistency has been demonstrated for crops in Britain, including horticultural species (Greenwood *et al.* (1977)) and for cereals and legumes in Central India (Monteith, 1987). When C = NAR × L is effectively constant, NAR must decrease as L increases.

Light and water capture

During the International Biological Program in the early 1960s, Professor G.E. Blackman was instrumental in setting up international trials to apply Watson-style growth analysis to stands of crops grown at a number of sites so that the influence of climate on productivity could be explored. This scheme was adopted with particular enthusiasm by the Japanese IBP Committee whose report (Murata, 1975) is a unique source of information about the dependence of NAR and RGR on radiation and temperature for a range of arable crops.

In a contribution that links traditional and contemporary forms of growth analysis, Murata (p.173) suggested that NAR should be correlated, not with incoming solar radiation as commonly attempted but with the mean flux of radiation captured per unit leaf area which he calculated as follows.

Suppose that the horizontal irradiance below a stand with a leaf area index L is given by the usual Beer's law expression:

$$S(L) = S(0) \exp(-KL) \tag{3}$$

where $S(0)$ is the irradiance incident on the stand and K is an extinction coefficient assumed constant through the canopy. Then the mean flux of radiation captured per unit of leaf area index is:

$$(1/L)\ S(L)\ \mathrm{d}L = S(0)\ \{1 - \exp(-KL)\}/KL \tag{4}$$

Murata's correlation of NAR = C/L with the left-hand of Equation 4 was therefore equivalent to correlating C with $S(0)\{1 - \exp(-KL)\}$, i.e. with the radiation *intercepted* by the canopy.

Earlier, Shibles and Weber (1965) growing soybean and Williams, Loomis and Lepley (1965) growing maize demonstrated tight linear relations between standing biomass measured at intervals throughout the growing season and the corresponding accumulated total of solar irradiance. The validity of this relation and its limitations has been examined in much later work reviewed by Campbell and Van Evert (Chapter 3).

Warren Wilson (1967) suggested that the conservative relation between biomass accumulation and the time integral of light capture could provide the basis for a new type of growth analysis. Hunt, Warren Wilson and Hand (1990) recently went a stage further, suggesting that for *all* types of resource capture, dry matter can usefully be expressed as the product of three terms: the resource available per unit ground area; the amount of resource captured per unit of available resource ('efficiency of resource capture'); and the rate of dry matter production per unit of captured resource ('efficiency of resource utilization'). This type of identity is convenient and logical for a resource such as light, received as a flux, but dimensional problems arise when it is applied to resources such nutrients or water present as 'pools of substance'. In this case, the 'efficiency' of capture is, for example, 'the rate of uptake of mineral nutrient relative to the amount of mineral nutrient available' (Hunt *et al.*, 1990) but this quantity is not an efficiency: it has dimensions of $(\mathrm{TIME})^{-1}$ and cannot be constrained to have a maximum value of unity. The difficulty could be partly resolved by rewriting the identity defining the set of efficiencies with standing biomass in place of the rate of dry matter production.

Assessment

Traditional growth analysis was an attempt to understand how the environment of plants influenced their size and growth rate. It stimulated quantitative thinking that eventually pointed to two related weaknesses. The first was the emphasis on the size of organs rather than on their ability to capture resources. For leaves, fortunately, the relation between area and light capture can be expressed in very simple terms, as outlined briefly above, although the details of light transmission through canopies and absorption by leaves are extremely complex (Chapters 2 and 3). It has proved much more difficult to relate the size of root systems to their ability to capture water and nutrients (Chapters 4 and 5).

The fact that root systems are always out of sight and often out of mind was responsible for the second weakness of traditional growth analysis: undue emphasis on the area of leaves as the prime determinant of growth rate. It can be argued that deficiencies in the soil environment are always manifest by a reduction in leaf area and that the investment of biomass in the capture of water and nutrients is usually much less than the corresponding investment in light capture. Nevertheless, our understanding of crop–environment interactions will not make significant further progress until we have developed better ways of observing, quantifying and modelling the behaviour of root systems. The problem remains formidable, despite the development of instruments such as the mini-rhizotron and the neutron probe. Until our understanding of water and nutrient capture by roots is on a par with our ability to describe how leaves capture light and carbon dioxide, attempts to integrate shoot and root activity (Chapters 6 and 7) are bound to be lopsided and therefore weak in predictive power. The same imbalance prevails in our understanding of the extent to which growth is impaired when plants are attacked by pests and diseases or damaged by pollutants (Chapters 9–11).

In the rest of this introductory chapter, some elementary ways of specifying resource capture by roots and leaves operating as co-ordinated components of the same system are examined.

Resource capture by whole plants

In its earliest form, exemplified by Equation 1, growth analysis was built on the hypothesis that isolated plants capture resources at a rate proportional to mass, implying a constant relative (specific) growth rate and an exponential increase of biomass with time (Equation 2). This is an acceptable description of growth during the establishment of a crop stand when mutual shading by leaves or competion for resources between roots are both trivial. From observations on young plants, Brouwer (1962) inferred that a 'functional balance' must exist between shoots and roots to ensure that the capture of resources was matched above and below ground. There are at least two ways in such a balance can be quantified. Davidson (1969) and others suggested that the uptake of water or nutrients per unit root mass should be proportional to the uptake of carbon per unit shoot mass and this hypothesis has been used to explore the allocation of carbon and nitrogen between shoots and roots (see Johnson and Thornley (1987) for example). Testing the hypothesis experimentally has provoked controversy (e.g. Hunt (1977), Thornley (1977)).

GROWTH OF A STAND

When competion in a stand of young plants becomes significant, either above or below ground, *relative* growth rate begins to decrease but *absolute* growth rate increases until it becomes limited by the availability of resources below ground, above ground, or both. The transition from a *maximum* relative growth rate (R_m) to a *maximum* absolute growth rate (C_m) is an implicit property of the empirical relation:

$$C = C_m \{1 - \exp(-aW)\} \tag{5}$$

where W is biomass per unit ground area and C tends to a maximum value of C_m when aW is large. For a very small initial value of aW, C tends to $C_m aW$ so that the initial relative growth rate can be identified as:

$$R_m = C_m aW/W = aC_m \tag{6}$$

Goudriaan and Monteith (1990) showed that Equation 5 can be integrated to give the standing biomass at time t as an 'expolinear' function of time, viz:

$$W = (C_m/R_m)\ [\ln(1 + \{f_0/(1 - f_0)\} \exp(R_m t)] \tag{7}$$

where f_0 is the fraction of incident radiation intercepted by the stand at $t = 0$, i.e. at emergence (see also Chapter 6). As this quantity cannot readily be measured, it is more convenient to rewrite the equation in terms of a time constant defined as:

$$t_b = -\ln \{f_0/(1 - f_0)\}/\ R_m \tag{8}$$

and called 'lost time' because $C_m t_b$ is the loss of biomass at the beginning of the growing season when the leaf canopy does not completely shade the ground. (Goudriaan (p.100) gives an alternative expression for t_b in terms of the biomass when $t = 0$).

Eliminating f_0 from Equations 7 and 8 gives:

$$W = (C_m/R_m) \ln [\ 1 + \exp\{\ R_m(t - t_b)\}] \tag{9}$$

which can be differentiated to give the absolute growth rate as:

$$C = \mathrm{d}W/\mathrm{d}t = \frac{C_m \exp\{\ R_m(t - t_b)\}}{1 + \exp\{R_m(t - t_b)\}} \tag{10}$$

From Equations 9 and 10, the absolute growth rate at time t_b is $C_m/2$ and the relative growth rate is:

$$C/W = (C_m/2)/\{\ln 2\ (C_m/R_m)\} = 0.72\ R_m \tag{11}$$

Figure 1.1 illustrates some implications of Equation 9 with f_0 set at 0.005. Values of $C_m = 20$ g m^{-2} d^{-1} and $R_m = 0.2$ d^{-1} (full) lines are characteristic of a healthy

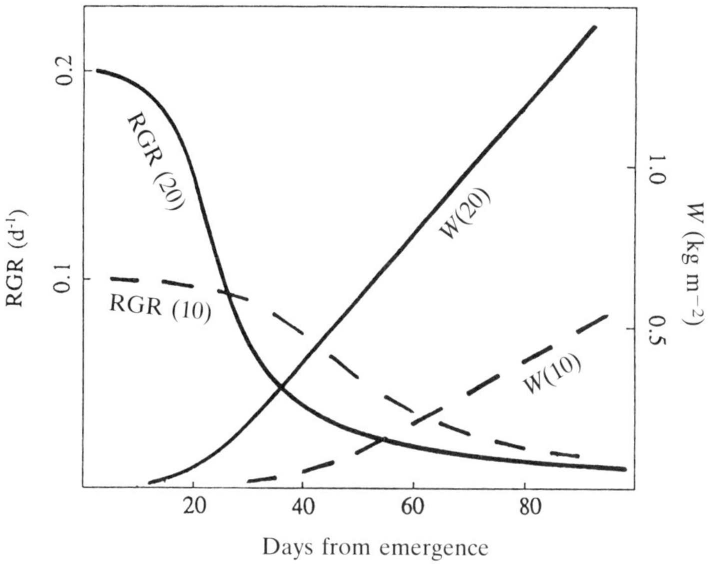

Figure 1.1 Change of biomass (W) and relative growth rate (RGR) with time estimated from Equation 7 for stands with maximum absolute growth rates of 10 g m^{-2} d^{-1} and 20 g m^{-2} d^{-1} (other details in text)

C_3 species in an environment where no resources are limiting. Halving both values (dashed lines) simulates the impact of resource shortage. Such a shortage leaves the term C_m/R_m in Equation 9 unchanged but increases 'lost time' from 26.5 to 53 days. In consequence, biomass 100 days after emergence is reduced by a factor of 3 when the maximum growth rate is reduced by a factor of 2.

Because expansion of the canopy is delayed by stress, the *relative* growth rate in the stressed stand exceeds that in the unstressed stand after 30 days. Other examples of this contrast are in Chapter 6. The relative growth rate of stands as distinct from isolated plants can therefore be a somewhat misleading index of resource capture.

Some quantitative aspects of resource capture during initial and final stages of growth will now be explored in more detail with the objective of comparing relative growth rates for leaves and roots when they are limited respectively by the absorption of light and the uptake of water.

INITIAL EXPONENTIAL GROWTH RATE

Capture of light by leaves

Equation 5 has the same form as the well known expression for the relation between crop growth rate and the capture of light referred to earlier, viz:

$$C = C_m \{ 1 - \exp(-KL)\} \tag{12}$$

When Equations 5 and 12 are both valid,

$$KL = R_m W / C_m \tag{13}$$

An expression for the relative growth rate of leaves can now be found by writing leaf biomass as $W_l = f_l W$ where f_l is the leaf weight ratio and $L = s_l W_l$ where s_l is the specific leaf area. Then from Equation 13 and by substituting $f_l s_l$ for L/W,

$$R_m = C_m KL/W = f_l C_m K s_l = f_l e S_a K s_l \tag{14}$$

where S_a is the absorption of radiation by the canopy when the interception of light is complete and e is the biomass produced per unit of absorbed radiation.

Capture of water by roots

The development of a comparable expression for R_m based on the uptake of water by roots is necessarily more speculative. Most models of water uptake by root systems are 'passive' in the sense that they deal with the diffusion of water towards a fixed root system, formalized as a set of cylinders. In practice, root systems are always 'dynamic', growing towards sources of water and nutrients. In most crop stands, the scope for horizontal movement is restricted by the proximity of neighbours and active capture is dominated by the descent of a 'root front'. When rainfall is slight or intermittent, the rate of descent can often be inferred from measurements of soil water content with a neutron probe. Strictly, probe measurements detect the arrival of an 'extraction front' but this usually follows close behind the root front and has virtually the same velocity (Gregory, McGowan and Biscoe, 1978).

During periods when no water is added to the soil profile, neutron probe measurements suggest that the rate of capture of water by roots at depth z and time t (referred to the first day of extraction) can be described by an exponential function:

$$\theta(z,t) = \theta(z,0) \exp(-t/\tau) \tag{15}$$

where τ is the time constant of the extraction process and $\theta(z,0)$ is the volume of water 'available' (i.e. extractable) per unit soil volume (Passioura, 1983). The value of τ is expected to depend on factors such as the density of the root system l_v (m of root per m^3 of soil) and the hydraulic properties of soil. Passioura suggested that τ could be assumed inversely proportional to l_v (i.e. $\tau = 1/(kl_v)$ where k is a constant accounting for soil hydraulic properties. This relation is difficult to establish in the field, because the error in measuring l_v for a crop stand is usually comparable with its variation within a soil profile. However, Robertson *et al.*, (1993a) were able to establish a loose relation between τ and l_v that appeared to be sensitive to evaporative demand. In the treatment that follows, l_v and k are are assumed independent of depth.

By differentiation, the rate of extraction of water at depth z is:

$$d\theta/dt = \{-\theta(z,0)/\tau\} \{\exp(-t/\tau)\tag{16}$$

At any time t much smaller than τ, the exponential term is close to unity and in a soil where the initial water content is constant with depth, the rate at which water is extracted to depth z is $\theta z/\tau$. If q is the biomass produced per unit *volume* of water transpired, the rate of (whole plant) biomass production will be:

$$dW/dt = q\ \theta\ z/\tau = l_v\ z\ k\ q\ \theta \qquad (17)$$

Root biomass to a depth z can be written as

$$W_r = l_v z/s_r \qquad (18)$$

where s_r is the specific root length (m per g), assumed, like l_v, to be constant throughout the profile. Then if f_r is the fraction of total biomass in the root system initially, Equations 17 and 18 give the initial relative growth rate of the roots as

$$R_m = f_r(dW/dt)/W_r = f_r\ k\ q\ \theta\ s_r \qquad (19)$$

Root:leaf ratio

In a young plant, balanced exponential growth is achieved when the relative growth rates of shoots and roots are equal. According to Equations 14 and 19, this condition is satisfied when the relative investment of biomass in roots and leaves is:

$$(f_r/f_l) = [K\ S_a/\ k\ \theta] \times [e/q] \times [s_l/\ s_r] \qquad (20)$$

The three terms in square brackets can be identified as follows:

- $K\ S_a/\ k\ \theta$: the *relative availability* of light and water. The quantity KS_a is the mean absorption of light energy by leaves exposed to a horizontal irradiance of S_a. Grouping k with the volume fraction of available water (θ) is speculative and based on the assumption that the maximum *rate* at which water becomes available to roots must depend on soil hydraulic properties as well as on water held per unit soil volume.
- s_l/s_r : the *relative resource capturing ability* of leaves and of roots per unit biomass.
- e/q : the *relative value* or equivalence of available resources for biomass production in terms of the mass of water extracted by roots per unit of solar radiation intercepted by foliage.

This is a more specific scheme than that proposed by Hunt, Warren Wilson and Hand (1990) but there are strong similarities. The first of the three terms is determined mainly by the environment and the second is a quantity found to be conservative in the absence of stress. An increase in the relative availability of resources between treatments or between seasons would therefore be expected to induce a balancing change *either* in the partitioning of biomass between roots and leaves (f_r/f_l); *or* by a change in the relative capturing ability per unit biomass (s_l/s_r), e.g. thinner leaves; *or* by a combination of both types of compensation.

Although Equation 20 was derived specifically for the very early stages of growth (and was, for example, subject to the condition that t should be much smaller than τ), qualitative inferences should remain valid beyond the time during which the relative growth rate is effectively constant. However, as the capture of resources becomes limited by their availability per unit field area rather than per plant, different relations must apply and these are now explored.

LIMITS TO ABSOLUTE GROWTH RATE

In principle, a maximum rate of growth eS is achieved by a canopy intercepting all incident light. In practice, the fraction intercepted by crop stands rarely exceeds 95–97% (Monteith and Elston, 1983) so that the effective maximum is a few per cent less than eS.

The maximum uptake of water by a root system whose extraction front is moving downwards with a velocity u in soil with a uniform water content θ is $u\ \theta$ (Monteith, 1986, 1988). Observations of the extraction front for seven species made by Angus *et al.* (1983) imply a range of u from about 1 to 2.5 cm d^{-1} and Robertson (1993 a,b) reported values in the range 2.7 – 3.3 cm d^{-1} for sorghum. The corresponding rate of biomass production is $q\ u\ \theta$.

As the rate of dry matter production based on light capture must be the same as the rate related to water capture, it follows that:

$$eS = q\ u\ \theta \qquad (21)$$

The available evidence suggests that q, the amount of dry matter accumulated per unit of water transpired, is more conservative than e when either water or light is limiting. Much less is known about the response of u to stress but the velocity presumably has a maximum value too. The maximum observed value of e (e_m), appears to be about 1.3 and 1.6 g MJ^{-1} for C_3 and C_4 species respectively (see Campbell and van Evert, p.35 this volume). From Equation 20, this value can be maintained only when $u\ \theta$ achieves a value of $e_m S/q$. Setting S at 15 MJ m^{-2} d^{-1}, e_m at 1.3 g MJ^{-1} and q at 5 g kg^{-1} gives $u\ \theta$ = 0.39 cm d^{-1}, compatible with values of u less than 2 cm d^{-1}, in soils for which θ exceeds 0.2 m^3 m^{-3}. Simple calculations of this kind can help to relate suboptimal values of e to soil water regimes and to estimate values of u when maximal values of e are achieved.

RESOURCE CAPTURE BY INDIVIDUAL ORGANS

When an organ such as a leaf or root starts to grow, the growth rate is determined, *inter alia*, by the size and activity of meristems and is expected to increase exponentially with time (Thornley, 1977). Eventually, however, when growth becomes limited by the supply of assimilate for the plant as a whole, the rate of growth attains a maximum value that may prevail until the onset of senescence or the operation of a biochemical switch initiating a new phase of development.

The transition from exponential to linear growth has rarely been reported in field trials because, in most environments, changes in weather and/or soil conditions usually obscure onogenetic trends. Exceptionally, a uniform environment exists during the first part of the post-rainy season in the semi-arid tropics, especially on deep clay soils holding enough water to sustain growth for two or three months. During this period, rainfall is usually trivial and radiation is rarely reduced by cloud.

A trial taking advantage of such an environment was conducted in 1988–89 at the International Crops Research Institute for the Semi-Arid Tropics (ICRISAT), Hyderabad, India. Treatments were imposed to explore the interaction of nitrogen (six levels of application) and water (two levels) on the growth and yield of sorghum. Irrespective of treatment, total dry matter increased exponentially with time initially, but roughly one month after emergence, growth achieved an almost constant rate maintained till anthesis. Dry matter allocated to leaves, stems and shoots behaved in the same way. It was therefore possible to estimate and compare, organ by organ, initial absolute and relative growth rates and the final absolute growth rate.

The initial *relative* growth rate (R_m) of the root system decreased with increasing nitrogen application whereas leaf R_m increased, a contrast consistent with the principle of functional balance. Similarly, at a given level of applied N, root R_m was larger in the unirrigated than in the irrigated treatment. Leaf R_m, on the other hand, showed no systematic response to water. It is unlikely that growth rate was limited by water initially even in the unirrigated treatment so the response of roots (but not leaves) appears to be related to the amount of water in the root environment rather than its rate of capture.

In leaves, the maximum *absolute* growth rate C_m responded strongly both to N and to water (at the higher levels of N) whereas root growth appeared to be almost independent of the availability of soil resources. Stems, in the dry treatment, responded moderately to N up to 60 kg ha^{-1} and the same response was observed in the irrigated treatment.

The analysis of growth for individual organs was therefore consistent with the form of the expolinear equation and the parameters of the equation could be correlated with the availability of soil resources.

Optimization of resource capture

The question of whether plants 'optimize' the capture of resources has been tackled in many ways over the past 20 years and has provoked argument about whether it is legitimate to attribute goal-seeking characteristics to plants. The word 'teleonomic' has been coined to describe the interpretation of processes in terms of *apparent* goal-seeking strategies such as the maximization of carbon uptake. However, the short-term maximization of dry matter production may not optimize the final biomass or yield of a crop if a resource becomes limiting before plants reach maturity.

Most optimization schemes are based on an explicit trade-off in terms of metabolism, allocation, or morphology. All three involve aspects of resource capture and aspects of allocation were discussed in the preceding section.

Optimization schemes themselves provide a means to optimize research efficiency. Far-reaching implications can be deduced from a relatively small number of initial conditions from which constant terms are removed by the process of differentiation to obtain the singular value of a parameter. However, it is rarely possible to optimize more than two or three processes simultaneously and we do not know whether those we have chosen because they happen to be amenable to algebraic treatment are given similar priority by species that survive from generation to generation without resort to calculus!

METABOLIC TRADE-OFF

Resource capture and metabolic consequences are often presented in economic terms such as cost:benefit ratios or the balance between revenue and expenditure. For example, Cowan and Farquhar (1977) investigated the optimal strategy for a leaf buying carbon at a rate A in exchange for water lost at a rate E. When stomatal aperture responded to an environmental change, A/E was maximized when water expenditure per unit marginal change of carbon assimilated remained constant. Givnish (1986) developed the analysis further by demonstrating that the value of the constant was proportional to the ratio of mesophyll to mesophyll plus stomatal resistance.

Friend (1991) described a model in which optimal stomatal aperture is achieved when the diffusion of carbon dioxide in the gaseous phase matches the rate at which it is reduced by the photochemical system. This rate is determined, *inter alia*, by the activity of the co-enzyme Rubisco which is assumed to depend on the water potential of leaf tissue. The model also predicts an optimum balance between the allocation of nitrogen to Rubisco and to chlorophyll.

The optimal allocation of nitrogen to roots and shoots was explored by Hilpert (1990) and more general models for partitioning carbon and nitrogen substrates were developed by Johnson and Thornley (1987).

MORPHOLOGICAL OR DIMENSIONAL TRADE-OFF

In this case, the application of dry matter to an organ is assumed fixed. For example, Fowkes and Landsberg (1981) explored the relation between the length, diameter and hydraulic resistance of individual roots of fixed biomass. They established an optimal relation between length and diameter that depended on a parameter linking the diameter and resistance of unit length of root. Although this analysis dealt only with roots of fixed length, they discussed implications for changes of soil water content and of inter-root competition as roots grew.

In a similar exercise, Robinson (1988) established an optimum relation between root length per unit soil volume and nutrient uptake,and demonstrated that the separation of neighbouring roots should be at least five times the root diameter to avoid competition. A comparable calculation could be attempted for leaves, based on the premise that a thin leaf with a large area gains by intercepting more light than a thick leaf with a smaller area but is likely to lose in terms of photosynthetic efficiency.

Conclusion

It is striking how quantitative analysis of crop growth did not begin to develop until well into the twentieth century, long after major developments in the physical sciences. In comparison with other fields of science, including many aspects of biology, we do not seem to have come very far since Blackman wrote his paper on compound interest.

Such relatively slow progress is not a consequence of the indolence or obtuseness of crop scientists! Rather it reflects the formidable complexity of plant responses to environmental factors in the field, with the consequence that it is impossible to observe more than a very limited number of responses in a single stand or to impose more than two or three treatments in a factorial design. Crop ecology is therefore likely to remain a good example of Tennyson's view that 'science moves but slowly, slowly, creeping on from point to point'! However, meetings of this kind with several major disciplines and research modes strongly represented are an encouraging sign that progress is accelerating as a consequence of improved synthesis between field experiments, work in controlled environments, and modelling. It may be some time before we discover how best to intercept and absorb all the information available from an extensive literature; but when that time comes, the subject will have reached its maximum *absolute* growth rate. For the time being, perhaps, we should strive to maximize the *relative* growth rate of our limited understanding!

References

Angus, J.F., Hasegawa, S., Hsiao, T.C., Liboon, S.P. and Zandstra, H.G. (1983) The water balance of post-monsoonal dryland crops. Journal of Agricultural Science, Cambridge, **101**, 699–710

Blackman, V.H.(1919) The compound interest law and plant growth. *Annals of Botany*, **33**, 353–360

Briggs, G.E., Kidd, F. and West, C.(1920) A quantitative analysis of plant growth. Part I. *Annals of Applied Biology*, **7**, 103–123

Brouwer, R. (1962) Distribution of dry matter in the plant. *Netherlands Journal of Agricultural Science*, **10**, 361–376

Cowan, I.R. and Farquhar, G.D. (1977) Stomatal function in relation to leaf metabolism and environment. *Symposium of the Society for Experimental Biology*, **31**, 471–505

Davidson, R.L. (1969) Effect of root/leaf temperature differentials on root/shoot ratios. *Annals of Botany*, **33**, 562–569

Evans, G.C. (1972) *The Quantitative Analysis of Plant Growth*. Oxford: Blackwell Scientific Publications

Fowkes, N.D. and Landsberg, J.J. (1981) Optimal root systems in terms of water uptake and movement. In *Mathematics and Plant Physiology*, pp. 109–125 (Edited by D.A. Rose and D.A. Charles-Edwards). London: Academic Press

Friend, A.D. (1991) Use of a model of photosynthesis and leaf microenvironment to predict stomatal conductance and leaf nitrogen partitioning. *Plant, Cell and Environment*, **14**, 895–905

Givnish, T.J. (1986) Optimal stomatal conductance, allocation of energy between leaves and roots and the marginal cost of transpiration. In *On the Econonmy of Plant Form and Function*, pp 171–213 Edited by T.J. Givnish. Cambridge: Cambridge University Press

Goudriaan, J. and Monteith, J.L. (1990) A mathematical function for crop growth. *Annals of Botany*, **66**, 695–701

Greenwood, D.J., Cleaver, T.J., Loquens, S.H.H., and Niendorf, K.B. (1977). Relationship between plant weight and growing period for vegetable crops in the United Kingdom. *Annals of Botany*, **41**, 987– 997

Gregory, F.G. (1926) The effect of climatic conditions on the growth of barley. *Annals of Botany*, **40**, 1–26

Gregory, P.J., McGowan, M. and Biscoe, P.V. (1978) Water relations of winter wheat. 2.Soil water relations. *Journal of Agricultural Science*, Cambridge, **91**, 103–116

Hilbert, D.W. (1990) Optimization of plant root:shoot ratios and internal nitrogen concentration. *Annals of Botany*, **66**, 91–99

Hunt, R. (1977) Significant relationships in the analysis of root–shoot equilibria. *Annals of Botany*, **41**, 657–659

Hunt, R., Warren Wilson, J. and Hand, D.W. (1990) Integrated analysis of resource capture and utilization. *Annals of Botany*, **65**, 643–648

Johnson, I.R. and Thornley, J.H.M. (1987) A model of shoot:root partitioning with optimal growth. *Annals of Botany*, **60**, 133–142

Kanemasu, E.T., Demetriades-Shah, T.H., Su, H. and Lang, A.R.G. (1990) Estimating grassland biomass using remotely sensed data. In *Remote Sensing In Agriculture*, (Edited by J.A. Clark). London: Butterworths

Monteith, J.L. (1986) How do crops manipulate water supply and demand? *Philosophical Transactions of the Royal Society, A*, **316**, 245–259

Monteith, J.L. (1987) Climate and productivity in dryland agriculture. *Indian Journal of Dryland Agricultural Research and Development*, **2**, 1–16

Monteith, J.L. (1988) Does transpiration limit the growth of vegetation or vice-versa? *Journal of Hydrology*, **100**, 57–68

Monteith, J.L. and Elston, J. (1983) Performance and productivity of foliage in the field. In *The Growth and Functioning of Leaves*, pp 499–518 Edited by J.E. Dale and F.L. Milthorpe. Cambridge: Cambridge University Press

Murata, Y. (1975). Crop productivity and solar energy utilization in various climates in Japan. *Japanese International Biological Programme*, Synthesis Volume 1. University of Tokyo Press

Passioura, J. (1983) Roots and drought resistance. *Agricultural Water Management*, **7**, 265–280

Robertson, M.J., Fukai, S., Ludlow M.M. and Hammer, G.L. (1993a) Water extraction by grain sorghum in a sub-humid environment. I Analysis of the water extraction pattern. *Field Crops Research*, **33**, 81–97

Robertson, M.J., Fukai, S., Ludlow M.M. and Hammer, G.L. (1993b) Water extraction by grain sorghum in a sub-humid environment. II Extraction in relation to root growth. *Field Crops Research*, **33**, 99–112

Robinson, D. (1988) Optimal relations between root length and nutrient inflow rates in plant root systems. *Journal of Theoretical Biology*, **135**, 359–370

Shibles, R.M. and C.R. Weber (1965) Interception of solar radiation and dry mattter production by various soybean planting patterns. *Crop Science*, **5**, 53–56

Sibma, L. (1968). Growth of closed green crop surfaces in the Netherlands. *Netherlands Journal of Agricultural Science*, **16**, 211–216

Thornley, J.(1977) Interpretation of shoot:root relationships. *Annals of Botany*, **41**, 461–464

Warren Wilson, J.W. (1967) Ecological data on dry-matter production by plants and plant communities. In *The Collection and Processing of Field Data*, pp 77–123 Edited by E.F. Bradley, and O.T. Denmead, New York: Interscience Publishers

Watson, D.J. (1947) Comparative physiological studies on the growth of field crops. I Variation in net assimilation rate and leaf area. *Annals of Botany*, **11**, 41–76

Williams, W.A., Loomis, R.S., and Lepley, C.R. (1965) Vegetative growth of corn as affected by population density. 1. Productivity in relation to interception of solar radiation. *Crop Science*, **5**, 211–215

Woodward, I. (1987) *Climate and Plant Distribution*. Cambridge: Cambridge University Press

2

RESOURCE CAPTURE BY SINGLE LEAVES

S.P. LONG*
Biosystems and Process Science Division, Department of Applied Science, Brookhaven National Laboratory, Upton, NY11973, USA

Introduction

The primary role of the leaf is the capture of light and carbon dioxide for photosynthesis and the provision of carbohydrate for the growth and reproduction of the plant. However, the fitness of an individual plant will not depend simply on the ability of individual leaves to capture these resources, but on the amount of CO_2 assimilated per unit of investment of the resources most limiting to survival in a given environment. Similarly, the contribution made by an individual leaf to the productivity of a crop stand will not only depend on the ability of the leaves to capture CO_2 and light, but also on the cost of their construction and maintenance.

While light is commonly recognized as a limiting resource for photosynthesis and crop production, doubts have been expressed as to whether CO_2 is a limiting resource for crop and terrestrial plants in general (Schimmel, 1990). Evidence that CO_2 is a limiting resource for crop production is indicated by studies of the effects of growth in elevated CO_2 concentration on dry matter formation. Among the 430 experiments reviewed by Kimball (1983), CO_2 enrichment to approximately double the present concentration increased dry matter production by 35%, on average, in C_3 crops. The selective importance of efficient CO_2 capture in evolution can be inferred from the affinity and specificity of Rubisco for CO_2. Rubisco extracted from a range of terrestrial C_3 plants shows a remarkable constancy in both its affinity for CO_2, as indicated by its Michaelis constant (K_c) and its specificity for CO_2 relative to O_2 (τ). This would be explained if naturally arising mutants with decreased τ are quickly eliminated from the population. Of course, τ may be constant simply because mutations affecting it are extremely rare. However, evidence that variation in τ does occur when selective pressures are removed is provided by C_4 plants. In C_4 plants, Rubisco occurs in the bundles sheath cells where CO_2 concentration is elevated by two to three orders of magnitude above atmospheric concentrations, so diminishing any selective advantage of a high τ and low K_c. C_4 plants in contrast to C_3 show considerable variability in K_c and τ, with specificities for CO_2 below, but not

*Permanent address: *Department of Biology, University of Essex, Colchester, CO4 3SQ, UK*

above, those found in C_3 species (Bowes, 1991). Both the conserved properties of Rubisco and studies of crop responses to elevated CO_2, suggest that CO_2 concentration is a major limitation to crop growth and that there have been strong evolutionary pressures to maximize CO_2 capture.

Capture of CO_2 and light in photosynthesis can be expressed in relation to other resources: water use efficiency has been expressed as the ratio of mol CO_2 absorbed per mol water transpired; nitrogen-use efficiency as mol CO_2 absorbed per mol N invested in the leaf; and energetic efficiency as chemical energy gained in photosynthesis per unit of chemical energy invested in the formation and maintenance of the leaf (see p. 30). In the context of returns relative to investment in the leaf, a range of strategies are indicated. Some features of leaves are obvious adaptations to maximizing resource capture, i.e. the typical flattened shape clearly maximizes light capture per unit of leaf dry mass. Properties such as the proportion of strengthening tissues, leaf longevity, and presence of spines have more subtle and less certain roles in determining resource capture and retention. This chapter examines how variation in leaf structure and metabolism may reflect compromises between the conflicting demands of CO_2 and light capture, and between maximizing photosynthesis while conserving water resources. Attention is also given to the conflict between maximizing light capture for photosynthesis and avoidance of photoinhibition. Only CO_2 and light capture are addressed here; some further roles of leaves in the capture and protection of plant resources are summarized in *Table 2.1*.

Table 2.1 MAJOR FUNCTIONS OF LEAVES IN CAPTURING AND PROTECTING PLANT RESOURCES

Function	*Mechanism*	*Benefit*
Light capture	Concentration of photosynthetic pigments and maximization of intercepting area	Increased photosynthesis
CO_2 capture	Stomata, concentration of primary carboxylase and acceptor for carboxylation, internal CO_2 concentrating mechanisms (C_4 and CAM metabolism)	Increased photosynthesis
Water capture	Channels on leaf surface and inclination of leaves toward the main stem, serve to direct captured precipitation toward stem base	Increases water capture by roots
Protection against predation and disease	Toxic secondary metabolites, physical deterrents, and decreased nitrogen content	Conserve resources
Protection against desiccation or frost damage	Insulation of buds by scale, developing and mature leaves	Conserve resources
Improved resource capture by roots	Shading by competitors	Increased capture of soil water and nutrients

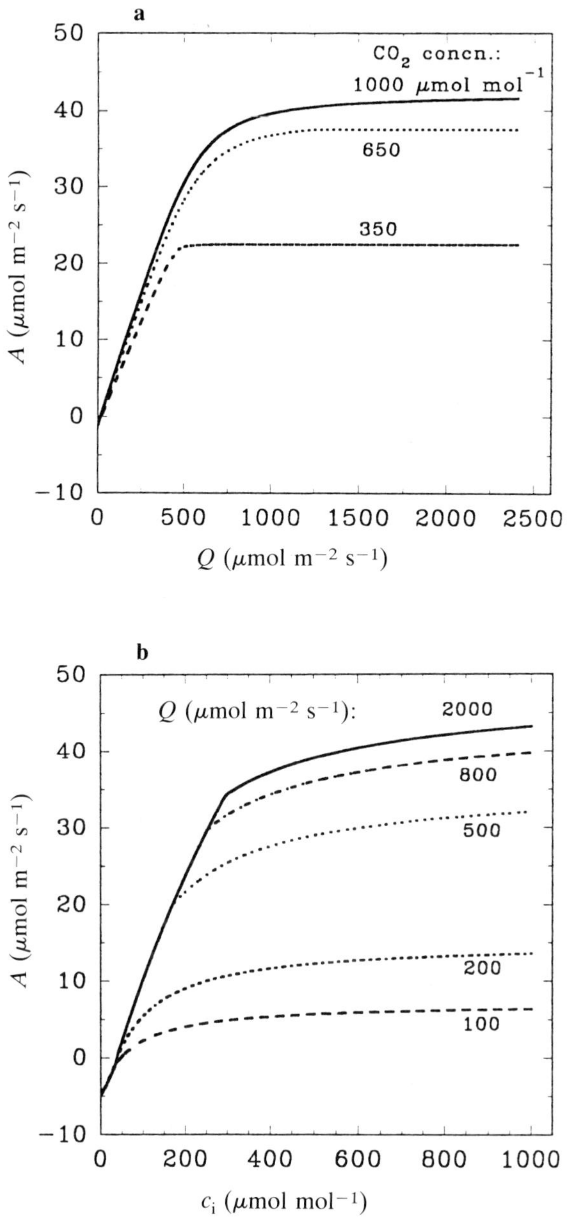

Figure 2.1 (**a**) The response of the photosynthetic rate of CO_2 uptake per unit leaf area (A) against incident photon flux (Q), at three atmospheric CO_2 concentrations. (**b**) The response of A to the intercellular CO_2 concentration (c_i) at five values of Q. Curves are predicted from the model of Farquhar, von Caemmerer and Berry (1980) as modified by Evans and Farquhar (1992), using average parameters for a C_3 leaf, and at a leaf temperature of 25°C (Long, 1991)

Light and CO_2, co-limiting resources for leaf photosynthesis

Figure 2.1 illustrates responses of photosynthesis to absorbed photon flux (Q) in different ambient CO_2 concentrations (c_a) and to c_a for different values of Q. Increase in CO_2 concentration increases the rate of CO_2 uptake, even at the light compensation point of photosynthesis. This results because the observed rate of CO_2 capture (A) in C_3 leaves is the net difference of two processes, photorespiration and photosynthesis, which compete for ribulose- 1:5–bisphosphate (RubP) at the primary carboxylase molecule; RubP carboxylase/oxygenase (Rubisco). If RubP is oxygenated, then adenosine triphosphate (ATP) and nicotinamide adenine dinucleotide phosphate (NADPH) generated on the photosynthetic membrane are utilized by the photosynthetic carbon oxidative cycle resulting in the evolution of CO_2 via photorespiration. Increased CO_2 concentration competitively inhibits oxygenation of RubP and hence photorespiration making a larger proportion of the ATP and NADPH available for CO_2 uptake into the photosynthetic C_3 reductive or Calvin cycle, so increasing quantum efficiency. Because a decrease in the ratio of photorespiration to photosynthesis increases the net efficiency of CO_2 uptake, an increase in A will occur regardless of whether other factors, including light, are limiting.

Inhibition of photorespiration by increased CO_2 concentration has been exploited by C_4 plants. In these plants, a C_4 photosynthetic cycle is imposed onto the Calvin cycle. This acts as a 'light-driven CO_2 pump', concentrating CO_2 in the bundle sheath cells which contain the leaf's complement of Calvin cycle enzymes, including Rubisco. The concentration of CO_2 within the bundle sheath competitively inhibits oxygenation of RubP and hence photorespiration. Because of the extended pathway of CO_2 assimilation, a minimum of two additional molecules of ATP are required for every molecule of CO_2 assimilated in C_4 photosynthesis. This however, is less than the additional cost imposed by photorespiration in C_3 photosynthesis in many environments (Long and Hutchin, 1991).

In canopies, light is continually changing with cloud, time of day and season. It might therefore be expected that the leaf will be well adapted for maximizing resource capture with variation in light quantity, quality and direction, but not to variation in atmospheric CO_2 concentration (c_a). Surprisingly, recent studies of leaves grown at elevated c_a show that in some instances leaves do adapt to variation in c_a by adjusting investment in different photosynthetic proteins, according to the changed balance between CO_2 supply and light flux (reviewed by Long and Drake, 1993). The following sections examine the internally controlled factors limiting CO_2 and light capture in the leaves of C_3 plants.

CO_2 capture

While the CO_2 concentration of the air around the leaf may be relatively constant during the day, the concentration within the leaf will vary because of the restriction of diffusion imposed by the stomata during photosynthesis. CO_2 capture by leaves is commonly described by the net rate of CO_2 uptake per unit of leaf area (A). Factors affecting A have been the subject of numerous reviews (e.g. Long, 1985). This account is limited to an outline of the mechanisms that underlie variation in

A. Within a C_3 leaf, CO_2 capture may be limited by (1) stomatal conductance; (2) the quantity of active Rubisco (carboxylation efficiency); and/or (3) the capacity for regeneration of RubP. How might these limitations be separated?

A MECHANISTIC MODEL OF CO_2 CAPTURE BY LEAVES

From a steady-state mathematical model of photosynthetic carbon metabolism, Farquhar, von Caemmerer and Berry (1980) demonstrated that the response of photosynthetic CO_2 uptake (A) to the intercellular CO_2 concentration (c_i) in a leaf consists of two phases. The initial and maximal slope of the response curve, termed the 'carboxylation efficiency', is determined by the quantity of active RubP carboxylase/oxygenase, in C_3 species. This initial phase is commonly linear up to a c_i value of about 250 μmol mol^{-1}. Beyond this inflexion, A is limited by capacity for the regeneration of RubP, which is commonly linked to the rate of whole chain electron transport (Farquhar, von Caemmerer and Berry, 1980). In the *absence* of photorespiration, photosynthesis is strictly limited by CO_2 concentration on the initial linear phase of the A/c_i response and strictly limited by photon flux beyond. Beyond the inflection of the curve, A shows no further increase with increase in

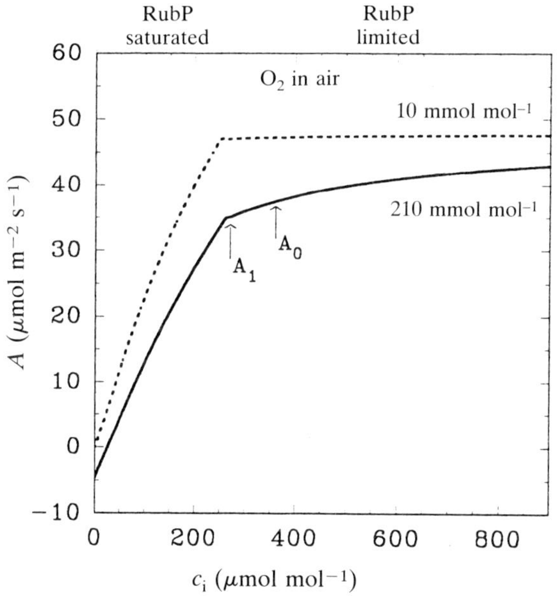

Figure 2.2 Lines indicate the dependence of the rate of CO_2 uptake per unit leaf area (A) on the intercellular CO_2 concentration (c_i), at two O_2 concentrations. 10 mmol mol^{-1} of O_2 in air (upper), where photorespiration would be inhibited, is compared with a normal atmosphere of 210 mmol mol^{-1}. The point A_1, indicated on the lower curve, is the value of A extrapolated from the typical c_i of a C_3 leaf photosynthesizing in a normal ambient CO_2 concentration of 360 μmol mol^{-1}. The decrease in c_i relative to c_a results from the diffusive limitation imposed by the stomata. In the absence of any limitation to diffusion into the leaf A would theoretically increase to A_0. Curves are derived from the model referred to in Figure 2.1

CO_2 concentration (*Figure 2.2*; upper curve). In the *presence* of photorespiration, A will continue to rise beyond the inflection of the curve because of a progressive inhibition of photorespiration.

Following the model of Farquhar, von Caemmerer and Berry, (1980) and incorporating some subsequent modifications (Evans and Farquhar, 1992) A is related to c_i and to Q as follows:

$$A = V_c - 0.5V_o - R_d = V_c\,(1 - \Gamma^*/c_i) - R_d \quad (1)$$

V_c and V_o are the velocities of carboxylation and oxygenation, R_d dark respiration, Γ^* the CO_2 compensation concentration of photosynthesis in the absence of mitochondrial respiration (R_d) and V_c is the velocity of carboxylation. Γ^* is related to the concentration of oxygen and kinetic constants for Rubisco at a given temperature by:

$$\Gamma^* = 0.5\,V_{o,max}\,K_c o_i\,/(V_{c,max}\,K_o) \quad (2a)$$

$V_{c,max}$ and $V_{o,max}$ are the maximum velocities of Rubisco catalysed carboxylation and oxygenation, respectively; K_c and K_o are the Michaelis constants of Rubisco for CO_2 and O_2, respectively; and o_i the intercellular O_2 concentration.

The velocity of carboxylation may be determined as:

$$V_c = \min\{W_c, W_j\}. \quad (2b)$$

where W_c is the Rubisco-limited rate of carboxylation and W_j is the RubP regeneration/electron transport limited rate of carboxylation given by:

$$W_c = V_{c,max} c_i/[c_i + K_c(1 + o_i/K_o)] \quad (3a)$$

$$W_j = J/(4.5c_i + 10.5\Gamma^*) \quad (3b)$$

where J is the rate of coupled electron transport. The second term in the denominator of Equation 3b represents the 'cost' of photorespiration. J depends on the absorbed photon flux available to photosystem II (Q_2) and on the light-saturated rate of whole chain electron transport (J_{max}) according to the non-rectangular hyperbolic response described by Evans and Farquhar (1992):

$$J = \{Q_2 + J_{max} - [(Q_2 + J_{max})2 - 4\Theta Q_2 J_{max}]^{1/2}\}/2\Theta \quad (4a)$$

where Θ is the convexity of the response (dimensionless) and Q_2 is related to the incident photon flux by:

$$Q_2 = Q(1-f)\alpha/2 \quad (4b)$$

The parameter f corrects for the spectral imbalance of the light (*c.* 0.15), α is the absorptance of the leaf (*c.* 0.12), and 2 accounts for the two photons required to drive 1 electron through both reaction centres (Evans and Farquhar, 1992).

At low values of c_i, A will be limited by W_c (Equation 2b), since there will be excess capacity for regeneration of RubP. Because the affinities of Rubisco for CO_2 and O_2 are considered to vary little between terrestrial C_3 species, the main determinant of A at low c_i is the maximum velocity of carboxylation of Rubisco ($V_{c,max}$) which is directly dependent on the quantity of active Rubisco *in vivo*.

Beyond the inflection of the A/c_i response (*Figure 2.2*) A is limited by the capacity of the leaf to regenerate RubP for carboxylation (W_j), but continues to increase on a lower gradient, as the proportion of the limiting supply of RubP which is carboxylated ($1-\Gamma^*/c_i$) increases as c_i rises (Equation 1). When the partial pressure of O_2 is reduced so that o_i = 10 mmol mol^{-1}, Γ^* and photorespiration approach zero, removing any increase in A when RubP regeneration is limiting (*Figure 2.2*, upper line).

OPTIMIZING THE USE OF NITROGEN IN CO_2 CAPTURE

Nitrogen is commonly a limiting resource for crop growth. A large proportion of plant nitrogen is allocated to leaves. Rubisco typically constitutes about half of the soluble protein of the leaf. The quantity of active Rubisco will determine the rate of CO_2 uptake, when RubP is saturating (*Figure 2.2*; Equation 3a). A decline in carboxylation efficiency represents a common, but not a universal, feature of acclimation of the photosynthetic apparatus to elevated CO_2 concentration (Stitt, 1991; Long and Drake, 1993). Why should such a decrease occur?

For a wide range of C_3 leaves in air at normal atmospheric concentrations

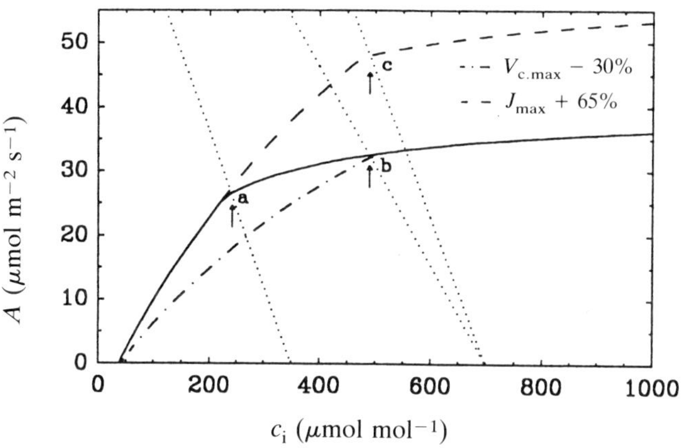

Figure 2.3 The solid line (——) illustrates a typical response of CO_2 uptake per unit leaf area (A) to the intercellular CO_2 concentration (c_i) for a C_3 leaf developed in a normal atmosphere. Arrow 'a' indicates the operating point, i.e. the coordinates of A and c_i when the external CO_2 concentration (c_a) is 360 μmol mol^{-1}. The dashed-dotted (–·–) and dashed line (– – –) indicate two theoretical ways in which photosynthesis could acclimate to growth in a c_a of 700 μmol mol^{-1} and optimize investment in Rubisco and apparatus for the regeneration of RubP. The dashed-dotted (–·–) line represents acclimatization by a 30% decrease in Rubisco activity, moving the inflexion of the curve and the operating point to 'b'. The dashed (– – –) line represents acclimatization in which Rubisco activity is unchanged and capacity for regeneration of RubP increased by 65% moving the inflexion of the curve and the operating point to 'c'. Curves are derived from the model referred to in Figure 2.1. The quantity of active Rubisco is described by its maximum catalytic activity ($V_{c,max}$) and the capacity for regeneration of RubP by J_{max}

of CO_2, the value of c_i during photosynthesis is close to the transition point between W_c and W_j limitation, i.e. the point where the quantity of active Rubisco and the capacity for regeneration of RubP are co-limiting (Evans and Farquhar, 1992; *Figure 2.3*). An environmental change that effects a change in capacity for carboxylation of RubP will often effect a parallel change in capacity for RubP regeneration, such that the two processes remain co-limiting. This suggests an 'optimization' of the distribution of resources within the chloroplast so that neither active Rubisco nor the apparatus for regeneration of RubP are in excess. If optimization of the distribution of resources between components of the photosynthetic apparatus is a widespread phenomenon, then adjustment must also be expected in plants grown in elevated CO_2 concentration which will increase carboxylation efficiency. Given the large investment of energy and nitrogen in Rubisco, strong selective pressures for optimizing carboxylase levels might be expected.

Figure 2.3 illustrates how such optimization might occur, in theory. Without any adjustment of the photosynthetic apparatus as the leaf developed in an elevated CO_2 concentration, the A/c_i curve would be unchanged. Considering a typical A/c_i curve for a leaf developed at current atmospheric CO_2 concentrations and based on the parameters of Long (1991), the operating value of c_i when c_a = 360 μmol mol^{-1} will be at the point of inflection of the curve (point 'a' on *Figure 2.3*), assuming $c_i/c_a = 0.7$. If c_a is now doubled, c_i moves to 490 μmol mol^{-1}, and is now on the upper portion of the A/c_i response (point 'b' on *Figure 2.3*). Here A would be limited by regeneration of RubP (W_j) and Rubisco activity (W_c) would be in considerable excess. Acclimation of the A/c_i curve so that the inflection of the curve moves to 490 μmol mol^{-1} could be achieved in two ways. Either the quantity of active Rubisco is decreased ($V_{c.max}$) so that W_c decreases to equal W_j at c_i = 490 μmol mol^{-1}; or the capacity for regeneration of RubP, which is assumed to be limited by the maximum rate of whole chain electron transport (J_{max}), could increase so that W_j increases to equal W_c at c_i = 490 μmol mol^{-1} (point 'c', *Figure 2.3*).

As light is decreased, capacity for regeneration of RubP (W_j) decreases. If a leaf is transferred from strong to weak light the operating point moves above the inflection of the A/c_i curve. This will occur even if stomatal conductance decreases in direct proportion to the decrease in A. Many crop leaves are formed in an open environment and later become shaded by younger leaves. Leaves will commonly respond to this decrease in mean photon flux by decreasing their Rubisco content (Field, 1983). This decrease in carboxylation efficiency may match the decrease in regeneration of RubP imposed by decreased photon fluxes (Equation 3b). When nitrogen supply is limiting this will be important in releasing N from Rubisco now in excess in the lower leaves for use in new leaves at the top of the canopy. Field (1983) has shown that redistribution of N within a plant canopy, in accordance with variation in the potential for limitation by Rubisco, can significantly increase canopy CO_2 capture.

In wheat, grown with three contrasting rates of nitrogen supply, Rubisco activity declined sharply with decreased available N. However, the decline was paralleled by decreased capacity for regeneration of RubP so that, irrespective of the N-supply rate Rubisco activity and regeneration of RubP were co-limiting (Evans and Farquhar, 1992). Thus at each level of N supply, distribution of resources between Rubisco and the apparatus for regeneration of RubP appears optimized so that capture of CO_2 per unit of N in the chloroplast is maximized.

STOMATAL *VERSUS* MESOPHYLL LIMITATIONS

An unavoidable cost of CO_2 capture by the leaf is water vapour loss. The compromise between the conflicting demands of CO_2 uptake and water conservation is governed by stomata. During photosynthesis, the stomata represent a significant diffusive resistance, such that c_i is significantly decreased below the concentration of the surrounding atmosphere.

Stomata clearly limit water loss, but to what extent do they limit CO_2 uptake? Before 1980, resistance analogues that ignored the non-linearity of the response of A to c_i were used in assessing limitations (Long, 1985). If A responded linearly to c_i then any decrease in c_i produced by a finite stomatal conductance would decrease A. By reference to *Figure 2.2* it can be seen that if c_i is below the inflexion of the A/c_i response, then any change in stomatal conductance would have a marked effect on A. However, beyond the inflexion, increases in stomatal conductance will have little effect on A. Transpiration (E) responds almost linearly to increase in conductance, at a given leaf–air water vapour concentration difference $(\chi_i - \chi_a)$:

$$E = g_s(\chi_i - \chi_a) \tag{5}$$

It follows that E will continue to increase at a similar rate with increase in g_s beyond the inflexion of the A/c_i response, while A shows a sharply diminished response. Thus, any increase in stomatal conductance that allowed c_i to move above the inflexion of the A/c_i response would result in a marked decrease in water use efficiency.

Stomatal conductance is affected by CO_2 concentration, such that the operating c_i has been found to be close to the point of inflection of the A/c_i response. The limitation (l) that stomata impose on CO_2 capture is given by:

$$l = (A_0 - A_1)/A_0 \tag{6}$$

A_1 is the assimilation rate in a normal atmosphere in which c_a = 360 μmol mol^{-1}, and for a given finite stomatal conductance. A_0 is the assimilation rate extrapolating from the A/c_i response (see *Figure 2.2*) that would be attained for the same leaf, if there was no diffusive limitation imposed by the stomata, i.e. $g_s = \infty$. Using this analysis, the limitation imposed by the stomata for crops grown under optimal conditions appears small and typically less than 0.2 (Farquhar and Sharkey, 1982).

Light capture

Although most leaves receive direct sunlight at some or at all stages of development, their capacity to use light usually matches the recent light history of their environment. Leaves of sun ecotypes within a species have long been known to show higher light-saturation points than leaves of shade ecotypes. Within a single plant, the exposed upper leaves typically show higher light-saturation points than those near the base (e.g. Field, 1983). As discussed earlier, the light-saturation point is commonly co-determined by the quantity of active Rubisco and the apparatus for regeneration of RubP.

ABSORPTANCE

The ability of a leaf to capture light depends partly on its absorptivity, which in turn is determined by light quality, the angular distribution of light, surface reflectivity and transmissivity. Chlorophyll in dilute solution shows a strong absorptance in the red and blue regions of the spectrum, with a pronounced dip in the green. However, the high concentration of chlorophylls and accessory carotenoids, coupled with multiple internal reflections within the leaf, result in a strong absorptance in all visible wavelengths. As a result, there is only a slight (*c.* 5%) depression of absorptance in the green in intact leaves, relative to the red and blue regions of the spectrum. Similarly, the spectral response of the quantum yield of photosynthesis for intact leaves is affected little by variation in wavelength between 400 nm and 700 nm (Long, 1985). Transmittance of light, estimated to be about 0.12 for crop leaves (Evans and Farquhar, 1992), depends on the path length and concentration of pigments within the leaf. The highly refractive nature of plant cell walls results in multiple reflections of light entering the leaf, greatly extending the mean path-length through the leaf and, hence, the probability of absorption (Vogelmann, Bornman and Josserand, 1989). Reflectance is affected predominantly by the smoothness of the cuticle, refractive properties of the epidermal cells, and density and form of epidermal hairs and glands. In many instances, reflectance appears to be an adaptation that reduces radiant heating at the expense of light capture. In this respect, cacti, in which the leaves are reduced to reflective spines, represent an extreme of adaptation to an environment where competition for light capture is minimal and survival is almost totally determined by minimizing water loss and radiation interception.

LIGHT GRADIENTS WITHIN THE LEAF

Although significant gradients of light must exist within the leaf (Terashima and Inoue, 1985a), their pattern is not easily predicted because of both the optical complexity within the leaf and the high degree of internal reflection. By the use of miniature fibre optic probes of 125 μm diameter, Vogelmann, Bornmann and Josserand (1989) described the attenuation of light within the leaf of *Medicago sativa*. In direct illumination, they found that the sum of upward and downward fluxes immediately below the leaf surface were some 1.5–2.0 times greater than that incident on the surface. This indicates both the efficiency of light capture and the effect of multiple internal reflections. Even within the 60 m deep palisade of *M. sativa* there was a steep light gradient, with more light being scattered upward than downward. Red light, strongly absorbed by the chlorophyll *a* absorption peak (664 nm) was rapidly depleted within the palisade so that less than 10% entered the spongy mesophyll, compared with 30% at 550 nm (Vogelmann, Bornmann and Josserand, 1989). Just as a differential allocation of resources between leaves within a canopy may increase the efficiency of CO_2 uptake at the canopy level, so a differential distribution of resources through a leaf could increase CO_2 capture. Terashima and Inoue (1985a) showed, by paradermal cryosectioning of leaves, that this expectation is fulfilled in *Spinacia oleracea*. They observed a decrease in Rubisco activity between the uppermost and lowest mesophyll cell layer of about 60%. Maximum rates of whole chain electron transport also decreased by *c.* 50% between the upper and lower layers of mesophyll cells, suggesting that both capacity for carboxylation and for regeneration of the acceptor may decline in parallel (Terashima and Inoue, 1985b). Such a pattern would help to optimize the use of resources in capturing CO_2.

Figure 2.4 uses Equations 1–3 to mimic the light response of photosynthesis, assuming a light gradient in which 80% of absorption is in the upper half and 20% in the lower. This simulation illustrates: (1) an equal distribution of Rubisco activity; and (2) an 80:20 distribution between the upper and lower mesophyll. This predicts that while the quantum yield and light saturated rates of photosynthesis for the whole leaf will be unaffected, the leaf with an equal distribution of Rubisco between the upper and lower halves of the mesophyll would show significantly lower rates of CO_2 uptake at intermediate light fluxes, for the same total investment of resources. This apparent advantage of increased investment into the cells of the upper mesophyll would prevail only when one surface receives higher photon fluxes.

In some amphistomatous grasses, the separation of an upper and lower air space system, allows separate measurement of CO_2 capture by the upper and lower mesophyll (Long *et al.*, 1989a). In *Zea mays* cv. LG11, the apparent responses of photosynthesis of the upper and lower mesophyll, were similar. However, with irradiance from above, the upper surface may suffer severe photoinhibitory

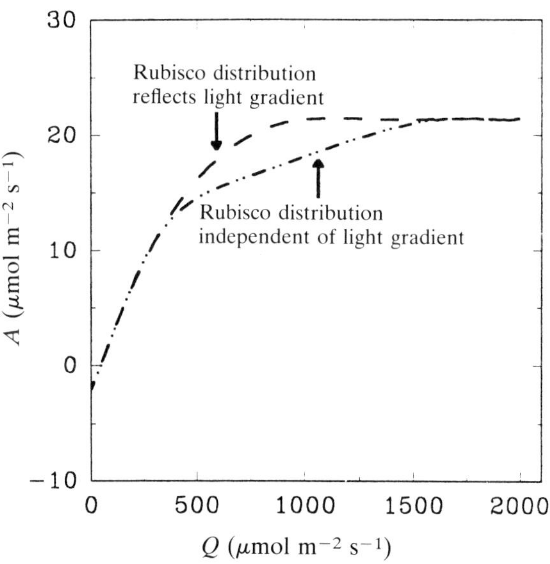

Figure 2.4 The effect of the distribution of active Rubisco within the leaf on the response of photosynthetic CO_2 uptake per unit leaf area (A) on incident photon flux (Q). Responses are simulated for a hypothetical leaf, illuminated from above with direct radiation, in which it is assumed that 80% of photon absorption occurs in the upper half of the mesophyll and 20% in the lower half. The upper curve simulates a leaf which contains 80% of its active Rubisco in the upper half of the mesophyll and 20% in the lower half. The lower curve simulates a leaf with the same total content of active Rubisco, but evenly distributed between the upper and lower halves of the mesophyll. Curves are derived from the model referred to in Figure 2.1

depression of photosynthetic capacity, while the lower surface is unaffected (Long *et al.*, 1989b).

INCREASING LIGHT CAPTURE

Increased leaf thickness and pigment concentration are obvious ways in which light capture, per unit leaf area, may be increased with the investment of more plant resources. How might light capture be increased, by more efficient use of the same resources? Chloroplast and leaf movements, provide two potential mechanisms.

Chloroplast movements

Senn (1908) recognized that chloroplast movements in a range of species could be related to the quantity and direction of light. For example, in the floating angiosperm *Lemna* the mesophyll chloroplasts are arranged at random in darkness. In weak illumination, perpendicular to the upper surface of the leaf, the chloroplasts move alongside and parallel to the paradermal walls. In strong illumination they move against and parallel to the anticlinal walls. For thin leaves these responses maximize light capture in the first case and minimize it in the second (Haupt and Scheuerlein, 1990). In other species, e.g. the fern *Adiantum capillus*, chloroplasts have been shown to move in response to heterogeneous illumination of cells. Where a weak beam of light passes through the cell, chloroplasts group within the beam. However, if an intense beam passes through the cell, the chloroplasts move out of the beam. Reports of chloroplast movements are largely limited to algae, the thin photosynthetic structures of bryophytes, and some thin leafed ferns and higher plants (Haupt and Scheuerlein, 1990). This may reflect the relative ease with which movements can be observed in these plants. Alternatively, the absence of reported movements in thicker-leafed plants suggest that the highly diffuse nature of light scattered by several cell layers may largely eliminate any advantage from chloroplast movements.

Leaf movements

Just as chloroplasts may move to adjust their efficiency of light capture, so the whole leaf may move. Leaf movements in response to light are of two types: (1) nastic, a non-directional response in relation to the quantity of light, e.g. the 'sleep' movements of leaflets of species of the Oxalidaceae; and (2) tropic, a directional response in which the direction of movement is determined by the direction of the light.

Two types of tropic response have been described: (1) growth responses in which the leaf turns parallel or perpendicular to the major direction of light as a result of growth of new tissues so that the response is irreversible without further growth; and (2) responses driven by reversible swelling and contraction of specialized cells within the petiole or leaf base which can move the whole leaf. This latter form of movement allows solar tracking in which a leaf may move through the course of a day to maintain its upper surface at 90° to the direct solar beam (diaphototropism). Diaphototropism has been recorded for a wide range of species, but is predominant within the Leguminosae and Malvaceae, including soybean and cotton (Koller, 1990). Leaves of Malvaceae species have been shown to track solar radiation from 1–2 h after sunrise to 1–2 h before sunset, with a precision of about 3%. Light interception by individual leaves is therefore maximized through most of the day and under clear skies maintained nearly constant. Depending on latitude and solar declination, this movement can more than double the amount of light captured per unit of leaf area. However, maximizing light capture by individual leaves will benefit the individual only under specific circumstances. Where individuals form a dense canopy, maximal interception by the outer leaves may decrease rather than increase canopy CO_2 capture, while maximizing light capture will also increase the incidence of photoinhibition of photosynthesis.

Other species show a paraphototropic response, i.e. leaves are displayed parallel to the solar beam (Rajendrudu and Rama Das, 1981), minimizing light capture and the probability of photoinhibition. Some species appear capable of both types of movement. This is achieved in many leguminous species by combining a nastic response with the diaphototropic response. At midday, leaflets at 90° to the solar beam may fold in a nastic response, around the midrib, moving the laminae parallel to the solar beam. This has been observed when leaves are in sub-optimal conditions, e.g. during periods of water stress (Forseth and Ehleringer, 1980).

CAN LEAVES CAPTURE TOO MUCH LIGHT?

If the light absorbed by the photosynthetic pigments of a leaf exceeds that which may be dissipated through the normal channels of photochemistry (including photosynthesis), thermal deactivation and fluorescence, then an apparent impairment of photosynthetic efficiency, termed photoinhibition ensues. Photoinhibition is defined here as a light-dependent depression of photosynthetic capacity irrespective of the molecular mechanism. Photoinhibition is promoted when other environmental factors are sub-optimal and inhibitory to CO_2 assimilation and hence photochemical dissipation of absorbed light energy, e.g. low temperature, water stress and nitrogen supply (Powles, 1984). However, transient photoinhibition may also be observed on days with clear sunshine when conditions otherwise appear optimal (Ögren, 1988). Photoinhibition, is commonly characterized by a decrease in the maximum quantum yield of CO_2 uptake (ϕ) and has been correlated with a decrease in the efficiency with which crop stands convert intercepted light into biomass (Farage and Long, 1991).

How might photoinhibition be avoided? The zeaxanthin/violaxanthin cycle may be one biochemical mechanisms for dissipating, as heat, an excess of excitation energy in the photosynthetic membrane (Adams and Demmig-Adams, 1992). However, such mechanisms will also lower the maximum photosynthetic efficiency. Paraheliotropic leaf movements, provide one alternative which does not impair the light utilization efficiency of the leaf.

The paraheliotropic response is particularly important to some shade species. Leaves of *Oxalis oregana* held horizontal during exposure to a sunfleck were strongly photoinhibited, while those allowed to develop their natural paraheliotropic movement avoided photoinhibition (Powles and Björkman, 1981). However, avoidance of high thermal radiation loads by leaf movements may be more important than avoidance of photoinhibition (Forseth and Ehleringer, 1980; Fu and Ehleringer, 1989).

The costs of forming and maintaining a leaf

The construction and maintenance of leaves impose significant costs on the plant's resources. With respect to carbon balance, the benefit of a leaf to the plant will be its total photosynthesis less the carbon used in the construction and maintenance of the leaf. What are the costs of leaf construction?

Nobel (1991) suggests that about 100 mol of ATP are consumed in the synthesis of 1 kg of biomass. If this is extrapolated to an average mass per unit area of leaves to be 50 g m^{-2}, then 5 mol ATP would be required in the synthesis of 1 m^2 of leaf. Assuming that the aerobic metabolism of one mol glucose to CO_2 will yield about 30 mol ATP, then approximately 1 mol of CO_2 would be released in the synthesis of 1 m^2 leaf. In healthy C_3 leaves, under optimal conditions, the average photosynthetic rate over a 12-h photoperiod will rarely exceed 12 $\mu mol\ m^{-2}\ s^{-1}$, in temperate climates, giving a daily total of 0.5 mol. If however it is assumed that respiratory costs in the further metabolism of this assimilate and its translocation removes about 25% and that maintenance respiration at 20°C will be 1.5% of the leaf's mass, then the daily total will be decreased to a net figure of 0.35 mol and about 3 days would be required to recover this cost. Daily net carbon gains of tree leaves are 0.02 to 0.2 mol (Long ,1985). Thus, even a healthy tree leaf in optimal conditions would require 5–50 days before the net gain of CO_2 equals the amount expended in its synthesis. In shady, dry or cold habitats, weeks, months or possibly years may pass before sufficient photosynthate is formed to offset the amount used in the synthesis of the leaf. Leaf construction costs, leaf metabolic activity, functional duration and the proportion of leaf resources recovered during senescence are the four determinants of net carbon capture by a leaf. Particular importance attaches to the additional loss of water imposed by maintaining a leaf during drought and the respiratory costs of maintaining a leaf both during drought and other prolonged periods in which photosynthesis may be negligible because of environmental restraints.

The longer a leaf lives and remains active, the greater the net capture of CO_2 . However, most arable crops have short-lived leaves. Taxonomically, long-lived leaves are most frequently found in the more 'primitive' vascular plant groups. Most flowerless vascular plants and the more 'primitive' angiosperms, e.g. Magnoliales, have long-lived leaves (Chabot and Hicks, 1982). Leaves of the gymnosperm, *Araucaria*, can live for 25 years, while leaves of the highly productive C_4 grass *Echinochloa polystachya* live for just 30 days (Piedade, Junk and Long, 1991). Leaves may die for several reasons: consumption by herbivores or pathogens; mechanical removal by wind or sand abrasion; death caused by extreme climatic events such as freezing or drought, or senescence at the end of a genetically determined life span in response to hormonal cues.

The benefits of longevity to CO_2 capture are obvious, but what costs favour short-lived leaves in the many successful angiosperm species of today? If trees of the temperate and boreal zones are considered, two contrasting strategies are apparent. Evergreen conifers dominate the boreal zone and are replaced in the milder cool temperate zone by deciduous species.

Trees of the boreal coniferous forests typically bear leaves with life spans of 2–7 years. These leaves must endure seasonal variation in climate and survive extreme weather events. This involves enduring a season of extreme cold when photosynthesis is negligible. The xeromorphic features of these leaves, i.e. thick cuticle, low boundary layer conductance and low stomatal conductance, all protect against desiccation during this cold period, but may inhibit CO_2 capture in the favourable summer months. Mechanical damage is most likely during the winter period and thus there must be sufficient strengthening tissue to survive storms. There is also a strong negative correlation between palatability and the longevity of leaves (Southwood, Brown and Reader, 1986). Epidemiologically the permanent canopy will increase the probability of loss to pathogens and insect grazers. Thus, evergreen leaves have an increased requirement for investment in adaptations deterring grazers, such as phenolics, a high proportion of structural material and a low nitrogen content. All of these requirements are met by coniferous trees of the boreal forest. However most of these adaptations militate against photosynthetic CO_2 capture and this is reflected in the negative correlation between the mean leaf longevity and maximum photosynthetic rate of a species (Chabot and Hicks, 1982; Diemer, Körner and Prock 1992).

Maintenance of an evergreen canopy with long-lived leaves, may have functions other than maximizing net CO_2 capture. It helps to exclude smaller competitors, particularly those exploiting soil resources, which could otherwise use the periods between leaf fall and bud burst for photosynthesis. It also intercepts precipitation throughout the year and increases protection of the soil surface against erosion (Reich, 1987).

In deciduous temperate forests, senescence during autumn allows the plant to recover much of its mineral investment and a part of the investment in carbon. Nevertheless significant losses of carbon occur in the strengthening materials of the leaf, cellulose, suberized and lignified cell walls, and in cuticularized surfaces. However, the deciduous habit requires only that the leaves are adapted to the warmer period of the year, which can sustain photosynthesis. Thus, these leaves commonly have higher diffusive conductances, higher nitrogen contents and a lower proportion of structural material, all are features which favour high photosynthetic rates of CO_2 capture (Chabot and Hicks, 1982; Long, 1985). The transition to a deciduous habit in cool temperate climates suggests that the increased photosynthetic rates that may be attained during the warm periods of the year by deciduous *versus* evergreen leaves more than outweigh the costs involved in replacing the entire canopy each year.

Most arable crops form canopies in which leaves are continuously replaced conferring some important advantages. Since leaves include no meristematic tissue they cannot adapt at the anatomical level to changes in the environment following completion of their growth. Thus a leaf developed in a high light environment has limited ability to adapt to a subsequent low light environment. A rapid turnover of leaves allows the plant to produce leaves adapted to the prevailing environment. Thus the capacity of the elements of the canopy for resource capture can change according to variation in climate, nutrient and water availability as the crop develops.

Conclusions

Leaves show a variety of strategies for CO_2 and light capture, in particular for maximizing capture relative to the use of water, nutrients and carbohydrate reserves. There is considerable variation between crops in the efficiency of CO_2 and light capture at the leaf level. Understanding of these mechanisms indicates some ways in which efficiency of resource capture could be increased. However, improvement of resource capture at the individual leaf level cannot be properly interpreted without simultaneous understanding of implications at the canopy level, as discussed by Campbell (see p. 35 *et seq.*).

References

Adams, W.W. and Demmig-Adams, B. (1992) Operation of the xanthophyll cycle in higher plants in response to diurnal changes in incident sunlight. *Planta,* **186**, 390–398

Bowes, G. (1991) Growth at elevated CO_2: photosynthetic responses mediated through Rubisco. *Plant Cell and Environment*, **14**, 795–806

Chabot, B.F. and Hicks, D.J. (1982) The ecology of leaf life spans. *Annual Review of Ecology and Systematics*, **13**, 229–259

Diemer, M., Körner, Ch., Prock, S. (1992) Leaf life spans in wild perennial herbaceous plants: a survey and attempts at a functional interpretation. *Oecologia,* **89**, 10–16

Evans, J.R. and Farquhar, G.D. (1992) Modeling canopy photosynthesis from the biochemistry of the C_3 chloroplast. In *Modeling Crop Photosynthesis – from Biochemistry to Canopy*, pp 1–15 Edited by K.J. Boote and R.S. Loomis. Madison: Crop Science Society of America Inc.

Farage, P.K. and Long, S.P. (1991) The occurrence of photoinhibition in an over-wintering oil-seed rape crop (*Brassica napus*) and its correlation with changes in crop growth. *Planta*, **185**, 279–285

Farquhar, G.D. and Sharkey, T.D. (1982) Stomata and photosynthesis. *Annual Review of Plant Physiology,* **33**, 317–345

Farquhar, G.D., von Caemmerer, S. and Berry, J.A. (1980) A biochemical model of photosynthetic CO_2 assimilation in leaves of C_3 species. *Planta,* **149**, 78–90

Field, C. (1983) Allocating leaf nitrogen for the maximization of carbon gain: leaf age as a control on the allocation program. *Oecologia*, **56**, 341–347

Forseth, I.N. and Ehleringer, J. (1980) Solar tracking response to drought in a desert annual. *Oecologia,* **44**, 159–163

Fu, Q.A. and Ehleringer, J. (1989) Heliotropic leaf movements in common beans controlled by leaf temperature. *Plant Physiology,* **91,** 1162–1167

Haupt, W. and Scheuerlein, R. (1990) Chloroplast movement. *Plant Cell and Environment,* **13**, 595–614

Kimball, B.A. (1983) Carbon dioxide and agricultural yield: an assemblage and analysis of 430 prior observations. *Agronomy Journal,* **75**, 779–782

Koller, D. (1990) Light-driven leaf movements. *Plant Cell and Environment,* **13**, 615–632

Long, S.P. (1985) Leaf gas exchange. In *Photosynthetic Mechanisms and the Environment*, pp 453–499 Edited by J. Baker and N.R. Baker. Elsevier: Amsterdam

Long, S.P. (1991) Modification of the response of photosynthetic productivity to rising temperature by atmospheric CO_2 concentrations. Has its importance been underestimated? – Opinion. *Plant Cell and Environment*, **14**, 729–739

Long, S.P., Bolhár-Nordenkampf, H.R., Croft, S.L., Farage, P.K., Lechner, E. and Nugawela, A. (1989b) Analysis of spatial variation in CO_2 uptake within the intact leaf and its significance in interpreting the effects of environmental stress on photosynthesis. *Philosophical Transactions of the Royal Society of London, Series B,* **323**, 385–395

Long, S.P. and Drake, B.G. (1993) Photosynthetic CO_2 assimilation and rising atmospheric CO_2 concentrations. In *Crop Photosynthesis: Spatial and Temporal Determinants*, pp 69–103 Edited by N.R. Baker and H. Thomas, Amsterdam: Elsevier.

Long, S.P., Farage, P.K., Bolhár-Nordenkampf, H.R. and Rohrofer, R. (1989a) Separating the contribution of the upper and lower mesophyll to photosynthesis *Zea mays* L. leaves. *Planta,* **177**, 207–216

Long, S.P. and Hutchin, P. (1991) Primary production in grasslands and coniferous forests in relation to climate change and rising CO_2 levels. *Ecological Applications*, **1**, 139–156

Nobel, P.S. (1991) *Physiochemical and Environmental Plant Physiology*. San Diego: Academic Press

Piedade, M.T., Junk, W.J. and Long, S.P. (1991) The productivity of the C_4 grass *Echinochloa polystachya* in the Amazon floodplain. *Ecology*, **72**, 1456–1463

Powles, S.B. (1984) Photoinhibition of photosynthesis induced by visible light. *Annual Reviews of Plant Physiology*, **35**, 15–44

Powles, S.B. and Björkman, O. (1981) Leaf movement in the shade species *Oxalis oregana*. II. Role in protection against injury by intense light. *Carnegie Institution of Washington Year Book,* **80**, 63–66

Ögren, E. (1988) Photoinhibition of photosynthesis of willow leaves under field conditions. *Planta*, **175**, 229–236

Rajendrudu, G. and Rama Das, V.S. (1981) Solar tracking and light interception by leaves of some dicot species. *Current Science,* **50**, 618–620

Reich, P.B. (1987) Quantifying plant response to ozone: a unifying theory. *Tree Physiology*, **3**, 63–91

Schimmel, D. (1990) Biogeochemical feedbacks in the Earth system. In *Global Warming. The Greenpeace Report*, pp 68–82 Edited by J. Leggett. Oxford: Oxford University Press

Senn, G. (1908) *Die Gestalts- und Lageveränderung der Pflanzen-Chromatophoren.* Stuttgart: Engelmann

Southwood, T.R.E., Brown, V.K. and Reader, P.M. (1986) Leaf palability, life expectancy and herbivore damage. *Oecologia*, **70**, 544–548

Stitt, M. (1991) Rising CO_2 levels and their potential significance for carbon flow in photosynthetic cells. *Plant Cell and Environment*, **14**, 741–762

Terashima, I. and Inoue, Y. (1985a) Palisade tissue chloroplasts and spongy tissue chloroplasts in spinach: Biochemical and ultrastructural differences. *Plant and Cell Physiology,* **26**, 63–75

Terashima, I. and Inoue, Y. (1985b) Vertical gradient in photosynthetic properties of spinach chloroplasts dependent on intra-leaf light environment. *Plant and Cell Physiology,* **26**, 781–785

Vogelmann, T.C., Bornman, J.F. and Josserand, S. (1989) Photosynthetic light gradients and spectral régime within leaves of *Medicago sativa. Philosophical Transactions of the Royal Society of London, Series B,* **323**, 411–421

Appendix 1: Abbreviations

A	net rate of CO_2 uptake per unit leaf area	$\mu mol\ m^{-2}\ s^{-1}$
A_{sat}	light saturated value of A	$\mu mol\ m^{-2}\ s^{-1}$
c_a	CO_2 concentration of the bulk air	$\mu mol\ mol^{-1}$
c_i	CO_2 concentration in the intercellular air spaces	$\mu mol\ mol^{-1}$
g_s	leaf conductance to water vapour diffusion	$mmol^{-2}\ m^{-2}\ s^{-1}$
J	the rate of coupled electron transport	$\mu mol\ m^{-2}\ s^{-1}$
J_{max}	the light saturated rate of coupled electron transport	$\mu mol\ m^{-2}\ s^{-1}$
K_c	Michaelis constant of Rubisco for CO_2	$\mu mol\ mol^{-1}$
K_o	Michaelis constant of Rubisco for O_2	$\mu mol\ mol^{-1}$
l	stomatal limitation to CO_2 uptake	dimensionless
o_i	intercellular O_2 concentration	$\mu mol\ mol^{-1}$
Q	photon flux	$\mu mol\ m^{-2}\ s^{-1}$
Rubisco	ribulose 1:5 bisphosphate carboxylase/oxygenase	
RubP	ribulose 1:5 bisphosphate	
$V_{c,max}$	the maximum Rubisco-catalysed velocity of carboxylation	$\mu mol\ m^{-2}\ s^{-1}$
$V_{o,max}$	the maximum Rubisco-catalysed velocity of oxygenation	$\mu mol\ m^{-2}\ s^{-1}$
W_c	the Rubisco-limited rate of carboxylation	$\mu mol\ m^{-2}\ s^{-1}$
W_j	the electron transport limited rate of regeneration of RubP	$\mu mol\ m^{-2}\ s^{-1}$
Γ^*	CO_2 compensation point of photosynthesis in the absence of dark respiration	$\mu mol\ mol^{-1}$
ϕ	the quantum yield of photosynthesis on an absorbed light basis	dimensionless
Θ	Convexity coefficient for the response to J to Q	dimensionless
χ_a	Water vapour concentration of the bulk air	$mol\ mol^{-1}$
χ_i	Water vapour concentration of the intercellular air space	$mol\ mol^{-1}$

3

LIGHT INTERCEPTION BY PLANT CANOPIES: EFFICIENCY AND ARCHITECTURE

G. S. CAMPBELL and F. K. VAN EVERT
Department of Crop and Soil Sciences, Washington State University, Pullman, WA, USA

Introduction

A plant canopy represents an investment of dry matter by the plant for the purpose of intercepting solar energy and CO_2. The capture of these resources allows the plant to produce additional dry matter. Monteith (1977) proposed a simple model that provides a useful framework for analysis of the efficiency of canopies in the capture and conversion of solar energy to dry matter. The model is based on the observation from field data that the cumulative dry matter production of a crop is linearly related to the integral of solar radiation absorbed by the crop. The dry matter produced on a given day can therefore be predicted from:

$$P = efS_t \tag{1}$$

where f is the fraction of incident solar energy intercepted or absorbed by the canopy, S_t is the flux density of incident radiation, and e is a conversion factor. Since ΣP always has a strong linear correlation with ΣfS_t (because of the induced correlation from taking sums), Equation 1 will always correctly predict dry matter production. It is most useful, however, when water is plentiful and solar radiation limits production, so that e is relatively constant and predictable.

Another simple model, proposed by Tanner and Sinclair (1983), relates dry matter production to transpiration. This model is also based on field observation, and states that:

$$P = kT/D \tag{2}$$

where T is the crop transpiration, k/D is the dry-matter:water ratio (dry mass produced per unit mass of water transpired), and D is the average vapour deficit of the air. The quantity, k is found to be conservative. When water is not limiting, T can be calculated from the product of f and E_p, the potential evaporation rate:

$$T = fE_p \tag{3}$$

Since E_p is mainly determined by the amount of absorbed solar radiation (Priestley and Taylor, 1972), Equations 2 and 3 also constitute a direct relationship between interception or absorption of radiation and accumulation of dry matter. Equation 2 works best in more arid environments where water tends to limit crop production.

Canopy architecture influences the efficiency of dry matter production in both models through f, the fractional interception or absorption. For most applications of these simple models, f has been arrived at through field measurements and simple correlations. However, a large body of literature exists relating interception and absorption of radiation to detailed descriptions of canopy structure and leaf area (Ross, 1975; Goudriaan, 1977; Campbell and Norman, 1989). These detailed models could help provide insight into relationships between canopy architecture and efficiency of radiant energy capture, but they tend to be quite complex. Another problem is that the detailed models are for interception at a particular time of day, while the time resolution of the simple models is, at best, one day. The purpose of this paper is to try to bridge that gap; to provide simple ways of computing daily averages of intercepted or absorbed radiation based on canopy structure parameters. These relationships will then be used to compare the performance and efficiency of different canopy structures.

Before going on to a consideration of canopy structure, brief mention needs to be made of some of the properties of solar radiation, which is the resource being captured. The total solar radiation, S_t, is made up of beam (S_b) and diffuse (S_d) components. The diffuse component is scattered by the atmosphere and by clouds, when present. For our purposes, it can be considered uniform over the entire hemisphere above the crop. Beam radiation is treated as a point source at the elevation and azimuth of the sun. On clear days, the ratio, S_d/S_t may be below 0.2, but data summarized by Monteith and Unsworth (1990) show monthly averages above 0.5 for a location in Britain, indicating that diffuse and beam radiation may be equally important as resources to capture.

The spectral distribution of radiation is also important. Leaves absorb most of the photosynthetically active radiation (PAR, 0.4–0.7 μm wavelength) incident upon them, and reflect most of the near infrared (NIR, 0.7–3 μm). Typical values of absorptivity for thin leaves are $a_{par} = 0.8$ and $a_{nir} = 0.2$ (Goudriaan, 1977). Half of the total energy in solar radiation is in each waveband (Stanhill and Fuchs, 1977), and this value changes only a few percent with time of day and atmospheric conditions.

Since the interaction of solar radiation with the canopy differs markedly depending on whether the radiation is beam or diffuse, or PAR or NIR, each of these components will be treated separately, and then combined to give total interception.

Interception by canopies of black leaves

The first increment of leaf area produced by a crop is unshaded, but as additional area is produced, some leaves shade others, making the interception per unit leaf area less and less effective as new increments are added. In order to describe this shading effect, we consider a canopy made up of black leaves (no reflection or transmission of radiation) which are randomly distributed in space, with random azimuthal distribution.

The basic equation describing transmission of a beam of light through such a plant canopy is similar to Bouguer's law:

$$\tau_b = \exp(-K_b L) \tag{4}$$

where τ_b is the transmission of the canopy, L is the leaf area index (area of leaves per unit area of ground), and K_b is a canopy extinction coefficient. Interception, for the beam, or the probability that the beam will strike a canopy element while passing through the canopy, is given by

$$p_b = 1 - \tau_b \tag{5}$$

The extinction coefficient depends, in general, on canopy structure and the zenith angle of the incident radiation. If the leaves are all horizontal, then $K_{bh} = 1$. The subscript, b, indicates black leaves, and h horizontal. Canopies of leaves with other inclination angles can be modelled by computing the projection of the inclined leaves on to a horizontal surface, and then finding τ_b for the equivalent canopy of horizontal projections. The K_b for non-horizontal canopies therefore becomes the ratio of the area of an average leaf projection to the area of the leaf.

For canopies whose leaf angle distributions approximate the angle distributions on the surfaces of solids of revolution, K_b can be calculated as the ratio of shadow area of the solid to its surface area (Monteith and Unsworth, 1990). An ellipsoidal angle distribution is particularly useful because it requires only a single structure parameter, x, to describe a wide range of canopy structures. The ratio of projected area to surface area for an ellipsoid is (Campbell, 1986):

$$K_{be} = \frac{\left(x^2 + \tan^2\theta\right)^{1/2}}{x + 1.774(x + 1.182)^{-0.733}} \tag{6}$$

The parameter, x, is the ratio of horizontal to vertical projections of canopy elements. When $x = 1$ the canopy has a spherical leaf angle distribution. When $x = 0$ the distribution is cylindrical (vertical leaves); as $x \rightarrow \infty$, the canopy becomes horizontal. Equation 6 therefore gives all of the simple 'K_b's as well as the ones in between.

Ross (1975) gives another inclination index, χ_L, as:

$$\chi_L = \pm\frac{1}{2}\int_o^{\pi/2} |\sin(\alpha) - g(\alpha)|\, d\alpha \tag{7}$$

where α is the leaf inclination angle, and $g(\alpha)$ is the leaf angle distribution function. The + sign is for distributions with inclinations more horizontal than spherical, and the − for inclinations more vertical than spherical. Horizontal leaves have $\chi_L = 1$, spherical have $\chi_L = 0$, and vertical leaves have $\chi_L = -1$.

A relation between the Ross inclination index and the ellipsoidal x value can be found by substituting the ellipsoidal angle distribution function from Campbell (1990) into Equation 7 and performing the integration. The integration was done numerically, and then an empirical function was fitted to the results, obtaining:

$$\chi_L = 0.85 - 1.85 \exp(-0.7778\, x^{0.75}). \tag{8}$$

Table 3.1 gives values of χ_L and their corresponding x values for several crops.

Table 3.1. THE ROSS INCLINATION INDEX, χ_L, AND THE ELLIPSOIDAL ANGLE DISTRIBUTION INDEX, x, FOR SEVERAL CROPS. THE FIRST SET OF VALUES IS FROM ROSS (1975); THE SECOND SET IS FROM CAMPBELL (1986) USING DATA BY LEMEUR (1973)

Crop	χ_L			x		
Ryegrass	–0.19	to	0.45	0.67	to	2.47
Maize	–0.13	to	0.46	0.76	to	2.52
Rye	–0.11	to	0.12	0.80	to	1.27
Wheat			–0.02			0.96
Barley			0.09			1.20
Timothy			0.06			1.13
Sorghum			0.18			1.43
Lucerne			0.22			1.54
Hybrid swede	0.13	to	0.30	1.29	to	1.81
Sugar beet	0.19	to	0.32	1.46	to	1.88
Rape	0.33	to	0.38	1.92	to	2.13
Cucumber			0.39			2.17
Tobacco	0.13	to	0.40	1.29	to	2.22
Potato	0.27	to	0.45	1.70	to	2.47
Horse bean	0.30	to	0.39	1.81	to	2.17
Sunflower	0.30	to	0.42	1.81	to	2.31
White clover	0.45	to	0.57	2.47	to	3.26
Strawberry			0.54			3.03
Soybean			–0.10			0.81
Maize			0.16			1.37
Jerusalem artichoke			0.39			2.16
Sunflower			0.65			4.10

Table 3.1 shows a strong tendency by most canopies toward a spherical to horizontal habit. In the analyses to be done later, a range of x of 0.5–5 (χ_L from –0.3 to 0.7) will be used, which covers all of the crops shown in *Table 3.1*.

Figure 3.1 shows how canopy interception of beam radiation varies with beam zenith angle and and the value of x. Note that horizontal canopies have no

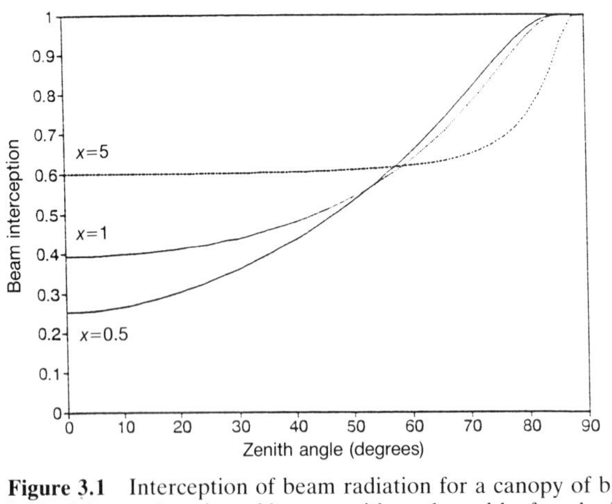

Figure 3.1 Interception of beam radiation for a canopy of black leaves with LAI of 1, as a function of beam zenith angle and leaf angle distribution

zenith angle effect, but for all other canopies, zenith angles below 60° have lower interception than horizontal, while at zenith angles greater than 60°, interception is higher than for a horizontal canopy.

Non-random foliage distribution in canopies

The equations so far have assumed that foliage is distributed randomly in space, and with a random distribution of leaf azimuth angles. These conditions are typical of closed canopies of agricultural crops and some dense forests, but are not typical of conditions that exist early in the growing season with agricultural crops, or when mixed species occur in a stand. Violation of these assumptions can occur in several ways. For example, some species have heliotropic leaves which orient themselves at fixed angles with respect to the sun. This behaviour invalidates the assumption that leaf azimuth angles are randomly distributed. Mann *et al.* (1980) show that, if leaves always orient themselves perpendicular to the sun, the extinction coefficient for the canopy is:

$$K_{bf} = 1/\cos\,\theta \tag{9}$$

which is twice the value for a spherical canopy. For a canopy made up partly of heliotropic leaves, the transmission coefficient could be calculated from:

$$\tau_b = \exp(-K_f L_f - K_r L_r) \tag{10}$$

where the subscript f indicates the extinction coefficient for sun follower leaves, and r indicates random azimuth leaves.

Other types of non-random canopies are those which have leaves grouped around some axis, rather than being randomly distributed in space. Examples are crops with row structure, where leaves are grouped around the row; orchards, where leaves are grouped into trees; and conifers, where needles are grouped around stems. In each case, there exists a level of organization within which the assumption of random distribution seems to hold, but the random distribution does not extend to the entire collection of sub-canopies. Several approaches have been used for modelling light interception in this type of canopy (Norman and Jarvis, 1975; Mann *et al.*, 1980; Norman and Welles, 1983). Methods for dealing with row crops will be discussed briefly.

Assume that the plants within a row can be surrounded by an ellipsoid having height, a and width, b. The distance between rows is d. The sunlit areas under the rows will be made up of irregular areas where the sun shines through holes in the vegetation within the rows, and possibly of long strips of sunlit area where the sun misses the row entirely. The width of the shaded strip can be calculated from (Mann *et al.*, 1980)

$$w = [b^2 + a^2(\sin z \tan\theta)^2]^{1/2} \tag{11}$$

where z is the azimuth angle between the direction of the sun and the direction of the row, and θ is the solar zenith angle. The ratio,

$$q = w/d \tag{12}$$

is the ratio of the width of the row shadow to the row spacing. If $q<1$, then a strip of sunlit area exists between rows, and the canopy transmission is the sum of the probability of the sun missing a row and the probability of hitting a row and getting through the holes (Fuchs and Stanhill, 1980):

$$\tau_b = (1 - q) + q \exp(-K_{be}L/q) \tag{13}$$

In Equation 13, L is divided by q because the leaf area, rather than being uniformly distributed throughout the canopy, is concentrated into the space within the row. When $q \geq 1$, the row shadows overlap, and the canopy transmits light according to the random canopy equations developed earlier. Fuchs and Stanhill (1980) used a similar approach to predict interception of radiation by a sorghum row canopy (though they used rectangles, rather than ellipsoids to describe the sub-canopy envelope). They found excellent agreement between predicted and measured interception using this type of model, while the random model was in error by almost 200% in some cases.

Other examples of non-random foliage distribution are clumping of vegetation around trees in an orchard, or individual plants that are randomly distributed in space. Models for these are given by Mann, Curry and Sharpe (1979) and Norman and Welles (1983).

Interception by canopies with several species

Sometimes canopies have mixed species, such as with weeds in crops, or with intercroping. In these cases it is important to determine how much light is intercepted by each component of the canopy so that dry matter production of each can be calculated. Spitters (1989) describes a simple model that can be used for canopies with more than one species. He assumes that the leaf area of each species is distributed evenly with height. The mixed canopy is then divided into as many horizontal layers as there are species, so that the top-most layer only has leaves from the tallest species; the second layer has leaves from the tallest and the next-tallest species, etc. The layers are numbered from the top as $i = 1. .n$. The transmission through layer i is:

$$\tau_i = \exp(-M_i) \tag{14}$$

where

$$M_i = \sum_{j=1}^{m} K_j L_{i,j}. \tag{15}$$

Here, K_j is the extinction coefficient for species j, and $L_{i,j}$ is the leaf area index for species j in layer i. The light intercepted by species j in layer i is calculated from:

$$I_{i,j} = S_i(1-\tau_i)\ K_j L_{i,j}/M_i \tag{16}$$

where S_i is the radiation incident on layer i. If all species have the same K, then the intercepted radiation in each layer is distributed over the species in proportion to $L_{i,j}$. The total light intercepted by each species is obtained by summing $I_{i,j}$ over all layers.

The height of each species is an important factor in calculating light interception by this model. Spitters (1989) uses a logistic equation in time, and a maximum height for each species to determine the height at any time. Best results are obtained when thermal time, rather than calendar time is used.

Transmission of diffuse radiation

The diffuse radiation comes from all directions, so is attenuated differently from beam radiation. Diffuse radiation can be thought of as many beams, so a diffuse interception coefficient for the canopy can be calculated by integrating beam interception over all zenith angles (assuming azimuthal symmetry of leaves in the canopy). For a uniform distribution of diffuse radiation in the sky hemisphere, the diffuse interception p_d is:

$$p_d = \int_o^{\pi/2} p_b(\theta) \sin\theta \cos\theta \, d\theta \tag{17}$$

where $p_b(\theta)$ is from Equation (5). For horizontal leaves, p_b is not dependent on θ and so $p_d = p_b$; but for the other leaf angle distributions the integration in Equation (17) must be carried out numerically.

Except for horizontal-leaf canopies, the transmission of diffuse radiation predicted by Equation 17 $(1 - p_d)$ is not exactly exponential with leaf area index. It is useful, however, to define an extinction coefficient for diffuse radiation such that the interception of diffuse radiation by a canopy with leaf area index, L, is the same as that given by Equation 17. This is given by:

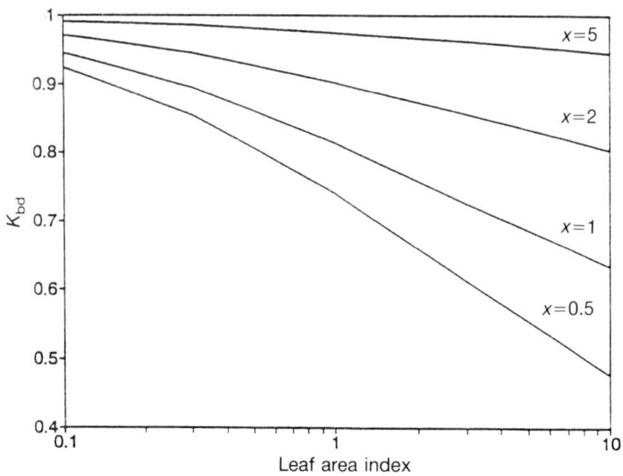

Figure 3.2 The apparent extinction coefficient for diffuse radiation for canopies of black leaves for a range of LAI and canopy structure

$$K_{bd} = -\ln(1-p_d)/L \ . \tag{18}$$

Figure 3.2 shows K_{bd} as a function of L for several values of the inclination index. For horizontal leaves, $K_{bd} = 1$.

Effects of light scattering on interception

Since leaves are not black, interception by a crop canopy will be less than that predicted by Equations 2 and 17. Several detailed light scattering models have been described (Norman and Jarvis, 1975; Goudriaan, 1977; Chen, 1985) which divide the canopy into layers and calculate the scattering by each layer and the interaction between the layers. These models are too complex for the type of analysis needed here, but several simplifications are possible. Goudriaan (1977) compared the results of his detailed light scattering model with the predictions from a simple exponential model. For transmission through the entire canopy, the simple model is:

$$T = f_b \exp(-\sqrt{a}K_{be}L) + (1-f_b)\ exp(-\sqrt{a}K_{bd}L) \tag{19}$$

where f_b is the fraction of incident radiation that is beam, and a is the absorptivity of the leaves for either PAR or NIR. It is assumed that the leaves are thin so that they reflect and transmit approximately equal amounts of radiation. The Goudriaan (1977) model was used to compute T for a range of values of L, a, x, θ and f_b. A comparison between the predictions of the layer model and Equation 19 is shown in *Figure 3.3*. The results are similar to those obtained by Goudriaan, in that the simple model approximates the predictions from the detailed model in almost all cases. Statistics for linear fits to the data are given in *Table 3.2*.

In order to compute absorption of radiation by the canopy, we also need to know how much radiation is reflected from the canopy. Fuchs, Stanhill and

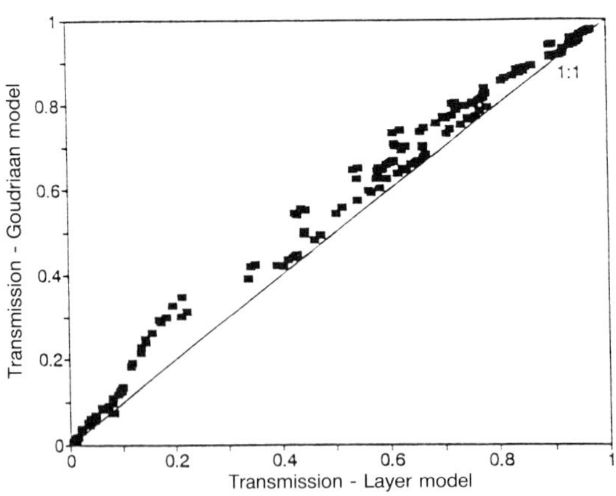

Figure 3.3 A comparison of predicted light transmission through canopies using a simple exponential model (Goudriaan, simple model) with predicted transmission using Goudriaan's detailed layer model. Ranges of parameters are θ:10–70°; LAI:0.1–5; f_b:0–0.8; x:0.8–4; a:0.2–0.8; ρ_s=0.2

Table 3.2. STATISTICS FOR LINEAR FITS OF DATA IN FIGURES 3.3–6

	Slope	*Intercept*	s_{yx}	*No. of Points*
Figure 3.3	0.948	0.002	0.016	192
Figure 3.4.	0.981	0.048	0.033	192
Figure 3.5	1.017	–0.002	0.016	288
Figure 3.6	0.968	0.030	0.030	288

Moreshet (1976) proposed a simple reflection model for canopies which is based on the scattering properties of the canopy, as a whole, rather than the properties of the individual leaves. Their equation is:

$$R = \left[\rho_c - \frac{\rho_s t(1-\tau_c)}{1 - p_d \rho_s \rho_c}\right] f_b(p_b - p_d) + p_d \rho_c + \frac{\rho_s t^2}{1 - p_d \rho_s \rho_c} \quad (20)$$

where ρ_c and ρ_s are canopy and soil reflectances, τ_c is the canopy transmittance, and $t = 1 - p_d(1-\tau_c)$. The canopy properties, ρ_c and τ_c are not the leaf reflectance and transmittance, because they involve multiple scattering, but they can be related to the leaf scattering properties. It was assumed, as with the leaf properties, that $\tau_c = \rho_c$, and they were related to the leaf properties by computing canopy reflectance with the detailed scattering model. Using this, the following relationship was obtained:

$$\tau_c = \rho_c = -0.388 \ln[1-0.75(1-a)] \quad (21)$$

where a is the absorptivity of leaves. The relationship between predictions from the detailed model and from Equation 20 is shown in *Figure 3.4* for the same range of parameters used in *Figure 3.3*. Statistics are given in *Table 3.2*. Again, the agreement is generally good, with the most significant departures coming at

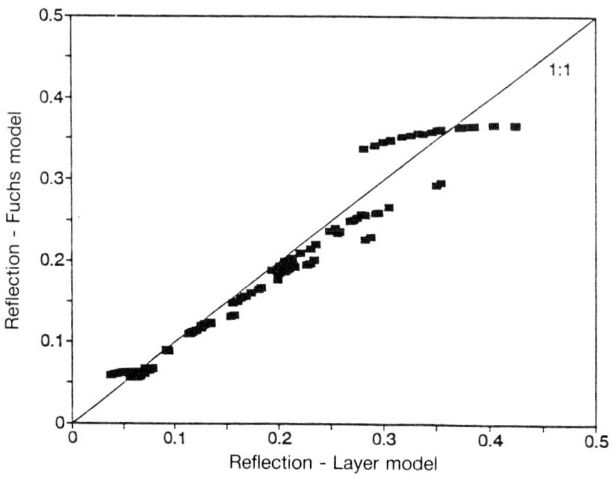

Figure 3.4 A comparison of predicted light reflection from canopies using the Fuchs model and Goudriaan's detailed layer model. Ranges of parameters are the same as in Figure 3.3

large values of L with low values of a. It appears, therefore, that Equations 19 and 20 will give satisfactory estimates of transmission and reflection in most cases.

Integration over days

In order to predict interception and absorption of radiation over whole days, it is necessary to integrate Equations 19 and 20 over the entire day. Fuchs, Stanhill and Moreshet (1976), however, suggested that this integral is closely approximated by

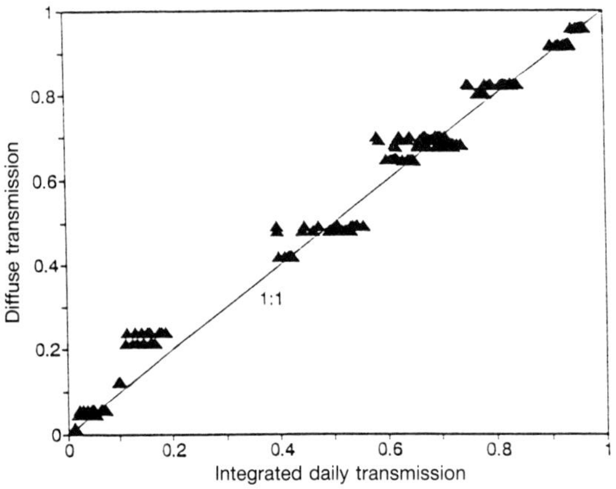

Figure 3.5 A comparison of diffuse transmission through a canopy computed using Goudriaan's simple exponential model with integrated daily transmission using the detailed layer model. Parameters are as in Figure 3.3 except zenith angles were calculated for solar declinations from 0° to 23° and latitudes from 10° to 50°. Values of f_b were appropriate for clear days at these solar elevations

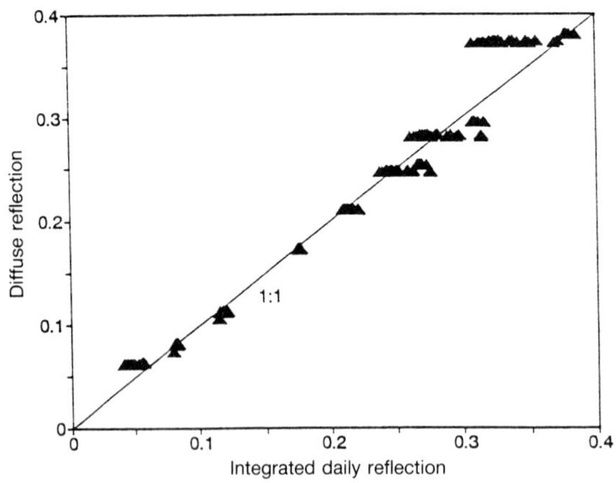

Figure 3.6 A comparison of diffuse reflection computed using Fuchs' simple reflection model with reflection from a canopy using the detailed layer model. Ranges of parameters are as in Figure 3.5

the transmission and reflection obtained if S_t were completely diffuse. *Figures 3.5* and *3.6* compare integrated values of T and R from the Goudriaan (1977) layer model with values of T and R obtained from Equations 19 and 20 with f_b=0. Statistics are shown in *Table 3.2*. Results are plotted for latitudes between 10° and 50°, and solar declinations between 0° and 23° (mid March to Mid September in the northern hemisphere). For overcast days, the integration and the direct calculation give similar results. *Figures 3.5* and *3.6* therefore represent the largest errors one is likely to encounter by using Fuchs' approximation. The approximation therefore seems adequate for most times during the normal growing season.

Using the Fuchs approximation, the average daily fractional transmission and reflection for the canopy are:

$$\overline{T} = \exp\left(-\sqrt{a}K_{bd}L\right) = \tau_d^{\sqrt{a}} \quad (22)$$

and

$$\overline{R} = p_d \rho_c + \frac{\rho_s t^2}{1 - p_d \rho_s \rho_c} \quad (23)$$

The fractional absorption of radiation over a day can now be calculated as:

$$\overline{A} = 1 - \overline{R} - \overline{T}(1 - \rho_s). \quad (24)$$

Daily interception and absorption of solar radiation by canopies

The following calculations will use typical values of leaf absorptance and soil reflectance for PAR and NIR suggested by Goudriaan (1977). These are a_p = 0.8,

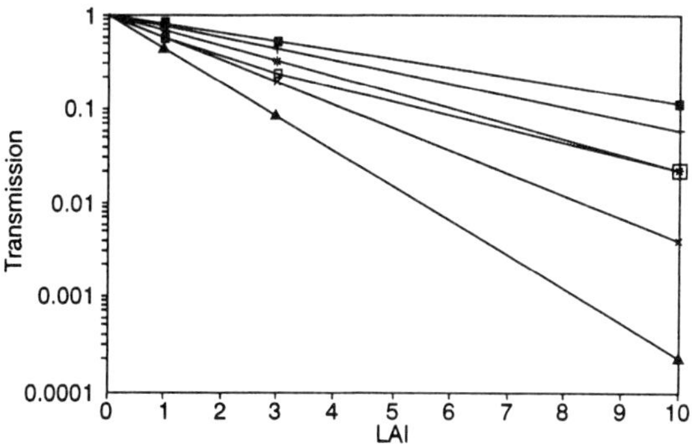

Figure 3.7 Average daily transmission of light through canopies having a range of structure parameters and leaf optical properties., –■–, a=0.2, x=0.5;–+–, a=0.2, x=1; –*–, a=0.2, x=5; –□– a=0.8, x=0.5; –×– a=0.8, x=1;–▲–, a=0.8, x=5

$a_n = 0.2$, $\rho_{sp} = 0.1$, $\rho_{sn} = 0.25$. Using these values, and Equation 22, gives the transmission for two wavebands and three leaf angle distributions shown in *Figure 3.7*. Transmission is influenced strongly by both the optical properties of the leaves and the leaf angle distribution of the canopy. A horizontal-tending canopy intercepts about the same amount of NIR as a vertical-tending canopy intercepts of PAR.

Calculation of the interception of total solar radiation, as is used in Equation 3, requires that an average be taken of the transmission for NIR and PAR in *Figure 3.7*. For L up to 3, at least, this is closely approximated by the transmission predicted by Equation 22 with $a_s = (a_p\, a_n)^{1/2}$, the geometric mean of the absorptances in the two wavebands. This gives a value of $a_s = 0.4$ for the absorptances used here. The product, $\bar{K}_s = \sqrt{a_s} K_{bd}$, should approximate the empirically determined extinction coefficient which is often used in simple crop models (Ritchie, 1972; Campbell and Diaz, 1988). *Figure 3.2* shows a value of K_{bd} around 0.7 for a spherical canopy with $L = 3$. This would give an average extinction coefficient for solar radiation of 0.44. This is similar to an extinction coefficient for net radiation of 0.4 reported by Ritchie (1972) for canopies of six different crops. For the range of angle distribution shown in *Figure 3.2*, K_s could range from around 0.35, for x=0.5 to 0.65 for x=5. The corresponding range for computing interception of PAR (for Equation 1) is 0.45 to 0.9, with the spherical value at 0.65.

The average reflection of these same canopies is shown in *Figure 3.8*. This shows a much smaller effect of canopy structure on reflection, with the main features being determined by the reflection of bare soil and of an infinitely thick canopy.

Using Equation 24 to combine the results of the transmission and reflection calculations gives the absorption. This is shown in *Figure 3.9* as a function of leaf area index. At intermediate LAI, horizontal leaves absorb a somewhat greater fraction of the incident radiation than do other canopy structures. Based on this observation, and Equation 1, one might be tempted to predict that horizontal leaf canopies would produce more dry matter per unit leaf area than other canopies. This would ignore, however, a probable effect of structure on the conversion factor *e*. The photosynthetic mechanism of a leaf tends to light saturate at high light levels. Inclined-leaf canopies, with high L and intermediate light levels on leaves,

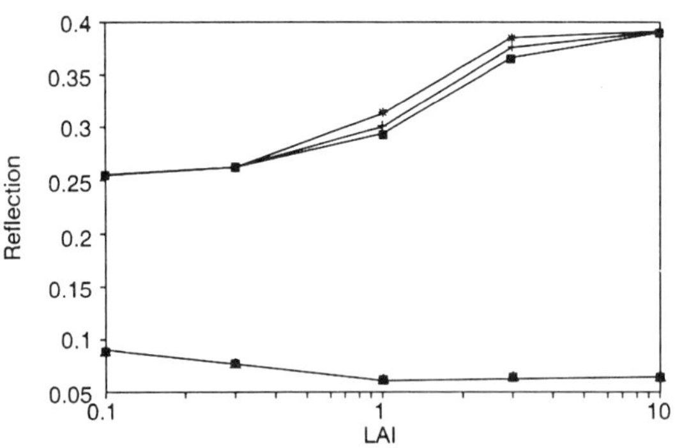

Figure 3.8 Average daily reflection of light from canopies having a range of structure parameters and leaf optical properties. Soil reflectances are ρ_s=0.1 for PAR and ρ_s=0.25 for NIR. –■–, a=0.2, x=0.5;–+–, a=0.2, x=1; –∗–, a=0.2, x=5; –□– a=0.8, x=0.5; –×– a=0.8, x=1; –▲–, a=0.8, x=5

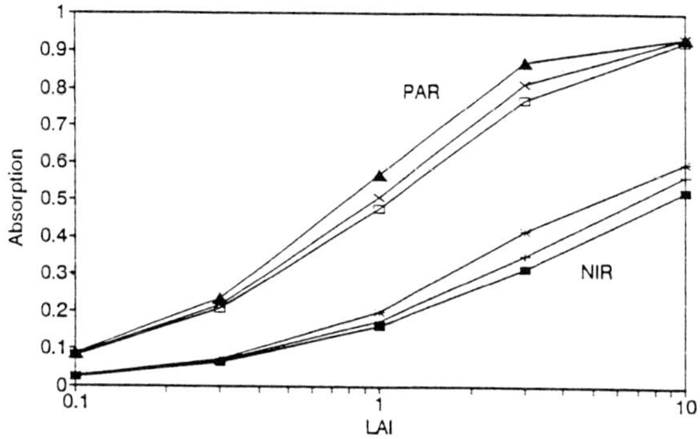

Figure 3.9 Absorption of radiation by canopies having a range of structure parameters and optical properties. –■–, a=0.2, x=0.5;–+–, a=0.2, x=1; –∗–, a=0.2, x=5; –□– a=0.8, x=0.5; –×– a=0.8, x=1;–▲–, a=0.8, x=5

can therefore fix more CO_2 per unit ground area than horizontal-leaf canopies with high light levels on top leaves, but less sunlit leaf area. This effect should result in larger values of e for inclined-leaf canopies.

In *Figure 3.10*, absorption is plotted as a function of interception. Now the values for all different canopy structure parameters fall on single lines, and only the optical properties of the leaves are important. The relationship between absorption and interception is almost linear, with a slope near 1 for PAR, 0.8 for total solar radiation, and 0.6 for NIR. Thus, even though f in Equation 1 should refer to absorbed PAR, it makes little difference whether one uses intercepted PAR (as is usually done) or absorbed PAR. If one is considering total solar radiation, then the value for e will depend on whether one has measured intercepted or absorbed solar radiation. It would be best to report e only in terms of absorbed or intercepted PAR, since it is only energy in this waveband that is useful for producing dry matter.

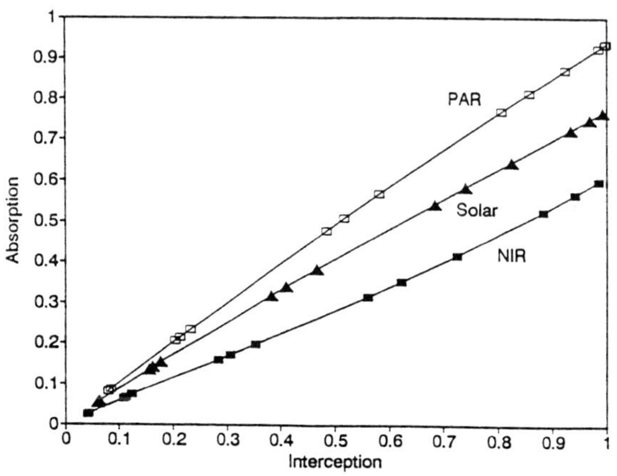

Figure 3.10 Absorption of PAR, NIR and solar radiation as a function of interception in these wavebands

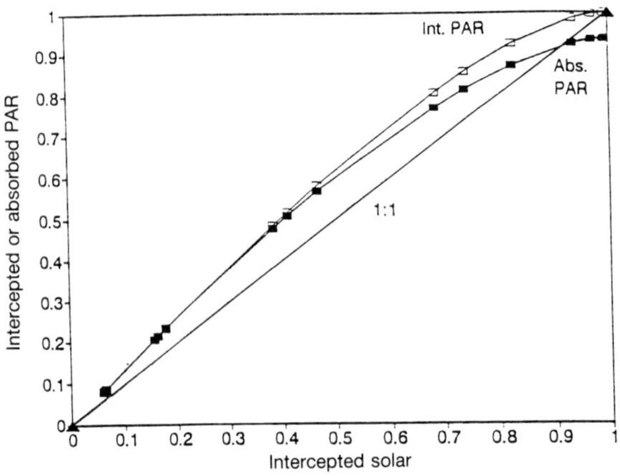

Figure 3.11 Interception and absorption of PAR as a function of intercepted solar radiation

The relationship between absorbed or intercepted PAR and intercepted total solar radiation is not linear (*Figure 3.11*), so some error could result from assuming solar and PAR interception are equal. At mid-range interception values, solar interception underestimates PAR interception by about 20%. Tube solarimeters, which are often used for determining interception, measure intercepted solar radiation. The information in Figure 3.11 should be used to convert these values to intercepted PAR, before e is computed for a canopy.

Efficiency of interception

Each increment in leaf area index produces a smaller and smaller increment in interception because of the shading of leaves by other leaves already in the canopy. Interception is $\bar{I} = 1 - \bar{T}$. A measure of interception efficiency, or the increase in interception per unit increase in L would be $\mathrm{d}I/\mathrm{d}L$. Using Equation 22 we obtain:

$$\mathrm{d}I/\mathrm{d}L = \sqrt{a}K_{\mathrm{bd}} \exp(\sqrt{a}K_{\mathrm{bd}}L). \quad (25)$$

Figure 3.12 shows values of $\mathrm{d}I/\mathrm{d}L$ as a function of leaf area index for the three canopy types used in previous analyses. At low leaf area index, the more horizontal leaves have somewhat higher interception efficiencies, but as L increases the efficiencies converge. At high L, the most horizontal leaves have the lowest efficiency.

It is interesting to determine the efficiency at which the cost of additional canopy dry matter just returns the initial investment. The energy cost of the increment in L is $(se)^{-1}$, where s is the area of leaf per unit dry mass of canopy. The return on the investment (also in terms of energy) is $\bar{S}\ \mathrm{d}I/\mathrm{d}L$, where $\bar{S}$ is the total solar energy the leaf increment can absorb in its lifetime. The value of $\mathrm{d}I/\mathrm{d}L$ at which the investment just equals the return is when $\mathrm{d}I/\mathrm{d}L = (\bar{S}se)^{-1}$. Taking very round numbers of $\bar{S}$ 600 MJ m^{-2} (30 days × 20 MJ m^{-2} day^{-1}), s = 13 m^2 kg^{-1}, (specific

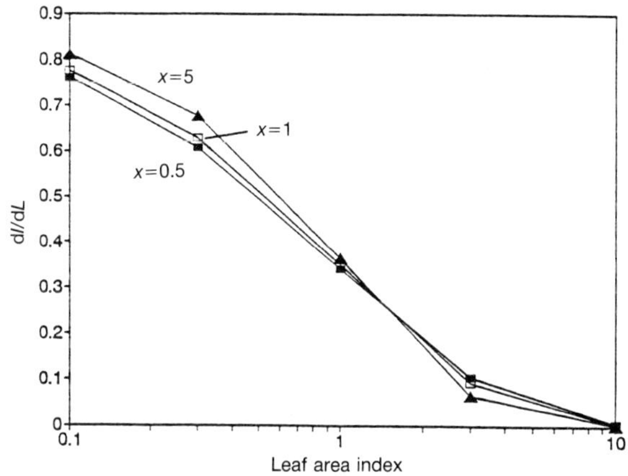

Figure 3.12 Efficiency of radiation capture of canopies as a function of LAI and canopy structure

leaf area of 26 $m^2\ kg^{-1}$; leaves are half of canopy biomass) and e = 1.5 kg GJ^{-1}, gives a break even value for dI/dL of 0.09. Even though this is a very rough analysis, a comparison of this number to the values in *Figure 3.12* indicates that new investments in leaf dry matter are likely to pay up to LAI around 3, with the payoff being a bit higher for vertical-tending canopies than for horizontal.

Applying the simple mixed canopy interception (Equations 14–16) model to a canopy composed of two species of differing height, and assuming that the two species have the same canopy light extinction coefficient, interception by the shorter of the two species is:

$$I_2 = [L_2/(L_{ib} + L_2)]\ (1-\tau_{bottom}) \qquad (26)$$

where L_{ib} is the leaf area index in the bottom layer of the taller species, and L_2 is the leaf area index of the shorter species. Total interception in the bottom layer is $1-\tau_{bottom}$, where:

$$\tau_{bottom} = \exp\ [-K(L_{ib} + L_2)] \qquad (27)$$

Taking the derivative of Equation 26 with respect to L_2 gives:

$$dI_2/dL_2 = L_{ib}/(L_{ib}/ + L_2)^2 + \exp\ [-K(L_{ib} + L_2)]$$
$$[KL_2/(L_{ib} + L_2) + L_{ib}/(L_{ib} + L_2)^2] \qquad (28)$$

Figure 3.13 shows the dependence of dI_2/dL_2 on L_{ib} and L_2, for K = 0.4.

The shorter species can increase its leaf area index profitably to the point where $dI_2/dL_2 = (S_{bottom}se)^{-1}$. Because the radiation incident on the bottom layer is a function of the leaf area index in the upper layer, the maximum leaf area index of the shorter species is a function of the leaf area index of the taller species and the ratio of heights of the species (*Figure 3.14*).

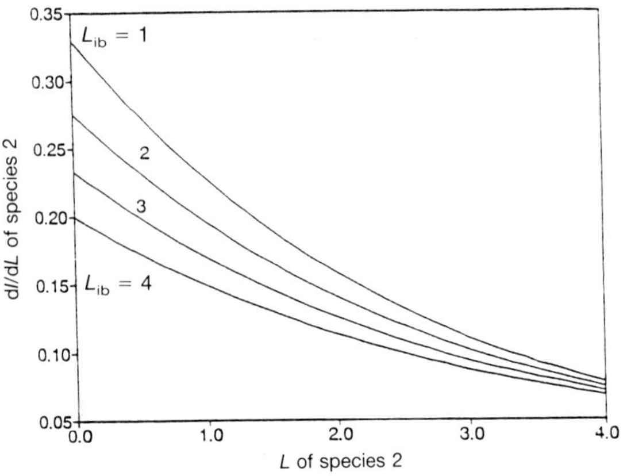

Figure 3.13 Efficiency of radiation capture by an understory canopy as a function of LAI of the understory and LAI in the bottom part of the overstory canopy

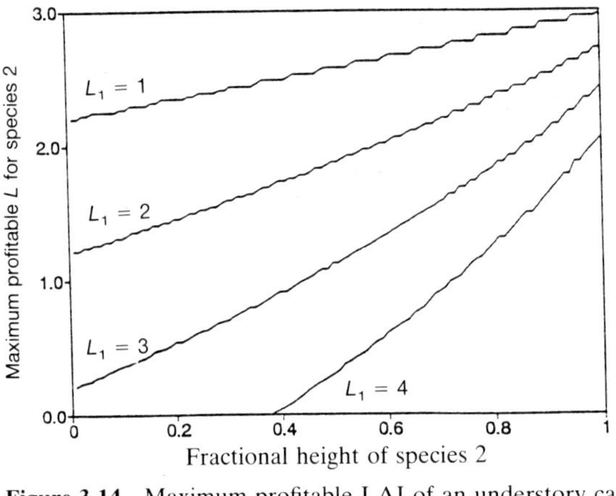

Figure 3.14 Maximum profitable LAI of an understory canopy as a function of its height (relative to the overstory) for four different overstory LAI values

Conclusions

This analysis has provided a link between the detailed models of light scattering and canopy architecture and the simple models for dry matter production based on daily interception of solar radiation or PAR. The analysis has shown the simple models to represent the detailed calculations adequately in most cases, and has confirmed Fuchs' claim that the integral of interception over days is equal to the diffuse interception for a canopy. Using these relationships, it was shown that the daily interception and absorption of PAR are about equal, while daily interception of solar radiation is about 80% of absorbed solar radiation. Intercepted solar radiation underestimates intercepted PAR by perhaps 20% at intermediate interception values.

A simple cost–benefit analysis indicates that investment in additional canopy dry matter is likely to pay up to LAI of around 3, with horizontal-tending canopies being most efficient at low LAI and vertical-tending canopies being more efficient at high LAI.

References

Campbell, G.S. (1986) Extinction coefficients for radiation in plant canopies calculated using an ellipsoidal inclination angle distribution. *Agricultural and Forest Meteorology*, **36**, 317–321

Campbell, G.S. (1990) Derivation of an angle density function for canopies with ellipsoidal leaf angle distributions. *Agricultural and Forest Meteorology*, **49**, 173–176

Campbell, G.S. and Diaz, R. (1988) Simplified soil–water balance models to predict crop transpiration. In *Drought Research Priorities for the Dryland Tropics*, pp 15–25 Edited by F.R. Bidinger and C. Johansen. Patancheru, A.P., India: ICRISAT

Campbell, G.S. and Norman, J.M. (1989) The description and measurement of plant canopy structure. In *Plant Canopies: Their Growth, Form and Function. SEB Seminar Series* **31**, 1–19. Edited by G. Russell, B. Marshall and P. Jarvis. Cambridge: Cambridge University Press

Chen, J. (1985) Kubelka–Munk equations in vector–matrix forms and the solution for bidirectional vegetative canopy reflectance. *Applied Optics*, **24**, 376–382

Fuchs, M. and Stanhill, G. (1980) Row structure and filiage geometry as determinants of the interception of light rays in a sorghum row canopy. *Plant, Cell Environment*, **3**, 175–182

Fuchs, M., Stanhill, G. and Moreshet, S. (1976) Effect of increasing foliage and soil reflectivity on the solar radiation balance of wide-row grain sorghum. *Agronomy Journal*, **68,** 865–871

Goudraan, J. (1977) *Crop Micrometeorology: A Simulation Study*. Wageningen: Pudoc, 249p

Lemeur, R. (1973) A method for simulating the direct solar radiation regime in sunflower, Jerusalem artichoke, corn and soybean canopies using actual stand structure data. *Agricultural Meteorology*, **12**, 229–247

Mann. J.E., Curry, G.L. and Sharpe, P.J.H. (1979) Light interception by isolated plants. *Agricultural Meteorology*, **20**, 205–214

Mann, J.E., Curry, G.L., DeMichele, D.W. and Baker, D.N. (1980) Light penetration in a row-crop with random plant spacing. *Agronomy Journal*, **72**, 131–142

Monteith, J.L. (1977) Climate and the efficiency of crop production in Britain. *Philosophical Transactions of The Royal Society B*, **281**, 277–294

Monteith, J.L. and Unsworth, M.H. (1990) *Principles of Environmental Physics*. 2nd Edn. London: Edward Arnold, 291p

Norman, J.M. and Jarvis, P.G. (1975) Photosynthesis in Sitka spruce (*Picea sitchensis* Bong.) Carr.) V. Radiation penetration theory and a test case. *Journal of Applied Ecology,* **12**, 839–878

Norman, J.M. and Welles, J.M. (1983) Radiative transfer in an array of canopies. *Agronomy Journal,* **75**, 481–488

Priestley, C.H.B. and Taylor R.J. (1972) On the assessment of surface heat flux and evaporation using large scale parameters. *Monthly Weather Review,* **100**, 1–8

Ritchie, J.T. (1972) Model for predicting evaporation from a row crop with incomplete cover. *Water Resources Research,* **8**, 1204–1213

Ross, J. (1975) Radiative transfer in plant communities. In *Vegetation and Atmosphere*, pp 13–56. Edited by J.L. Monteith. New York: Academic Press

Spitters, C.J.T. (1989) Weeds: population dynamics, germination and competition. In *Simulation and Systems Management in Crop Protection*. pp 217–239. Edited by R. Rabbinge, S.A. Ward and H.H. van Laar. Wageningen: Pudoc

Stanhill, G. and Fuchs, M. (1977) The relative flux density of photosynthetically active radiation. *Journal of Applied Ecology*. **14**, 317–322

Tanner, C.B. and Sinclair, T.R. (1983) Efficient water use in crop production: research or re-search? In *Limitations to Efficient Water Use*, pp 1–27 Edited by H.M. Taylor, W.R. Jordan and T.R. Sinclair. Madison, Wisconsin: American Society of Agronomy

4

RESOURCE CAPTURE BY SINGLE ROOTS

D. ROBINSON
Soil-Plant Dynamics Group, Cellular & Environmental Physiology Department, Scottish Crop Research Institute, Dundee DD2 5DA, UK

Introduction

This chapter is about the uptake of nutrients and water by *discrete* parts of *intact* root systems growing in *soil*. Why this emphasis?

Roots in *soil* have subtle relations with microbes, and with the physical and chemical features of the soil through which they grow. These relations are lost in nutrient solutions. Resource capture by roots in nutrient solutions has been reviewed many times, by Clarkson (1981), Ingestad (1982), Pitman (1982), Robinson (1986), Wild, Jones and Macduff (1987), among others. Nutrient solution experiments have been valuable but, at the same time, conspicuous in their failure to explain how roots operate in their natural environment.

The use of excised roots and root systems is, for many physiologists, the standard way to study nutrient and water uptake by plants. But this introduces nutritional and hydraulic artefacts (Pitman, 1982; Zimmermann *et al.*, 1991; Saliendra and Meinzer, 1992). The use of *intact* material avoids this problem. But for a whole root system this means treating it as a uniform entity, averaging uptake rates over the whole of its surface and over the interval between measurements. This ignores differences in maturity, geometry and anatomy between roots, and takes no account of local environmental variations in the soil around the root system. Once obtained, such averages are of limited use. Their relation to actual rates at specific points, for which mechanistic interpretations may be available, can be tenuous and misleading (Robinson, 1986; Robinson, Linehan and Caul, 1991).

Alternatively, actual rates of uptake (or close approximations to them) that are not averaged over the whole root system can be obtained by restricting measurements to *discrete* parts of intact roots. This is difficult, even with plants grown in nutrient solutions (Clarkson, 1981). Obviously, the difficulties are multiplied in soil. But a few important experiments of this type have been reported for nutrient uptake, and an influential literature has built up around them (Nye and Tinker, 1977). Similar, more recent experiments have measured water uptake *in situ*. What follows is based largely on these pieces of work.

Nutrient uptake

NITRATE AND AMMONIUM UPTAKE BY WHEAT

Clarke and Barley (1968) measured the uptake of nitrate and ammonium by roots of wheat (*Triticum aestivum*) in either a sandy or a loamy soil. Their aim was to compare uptake rates with those predicted by a simple model of diffusion to a cylindrical sink. Seedlings were grown individually in 25 mm diameter, 50 mm long cylinders containing leached, sterilized sand moistened with a nitrogen-free nutrient solution. On the appearance of a root through a small hole at the base of this cylinder, it was encouraged to grow vertically through a 5 mm layer of the soil, and into another cylinder of nitrogen-free sand. The 5 mm thick layer had previously been loaded with either a nitrate or ammonium solution containing nitrogen at a concentration of 14 mol m^{-3}. These solutions established volumetric moisture fractions in the soil between 0.04 and 0.25. Transpiration and water movements to the plant's roots were minimized by enclosing its leaves in a transparent chamber.

After 65 or 75 h, nitrate or ammonium was extracted from the soil. Uptake was estimated by the difference between the amounts of ammonium or nitrate present in the soil at the start of the experiment and those measured at the end. Because only the 5 mm length of root crossing the central layer of soil had access to it, nitrogen uptake must have taken place *via* that part of the root system. Uptake rates referred only to the 5 mm length of root, avoiding the problem of spatial averaging. These rates were not instantaneous, but were averaged over the 65 or 75 h that the experiments lasted. Some results are in *Table 4.1*.

Table 4.1 NITROGEN INFLOWS TO SINGLE, INTACT ROOTS OF WHEAT IN SOIL (CLARKE AND BARLEY, 1968)

Soil type	*Ion*	*Volumetric moisture fraction of the soil*	*Inflow* (nmol m^{-1} s^{-1})	
			Measured	*Predicted*
Loam	Nitrate	0.25	1.03	1.29
	Ammonium	0.25	0.42	n.d.*
	Nitrate	0.24	1.39	1.27
	Nitrate	0.08	0.14	0.27
Sand	Nitrate	0.17	1.31	1.13
	Nitrate	0.04	0.01	0.18

* Not determined

Net rates of nitrate uptake per unit length of root (conventionally termed *inflows*: Nye and Tinker, 1977, p. 110) in moist loam or sand were predicted to within ±9 or 16%, good for only duplicate measurements. Ammonium is absorbed by roots in nutrient solutions up to an order of magnitude faster than nitrate. But Clarke and Barley found that nitrate was absorbed by roots in soil more than twice as fast as ammonium. Ammonium's diffusivity through soil, and its concentration in the soil solution from which ions are taken up by roots, are limited by sorption reactions with negatively charged surfaces. Because nitrate does not normally react with static charges in soils, it diffuses about an order of magnitude faster than ammonium. If the rate at which a root takes up an ion from the soil solution is dominated by the rate at which the ion arrives at its surface, it follows that nitrate inflows will exceed those of ammonium, other things being equal. The sorption reactions that have such a great influence on ammonium's mobility in soil were not built into Clarke and Barley's model. Comparisons between theory and experimental data were not possible for ammonium uptake.

When the moisture contents of the loam and the sand were reduced to 0.08 and 0.04, respectively, the corresponding rates of nitrate inflow were reduced to 42% and 8% of their values in moist loam or sand. This illustrates the profound influence of soil moisture content on transport and uptake, a factor obviously irrelevant in solution cultures! Even though an ion's concentration in solution increases as soil dries, its diffusivity declines rapidly. This is because less of the soil's cross sectional area is occupied by solution-filled pores. The pathways in the soil in which diffusion can occur become more tortuous. If this was the only effect of soil moisture content on uptake rate, it could easily be accounted for in transport models by making diffusivity moisture-dependent. Clarke and Barley did this empirically. Yet, their model still failed to predict nitrate inflows in dry soils: it over-estimated the measured values by almost 90%. Factors *not* included in simple transport models come into play in dry soils. One is decreased cortical permeability in droughted roots, as found by Clarkson, Sanderson and Russell (1968) in barley (*Hordeum vulgare*) and by Jupp and Newman (1987) in rye-grass (*Lolium perenne*). Another is a decrease in contact between a droughted root and the soil around it. This may be caused by shrinkage of the cortex, the soil, or both (Nobel and Cui, 1992).

Clarke and Barley measured the depletion of ammonium and nitrate from the middle zone of soil as a fraction of that present initially. This fraction represents an *efficiency* of nitrogen capture by a single wheat root. Even after relatively short periods, and before any significant morphological responses to the localization of inorganic nitrogen would be expressed, efficiencies of nitrate capture from moist soil ranged from 12 to 33%; that of ammonium was less than 3%. In drier soil, efficiencies were less than half those in moist soil. But here, a puzzle arises. Although inflows in dry soil were reduced to 10% or less of those in the moist, *efficiencies* of nitrate capture from dry soil were 40–50% of those from moist soil. In other words, capture efficiency was maintained to a greater extent in dry soil than would be expected had it depended directly and solely on inflow. It is not obvious what other processes could explain this discrepancy. Efficiency as defined here will be a function of time: the longer a root is absorbing, the more resource it will capture.

Many solution culture experiments (Robinson, 1986) have shown that if a plant is starved of a nutrient, inflows following re-supply can for some time exceed those of plants that had received a constant supply of the nutrient. Similarly, increases in inflow occur if only part of the root system has access to the nutrient. Inflows in that part exceed those in an *equivalent* zone of a root system wholly supplied with the nutrient. Such increases are thought to *compensate* temporarily for the spatial or temporal non-uniformities in supply. They are part of the 'ecological mechanisms' that slow-growing perennials tolerant of nutrient-deficient soils express, allowing them to exploit localized, transient inputs of nutrients (such as from organic residues) in otherwise oligotrophic conditions (Robinson and Rorison, 1983; Campbell and Grime, 1989; Jackson, Manwaring and Caldwell, 1990). Most crops, bred from fast-growing annuals, are not likely to experience such local discontinuities of their nutrient supplies in fertilized, well-structured soil. Nevertheless, they have the capacity to adjust local inflow rates and, if so challenged, 'buffer' their capture of nutrients against local variation in supplies. The *physiological* mechanisms of increased inflows are not known. The inflows Clarke and Barley measured were those of plants subjected to a serious non-uniformity in nitrogen availability. They made no measurements of nitrogen uptake by discrete zones when *all* of the root system had access to nitrate or ammonium. Their work unfortunately could shed no light on the capacities of roots in soil to express any adjustments in inflow. For such information another set of experiments must be examined.

POTASSIUM UPTAKE BY ONION, LEEK AND RYE-GRASS

A technique similar to that of Clarke and Barley was used by another group to measure potassium uptake: Drew and Nye (1969), and Drew, Nye and Vaidyanathan (1969). They grew roots of onion (*Allium cepa*), leek (*A. porrum*) and rye-grass (*Lolium multiflorum*) through a 10 mm thick layer of soil loaded with $^{42}K^{+}$. The roots above and below this layer occupied either soil with the same potassium (but not $^{42}K^{+}$) availability as the central layer, or a nutrient-free acid-washed sand. The 'sand/soil/sand' sandwich was equivalent to Clarke and Barley's arrangement in which only the root crossing the central layer had access to the nutrient. The 'soil/soil/soil' treatment was a control in which all roots had access to the same nutrient supplies, but only the root in the central layer had access to the radioisotope. As in Clarke and Barley's work, transpiration and water movements in soil were minimized by keeping the plants in a humid atmosphere. Potassium uptake was measured by destructive harvesting at regular intervals up to 16 days for onion and leek, and 5 days for rye-grass.

The cumulative uptake of potassium by an onion root followed that predicted by a diffusion model (*Figure 4.1a*). Translated into instantaneous inflows (*Figure 4.1b*), the pattern was one of an initially fast inflow followed by a gradual decline and levelling-off. Initially, the concentration of potassium in the soil solution adjacent to the root from the soil would have been adequate to drive rapid inflow. But as potassium diffusivity in soil, like that of ammonium, is limited by sorption reactions, these fast inflows occurred more rapidly than the potassium could diffuse towards the root. The result was that at the root surface, potassium concentration in the soil solution declined as demand outstripped supply. Inflow, if solution culture experiments are to be believed, is a non-linear function of concentration. As potassium concentration in the soil solution declined, so did inflow. In the soils used in these experiments, the diffusivity of potassium varied with its concentration per unit volume of soil, and probably contributed to the progressive decline in inflow. These events would have continued in a self-amplifying manner until the diffusive flow of potassium through the rhizosphere from the bulk soil matched the inflow at the root surface. The approach to this steady-state corresponds to the levelling-off of the curves in *Figure 4.1b*.

In soil, such time-dependent reductions in inflow will occur as an inevitable consequence of the root's demand for a nutrient exceeding the supply from the soil. An older zone is more likely to have reached a steady-state and to absorb nutrients slower than a younger one that is extending into undepleted soil. Age-related reductions in inflow in *nutrient solution* experiments (in which demand is

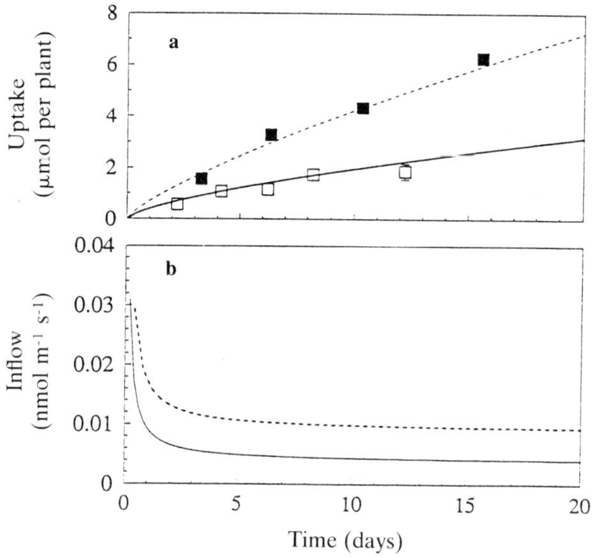

Figure 4.1(a) Time-course of potassium uptake by a 10 mm length of onion root when only one 10 mm length (■) or all the roots (□) were supplied with potassium from a sandy soil (from Drew, Nye and Vaidyanathan, 1969). Curves are theoretical values calculated from a solution of Equation 1 (see p. 67), assuming zero convection. **b** Instantaneous inflows of potassium derived from the curves in **a**

never likely to exceed supply) have been attributed to developmental changes that influence the permeability of the root's symplast and apoplast. These must also be taken into account when interpreting the constraints operating on roots in soil. As a root of a monocot species ages, the cell walls of its endodermis become suberized. Environmental conditions such as nutrition, drought, or oxygen starvation also influence anatomical development (e.g. aerenchyma formation, loss of cross-walls in large metaxylem vessels, cortical senescence) (Deacon, 1987; Justin and Armstrong, 1987; McCully and Canny, 1988; Cruz, Jordan and Drew, 1991; Saliendra and Meinzer, 1992). Endodermal suberization restricts apoplastic transport from the cortex into the stele. Substances transported apoplastically are then taken up faster close to the root's apex than in mature regions (Clarkson, 1981). Both symplastic and apoplastic transport may be reduced if cortical cells become senescent. Those parts of the root would, presumably, become more susceptible to shrinkage and loss of contact with the soil during drying making them less permeable to nutrients and water. This reduces progressively the uptake rates of substances transported apoplastically (e.g. calcium and water) *and* those transported primarily symplastically (e.g. nitrate and phosphate). However, the precise reduction associated with developmental events must depend on species and on the physical properties of the soil.

The role of suberized cell walls in the endodermis (and exodermis, too, in some species: Peterson, 1992) has been likened to that of a damp-proof course in a building, preventing the flow of water to an undesired destination (Clarkson, 1981). In a root, this means regulating the passage of materials from the soil into the xylem by directing their flow across at least two membranes. An alternative analogy would be that of the butyl rubber lining of a garden pond, minimizing the flow of water (and nutrients) *from* the root, especially its stele. This is likely to be most important in dry soil. There, the gradient in water potential driving the flow can be from root to soil. The flow can be as liquid or vapour if air gaps develop around a shrunken root (Nobel and Cui, 1992). The stele contains the plant's pathways of long-distance transport, whereas the cortex is a relatively ephemeral tissue whose loss does not necessarily affect the root's uptake nor its capacity for long-distance transport. A point of high resistance to water flow around the stele would entrain the stele's water potential with that of the xylem and shoot, while that of the root's cortex would tend to follow the soil's water potential (Passioura, 1985). The loss of water from the stele would then be minimized should low water potentials become established in the cortex and beyond. A similar barrier below the epidermis, in the exodermis, may be a second line of hydraulic defence, and (or) it may stiffen the root's circumference against massive shrinkage, so preserving good contact with the soil.

From Drew and Nye's (1969) experiment it was possible to obtain concentration-inflow relations, 'concentration' being estimated by the diffusion model for the soil solution at the root surface, and 'inflow' taken as the mean over the 10-day experiment. In the same soil, leek absorbed potassium almost twice as fast as did onion at concentrations > 0.2 mol m^{-3} (*Figure 4.2a*). The reason for this superiority is not known. Rye-grass absorbed potassium even faster, about three times faster than onion, but because only two data points are available, not too much should be read into precise comparisons. One reason for the superior potassium inflow of rye-grass could have been the extra surface area afforded by root hairs. Root hair production does depend on, among other things, nutrient availability. Roots in nutrient-deficient plants tend to be hairier (Robinson and Rorison, 1987), a good example of the coupling between current resource availability in the soil and the facultative development of certain features that can influence the future availability of the resource. Only a fraction of the observed uptake of potassium by rye-grass could have originated from the volume of soil enclosed by its root hairs. The rest must have come from the soil beyond, by diffusion in response to the depletion in its concentration at the root surface.

Both onion (*Figures 4.1 and 4.2b*) and rye-grass (*Figure 4.2b*) roots expressed a potentially compensatory response to the spatial localization of potassium. The inflow in onion increased 2.4-fold in the 'sand/soil/sand' system compared with the control; that of rye-grass increased by almost 8.3-fold. Whereas wheat's efficiencies of nitrate capture from dry or moist soil were *not* related directly to its corresponding inflows (above), those of potassium capture by onion and rye-grass with restricted access to the nutrient were. (For potassium, efficiency is expressed as the quotient of the amount taken up during the experiment and that which could potentially have been taken up after an infinite time.) The increases in capture efficiency corresponding to those in inflow quoted above were 2.3- and 7.1-fold.

Efficiencies of potassium capture by these three species are plotted in *Figure 4.3* in relation to the initial concentrations of potassium in the soil solutions. These data do not suggest that any of these species is more or less efficient than another in potassium capture.

To summarize, the inflow of potassium varies with its concentration in the soil solution at the root surface. When access to potassium is restricted to a discrete part of a single root, inflow can increase and is associated with a proportionate increase in capture efficiency.

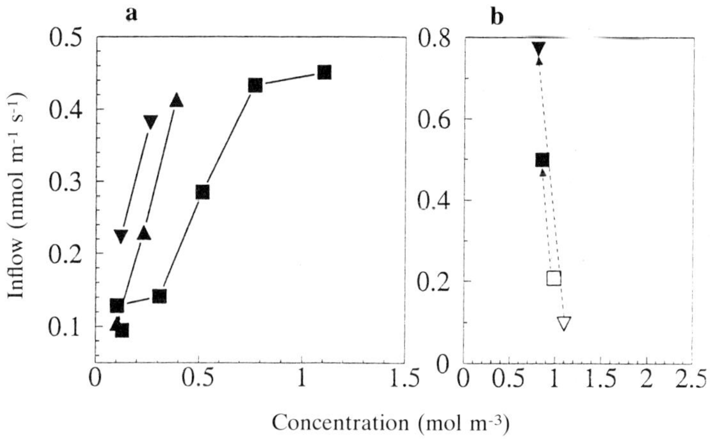

Figure 4.2 Potassium inflows in relation to its calculated concentration in the soil solution at the root surface. **a** Rye-grass (▼), leek (▲) and onion (■), in which only one 10 mm length of root was supplied with potassium from a clay soil **b** Closed symbols as for **a**, for plants grown in sandy soil, compared with their inflows when all their roots were supplied with potassium (open symbols). Arrows indicate the extent of the 'compensatory' response (Drew, Nye and Vaidyanathan (1969) and Drew and Nye (1969))

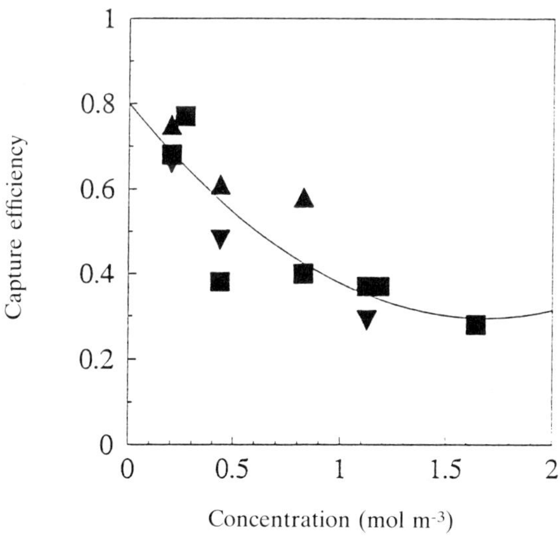

Figure 4.3 Efficiencies of potassium uptake by leek (▲), onion (■) and rye-grass (▼) roots in relation to the calculated concentration of potassium in the soil solution at the root surface. Efficiency is the quotient of the amount of potassium taken up during the experiment (4 days for rye-grass, 10 days for onion and leek) and that which could have been taken up after an infinite time (Drew, Nye and Vaidyanathan (1969) and Drew and Nye (1969))

PHOSPHATE UPTAKE BY ONION, LEEK AND RYE-GRASS

Drew and Nye (1970) followed up their work on potassium with a similar series of experiments on phosphate uptake. The middle layer of soil was loaded with ^{32}P-labelled phosphate, but otherwise the techniques they used were the same as for potassium.

The most striking difference they found between the uptake of potassium and that of phosphate, and between phosphate uptake by roots in soil and those in nutrient solutions, was that *no* potentially compensatory responses to the localized supply were expressed (*Figure 4.4a*). This was because phosphate diffuses in soil an order of magnitude slower than potassium, for example. The initially fast inflow of phosphate (*Figure 4.4b*) depleted its concentration in the soil solution close to zero. This dilution was maintained thereafter because phosphate could not diffuse through the soil fast enough to restore its original concentration close to the root. Such extreme, local depletions of phosphate around living roots have been visualized auto-radiographically (Nye and Tinker, 1977, p. 130). After very short times, therefore, the rate of phosphate inflow is dominated by its rate of diffusion towards the root. Inflow rate is probably always close to a maximum set by the soil, irrespective of the extent to which phosphate is spatially localized.

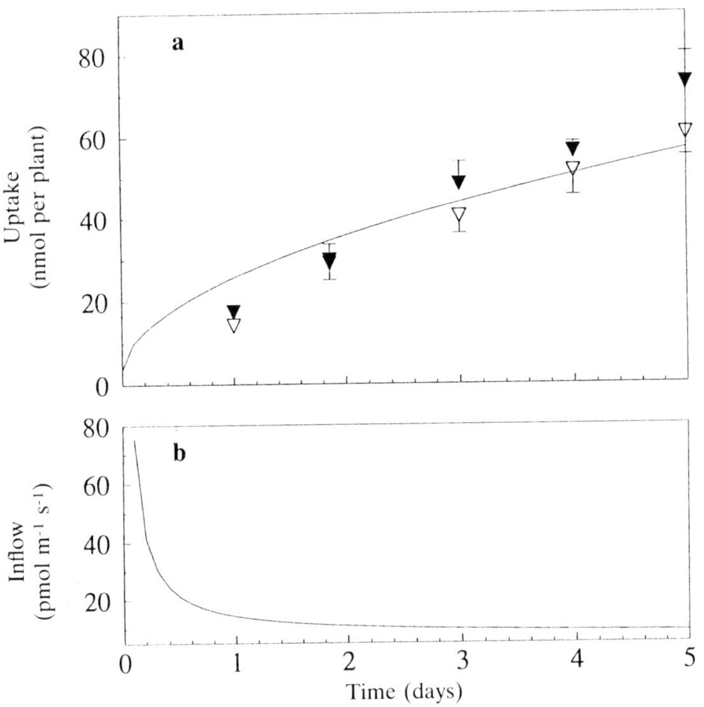

Figure 4.4 As for *Figure 4.1*, for phosphate uptake by rye-grass (Drew and Nye, 1970). Only one curve is shown in each graph because there was no significant difference between the treatments

No potentially compensatory increases in inflow can be expressed. For the same reason, Drew and Nye also found no difference between the phosphate inflows of onion, leek, and rye-grass, again in contrast to potassium (above).

The concentration-inflow relation that Drew and Nye measured for onion in a clay soil was strongly non-linear, reminiscent of many reported by others using solution cultures. It is reassuring that there is *some* agreement between roots in soil and those in solution!

Efficiencies of phosphate capture by onion and leek roots ranged from 65% to 98% over 10 days; those by rye-grass roots were 55%, for only a 4-day uptake. Again, at a given phosphate concentration in the soil solution, there was little evidence to suggest that any of these species had a significantly superior efficiency of phosphate capture. Some efficiencies Drew and Nye measured for onion exceeded 100%. This illustrates the difficulty of estimating reliably the potentially available amount of a nutrient that undergoes complex chemical reactions, exists in several interchangeable forms, and occupies both solid and liquid phases in the soil.

To summarize, there is no evidence that, in soil, roots of any of the species for which data are available can express potentially compensatory inflows of phosphate. This is because of phosphate's poor mobility in soil.

Ways of mitigating the constraint on a plant's capacity to capture phosphate (and other poorly mobile and available nutrients) have evolved. The most well-known is the sometimes-symbiotic relation between plants and mycorrhiza-forming fungi. Those who work on mycorrhizas are fond of pointing out that there is no such thing as a 'single root'. The mycorrhizal condition is far more common among crop species than the non-mycorrhizal state studied in many physiology laboratories. Hyphae that link the root to the soil beyond the root hair zone can 'short-circuit' the diffusive pathways through the soil to the root surface (Li, George and Marschner, 1991). They also increase the root's absorptive area per unit length. More subtle mechanisms of resource capture have been attributed to mycorrhizas. For example, certain ectomycorrhizal fungi can translocate to trees, whose roots they infect, nitrogen derived from proteins in leaf litter (Abuzinadah and Read, 1986). These trees have access to this source of nitrogen *before* it is broken down into ammonium and nitrate by free-living saprophytes. This again is a kind of 'short-circuit' of the generally accepted routes by which plants capture nutrients. It is not likely to be a significant route for resource capture by annual crops, but may be important in forests. A plant cannot be presumed to benefit from mycorrhizal associations. The effect of the association on the plant may be positive, negative or neutral, depending on species, soil and environment, the latter including neighbouring plants (Newman, 1991). Another way of capturing poorly mobile nutrients is the release from roots of substances that render the nutrients more soluble and more mobile. Examples include the acidification in the rhizospheres of white lupin (*Lupinus albus*) and rape (*Brassica napus*). This shifts the sorption equilibria of phosphate (and probably of some micronutrient metal cations also), so increasing its equilibrium concentration in the soil solution (Marschner, 1991). Ligands are also released from roots. These form mobile complexes with iron, for example, which allows this element to be captured effectively from soils in which it is otherwise barely available (Römheld, 1987). Other, less familiar mechanisms were described by Lamont (1982).

Water uptake

ONION

Wray and Tinker (1969; see also Nye and Tinker, 1977, pp. 153–156) grew an onion root through a 1 mm layer of continually moistened soil. They passed a stream of dry air over the plant's leaves, trapping the water transpired and thereby obtaining a direct measurement of the rate at which water entered the root. From these measurements, inflows in the range 0.036–0.072 $mm^3\ m^{-1}\ s^{-1}$ can be calculated. They are comparable to inflows measured by micropotometry on various species grown in nutrient solutions (*Table 4.2*). This suggests that when the soil is close to saturation and the contact between it and the root is good, resistances to the flow of water from soil to root are low – so low, in fact, that inflow is not determined by rates of flow to the root, but by the conductance of water through the plant. This would have been the case in the solution culture experiments, in which resistances to the flow of water to the roots were, by definition, zero.

Table 4.2 WATER INFLOWS TO DISCRETE ZONES OF INTACT ROOTS

Species	*Methodology*	*Inflow* ($mm^3\ m^{-1}\ s^{-1}$)	*Reference*
Solution culture experiments			
Vicia faba	Micropotometry	0.056–0.23	1
Hordeum vulgare	Micropotometry	0.016–0.11	2
Curcubita pepo	Micropotometry	0.016–0.10	2
Zea mays	Micropotometry	0.0092–0.037	3
Soil experiments			
Allium cepa	β scanning	0.036 to 0.072	4
Raphanus sativus	CT scanning	0.23 to 0.56	5

1, Brouwer (1953)
2, Graham, Clarkson and Sanderson (1974), calculated assuming a root radius of 0.1 mm
3, Hayward, Blair and Skaling (1942), calculated assuming a root radius of 0.1 mm
4, F. J. Wray, quoted in Nye and Tinker (1977, p. 155), calculated assuming a root radius of 0.26 mm (Drew, Nye and Vaidyanathan, 1969)
5, Hainsworth and Aylmore (1986)

Dunham and Nye (1973) measured water uptake by 10 mm lengths of onion roots arranged in a plane adjacent to a block of a sandy loam soil, rather than by an individual root at the centre of a soil cylinder as for nutrients. Three volumetric moisture fractions were established in the soil blocks. After 2.5 or 6 days, the blocks were cut at successive 1.5 mm distances from the root plane and their moisture contents determined. Uptake was calculated as a flux rather than an inflow, i.e. as volume of water per unit area of the root plane per unit time. Flux increased linearly as the soil's moisture content increased. Significant gradients in moisture content were measured adjacent to the roots in the two driest soils. Drier soil has a lower hydraulic conductivity than when wet. This limited the speed at which water extracted from the soils in Dunham and Nye's experiment was replaced by the movement of water towards the roots, so reducing the flux into the root. This effect was not seen in the wettest soil, in which gradients were not detected and flux increased with time.

It is not possible to estimate efficiency of water capture from the data presented in these papers. Nor is it possible to deduce how inflow varied precisely with the plant's transpiration rate.

WHEAT

No one has measured potentially compensatory increases in water inflow comparable to those described above for potassium. Passioura's (1972) experiment is the nearest equivalent. He grew wheat plants in which only one or all of their seminal roots, plus laterals, had access only to water stored in the soil during 70 days' growth.

Single rooted plants initially took up water slower than the others, but their total lengths of root were similar. If anything, the inflows in the single rooted plants were less than those of the fully rooted ones. Passioura attributed this to the former plants having high resistances to water flow through their xylem to the shoot. The xylem connection between the single root and the shoot consisted, essentially, of a single large vessel 50 μm in diameter. Fully rooted plants had at least three such vessels and correspondingly lower resistances. Here is an example of an anatomical constraint on resource capture by a single root. The result of this constraint in Passioura's experiment was that the single rooted plants apparently conserved water in the soil, so that by anthesis, they had more water available. They out-yielded the fully rooted plants' grain production by 100%, yield being positively correlated with the amounts of water available to the plants at anthesis.

The interpretation of this result is not straightforward, however. Different shoot dry weights were established in the two treatments, and this would have fed back onto the plants' respective demands for water. The treatments appear to have altered allocation patterns between and within the roots and shoot, affecting tillering and the trade-off between grain number per ear and weight per grain. The sequence of causes and effects leading from the restriction of the root systems of some plants to their conservative use of water and higher yield is far from clear.

RADISH

Recent experiments have used modern instrumentation to measure water uptake non-destructively. These include computerized tomography (CT) scanning and nuclear magnetic resonance (NMR) imaging. Using them, near-instantaneous inflows can be obtained, removing any lingering uncertainties associated with time-averaging these rates, while adding some new ones.

Hainsworth and Aylmore (1986) and Hamza and Aylmore (1992a, b) used CT scanning to measure moisture gradients around a radish (*Raphanus sativus*) root growing in a sand–kaolinite mixture. At first sight, the most surprising thing about their results was that any gradients were measured at all. When water is being removed from moist sand, its initially high mobility should prevent the formation of sharp gradients in moisture content around the point of extraction. Instead, a concentric reduction in moisture content might be expected around it with (if the volume of sand is large enough) the maintenance of the initial content in sand some distance away (Passioura, 1988; Grose and Hainsworth, 1992). Once the uniform, relatively dry zone is established, movement of water into it would be severely restricted by the very low hydraulic conductivities of dry sand, compared with the rates at which roots can take up water. Instead, Aylmore and his colleagues detected gradients very similar to those which Dunham and Nye (1973) measured in their blocks of sandy loam. Subsequent direct measurements (P. J. Gregory, personal communication) of the sand's moisture content confirmed the reliability of Aylmore's results. The answer must lie in the kaolinite. When Mullins and Panayiotopoulos (1984) added kaolinite (between 2 and 8% by weight) to sand, the mixture's moisture release characteristic shifted substantially from that of pure sand. If applicable to Hainsworth and Aylmore's mixture (15% kaolinite by weight), the moisture gradients should be similarly different from those expected around roots in pure sand: see the comments below about Grose and Hainsworth's (1992) experiments.

The inflows of water that Hainsworth and Aylmore calculated from their data (0.27–0.56 $mm^3\ m^{-1}\ s^{-1}$) were unusual in that they were an order of magnitude faster than the only directly comparable data (calculated from those of Wray, quoted in Nye and Tinker, 1977, p. 155). They were also significantly faster than inflows measured by micropotometry on solution-grown plants (*Table 4.2*). Hainsworth and Aylmore provided qualitative information on how inflow was coupled to transpiration. At a fast transpiration rate, inflow was increased compared with that at a slow one, in both wet and relatively dry sand. Unfortunately, they did not report the vapour pressure deficits that the plants experienced. It was not possible to quantify how inflow and vapour pressure deficit co-varied.

Inflow measured along the root increased as the root aged and as its diameter increased. When inflow is corrected for diameter, i.e. converted to a flux, rates of water uptake close to the root's tip were twice as fast as those further back. Rates of change (as a result of uptake and subsequent transport) in the amount of substance per unit volume of space (i.e. the substance's concentration) around a cylindrical sink are related *directly* to the rate at which the substance is removed by the sink per unit of its length, i.e. to the inflow. This can be shown by deriving from first principles the continuity equation for transport in a cylinder (see below). It is related less directly to the flux into the sink. This is why inflow, not flux, is the conventional measure of uptake rate by roots. But these relations are true only if the sink is a cylinder, i.e. it has a *constant* diameter. Roots of members of the Poaceae, Liliaceae and Leguminosae, including many of the crops favoured by plant physiologists, are usually more-or-less cylindrical. Those of many dicots thicken secondarily and are evidently not cylindrical, as Hainsworth and Aylmore reported for radish. For such plants, inflow is a less meaningful measure of a root's activity.

LUPIN

Hamza and Aylmore (1992a) also measured unexpectedly sharp gradients in moisture content around a lupin (*Lupinus angustifolius*) root in the sand–kaolinite mixture, using the same CT scanning technique. But when Grose and Hainsworth (1992) measured them in pure sand, the gradients were not unusual (*Figure 4.5*).

Grose and Hainsworth showed that only 8 days after a lupin root was infected with a pathogenic fungus, *Phytophthora cinnamoni*, its capacity to take up water, as determined from the moisture gradients established around it, was virtually non-existent (*Figure 4.5*). The rapidity of this effect argues against the pathogen impairing water uptake by inducing developmental changes in the root. It points instead to a biochemical or ultra-structural change which modifies the root's permeability to water. Of these changes, variable plasmodesmatal conductance is the likeliest, but unproven, candidate (Passioura, 1991a).

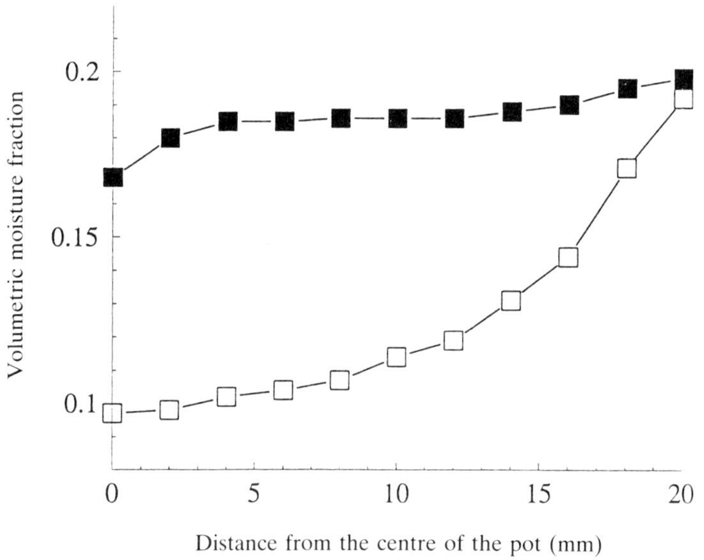

Figure 4.5 Volumetric moisture fractions of sand containing a single root of lupin at the centre of a 20 mm diameter pot, measured at a depth of 10 mm by CT-scanning, when healthy (□) and 8 days after being infected with *Phytophthora cinnamoni* (■) (Grose and Hainsworth, 1992)

PINE

Strange results were obtained by MacFall, Johnson and Kramer (1991) who used NMR imaging to 'visualize' the depletion of water around individual roots of a loblolly pine (*Pinus taeda*) seedling growing in initially moist sand. Over 9 hours, water was apparently not removed concentrically from the sand around individual roots. Certain zones were drained rapidly, while adjacent ones were not. Even if water was being absorbed eccentrically by the pine roots because of localized suberization or wounding (Caldwell, 1976), asymmetrical moisture gradients in wet sand would disappear very quickly. This would then establish the concentric gradients expected around a cylindrical sink. MacFall *et al.*'s results currently defy conventional explanations. Mechanisms involving roots modifying the rhizosphere's viscosity or compaction are all possible, but speculative.

They also measured the efficiency of water uptake by different classes of root (taproot, lateral and fine root), 'efficiency' here being the quotient of the amount of water absorbed by a root of one class and that absorbed by another. Roots of individual seedlings expressed very different efficiencies, depending on how many laterals or fine roots they had. Tap roots were usually the most efficient per unit length, despite their secondary thickening. Presumably this was because they contained more xylem and were therefore more conductive for a given suction, compared with the narrower roots. This interesting observation shows that the 'division of labour' in resource capture can be varied between members of a root system even if they have different ages and anatomies.

To summarize, techniques allowing non-destructive measurements of water uptake are becoming available, but require careful evaluation before the information they produce can be reconciled with established theories. The potential of these methods to produce instantaneous measurements and images of roots in soil will probably be the best way of answering some long-standing questions about the interactions between soil and plant resistances to uptake, and between the uptake of water and nutrients.

Interactions between the uptake of water and nutrients

EFFECTS OF WATER ON NUTRIENT UPTAKE

Diffusion and convection in relation to moisture

One interaction has been mentioned already: in dry soil, ions diffuse more slowly than in wet. Dry soil's hydraulic conductivity is also less than when it is wetter, so reducing the rate at which the soil solution arrives at the root surface by convection. Both effects will reduce inflow of a nutrient, other things being equal. The effect of ions becoming more concentrated in solution in a drying soil is unlikely to offset the simultaneous fall in hydraulic conductivity (which may be by orders of magnitude: Passioura, 1988).

Debates among plant physiologists about the importance of transpiration for nutrient uptake could be resolved with a deeper appreciation of the physical interactions between the supplies of the two resources. For example, Tanner and Beevers (1990) grew maize in atmospheres with different relative humidities. The resulting 3-fold difference in water uptake during 4 or 5 weeks' growth was, to their surprise, associated with no difference between the plants' production of dry matter or uptake of nutrients. But mobile nutrients (e.g. nitrate, sulphate) in moist soil will diffuse almost as fast as the plant's roots can take them up. Reducing water uptake without reducing the soil's moisture content should have no effect on potential supplies of mobile nutrients. The supplies of less mobile nutrients (e.g. potassium, phosphate, calcium) are independent of water uptake whatever the plants' uptake of water, provided that the soil does not dry.

Physical coupling between water and nutrient uptake

There are three extremes of physical interaction between water and nutrient inflows. First, when water inflow exceeds that of the nutrient (i.e. there is an exclusion of the nutrient by the root), the nutrient's concentration at the root surface tends to increase relative to that some distance from the root. The nutrient then diffuses down its concentration gradient against the convective flow. Fast inflows of water are likely if the plant is transpiring rapidly and if its roots are in moist soil. But back-diffusion in such soils will occur rapidly to disperse any build-up in concentration (Passioura and Frere, 1967; Nye and Tinker, 1977, p. 154; Hamza and Aylmore, 1992a).

Second, when the inflows of water and a nutrient are matched, if, for example, the concentration of a nutrient in the soil solution is 10^{-3} mol m^{-3} (1 mmol m^{-3}), the nutrient's inflow rate is 10^{-12} mol m^{-1} s^{-1} (1 pmol m^{-1} s^{-1}) and that of water 10^{-9} m^3 m^{-1} s^{-1} (1 mm^3 m^{-1} s^{-1}). Then, no concentration gradient of the nutrient could exist around the root, so no net diffusive inflow could occur. The root's entire supply of that nutrient would be convective.

Third, when the inflow of nutrient exceeds that of water. The nutrient's concentration at the root surface falls because its rate of replacement from the soil is less than that of its removal by the root. Diffusion towards the root then occurs. The total supply to the root is a combination of diffusion and convection. As mentioned earlier, nitrate and sulphate are so mobile in moist soil that their diffusive supplies virtually match those by convection. Their concentrations at the root surface are depleted only slightly compared with those of phosphate and potassium. These three extremes tend to merge into one another daily as the rate of transpiration rises and falls.

Modelling interactions

The physical interactions between water and nutrient uptake have been formalized mathematically for many years (Passioura and Frere, 1967; Nye and Tinker, 1977, Ch. 6). The basis of all such models is the continuity equation for a nutrient ion which, in its simplest form for cylindrical geometry, is:

$$\partial C/\partial t = (1/r) \,.\, (\partial[rD(\partial C/\partial r) + (rI_w C_l/2\pi a)]/\partial r) \qquad (1)$$

C is the ion's concentration per unit volume of *soil* (mol m^{-3}). C_l is the ion's concentration per unit volume of soil *solution* (mol m^{-3}). D is the ion's diffusion coefficient in the soil (m^2 s^{-1}). For mathematical convenience, D is usually assumed to be independent of the ion's concentration, but this is often not true (e.g. Drew and Nye, 1969, 1970; Drew *et al.*, 1969); it certainly depends strongly on the soil's moisture content (Nye and Tinker, 1977, p. 78). r is the radial distance (m) from the root's axis. t is time (s). I_w is the volume inflow of soil solution to the root (m^3 m^{-1} s^{-1}). a is the root's radius (m); this is usually assumed to be measured from the root's axis to its epidermis, even though uptake may occur from within the cortex, at the endodermis if the cortex is missing or damaged, or within the root hair cylinder, and appropriate corrections to a can be made to account for these features.

Zero inflows of nutrients and water at a certain time-dependent value of r are often assumed in order to define the volume of soil from which the single root extracts resources. This value is approximated by $a + 2\,(Dt)^{1/2}$ for as long as it is less than half the distance between neighbouring single roots. The latter is $(\pi L_v)^{-1/2}$ in a system of upright, parallel, regularly spaced cylinders, where L_v is the length of those cylinders (i.e. the roots) per unit volume of soil (m m^{-3}) (Nye and Tinker, 1977, p. 231). This boundary condition defines a 'watershed' for each root, beyond which it has no access to either water or nutrients, but within which it is the sole sink for them. This is a gross caricature of the configurations of roots in soil, and it can produce wildly inaccurate predictions of uptake, but it is the simplest geometry that can be devised. Inner boundary conditions, i.e. at the root surface, $r = a$, include: the assumption that a zero concentration is maintained there, when inflow exceeds supply (Gardner, 1965; Robinson, 1986); a direct proportionality between inflow and concentration (Nye and Tinker, 1977, p. 110); or a non-linear relation between inflow and concentration, as in Michaelis–Menten kinetics (Nye and Tinker, 1977, p. 106).

Assuming that a steady-state has been reached and that appropriate boundary conditions apply, Equation 1 can be solved analytically to give the nutrient inflow by a combination of diffusion and convection (Gardner, 1965). When water uptake is zero, Equation 1 reduces to the 'diffusion equation' (Nye and Tinker, 1977, p. 132), from which a simple equation describing nutrient inflow by diffusion alone can be derived (Gardner, 1965; Robinson, 1986). At the other extreme, if the rate of water uptake far exceeds that of the nutrient and $\partial C/\partial r$ is zero (i.e. there is no concentration gradient around the root: see above), Equation 1 then reduces to another, simpler form. From that, an equation describing nutrient inflow by convection can be obtained (Gardner, 1965).

I_w can be obtained as a solution to an analogous continuity equation for water:

$$\partial \theta_v/\partial t = (1/r).(\partial[rD_w(\partial \theta_v/\partial r)]/\partial r) \qquad (2)$$

θ_v is the volumetric moisture fraction of the soil (m^3 m^{-3}). D_w is the apparent diffusivity of water in the soil (m^2 s^{-1}); D_w is often assumed independent of θ_v, and although it is in soil close to saturation, it depends strongly on θ_v when the soil is unsaturated (Hainsworth and Aylmore, 1986). D_w is related to hydraulic conductivity, but not uniquely so (Campbell, 1985, p. 57).

Some of the most useful solutions to these equations have been given by Gardner (1965), Passioura and Cowan (1968), Tinker (1976), Nye and Tinker (1977), Passioura (1980) and Campbell (1985). Inevitably, analytical solutions will be approximations. But an approximate analytical solution is often preferable to an exact, numerical solution to get a feeling for how the important variables interact, or to test the effects of explicit assumptions (e.g. steady-state or steady-rate conditions: Passioura and Cowan, 1968) against experimental data. Even though 'the pursuit of exact [analytical] solutions at considerable effort is hardly warranted in a situation where our understanding of the detailed processes is so incomplete' (Tinker, 1976), considerable effort *has* been expended on producing such solutions. Their gruesome mathematics does not make many of these solutions easily accessible to, and usable by, other scientists. Numerical or approximate analytical solutions are adequate for most applications, given the imprecisions and uncertainties in experimental data.

How well are interactions between water and nutrient uptake modelled?

Wray and Tinker (1969; see above) induced a build-up in the concentration of ^{35}S-labelled sulphate around a single root of onion. The radiation emitted from

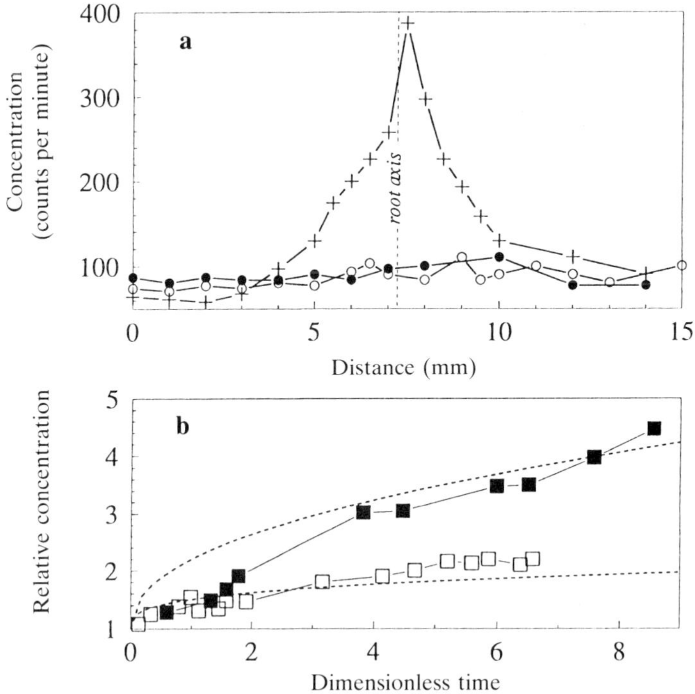

Figure 4.6(a) Concentrations (in counts per minute) of ^{35}S-labelled sulphate in soil around a single root of onion, initially (●), 18 h after inducing a fast rate of transpiration in a flow of dry air (+), and 16 h after stopping the airflow (○). **(b)** Concentrations of ^{35}S-labelled sulphate at the root surface relative to the initial concentration, in relation to time scaled as a function of the rate of water uptake and of the diffusivity of sulphate in soil. Symbols are experimental data for fast (■) and slow (□) rates of water uptake relative to the corresponding rates of sulphate uptake. Curves are theoretical values calculated from an approximate solution to Equation 1 appropriate for planar coordinates (Wray, in Nye and Tinker, 1977, pp. 153–155)

the soil was monitored non-destructively using a β-detector mounted behind a moving slit. A 4-fold increase in radioactivity at the root surface was seen after dry air had been flowing across the leaves for 18 h to induce a fast rate of transpiration, compared with that detected 5 mm away from the root (*Figure 4.6a*). The sulphate concentration declined approximately exponentially with distance from the root. About 14 h after the air flow was stopped, the concentration gradient had disappeared. Back-diffusion of sulphate in the moist soil had occurred in that time. The geometry of the experimental system used by Wray and Tinker was effectively planar rather than cylindrical as the root was embedded in a wide slab of soil only 1 mm thick. Movements of water and ions to the root were approximately linear. An analytical solution to a planar form of Equation 1 was available with which they could compare their results. The concentration of sulphate at the root surface changed with time and with the rate of water uptake generally as the theory predicted, but with some systematic deviations between them (*Figure 4.6b*).

Hamza and Aylmore (1992a, b) used a combination of CT-scanning and ion-specific microelectrodes to investigate the complex interactions between rates of water uptake by single roots of lupin and radish and the concentrations of sodium in a moist sand–kaolinite mixture. They separated the matric and osmotic potentials driving the movement of water to or from the roots. Osmotic potential was the major influence on water fluxes. The dynamics of water uptake involved opposing contributions of matric and osmotic potentials to the net potential difference in free energy between the root surface and the bulk soil. These effects depended on the initial concentration of sodium (in this case) in the soil solution. The moist soil allowed substantial back-diffusion of sodium to occur, so that the build-up in concentration was not exponential with time, as expected, but linear.

Another test was performed by Dunham and Nye (1974). They made destructive measurements of concentration gradients of phosphate, potassium, calcium and magnesium in blocks of soil adjacent to a plane of onion roots (see above). A range of soil moisture contents was used. They compared their measurements with the predictions of a numerical solution to a planar continuity equation. The model predicted the general patterns of the concentration gradients of phosphate and potassium, but under-estimated the distance to which the latter spread. This was attributed to inadequate understanding of desorption processes. Phosphate uptake was more sensitive to the soil's moisture content than that of potassium; magnesium uptake was more sensitive than that of calcium.

While it is good to see some agreement between theory and experiment, one of the most valuable applications of models is to check for *discrepancies*. These can provide clues about the constraints under which real roots operate, provided that the model is constructed simply enough for its underlying assumptions and internal workings to be apparent. An example is Brewster *et al.*'s (1976) discovery that rape plants took up more phosphorus from phosphate deficient soil than theory predicted. This suggested the possibility that rape could modify phosphate availability in its rhizosphere in ways not included in the model, a suggestion that was subsequently proven. Another is the disagreement between single root models and uptake by whole root systems; these are discussed below.

EFFECTS OF NUTRIENTS ON WATER UPTAKE

Nutrient deficiency can reduce a root's hydraulic conductivity in nutrient solutions, but this effect has never been examined in soil. In cotton (*Gossypium hirsutum*), Radin and Matthews (1989) measured a 50% decrease in the hydraulic conductivities of whole roots and of their cortices within 3 days of the plants being deprived of nitrogen or phosphate. The same was found in the roots of tomato (*Lycopersicon esculentum*) starved of nitrogen for 8 days, but not in those of barley, whose hydraulic conductivity *increased* (Chapin *et al.*, 1988). It is tempting to speculate that, because the response of hydraulic conductivity to nutrient availability was so rapid, an important factor might again be variable plasmodesmatal conductance, especially in the cortex. Changes in it can precede certain developmental events such as flowering in some species, but it is not known to what extent the two are linked causally.

Drew, He and Morgan (1989) and He, Morgan and Drew (1992) found that the temporary removal of phosphate or nitrogen from solutions containing maize

(*Zea mays*) plants increased the sensitivity of their cortical cells to ethylene. This substance is synthesized in oxygen-starved roots, and precedes the formation of aerenchyma in them. Aerenchyma allows tissues far from an oxygen supply to be ventilated by the movement of large volumes of air through the root. Presumably its formation in nutrient-deficient roots does not occur primarily to ventilate them. Partial pressures of ethylene in well-aerated soil may still be sufficient to induce aerenchyma in roots sensitized by nutrient deficiency. The discontinuity of the cell wall network in an aerenchymatous cortex may explain why nutrient deficiency reduces hydraulic conductivity. If so, that would have to be reconciled with Drew and Saker's (1986) finding that aerenchymatous roots of maize absorbed phosphate and rubidium (an analogue of potassium) from solution as fast as non-aerenchymatous ones. Phosphate and potassium are transported across the cortex primarily symplastically (Clarkson, 1981). Water, in contrast, moves mainly through the cortical apoplast. Drew and Saker's experiment ought to be repeated with apoplastically transported nutrients (e.g. calcium) as well as with water itself.

To summarize, interactions between the capture of water and nutrients are likely to depend on local moisture and solute availabilities, transpiration, time, the solute in question and the plant's demand for it. Simple experimental systems can be devised to show particular interactions that are explicable in terms of equally simple physical and chemical models. Adding *biology* to such models to allow more subtle interactions to be included demands information about developmental changes in roots, and about any associations with microbes that result (e.g. infection of senescent cortical cells by mycorrhizal or pathogenic fungi, or nitrogen-fixing organisms; the costs to the plant in terms of carbon substrates 'lost' from the root to the microbe(s)). It also demands information about how the roots of genotypes and species differ quantitatively and qualitatively in their capture of resources. Such information that is of sufficient detail to allow mechanistic interpretations to be made is extremely thin on the ground. Another aspect in need of much more attention is the functioning of single roots when operating not in isolation from, but as part of a root system.

Scaling up from root to root system: the 'big root'

The close agreement between the models and the experiments described above was possible because techniques were devised to allow uptake to be measured at a scale appropriate to the processes described in the models. The models describe movements of solutes and solutions over distances of several millimetres or centimetres to uniform cylindrical or planar sinks up to 1 mm thick. They cannot be expected to describe accurately the transport of materials over metres of soil profile, or through several nanometres of mineral lattice or cellulose matrix. Problems have been found in describing the behaviour of root systems in terms of these models (Brewster, Bhat and Nye, 1976; Herkelrath, Miller and Gardner, 1977; Passioura, 1980; Robinson *et al.*, 1991; but see also Baldwin, 1975, and Taylor and Klepper, 1975, for examples of good agreement between single root models and real root systems). The reason for this is that root systems are not collections of identical, cylindrical sinks, distributed uniformly in a uniform soil, the usual assumption when the 'single root' model is applied to root systems. They are clumped, have an age-structure and demography like any population, alter the local chemistry and microbiology of the soil, and may have genotypic peculiarities.

Scaling up from root to root system is possible in principle, but difficult in practice. Jarvis and McNaughton (1986) explored the analogous problems of interpreting gas exchange by vegetation in terms of stomatal conductances in individual leaves. One way this is done is to treat a collection of leaves in the same canopy as if it was one 'big leaf'. Passioura (1985, 1991b) suggested a way of doing something similar for a collection of roots, assuming it behaved as a 'big root'. The problem was that when applied naively, the single root model predicted that a root system of typical density in the soil should remove all the available water from the profile in a matter of days. Such crops never manage to extract all the potentially available water during their lives. Passioura assumed that the most important point of transfer of water into the plant was not from the cylinder of soil around every root, but from a volume of soil enclosing every *clumped group* of roots (*Figure 4.7*). The single root model was then applied to these 'big roots'. They were predicted to take many weeks to deplete the soil's available water, a much more realistic time-scale.

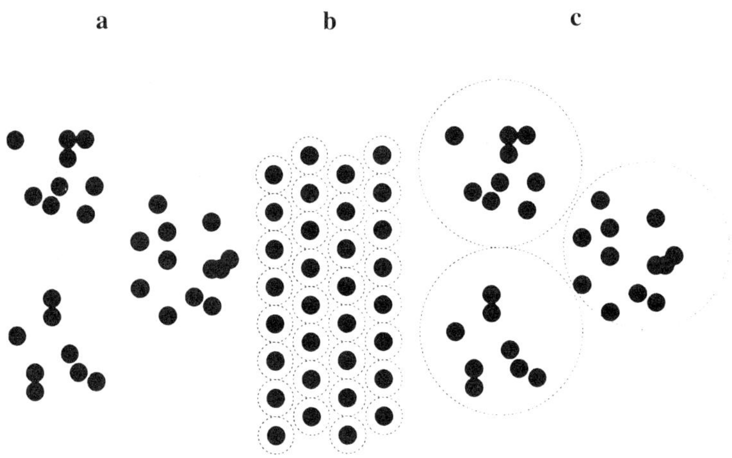

Figure 4.7(a) A map of 32 roots in soil. **(b)** A regular array of 32 roots each with its unique zone of depletion (broken line) from which it withdraws water and nutrients. The radius of each zone is defined by half the distance between neighbouring roots. **(c)** The same map as in **(a)**, but the zones of depletion are defined by half the distance between adjacent *clumps* of roots

More precise, detailed models of resource capture by root systems that include information about branching, demography or environment have not been produced. It is hard to see how they could be any better at the moment than the 'big root' model, given the current inadequacies of the data needed to construct them. Theories and models have dominated this field. The emphasis should now be shifted towards experiments in which as many of the 'unknowns' as possible are removed or controlled. New techniques and instrumentation, if handled critically, will probably offer the best opportunities to do this. Understanding how root systems operate demands *some* knowledge of how water and nutrients are transported to, and are taken up by, single roots. But it also requires information about the distributions of roots laterally and vertically in the soil, and about how roots interact with the shoot and through it, with the environment above-ground. These aspects are dealt with in the next chapter.

Acknowledgements

The author is grateful to Denis Linehan, Bruce Marshall, John Passioura, John Monteith and, especially, Margret van Vuuren for their helpful comments and discussions.

References

Abuzinadah, R. A. and Read, D. J. (1986) The role of proteins in the nitrogen nutrition of ectomycorrhizal plants. III. Protein utilization by *Betula, Picea* and *Pinus* in mycorrhizal association with *Hebeloma crustuliniforme. New Phytologist,* **103**, 507–514

Baldwin, J. P. (1975) A quantitative analysis of the factors affecting plant nutrient uptake from soils. *Journal of Soil Science,* **26**, 195–206

Brewster, J. L., Bhat, K. K. S. and Nye, P. H. (1976) The possibility of predicting solute uptake and plant growth response from independently measured soil and plant characteristics. V. The growth and phosphorus uptake of rape in soil at a range of phosphorus concentrations and a comparison of the results with the predictions of a simulation model. *Plant and Soil,* **44**, 295–328

Brouwer, R. (1953) Water absorption by *Vicia faba* at various transpiration strengths. I. Analysis of the uptake and the factors determining it. *Proceedings Koninklijke Nederlandse Akademie van Wetenschappen*, **C56**, 106–115

Caldwell, M. M. (1976) Root extension and water absorption. In *Water and Plant Life*, pp 63–85 Edited by O. L. Lange, L. Kappen and E.-D. Schulze. Heidelberg: Springer-Verlag

Campbell, B. D. and Grime, J. P. (1989) A comparative study of plant responsiveness to the duration of episodes of mineral nutrient stress. *New Phytologist,* **112**, 261–267

Campbell, G. S. (1985) *Soil Physics with BASIC*. London: Elsevier

Chapin, F. S., Clarkson, D. T., Lenton, J. R. and Walter, C. H. S. (1988) Effect of nitrogen stress and abscisic acid on nitrate absorption and transport in barley and tomato. *Planta,* **173**, 340–351

Clarke, A. L. and Barley, K. P. (1968) The uptake of nitrogen from soils in relation to solute diffusion. *Australian Journal of Soil Research,* **6**, 75–92

Clarkson, D. T. (1981) Nutrient interception and transport by root systems. In *Physiological Processes Limiting Plant Productivity*, pp 307–330 Edited by C. B. Johnson. London: Butterworths

Clarkson, D. T., Sanderson, J. and Russell, R. S. (1968) Ion uptake and root age. *Nature,* **220**, 805–806

Cruz, R. T., Jordan, W. R. and Drew, M. C. (1991) Structural changes and associated reduction of hydraulic conductance in roots of *Sorghum bicolor* L. following exposure to water deficit. *Plant Physiology,* **99**, 203–212

Deacon, J. W. (1987) Programmed cortical senescence: a basis for understanding root infection. In *Fungal Infection of Plants*, pp 285–297 Edited by G. F. Pegg and P. G. Ayres. Cambridge: Cambridge University Press

Drew, M. C., He, C. J. and Morgan, P. W. (1989) Decreased ethylene biosynthesis and induction of aerenchyma by nitrogen- or phosphate-starvation in adventitious roots of *Zea mays* L. *Plant Physiology,* **91**, 266–271

Drew, M. C. and Nye, P. H. (1969) The supply of nutrient ions by diffusion to plant roots in soil. II. The effect of root hairs on the uptake of potassium by roots of rye-grass (*Lolium multiflorum*). *Plant and Soil,* **31**, 407–424

Drew, M. C. and Nye, P. H. (1970) The supply of nutrient ions by diffusion to plant roots in soil. III. Uptake of phosphate by roots of onion, leek and rye-grass. *Plant and Soil,* **33**, 545–563

Drew, M. C., Nye, P. H. and Vaidyanathan, L. V. (1969) The supply of nutrient ions by diffusion to plant roots in soil. I. Absorption of potassium by cylindrical roots of onion and leek. *Plant and Soil,* **30**, 252–270

Drew, M. C. and Saker, L. R. (1986) Ion transport to the xylem in aerenchymatous roots of *Zea mays* L. *Journal of Experimental Botany,* **37**, 22–33

Dunham, R. J. and Nye, P. H. (1973) The influence of soil water content on the uptake of ions by roots. I. Soil water content gradients near a plane of onion roots. *Journal of Applied Ecology,* **10**, 585–598

Dunham, R. J. and Nye, P. H. (1974) The influence of soil water content on the uptake of ions by roots. III. Phosphate, potassium, calcium and magnesium uptake and concentration gradients in soil. *Journal of Applied Ecology,* **13**, 967–984

Gardner, W. R. (1965) Movement of nitrogen in soil. In *Soil Nitrogen*, pp 550–575 Edited by W. V. Bartholomew and F. E. Clark. Madison: American Society of Agronomy

Graham, J. P., Clarkson, D. T. and Sanderson, J. (1974) Water uptake by the roots of marrow and barley plants. *Agricultural Research Council Letcombe Laboratory Annual Report for 1973*, pp 9–12 Wantage, UK: Letcombe Laboratory

Grose, M. J. and Hainsworth, M. (1992) Soil water extraction, measured by computer-assisted tomography, in seedling *Lupinus angustifolius* cv. Yandee when healthy and when infected with *Phytophthora cinnamoni. Journal of Experimental Botany,* **43**, 121–127

Hainsworth, M. and Aylmore, L. A. G. (1986) Water extraction by single plant roots. *Soil Science Society of America Journal,* **50**, 841–848

Hamza, M. A. and Aylmore, L. A. G. (1992a) Soil solute concentration and water uptake by single lupin and radish plant roots. I. Water extraction and solute accumulation. *Plant and Soil,* **145**, 187–196

Hamza, M. A. and Aylmore, L. A. G. (1992b) Soil solute concentration and water uptake by single lupin and radish plant roots. II. Driving forces and resistances. *Plant and Soil,* **145**, 197–205

Hayward, H. E., Blair, W. M. and Skaling, P. E. (1942) Device for measuring entry of water into roots. *Botanical Gazette,* **104**, 152–160

He, C. J., Morgan, P. W. and Drew, M. C. (1992) Enhanced sensitivity to ethylene in nitrogen- or phosphate-starved roots of *Zea mays* L. during aerenchyma formation. *Plant Physiology,* **98**, 137–142

Herkelrath, W. N., Miller, E. E. and Gardner, W. R. (1977) Water uptake by plants. II. The root contact model. *Soil Science Society of America Journal,* **41**, 1039–1042

Ingestad, T. (1982) Relative addition rate and external concentration: driving variables used in plant nutrition research. *Plant, Cell and Environment,* **5**, 443–453

Jackson, R. B., Manwaring, J. H. and Caldwell, M. M. (1990) Rapid physiological adjustment to localized soil enrichment. *Nature,* **344**, 58–60

Jarvis, P. G. and MacNaughton, K. G. (1986) Stomatal control of transpiration: scaling up from leaf to region. *Advances in Ecological Research,* **15**, 1–49

Jupp, A. P. and Newman, E. I. (1987) Morphological and anatomical effects of severe drought on the roots of *Lolium perenne. New Phytologist,* **105**, 393–402

Justin, S. H. F. W. and Armstrong, W. (1987) The anatomical characteristics of roots and plant response to soil flooding. *New Phytologist,* **106**, 465–495

Lamont, B. (1982) Mechanisms for enhancing nutrient uptake in plants, with particular reference to Mediterranean South Africa and Western Australia. *Botanical Reviews,* **48**, 597–689

Li, X.-L., George, E. and Marschner, H. (1991) Phosphorus depletion and pH decrease at the root–soil and hyphae–soil interfaces of VA mycorrhizal white clover fertilized with ammonium. *New Phytologist,* **119**, 397–404

McCully, M. E. and Canny, M. J. (1988) Pathways of water and nutrient movement in roots. *Plant and Soil,* **111**, 159–170

MacFall, J. S., Johnson, G. A. and Kramer, P. J. (1991) Comparative water uptake by roots of different ages in seedlings of loblolly pine (*Pinus taeda* L.). *New Phytologist,* **119**, 551–560

Marschner, H. (1991) Plant–soil relationships: acquisition of mineral nutrients by roots from soil. In *Plant Growth: Interactions with Nutrition and the Environment*, pp 125–155 Edited by J. R. Porter and D. W. Lawlor. Cambridge: Cambridge University Press

Mullins, C. E. and Panayiotopoulos, K. P. (1984) The strength of unsaturated mixtures of sand and kaolin and the concept of effective stress. *Journal of Soil Science,* **35**, 459–468

Newman, E. I. (1991) The role of root-inhabiting fungi in mixed-species communities. In *Plant Root Growth: an Ecological Perspective*, pp 437–451 Edited by D. Atkinson. Oxford: Blackwell Scientific Publications

Nobel, P. S. and Cui, M. (1992) Hydraulic conductances of the soil, the root-soil air gap, and the root: changes for desert succulents in drying soil. *Journal of Experimental Botany,* **43**, 319–326

Nye, P. H. and Tinker, P. B. (1977) *Solute Movement in the Soil–Root System*. Oxford: Blackwell Scientific Publications

Passioura, J. B. (1972) The effect of root geometry on the yield of wheat growing on stored water. *Australian Journal of Agricultural Research,* **23**, 745–752

Passioura, J. B. (1980) The transport of water from root to shoot in wheat seedlings. *Journal of Experimental Botany,* **31**, 333–339

Passioura, J. B. (1985) Roots and water economy of wheat. In *Wheat Growth and Modelling*, pp 185–198 Edited by W. Day and R. K. Atkin. New York: Plenum

Passioura, J. B. (1988) Water transport in and to roots. *Annual Reviews of Plant Physiology and Plant Molecular Biology,* **39**, 245–265

Passioura, J. B. (1991a) An impasse in plant water relations? *Botanica Acta,* **104**, 405–411

Passioura, J. B. (1991b) Soil structure and plant growth. *Australian Journal of Soil Research,* **29**, 717–728

Passioura, J. B. and Cowan, I. R. (1968) On solving the non-linear diffusion equation for the radial flow of water to roots. *Agricultural Meteorology,* **5**, 129–134

Passioura, J. B. and Frere, M. H. (1967) Numerical analysis of the convection and diffusion of solute to roots. *Australian Journal of Soil Research,* **5**, 149–159

Peterson, R. L. (1992) Adaptations of root structure in relation to biotic and abiotic factors. *Canadian Journal of Botany,* **70**, 661–675

Pitman, M. G. (1982) Transport across plant roots. *Quarterly Reviews of Biophysics,* **15**, 481–554

Radin, J. W. and Matthews, M. A. (1989) Water transport properties of cortical cells in roots of nitrogen- and phosphorus-deficient cotton seedlings. *Plant Physiology,* **97**, 1087–1093

Robinson, D. (1986) Limits to nutrient inflow rates in roots and root systems. *Physiologia Plantarum,* **68**, 551–559

Robinson, D., Linehan, D. J. and Caul, S. (1991) What limits nitrate uptake from soil? *Plant, Cell and Environment,* **14**, 77–85

Robinson, D. and Rorison, I. H. (1983) A comparison of the responses of *Lolium perenne* L., *Holcus lanatus* L. and *Deschampsia flexuosa* (L.) Trin. to a localized supply of nitrogen. *New Phytologist,* **94**, 263–273

Robinson, D. and Rorison, I. H. (1987) Root hairs and plant growth at low nitrogen availabilities. *New Phytologist,* **107**, 681–693

Römheld, V. (1987) Existence of two different strategies for the acquisition of iron in higher plants. In *Iron Transport in Microbes, Plants and Animals*, pp 353–374 Edited by G. W. Winkelmann, D. van der Helm and J. B. Neilands. Weinheim: VCH Verlag

Saliendra, N. Z. and Meinzer, F. C. (1992) Genotypic, developmental and drought-induced differences in root hydraulic conductance of contrasting sugarcane cultivars. *Journal of Experimental Botany,* **43**, 1209–1217

Tanner, W. and Beevers, H. (1990) Does transpiration have an essential function in long distance ion transport in plants? *Plant, Cell and Environment,* **13**, 745–750

Taylor, H. M. and Klepper, B. (1975) Water uptake by cotton root systems: an examination of assumptions in the single root model. *Soil Science,* **120**, 57–67

Tinker, P. B. (1976) Transport of water to plant roots in soil. *Philosophical Transactions of the Royal Society,* **B272**, 445–461

Wild, A., Jones, L. H. P. and Macduff, J. H. (1987) Uptake of mineral nutrients and crop growth: the use of flowing nutrient solution. *Advances in Agronomy,* **41**, 171–219

Wray, F. J. and Tinker, P. B. (1969) A scanning apparatus for detecting concentration gradients around single plant roots. In *Root Growth*, pp 418–422 Edited by W. J. Whittington. London: Butterworths

Zimmermann, U., Rygol, J., Balling, A., Klock, G., Metzler, A. and Haase, A. (1991) Radial turgor and osmotic pressure profiles in intact and excised roots of *Aster tripolium*. *Plant Physiology,* **99**, 186–196

5

RESOURCE CAPTURE BY ROOT NETWORKS

P.J. GREGORY*
CSIRO, Division of Plant Industry, Dryland Crops and Soils Research Unit, Private Bag, PO, Wembley, WA 6014, Australia

Introduction

Roots are the organs that, in most crops, absorb nearly all of the water and nutrients taken up; they are sites of production for chemicals that maintain root: shoot integrity and also provide anchorage so that the shoot can expand leaf area to capture radiation without the plant toppling. It is now well recognized that the growth of root and shoot systems is an integrated process in which homeostasis is maintained as a consequence of both the size and activity of the two systems.

Brouwer (1963) was among the first to realize that the relative size of the root and shoot systems changes in response to the environment depending upon the comparative ease with which necessary resources can be obtained from the environment. He suggested that a 'functional equilibrium' exists such that photosynthate is partitioned to roots and shoots in inverse proportion to the rates at which they capture resources. Davidson (1969) expressed this condition as:

$$\underset{\text{(absorption)}}{\text{Root mass} \times \text{specific root activity}} = \underset{\text{(photosynthesis).}}{\text{shoot mass} \times \text{specific shoot activity}} \qquad (1)$$

The consequence of this statement is that the root:shoot mass of a crop will vary to compensate for changes in root and shoot activity induced by changes in the edaphic and atmospheric environments. Most investigation of this hypothesis has explored nitrogen uptake and carbohydrate synthesis of young, vegetative plants (Thornley, 1972) and 'goal-setting' models have been developed subsequently to describe the response of crops to nitrogen availability. However, although the right-hand side of Equation 1 can be quantitatively evaluated relatively easily, measurement of the capture of below ground resources poses many difficulties in practice. In their analysis of resource capture, Hunt, Warren Wilson and Hand (1990) indicate some of the problems in defining the amount of mineral nutrient available per unit ground area to a specified rooting depth. These include the difficulties of first, quantifying the fraction of a nutrient that is available at a particular time; second, allowing for the different nutrient contents of different soil layers and horizons; third, allowing for the differential rates of nutrient transport to different roots depending upon extant gradients of concentration; and finally, allowing for the different depths and spatial distributions of roots when plants are

*Present address: *Department of Soil Science, University of Reading, London Road, Reading, RG1 5AQ, UK*

grown in mixed communities. A similar list of difficulties can also be drawn up in relation to the capture of water.

This chapter will examine the capture of resources by root systems of field crops using Equation 1 as a convenient framework for discussion. The aims of this chapter are to examine first, typical patterns of biomass accumulation by root systems of crops; second, the allocation of dry matter and resource capturing surfaces between roots and shoots; and third, the capture of water and nutrients by root systems. Finally, some means whereby roots are able to change their environment to aid the acquisition of resources will be indicated briefly. Examples will be given from published data and particularly from some of the author's recent work in the mediterranean-type environment of Western Australia.

Biomass production by root systems

PATTERN OF GROWTH

The pattern of root development and growth is closely allied to that of the shoot and several workers have described the resulting 'developmental harmony'. In cereals, the development of root axes is closely allied to the production of leaves and tillers so that leaf number can be used to predict the potential number of axes. Klepper, Belford and Rickman (1984) grew winter wheat in field conditions and found that the number of root axes (R_n) was related to the number of leaves (L_n) by $R_n = 1.95L_n - 3.06$ and this relation has provided the basis for a model of root growth (Porter, Klepper and Belford, 1986). Similarly, Gregory (1983) determined that for pearl millet $R_n = 1.42L_n - 2.26$. Such relations describe the potential production of axes well although the actual production depends on environmental conditions, particularly the surface water content for the production of nodal axes (Gregory, 1987) and the radiative and nutritional environments for the production of tillers (Vincent and Gregory, 1989) and their accompanying nodal axes.

Dry matter accumulation by root systems typically follows the sigmoidal pattern commonly observed with shoots although the phases of rapid growth and decay may not entirely coincide (*Figure 5.1*). In particular, flowering appears as a particularly important developmental stage after which assimilates are required predominantly to fill the growing grain leaving little for roots (Gregory and Atwell, 1991). In most studies of cereals, the mass of the root system rarely increases after flowering and may decrease substantially depending on soil conditions. For legumes, the situation is less certain. *Figure 5.1d* shows continued increase in mass of the lupin root system during early grain-filling, a result also obtained by Sivakumar, Taylor and Shaw (1977) and Kaspar, Stanley and Taylor (1978) with soybean. However, the degree of determinancy may influence the pattern of carbohydrate allocation to the root system (Mayaki, Teare and Stone, 1976). The preceding description and *Figure 5.1* highlight a major deficiency in most estimates of root biomass: the technique commonly used (soil coring followed by extraction of the roots by washing with water) gives a measure only of the balance between new roots produced and older roots that have senesced (Huck, Hoogenboom and Peterson, 1987). Furthermore, the washing and storage techniques can result in substantial losses of fine root material and soluble carbohydrates resulting in substantial underestimates of root biomass (Logsdon and Reneau, 1988; Grzebisz, Floris and Noordwijk, 1989).

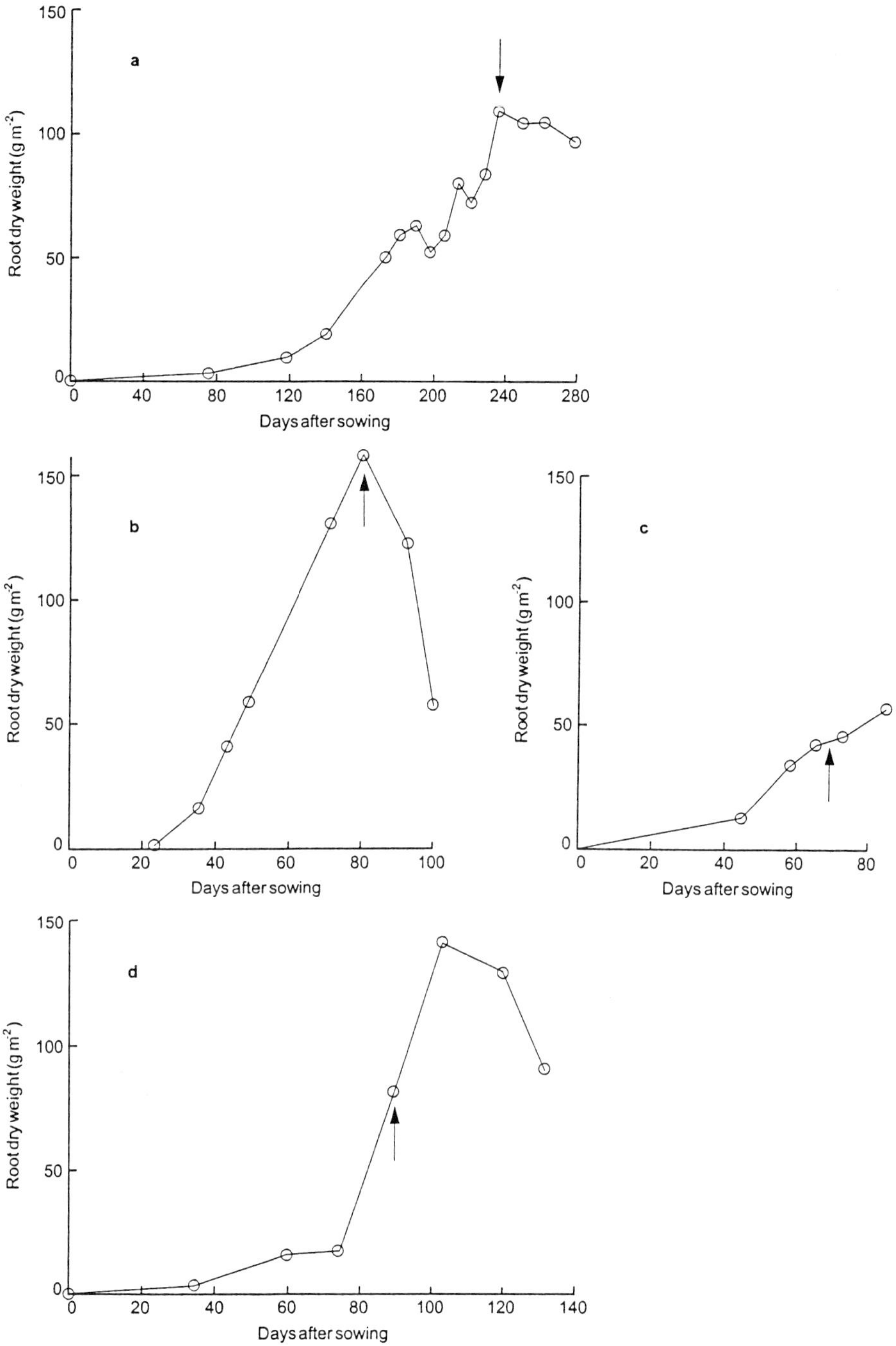

Figure 5.1 Changes in root dry weight with time for crops of (**a**) winter wheat (Sutton Bonington, UK, Gregory *et al.*, 1978); (**b**) maize (Lafayette, USA, Mengel and Barber, 1974; (**c**) soybean (Castana, lowa, Sivakumar, 1977); (**d**) lupin (East Beverley, Western Australia). The approximate time of flowering is indicated by an arrow

Generally, soil conditions early in the growing season are favourable and there is only a small amount of root death. However, as the season proceeds both natural root senescence and the incidence of unfavourable soil conditions increase so that it is impossible to state unequivocally, the amount of dry matter that a crop has invested in its root system. This 'turnover' of roots has not yet been reliably quantified so that it is difficult to be certain that the apparent differences in growth patterns between cereals and legumes discussed previously are real, or a consequence of differences in rates of root decay, determined genetically or environmentally. An appreciation of the possible error introduced into estimates of root biomass can be seen in some results obtained using washed soil cores for wheat and lupin crops grown in Western Australia. *Table 5.1* shows that the overall size of both root systems either increased or remained almost constant between 27 August and 24 September but the weight of lupin roots in the 0–10 cm layer decreased by 20% during the same period. This was a period of rapid shoot growth but limited rainfall meant that the volumetric soil water content at 0–10 cm decreased from 12% to 4%. It is also noteworthy, that the amount of 'dead' root material collected from the 0–10 cm layer of the wheat crop increased 2.5–fold over the period.

Table 5.1 ROOT MASS MEASURED AFTER WASHING SOIL CORES FOR WHEAT AND LUPIN CROPS GROWN AT EAST BEVERLEY, WESTERN AUSTRALIA, 1991

Crop	*Date*	*Total root mass* ($g\ m^2$)	*Live root mass 0–10 cm* ($g\ m^{-2}$)	*Dead root masss 0–10 cm* ($g\ m^{-2}$)
Wheat	27/08	63.8	44.3	1.1
	24/09	94.9	59.0	2.7
Lupin	27/08	86.2	61.4	
	24/09	83.1	48.3	

Several attempts have been made to demonstrate the amount of root turnover (e.g. Merckx, den Hartog and van Veen, 1985 with wheat; Cheng, Coleman and Box, 1990 with sorghum) using mesh bags, ^{14}C labelling and minirhizotron techniques. Cheng, Coleman and Box, 1990 used minirhizotrons and estimated that only about 60% of the total root production was accounted for if the peak values of live roots were used and dead roots ignored. However, despite its ability to examine individual roots in detail (assuming that the soil–glass interface is representative of growth and decay rates in the bulk soil) the minirhizotron technique consistently underestimates root growth in the upper 20 cm or so; this is typically the layer containing most roots (often >50% of the total length) and experiencing greatest environmental fluctuations. It is unlikely, then, that the minihizotron technique alone will give true estimates of root system productivity.

Pulse-labelling with radioactive carbon can provide estimates of carbon losses via respiration and rhizodeposition (Keith, Oades and Martin, 1986; Meharg and Killham, 1990; Gregory and Atwell, 1991). There has been much speculation about the magnitude of carbon losses from roots to the soil and their impact on the soil biomass (Martin, 1977; Wood, 1987). Most estimates of the fate of carbon have been made from young plants grown in controlled environments. Measurements by Martin and Kemp (1986) with wheat grown in field plots showed that a high proportion of ^{14}C translocated to the roots was released as rhizosphere CO_2 (65% at 7 weeks after emergence decreasing to 45% at 10 weeks). Similarly Keith, Oades and Martin (1986) found that at early tillering about 50% of photosynthate was translocated below ground of which about one-half was respired and one-quarter each was recovered in the soil and the roots. The pattern of partitioning and loss changed during growth so that < 5% of the photosynthate was translocated below-ground after anthesis. Gregory and Atwell (1991) conducted a similar field experiment in Western Australia and found that in young wheat plants about 33% of the photosynthate was recovered in below-ground components of which about one-half was as respiration. However, their results suggested that loss to the rhizosphere by exudation was small. After anthesis, only 9% of photosynthate was partitioned below ground. Over the whole season, the input of carbon to the below ground components was 48 gC m^{-2} of which only 3.3% was collected as exudate/sloughed material and 28% was collected as respired carbon (*Figure 5.2*). To determine the full carbon costs of root biomass production, however, it would be necessary to feed plants continuously with labelled carbon.

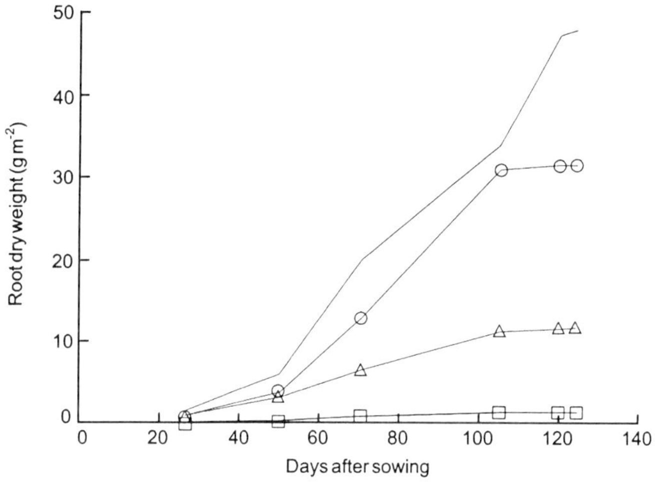

Figure 5.2 Cumulative estimated carbon accumulation by roots (○) and losses by exudation/sloughing of cells (□) and respiration (△) for a wheat crop grown at East Beverley, Western Australia (from Gregory and Atwell, 1991). The total is shown by a line without symbols

ROOT MASS AND LENGTH

Table 5.2 summarizes information from several experiments to show the maximum values of root dry weight and length recorded for various crops. Generally, the results were obtained from washed soil cores so that the reservations expressed previously apply and these values should be regarded only as the maximum values of live roots recorded rather than true indicators of total root biomass for the growing season. It is clear that cereals have heavier and longer root systems than legumes and other crops. The range of root weights is 170–45 g m^{-2} with

Table 5.2 MAXIMUM ROOT DRY WEIGHT AND LENGTH OF SOME COMMON CROPS

Crop	*Soil*	*Maximum root*		*Author*
		Dry weight (g m^{-2})	*Length* (km m^{-2})	
Sorghum	Tindall clay loam, NT, Australia	100	26.5	Myers (1980)
Winter wheat	Batcombe series, UK	170	32.0	Barraclough and Leigh (1984)
	Blithe series, UK	150	27.0	
Winter wheat	Astley Hall series, UK	105 113 97	23.5 26.8 15.4	Gregory *et al.*, (1978) McGowan *et al.* (1984)
Spring wheat	Typic Natrixeralf, Western Australia	77	3.8	Gregory, Tennant and Belford (1992)
Maize	Typic Argiaquoll, USA	160	15.1	Mengel and Barber (1974)
Soybean	Typic Udorthent, USA	58	5.5	Sivakumar, Taylor and Shaw (1977)
Groundnut	Alfisol, ICRISAT, India	143	6.6	Nageswara Rao *et al.* (1989)
Chickpea	Palexerollic, Chromoxerert, Syria	45	4.5	Brown *et al.* (1989)
Potato	Polder soil, Netherlands	77	7.1	Vos and Groenwold (1986)
Sugarbeet	Ludford series, UK	76	9.7	Brown and Biscoe (1985)

legumes typically having about half the root mass of cereals (note that lupins *Table 5.3* in Western Australia are exceptional). Similarly the range of root length is 32–45 km m^{-2} with legumes having about one-fifth the root length of cereals; the greater mass per unit length of legume roots accounting for the difference in the comparisons.

The values in *Table 5.2* must be treated with caution since comparisons involve differences in soils, climates and management as well as crops. There are few studies of roots of different species grown under comparable conditions and only a handful of studies in which a species has been grown on the same site over several seasons. This makes comparative statements about species difficult.

Hamblin and Hamblin (1985) compared the root growth of several species and varieties on a deep loamy yellow earth (Typic xeric psamment) at two locations in Western Australia and found that ranking of total root length was: clovers and medics > wheat > lupins and peas. The pasture species had total root lengths about twice to thrice that of wheat but 70% of their roots were in the upper 20 cm compared with only 50% for wheat and lupins. Other comparisons of crop species also confirm the general impression that the temperate cereals (barley and wheat) have greater root length than legumes (*Table 5.3*). However, although total length may be shorter, the root dry weight may be greater (e.g. lupin at East Beverley, Western Australia) depending on the growing conditions and the

Table 5.3 GROWTH OF ROOT SYSTEMS OF CEREAL AND LEGUME CROPS GROWN AT THE SAME SITE

Location	*Year*	*Crop*	*Maximum root*		*Author*
			Dry weight ($g\ m^{-2}$)	*Length* ($km\ m^{-2}$)	
E. Beverley, W. Australia	1990	Lupin	165	1.6	Gregory (unpublished)
		Wheat	107	4.8	
	1991	Lupin	135	1.7	
		Wheat	95	3.5	
Wongan Hills, W. Australia	1983	Lupin		2.4	Hamblin and Tennant (1987)
		Pea		3.0	
		Wheat		12.3	
		Barley		14.0	
Merredin, W. Australia	1984	Lupin		1.2	
		Pea		1.8	
		Wheat		7.7	
		Barley		5.3	
Jindiress, Syria	1983	Chickpea	45	4.5	Brown *et al.* (1989)
		Barley	83	13.3	Brown *et al.* (1987)

influence these have on both the total size of the root system and the length:weight ratio of the roots.

Welbank *et al.*, (1974) compared the growth of barley roots (cv. Maris Badger) over a 4–year period (1966–1969) on Cottenham series soil, Woburn Farm, Woburn, UK. Their results showed little interseasonal difference in the maximum dry weight of roots produced (95 $g\ m^{-2}$ in year 1, 80 $g\ m^{-2}$ in the other years) although there were differences in the rates of root growth from year to year.

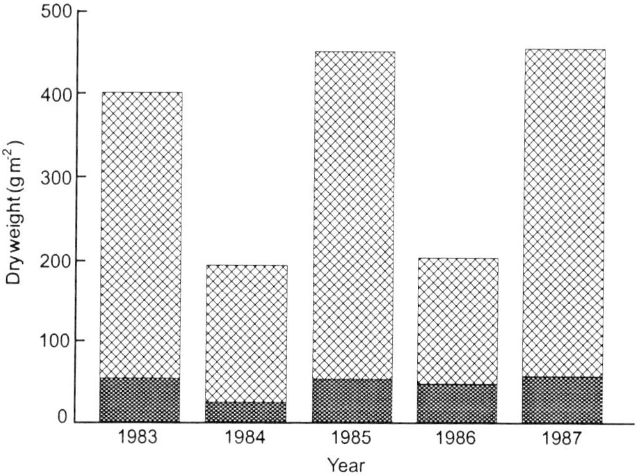

Figure 5.3 Year to year variation in root and shoot dry weight at anthesis for crops of barley (cv. Arabic abiad) given N and P fertilizer grown at Breda, northern Syria. ▩, root; ⊠, shoot

Root length was measured only in the last 2 years and reached maxima of 13.8 km m^{-2} and 12.6 km m^{-2}. *Figure 5.3* shows the year-to-year variation in root and shoot growth at anthesis of barley (Arabic abiad landrace) grown at Breda, northern Syria. All crops were given N and P fertilizers and seasonal rainfall was the principal factor causing the differences in growth. Root length ranged in a similar manner to root weight (from 3.2 km m^{-2} in 1984 to 13.9 km m^{-2} in 1985) and there was a linear relation between root length and root weight at anthesis; specific root length, then, was constant (174 m g^{-1}) despite the large, seasonal variation in rainfall.

ROOT:SHOOT ALLOCATION OF DRY MATTER

The partitioning of dry matter between root and shoot systems has frequently been used as an indication of the cost to a crop of the root system. Almost invariably the partitioning to roots is regarded as wasteful since it apparently detracts from the marketable yield which, with few exceptions, is part of the shoot system. Frequently, the assertions of wastefulness are based on an analysis of water uptake for which there is evidence that cereal root systems, in particular, are excessively large. However, such single factor analyses ignore the other functions of root systems and the requirement to take up less mobile resources such as phosphate. For example, the calculations of van Noordwijck and de Willigen (1987) indicate an approximate five- to ten-fold difference in optimal root length for P uptake compared with that for water uptake. This notion of the wasteful partitioning of dry matter to roots also appears to fly in the face of the 'functional equilibrium' of roots and shoots discussed earlier. Equation 1 makes clear that the relative size of the root and shoot systems will be a function of their relative activities.

The allocation of carbon to roots changes with crop ontogeny (e.g. Keith, Oades and Martin, 1986 for wheat) and with growing conditions. In general, any shortage of resources in the root environment causes a shift of assimilates in favour of an increased size of root system relative to the shoot and *vice versa* (Brouwer, 1983; Wild, Jones and MacDuff, 1987). *Figure 5.4* shows the changes in root weight as a fraction of total plant weight for the same crops shown in *Figure 5.1*. Typically root weight fraction was about 0.4 during early growth decreasing as the crop started to grow rapidly to a value of about 0.1–0.2 at flowering. However, the allocation of dry matter to roots may vary seasonally depending upon the availability of resources. For example, Hamblin, Tennant and Perry (1990) grew wheat in two seasons at Merredin, Western Australia (mean annual rainfall 310 mm) subjected to a range of treatments including nitrogen fertilizer, rotational cropping with lupins and sub-soiling to break up a traffic pan. Their results showed that root:total-biomass ratio at anthesis ranged from 0.1 to 0.6 as the conditions for growth became more harsh. Similarly, the root:total-biomass ratios of the barley crops shown in *Figure 5.3* varied from 0.18 in dry years to 0.13 in the wettest year. There are many examples in the literature where applications of fertilizer have not only increased the overall dry matter production of a crop but also reduced the proportion of total dry weight in roots during vegetative growth (e.g. Brown *et al.*, 1987).

While root and shoot mass provide a measure of the end product of resource utilization, they provide only a poor indication of activity and of the ability

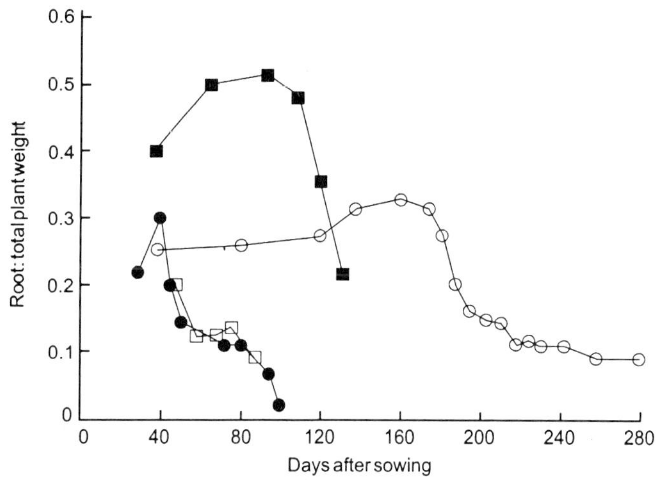

Figure 5.4 Changes in root:total-plant weight with time for the crops shown in Figure 5.1; winter wheat (○), maize (●), chickpea (□), and lupin (■)

to capture resources. Simmonds and Azam Ali (1989) suggested that in terms of water capture and utilization the ratio of L_A (the length of root per unit land area) to f (the fraction of incident radiation intercepted by foliage) was an appropriate index. However, this ratio varied considerably for groundnut crops grown at different spacings and did not provide a unique measure of capturing and utilizing structures.

Water capture by root systems

The supply of water to plants from soils is of considerable importance to terrestrial plants because it determines for how long they can grow and survive on the reserve held in the soil. The concept of available water is helpful in providing an indication of the water that is potentially extractable but the rate at which water is delivered to roots is a function of soil, plant and atmospheric factors. Considerable attention has been given to defining the factors affecting the rate of water supply to a root, with much less attention given to the factors determining how roots secure new supplies of water by extension into new soil zones and by subsequent root proliferation.

Most approaches to defining the uptake of water are based on the work of Philip (1957) and Gardner (1960) who described the system in terms of water flowing radially to an isolated, cylindrical root. Combining Darcy's Law and the continuity Equation gives:

$$\left(\frac{\partial \theta}{\partial t}\right) = \frac{1}{r}\frac{\partial}{\partial r}\left(r\, D_{\mathrm{W}} \frac{\partial \theta}{\partial r}\right) \qquad (2)$$

where θ is the water content, t is time, r is the radial distance from the root and D_w is the soil water diffusivity. This simple model assumes that there is no competition between roots for the water available so that each root can draw from an infinite soil volume. However the analysis can be extended to root systems by the simple expedient of apportioning to each root exclusive access to a cylinder of soil whose outer radius, b, is half the average distance between roots (Tinker, 1976). The radius, b, is related to the root length density (L_v) by:

$$b = 1/(\pi L_v)^{0.5} \tag{3}$$

If it is assumed that water is injected into the system at distance b, the steady state flow in a cylinder can be approximated as (Passioura and Cowan, 1969):

$$\theta_b - \theta_a = \frac{I_w}{2\pi D_w} \ln \frac{b}{a} \tag{4}$$

where a is the radius of the root and I_w is the rate of water uptake per unit length of root. To a first approximation, then, the rate of uptake from a given soil volume is proportional to the root length density (L_v), the diffusivity of soil water (D_w) and the difference in soil water content between the bulk soil and the root surface. The parameters θ_b and I_w change with time, but θ_a remains essentially constant so that I_w as a function of time is proportional to the amount of water unextracted. If D_w is assumed to be constant, then Equation 4 can be simplified to (Passioura, 1983):

$$\theta = \theta^* \exp(-kL_v t) \tag{5}$$

where k is a constant (with dimensions of a diffusion coefficient). Because $\theta = \theta^*$ at $t = 0$ and $\theta = 0$ when t is infinite, θ^* is the maximum amount of water that roots are capable of extracting from the surrounding soil. Passioura (1983) showed that Equation 5 did not adequately describe the uptake of water from the deeper soil layers beneath a sorghum crop detailed by Jordan and Miller (1980). Jordan and Miller (1980) showed that a large amount of 'available water' was left in the profile below 50 cm at the end of season and suggested that L_v was too low for this extraction to occur. Numerical solution of Equation 5 by Passioura suggests that L_v should have been large enough to extract the water so that there must have been other reasons why the extraction did not occur. It has been recognized for a long time that the expedient of assuming uniform root distribution (i.e. each root having the same size of equivalent cylinder of soil to exploit) is unlikely to be realized in soils with irregular structure, and this, together with our poor understanding of the physiology of water uptake, may render the assumptions implicit in Equation 5 invalid. Passioura (1983 and 1988) has discussed several reasons why such a simple model may fail including the clumping of roots in continuous, large pores, the development of an interfacial resistance between the root and the soil, and the impermeability of portions of the root system.

Despite its evident flaws, Equation 5 has been used as the basis of a model to describe water uptake by root systems (Monteith, Huda and Midya, 1989). Monteith (1986) combined Equation 5 with a mathematical function describing the descent of the root system to describe water extraction by crops of sorghum and millet grown on stored soil moisture. The quantity $1/kL_v$ can be regarded as a time constant and roots move downward to arrive at depth z at time $t_{(z)}$. Monteith assumed that when $t < t_{(z)}$, $L_v = 0$ and $\theta_{(z)} = \theta^*_{(z)}$; when $t > t_{(z)}$, L_v has a fixed value and $\theta_{(z)}$ decreases according to Equation 5.

Robertson *et al.*, (1993b) used Monteith's (1986) analytical framework to examine the water extraction pattern of six crops of sorghum under conditions of continuous soil drying. They found that the rate of downward progress of the extraction front was a conservative quantity (mean value 3.4 cm d^{-1}) across a range of soil types and agronomic conditions. However, the value of $1/kL_v$ varied considerably between crops depending upon L_v and atmospheric conditions such that it could not be used to predict the potential rate of uptake.

In the mediterranean-type environment of Western Australia, crops are grown on the current season's rainfall (plus a small amount of stored water from summer rains) so that the development of an observable drying front may be several weeks behind that of the actual rooting depth (*Figure 5.5*). Robertson *et al*, (1993a) noted a similar disparity between rooting and drying depths in the upper metre of soil beneath sorghum crops. It is noteworthy that the rooting depth increases linearly with time (9.1 mm d^{-1} for wheat and 6.9 mm d^{-1} for lupin) until about anthesis but the clear development of a drying front is only visible below 40 cm and linear extrapolation does not pass through the origin. These crops were grown on duplex soils (about 40 cm sand overlying kaolinitic clay – see Gregory, Tennant and

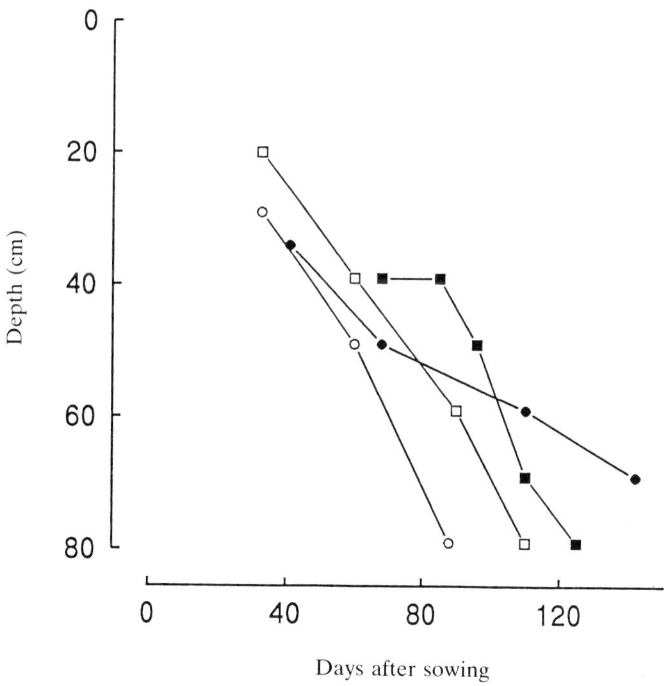

Figure 5.5 Rooting depth (open symbols) and depth of the water extraction front (solid symbols) for crops of wheat (circles) and lupin (squares) grown at East Beverley, Western Australia in 1990

Belford (1992) for details) where rooting depth and depth of water extraction are limited by soil properties.

Monteith (1986) and Robertson *et al.* (1993b) were unable to use Equation 5 to describe changes of water content in the upper 30 cm and 50 cm of soil respectively presumably because of substantial evaporation directly from the soil surface. However, in Western Australia where crops have reached an appreciable size before depletion of the soil water reserve occurs, Equation 5 described the depletion of soil water from layers to within 10 cm of the soil surface (*Table 5.4*). *Table 5.4* shows that $1/kL_v$ did not change consistently with depth but was larger in the clay horizon (40 and 50 cm) than the sand layer. Two other points arise from the analysis. First, the depletion from each layer was of the form described by Equation 5 despite the fact that L_v was not constant during the period of depletion (*Table 5.4*). Monteith (1986) assumed that L_v was constant once the drying front had arrived (an unlikely assumption unless there is no branching). Presumably in the present results, any increase of L_v was compensated for by a decrease in k because as the soil dried, D_w did not remain constant as initially assumed by Passioura (1983). It seems odd, however, that the compensation should be so exact. Second, the distribution of values of $1/kL_v$ down the soil profile was very similar for both crops despite the almost ten-fold greater length of the wheat root system. Several workers (Hamblin and Tennant, 1987; Gregory and Brown, 1989) have commented upon the ability of legume root systems to extract water at the same rate as cereals despite the smaller root length. This is difficult to resolve in relation to Equation 5 unless there are substantial differences in the gradients of water potential generated by the two crop types. Hamblin and Tennant (1987) demonstrated that the bulk soil water potential of wheat and lupins was similar (as was also the case in the 1990 crops) and hypothesized that the difference between crops was due to the difference in axial resistance of the root. They estimated that the axial resistance of wheat roots was some 1000 times greater than that of lupin roots.

A different approach to examining water uptake by root systems has been adopted by Gardner (1983 and 1991). One of the problems with models of water uptake based on Equation 4 is that when a soil profile has been wetted, uptake

Table 5.4 RESULTS OF REGRESSIONS OF ln θ WITH THE TIME (DAS = DAYS AFTER SOWING) FOR CROPS OF WHEAT AND LUPIN GROWN AT EAST BEVERLEY, WESTERN AUSTRALIA IN 1990

Crop	*Depth* (cm)	*Time interval* (das)	r^2	$^1/kL_v$ (days)	L_v (cm cm^3) *88 das*	*116 das*
Wheat	10	88–130	0.99	32.6	9.91	2.67
	20	88–116	0.85	22.2	2.91	1.34
	30	88–116	0.97	11.3	0.26	0.39
	40	88–158	1.00	53.7	0.17	0.21
	50	102–158	1.00	273	0.12	0.16
Lupin	10	88–130	0.98	43.1	0.46	0.46
	20	88–130	0.88	17.0	0.38	0.51
	30	88–144	0.95	24.5	0.06	0.13
	40	88–158	0.98	97.7	0.02	0.13
	50	88–158	0.99	275	0	0.05

should be related to the root length and gradient of hydraulic head in each layer so that water uptake will occur in each layer. This is not usually observed in practice and it is common that water uptake occurs preferentially from the surface layers with uptake from depth occurring only after substantial quantities of water have been removed from the upper layers. Gardner (1983) compared patterns of water uptake for different crops and soils at the time at which the crop first started to wilt. When the water uptake for each crop was nomalized on a scale of 0 to 1 and the depth was similarly scaled so that all curves passed through the same point at a water uptake value of 0.5, a surprising degree of uniformity was obtained.

Gardner (1991) has extended this analysis to develop the hypothesis that water uptake can be described by a distributed sink moving downward. He argued that the root system tends to linearize the otherwise highly non-linear soil–water flow system so that the upper portion of the root system is characterized by a soil–plant water diffusivity that is independent of soil water content. In the lower portion of the root system there is a transitional zone before non-linear flow again applies. Such a model gives good agreement with field-derived data and describes well the situation when the surface is re-wetted because the sink moves back to the surface and moves downward at a rate dependent on the water content of each soil layer.

Nutrient capture by root systems

The basic principles defining the ability of root systems to acquire nutrients were first outlined qualitatively by Bray (1954). He recognized that the ability to secure nutrient resources was dependent not only on the amount of nutrient present in the soil but also on its mobility relative to the morphology of the root system. This means that in a crop or natural ecosystem, the zone of competition between roots for mobile nutrients (e.g. nitrate) is much larger than that for immobile nutrients such as phosphate. These ideas have been developed quantitatively by several workers to provide computer-based models of nutrient uptake by crops (e.g. Baldwin, 1975; Barber and Cushman, 1981; Claassen, Syring and Jungk, 1986.) Such models differ in their specific details but contain three common components. First, nutrients are supplied to the roots by mass flow and diffusion and the concentration at the root surface is calculated. As with water flow, a uniform cylindrical root surface is assumed (Nye and Marriott, 1969):

$$\frac{\partial C_l}{\partial t} = \frac{1}{r}\frac{\partial}{\partial r}\left(r\,D_e\frac{\partial C_l}{\partial r} + \frac{v\,a\,C_l}{b*}\right) \qquad (6)$$

where C_l is the concentration of nutrient in the soil solution, D_e is the effective diffusion coefficient of the nutrient in the soil, v is the water flux to the root and b^* is the buffer power. A great deal of attention has been paid to this component of nutrient uptake and there are several experiments demonstrating its robustness (e.g. Brewster, Bhat and Nye, 1975; Barber, 1984).

Second, the rate of uptake by the root is assumed to be dependent on the concentration at the root surface and a Michaelis–Menten function is adopted to define the relation. This approach is open to many criticisms (Seward, Barraclough and Gregory, 1990; Wild, Jones and MacDuff, 1987) but describes some of the observed uptake reasonably well. Finally, uptake is calculated per unit of root length and summed over the whole root system. Several assumptions are involved in this step which may, in practice, not be realized. Equation 6 assumes that roots are uniform cylinders with smooth surfaces; this is unlikely for soil-grown roots. As with water uptake, the roots are assumed to be parallel and uniformly distributed in the soil and assigned a cylinder of soil according to Equation 3. It is further assumed that all roots are equally physiologically capable of taking up nutrients so that any differences in uptake between sections of the root system are a reflection of differences in concentration at the root surface.

Silberbush and Barber (1984) successfully used such a model to predict uptake of potassium by cultivars of soybean grown in the field. However, potassium uptake by crops of wheat growing at Woburn experimental farm was substantially overestimated (*Figure 5.6*) except for the crop grown under low nutrient conditions where nutrient supply may be expected to limit growth (Seward, Barraclough and Gregory 1990). Overestimation of uptake would occur if either the roots were not all active in uptake or the parameter relating rate of uptake to concentration at the root surface were wrong.

Robinson, Linehan and Caul (1991) have argued that not all roots are equally active in nitrate uptake and calculated that only 11% and 3.5% of the total root length of spring wheat roots were involved in uptake in soils given 0 or 200 kgN ha^{-1} respectively. There is a need for direct measurements of the physiological

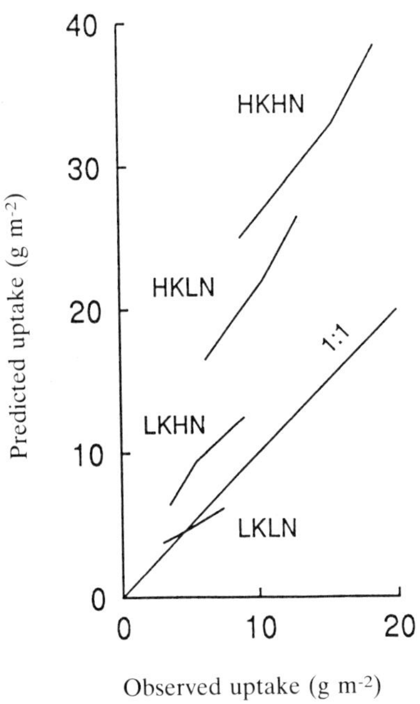

Figure 5.6 Comparison of measured potassium uptake of four crops of spring wheat with that predicted by the model of Claassen, Syring and Jungk (1986). Crops were grown with combinations of high (H) and low (L) soil potassium (K) and nitrogen (N) (from Seward, Barraclough and Gregory, 1990)

ability of components of root systems to take up nutrients (e.g. Ernst, Römheld and Marschner, 1989) so that this problem can be resolved.

An alternative (but complementary) approach to quantifying nutrient uptake is to specify the relationship between nutrient concentration in the dry matter and growth rate. In this way uptake can proceed according to plant demand while the potential supply is adequate but when the supply can no longer meet demand, supply then controls the rate of growth. This approach has been successfully adopted by Seward (1991).

Modification of the root environment to aid resource capture

The image conveyed by the previous discussion of resource capture is of inactive root surfaces receiving water and nutrients by well-defined physical and chemical processes. However, it is increasingly realized that the roots of many crop plants interact with their soil environment to enhance the availability and capture of resources. It has been known for many decades that the soil immediately surrounding a root (the rhizosphere) has a different microbial population from the bulk soil as a consequence of carbon materials exuded by roots and cellular material that is sloughed off during growth and decay (Foster, 1986). Other chemical and physical changes in the rhizosphere environment combine with these microbial changes to aid resource capture.

THE CHEMICAL ENVIRONMENT

Changes in the pH surrounding roots have been observed for several crop species including beans (Sarkar and Wyn Jones, 1982), rape (Moorby, White and Nye 1988) maize, wheat, chickpea and white clover (Marschner and Römheld, 1983). Nye (1984) reviewed the sources of acids and bases in the rhizosphere and assessed the four possible sources: first, the imbalance of cation and anion uptakes of nutrients requires the release of acids (H^+) or bases (usually HCO_3^-) to maintain electrical neutrality; second, respiration producing CO_2; third, excretion of organic acids by roots; and fourth, the microbial production of acids from carbon sources released by roots. He concluded that the first source was the most important. The source of nitrogen has a particularly important effect in relation to cation–anion balance. For example, Marschner and Römheld (1983) found that the rhizosphere pH of maize plants supplied with NO_3-N increased to about 7.5 compared with the bulk soil pH of 6.0 whereas supplying NH_4^+-N caused a reduction to pH 4. Their results also indicate that pH changes in the rhizosphere may vary depending upon the concentration of nitrate in the bulk soil and whether the whole of the system is actively taking up N. In legumes, the production and efflux of protons is an inevitable consequence of N_2-fixation and this, together with the large cation exchange capacity of dicotyledonous roots, results in legumes being very effective at lowering rhizosphere pH.

Acidification of the rhizosphere has the benefit of solubilizing sparingly-soluble Ca phosphates (Aguilar and van Diest, 1981) but in soils high in free iron and aluminium oxides, the effect on phosphate availability may be the reverse.

Ae *et al.*, (1990) have demonstrated that pigeonpea is able to utilize iron-bound phosphorus by the exudation of piscidic acid which releases P by chelating Fe^{3+}.

The practical consequence of this is that pigeonpea has substantial residual benefits for subsequent crops (particularly cereals). Arihara, Ae and Okada (1991) have demonstrated that these residual benefits cannot be explained solely in terms of N additions by N_2-fixation and suggest that the increased availability of P is a major contributor to improved cereal growth.

THE PHYSICAL ENVIRONMENT

Besides the indirect effects on the physical environment that carbon materials have through their effects on stabilizing aggregates and hence allowing more thorough exploitation of the soil by roots, exudates may also play a rôle in securing water supplies. There has long been an interest in the adherence of soil particles to roots by means of mucilage and root hairs and their role during periods of water stress (Sprent, 1975). Several roles have been ascribed to this root-cap mucilage including the stabilization of soil aggregates, the selective accumulation of ions, the reduction of friction between the moving root tip and the soil, and the maintenance of root–soil hydraulic contact (Oades, 1978). Detailed examination of roots by McCully (1987) and Guinel and McCully (1986) has demonstrated the presence of such soil sheaths on many young roots of grass species even in moist conditions and shown that the water potential of fully expanded mucilage gel from maize roots is high (–7.3 kPa). The high water potential makes its function as a lubricant unlikely since it will lose water in any environment with a water potential that is more negative (i.e. almost any soil drier than field capacity). Guinel and McCully (1986) conclude that 'the relationship of mucilage hydration and the maintenance of contact between root and soil is still unclear'.

Passioura (1988) has drawn attention to a function of root exudates that has, so far, remained almost unexplored. Some soils contain surface-active agents that reduce the surface tension of water; presumably these are of biological origin but their source has yet to be identified. The effect of these, if present in mucilage adjacent to the root, would be to reduce the surface tension (from 70 mN m^{-1} to say 30 mN m^{-1}) which would cause a proportionate reduction in the soil water tension at a particular water content. As Passioura (1988) states, although this effect would be negligible at low tensions, it could be substantial at higher tensions (say 500–1000 kPa).

Conclusion

This brief review has attempted to highlight those areas in which the means whereby roots acquire resources are poorly comprehended. Using Davidson's (1969) Equation as a base, it is evident that much still remains to be done to quantify the left-hand side of the Equation. Specifically, the following points require resolution:

1. Although there are now several measurements of root dry weight through the season, the full carbon costs of the turnover of root material (and therefore of the root system) are unknown especially in environments where soil stresses are common.
2. Although single-root models of water uptake have proved useful in specific circumstances, they do not adequately describe the uptake of water by root

systems. In particular, current models are generally inadequate at describing the preferential uptake of water from surface layers in environments where the surface is frequently re-wetted and at describing the restricted uptake from deeper layers when roots are confined to gaps around peds.

3. To date, models of nutrient uptake have largely been made by soil scientists with the emphasis on the supply from the soil. There is still a need to incorporate a plant demand characteristic, to define which roots are active in uptake and to characterize the long-term supply of nutrients.
4. There is increasing evidence that roots are able to modify their environment to secure resources. The role of organic substances in the rhizosphere for water uptake requires resolution and the prospects for genotypic variation, coupled with molecular biology approaches, offer future prospects for manipulation of resource acquisition.

References

Ae, N., Arihara, J., Okada, K., Yoshihara, T. and Johansen, C. (1990) Phosphorus uptake by pigeon pea and its role in cropping systems of the Indian subcontinent. *Science*, **248**, 477–480

Aguilar, A.S. and van Diest, A. (1981) Rock-phosphate mobilization induced by the alkaline uptake pattern of legumes utilizing symbiotically fixed nitrogen. *Plant and Soil*, **61**, 27–42

Arihara, J., Ae, N. and Okada, K. (1991) Root development of pigeonpea and chickpea and its significance in different cropping systems. In *Phosphorus Nutrition of Grain Legumes in the Semi-arid tropics*, pp 183–194. Edited by C. Johansen, K.K. Lee and K.L. Sahrawat. Patancheru, India: ICRISAT

Baldwin, J.P. (1975) A quantitative analysis of the factors affecting plant nutrient uptake from soils. *Journal of Soil Science*, **26**, 195–206

Barber, S.A. (1984) *Soil Nutrient Bioavailability: a Mechanistic Approach*. New York: Wiley Interscience

Barber, S.A. and Cushman, J.H. (1981) Nitrogen uptake model for agronomic crops. In *Modelling Waste Water Renovation – Land Treatment*, pp 382–409 Edited by J.K. Iskander. New York: Wiley Interscience

Barraclough, P.B. and Leigh, R.A. (1984) The growth and activity of winter wheat roots in the field: the effect of sowing date and soil type on root growth of high-yielding crops. *Journal of Agricultural Science, Cambridge*, **103**, 59–74

Bray, R.H. (1954) A nutrient mobility concept of soil–plant relationships. *Soil Science*, **78**, 9–22

Brewster, J.L., Bhat, K.K.S. and Nye, P.H. (1975) The possibility of predicting solute uptake on plant growth response from independently measured soil and plant characteristics. III. The growth and uptake of anions in a soil fertilized to different initial levels of phosphate and a comparison of the results with model predictions. *Plant and Soil*, **42**, 197–226

Brouwer, R. (1963) Some aspects of the equilibrium between overground and underground plant parts. *Jaarboek IBS*, Wageningen 1963, 31–39

Brower, R. (1983) Functional equilibrium: sense or nonsense? *Netherlands Journal of Agricultural Science*, **31**, 335–348

Brown, K.F. and Biscoe, P.V. (1985) Fibrous root growth and water use of sugar beet. *Journal of Agricultural Science, Cambridge* **105**, 679–691

Brown, S.C., Gregory, P.J., Cooper, P.J.M. and Keatinge J.D.H. (1989) Root and shoot growth and water use of chickpea (*Cicer arietinum*) grown in dryland conditions: effects of sowing date and genotype. *Journal of Agricultural Science, Cambridge*, **113**, 41–49

Brown, S.C., Keatinge, J.D.H., Gregory, P.J. and Cooper, P.J.M. (1987) Effects of fertilizer, variety and location on barley production under rainfed conditions in northern Syria. 1. Root and shoot growth. *Field Crops Research*, **16**, 53–66

Cheng, W., Coleman, D.C. and Box, J.E. (1990) Root dynamics, production and distribution in agroecosystems on the Georgia Piedmont using minirhizotrons. *Journal of Applied Ecology*, **27**, 592–604

Claassen, N., Syring, K.M. and Jungk, A. (1986) Verification of a mathematical model by simulating potassium uptake from soil. *Plant and Soil*, **95**, 209–220

Davidson, R.L. (1969) Effect of root/leaf temperature differentials on root/shoot ratios in some pasture grasses and clover. *Annals of Botany*, **33**, 561–569

Ernst, M., Römheld, V. and Marschner, H. (1989) Estimation of phosphorus uptake capacity by different zones of the primary root of soil-grown maize (*Zea mays* L.). *Zeitschrift für Pflanzenernährung und Bodenkunde*, **152**, 21–25

Foster, R.C. (1986) The ultrastructure of the rhizoplane and rhizosphere. *Annual Review of Phytopathology*, **24**, 211–234

Gardner, W.R. (1960) Dynamic aspects of water availability to plants. *Soil Science*, **89**, 63–73

Gardner, W.R. (1983) Soil properties and efficient water use: an overview. In *Limitations to efficient water use in crop production*, pp 45–64 Edited by H.M. Taylor, W.R. Jordan and T.R. Sinclair. Madison USA: American Society of Agronomy

Gardner, W.R. (1991) Modeling water uptake by roots. *Irrigation Science*, **12**, 109–114

Gregory, P.J. (1983) Response to temperature in a stand of pearl millet (*Pennisetum typhoides* S&H). III. Root development. *Journal of Experimental Botany*, **34**, 744–756

Gregory, P.J. (1987) Development and growth of root systems in plant communities. In *Root Development and Function*, pp 147–166 Edited by P.J. Gregory, J.V. Lake and D.A. Rose. Cambridge: Cambridge University Press

Gregory, P.J. and Atwell, B.J. (1991) The fate of carbon in pulse-labelled crops of barley and wheat. *Plant and Soil*, **136**, 205–213

Gregory, P.J. and Brown, S.C. (1989) Root growth, water use and yield of crops in dry environments: what characteristics are desirable? *Aspects of Applied Biology*, **22**, 235–243

Gregory, P.J., McGowan, M., Biscoe, P.V. and Hunter, B. (1978) Water relations of winter wheat. I. Growth of the root system. *Journal of Agricultural Science, Cambridge*, **91**, 91–102

Gregory, P.J., Tennant, D. and Belford, R.K. (1992) Root and shoot growth, and water and light use efficiency of barley and wheat crops grown on a shallow duplex soil in a mediterranean-type environment. *Australian Journal of Agricultural Research*, **43**, 555–573

Grzebisz, W., Floris, J. and van Noordwijk, M. (1989) Loss of dry matter and cell contents from fibrous roots of sugar beet due to sampling, storage and washing. *Plant and Soil*, **113**, 53–57

Guinel, F.C. and McCully, M.E. (1986) Some water-related physical properties of maize root-cap mucilage. *Plant, Cell and Environment*, **9**, 657–666

Hamblin, A.P. and Hamblin J. (1985) Root characteristics of some temperate legume species and varieties on deep, free-draining entisols. *Australian Journal of Agricultural Research*, **36**, 63–72

Hamblin, A. and Tennant, D. (1987) Root length density and water uptake in cereals and grain legumes: how well are they correlated? *Australian Journal of Agricultural Research*, **38**, 513–527

Hamblin, A., Tennant, D. and Perry, M.W. (1990) The cost of stress: dry matter partitioning changes with seasonal supply of water and nitrogen to dryland wheat. *Plant and Soil*, **122**, 47–58

Huck, M.G., Hoogenboom, G. and Peterson, C.M. (1987) Soybean root senescence under drought stress. In *Minirhizotron Observation Tubes: Methods and Applications for Measuring Rhizosphere Dynamics*, pp 109–121 Edited by H.M. Taylor. American Society of Agronomy Special Publication Number 50. Madison, USA: American Society of Agronomy

Hunt, R., Warren Wilson, J. and Hand D.W. (1990) Integrated analysis of resource capture and utilization. *Annals of Botany*, **65**, 643–648

Jordan, W.R. and Miller, F.R. (1980) Genetic variability in sorghum root systems: implications for drought tolerance. In *Adaptation of Plants to Water and High Temperature Stresses*, pp. 383–399 Edited by N.C. Turner and P.J. Kramer. New York: Wiley Interscience

Kaspar, T.C., Stanley, C.D. and Taylor, H.M. (1978) Soybean root growth during the reproductive stages of development. *Agronomy Journal*, **70**, 1105–1107

Keith, H., Oades, J.M. and Martin, J.K. (1986) Input of carbon to soil from wheat plants. *Soil Biology and Biochemistry*, **18**, 445–449

Klepper, B., Belford, R.K. and Rickman, R.W. (1984) Root and shoot development in winter wheat. *Agronomy Journal*, **76** 117–122

Logsdon, S.D. and Reneau, R.B. (1988) Influence of storage methods on corn root length measurements. *Plant and Soil*, **111**, 155–157

McCully, M.E. (1987) Selected aspects of the structure and development of field-grown roots with special reference to maize. In *Root Development and Function*, pp 53–70 Edited by P.J. Gregory, J.V. Lake and D.A. Rose. Cambridge: Cambridge University Press

McGowan, M., Blanch, P., Gregory, P.J. and Haycock, D. (1984) Water relations of winter wheat. 5. The root system and osmotic adjustment in relation to crop evaporation. *Journal of Agricultural Science, Cambridge*, **102**, 415–425

Marschner, H. and Römheld, V. (1983) *In vivo* measurement of root-induced pH changes at the soil–root interface: effect of plant species and nitrogen source. *Zeitschrift für Pflanzenphysiologie*, **111**, 241–251

Martin, J.K. (1977) Factors affecting the loss of organic carbon from wheat roots. *Soil Biology and Biochemistry*, **9**, 1–7

Martin, J.K. and Kemp, J.R. (1986) The measurement of C transfers within the rhizosphere of wheat grown in field plots. *Soil Biology and Biochemistry*, **18**, 103–107

Mayaki, W.C., Teare, I.D. and Stone, L.R. (1976) Top and root growth of irrigated and non-irrigated soybeans. *Crop Science*, **16**, 92–94

Meharg, A.A. and Killham, K. (1990) Carbon distribution within the plant and rhizosphere in laboratory and field-grown *Lolium perenne* at different stages of development. *Soil Biology and Biochemistry*, **22**, 471–477

Mengel, D.B. and Barber, S.A. (1974) Development and distribution of the corn root system under field conditions. *Agronomy Journal*, **66**, 341–344

Merckx, R., den Hartog, A. and van Veen, J.A. (1985) Turnover of root-derived material and related microbial biomass formation in soils of different texture. *Soil Biology and Biochemistry*, **17**, 565–569

Monteith, J.L. (1986) How do crops manipulate water supply and demand. *Philosophical Transactions of the Royal Society, London*, **A316**, 245–259

Monteith, J.L., Huda, A.K.S. and Midya, D. (1989) RESCAP: A resource capture model for sorghum and pearl millet. In *Modelling the Growth and Development of Sorghum and Pearl Millet*, pp 30–34 Edited by S.M. Virmani, H.L.S. Tandon and G. Alagarswarmy. Research Bulletin No 12, Patancheru, India: ICRISAT

Moorby, H., White, R.E. and Nye, P.H. (1988) The influence of phosphate nutrition on H ion efflux from the roots of young rape plants. *Plant and Soil*, **105**, 247–256

Myers, R.J.K. (1980) The root system of a grain sorghum crop. *Field Crops Research*, **3**, 53–64

Nageswara Rao, R.C., Simmonds, L.P., Azam-Ali, S.N. and Williams, J.H. (1989) Population, growth and water use of ground nut maintained on stored water. I. Root and shoot growth. *Experimental Agriculture*, **25**, 51–61

Nye, P.H. (1984) Acid-base changes in the rhizosphere. *Advances in Plant Nutrition*, **2**, 129–153

Nye, P.H. and Marriott, F.H.C. (1969) A theoretical study of the distribution of substances around roots resulting from simultaneous mass flow and diffusion. *Plant and Soil*, **30**, 451–472

Oades, J.M. (1978) Mucilages at the root surface. *Journal of Soil Science*, **29**, 1–16

Passioura, J.B. (1983) Roots and drought resistance. *Agricultural Water Management*, **7**, 265–280

Passioura, J.B. (1988) Water transport in and to roots. *Annual Review of Plant Physiology and Plant Molecular Biology*, **39**, 245–265

Passioura, J.B. and Cowan, I.R. (1968) On solving the non-linear diffusion Equation for the radial flow of water to roots. *Agricultural Meteorology*, **5**, 129–134

Philip, J.R. (1957) The physical principles of soil water movement during the irrigation cycle. *Proceedings of the 3rd Congress of the International Committee on Irrigation and Drainage*, **8**, 125–154

Porter, J.R., Klepper, B. and Belford, R.K. (1986) A model (WHTROOT) which synchronizes root growth and development with shoot development for winter wheat. *Plant and Soil*, **92**, 133–145

Robertson, M.J., Fukai, S., Hammer, G.L. and Ludlow, M.N. (1993a) Water extraction by grain sorghum in a sub-humid environment. II. Extraction in relation to root growth. *Field Crops Research*, **33**, 99–112

Robertson, M.J., Fukai, S., Ludlow, M.M. and Hammer, G.L. (1993b) Water extraction by grain sorghum in a sub-humid environment. I. Analysis of the water extraction pattern. *Field Crops Research*, **33**, 81–97

Robinson, D., Linehan, D.J. and Caul, S. (1991) What limits nitrate uptake from soil? *Plant, Cell and Environment*, **14**, 77–85

Sarkar, A.N., and Wyn Jones, R.G. (1982) Influence of rhizosphere on the nutrient status of dwarf French beans. *Plant and Soil*, **64**, 362–380

Seward, P. (1991) *The Development of a Mechanistic Mathematical Model to Predict the Uptake of Potassium by Wheat Plants from Soil*. Ph.D. Thesis, University of Reading

Seward, P., Barraclough, P.B. and Gregory, P.J. (1990) Modelling potassium uptake by wheat (*Triticum aestivum*) crops. *Plant and Soil*, **124**, 303–307

Silberbush, M. and Barber, S.A. (1984) Phosphorus and potassium uptake of field-grown soybean cultivars predicted by a simulation model. *Soil Science Society of America Journal*, **48**, 592–596

Simmonds, L.P. and Azam-Ali, S.N. (1989) Population, growth and water use of groundnut maintained on stored water. The influence of population on water supply and demand. *Experimental Agriculture*, **25**, 87–98

Sivakumar, M.V.K., Taylor, H.M. and Shaw, R.H. (1977) Top and root relations of field-grown soybeans. *Agronomy Journal*, **69**, 470–473

Sprent, J.I. (1975) Adherence of sand particles to soybean roots under water stress. *New Phytologist*, **74**, 461–463

Thornley, J.H.M. (1972) A balanced quantitative model for root:shoot ratios in vegetative plants. *Annals of Botany*, **36**, 431–441

Tinker, P.B. (1976) Transport of water to plant roots in soil. *Philosophical Transactions of the Royal Society, London*, **B273**, 445–461

Vincent, C.D. and Gregory, P.J. (1989) Effects of temperature on the development and growth of winter wheat roots. I. Controlled glasshouse studies of temperative, nitrogen and irradiance. *Plant and Soil*, **119**, 87–97

Van Noordwijck, M. and de Willigen, P. (1987) *Roots, Plant Production and Nutrient Use Efficiency*. Ph.D. Thesis, Agricultural University, Wageningen, The Netherlands

Vos, J. and Groenwold, J. (1986) Root growth of potato crops on a marine-clay soil. *Plant and Soil*, **94**, 17–33

Welbank, P.J., Gibb, M.J., Taylor, P.J. and Williams, E.D. (1974) Root growth of cereal crops. *Report of Rothamsted Experimental Station for 1973*, Part 2, 26–66

Wild, A., Jones, L.H.P. and MacDuff, J.H. (1987) Uptake of mineral nutrients and crop growth: the use of flowing nutrient solutions. *Advances in Agronomy*, **41**, 171–219

Wood, M. (1987) Predicted microbial biomass in the rhizosphere of barley in the field. *Plant and Soil*, **97**, 303–314

6

USING THE EXPOLINEAR GROWTH EQUATION TO ANALYSE RESOURCE CAPTURE

J. GOUDRIAAN
Department of Theoretical Production-Ecology, Agricultural University, POB 430, 6700AA Wageningen, The Netherlands

Introduction

Complex growth models have the potential to synthesize knowledge from different disciplinary fields, and to evaluate their combined behaviour. This process of putting models together, running them and studying their output, comprises the activity of 'simulation' (Rabbinge and De Wit, 1989), and will enhance our understanding if we do it with an open mind and a some sensible amount of self-criticism. Even in relatively simple models the complexity of interactions is baffling. Yet sometimes it is possible to discern guiding principles in the behaviour of the models. Being a simplified representation of reality, models should reflect nature also in this respect. Both careful real world experimentation and model analysis will be needed to find such guiding principles.

This quest for guiding principles is essential in science. Models can serve a useful role in assisting the scientist in this task, but in addition to the 'simulation' approach in the traditional sense, there is also a line of model research which explores highly simplified models (Monteith, 1981a; Waggoner, 1990). These models will perhaps suffer from lack of realism, but they do have the advantage of highlighting those features of dominating processes that we want to emphasize. In this respect these models are more subjective than large, comprehensive models.

Whereas the individual scientist may prefer one type of modelling or the other, it appears to the author that in general the use of complex models and of simplified models should go hand in hand (Spitters, 1990).

A simplified model will be used in this chapter to aid our understanding of a few features of crop growth and resource capture, rather than for a well-validated and generalized description of many of its aspects.

The expolinear growth equation will be used to investigate the phenomenon of apparent down-regulation of relative growth rate under increased CO_2, and the theory will be extended to allow for constrained leaf area growth, such as occurs under shortage of nitrogen.

A simple growth model: the expolinear growth equation

The basis of the analysis here is the assumption that plant growth follows the expolinear growth curve (Goudriaan and Monteith, 1990). The expolinear equation provides a continuous transition from the exponential growth phase, observed in young relatively widely spaced plants, to the linear growth phase that is seen later when the leaves have filled the space between plants. At that stage the plants have become so large that they fully compete with each other, so that their combined growth rate will be constrained, and approaches a maximum value. The equation for the expolinear growth curve is:

$$w = \frac{c_m}{r_m} \ln\left\{1 + \exp\left(r_m (t - t_b)\right)\right\} \qquad \left[\mathrm{g\ m^{-2}}\right] \qquad (1)$$

where:

w is the crop biomass at time t [g m^{-2}]
c_m is the maximum absolute growth rate in the linear phase [g m^{-2} d^{-1}]
r_m is the maximum relative growth rate in the exponential phase [d^{-1}]
t_b is a timing parameter [d]

These three parameters completely describe the growth curve. The initial biomass (w_0) at time zero can be expressed by these three parameters by setting $t = 0$ in the equation, but it is often a more natural sequence to express the timing parameter t_b using initial biomass, i.e.:

$$t_b = -\frac{\ln\left\{\exp\left(\frac{r_m\, w_0}{c_m}\right) - 1\right\}}{r_m} \qquad [\mathrm{d}] \qquad (2)$$

This may be useful when we compare treatments that start at the same initial biomass.

The expolinear growth equation is consistent with a few simple and plausible assumptions:

1. Exponential extinction of radiation with leaf area index L, and a constant extinction coefficient k, according to the well-known expression:

$$f = 1 - \exp(-k\, L) \qquad [-] \qquad (3)$$

2. A direct proportionality between this fraction f and actual growth rate, c ($c = \mathrm{d}w/\mathrm{d}t$), of biomass:

$$c = c_m\, f \qquad [\mathrm{g\ m^{-2}\ d^{-1}}] \qquad (4)$$

3. A constant ratio between leaf area and plant dry matter, leaf area ratio (LAR), that is further separated into leaf weight ratio (LWR) (partitioning to leaves) p_l [-] and specific leaf area (SLA) s [m^2 g^{-1}]:

$$\mathrm{d}L/\mathrm{d}t = p_l\, s\, c \qquad [\mathrm{d^{-1}}] \qquad (5)$$

This is the weakest assumption. Later we shall develop an equation for a situation in which partitioning to the leaves decreases after a while.

The potential relative growth rate (RGR) r_m for the early exponential phase can now be expressed as:

$$r_m = k\ p_l\ s\ c_m \qquad [d^{-1}] \tag{6}$$

Typical parameter values used in this paper are given in *Table 6.1*.

Table 6.1 PARAMETER SET USED FOR CONTROL CALCULATIONS WITH THE EXPOLINEAR EQUATION

Description	*Symbol*	*Value*	*Unit*
Maximum absolute growth of crop biomass	c_m	20	[g m^{-2}]
Partitioning to leaf dry weight	p_l	0.5	[-]
Specific leaf area	s	0.02	[m^2 g^{-1}]
Initial crop biomass	w_0	1	[g m^{-2}]
Extinction coefficient	k	0.7	[-]
Derived variables:			
Maximum (potential) relative growth rate	r_m	0.14	[d^{-1}]
Timing parameter ('lost time')	t_b	35.4	[d]

Down-regulation of RGR, a functional adaptation?

Plant growth experiments with increased CO_2 have shown that biomass increased by 30–40% upon CO_2 doubling, at least when the plants were grown under non-limiting nutrients (Kimball, 1983). This increase is usually attributed to enhanced photosynthesis (Lawlor and Mitchell, 1991), although decreased respiration may also be a contributing factor (Amthor, 1991). The method of growth analysis can be used to discern between functional and morphological factors that give rise to growth stimulation. The classical growth analysis, developed for young plants, emphasizes *relative growth rate* and separates it into *growth per leaf area* (the functional component net assimilation rate ('NAR') or unit leaf rate ('ULR')) and in *leaf area per unit of biomass* (the morphological component 'LAR'). In a comparison of relative growth rate in Equation 6 with the familiar expression from classical growth analysis RGR=NAR*LWR*SLA, the term 'NAR' becomes equal to $k\ c_m$ in the unshaded exponential phase.

The difference between fast and slow growing species is mainly caused by a difference in morphology (LWR and SLA) (Poorter, 1993), but the effect of increased CO_2 is caused by an increased rate of growth per unit leaf area ('NAR'). Increased CO_2 may alter morphology as well, but if it does it will suppress rather than enhance the effect of increased growth rate per unit leaf area. This morphological adaptation to higher CO_2 consists of increased thickening of leaves, so that more biomass is needed to produce the same leaf area (Lemon, 1983).

When the time courses of relative growth rate of control plants and of plants grown under higher CO_2 are plotted together *versus* time, it seems that the initial stimulus of RGR disappears quickly, and may even turn into a reduction

(Poorter, 1993). The question now is, whether this experimental finding is largely a result of functional down-regulation, such as a reduction in photosynthetic capacity (Besford and Hand, 1989; Besford, Ludwig and Withers, 1990; Bowes, 1991), or whether there is a simple mechanistic explanation.

The methodology chosen here is to investigate what happens with RGR

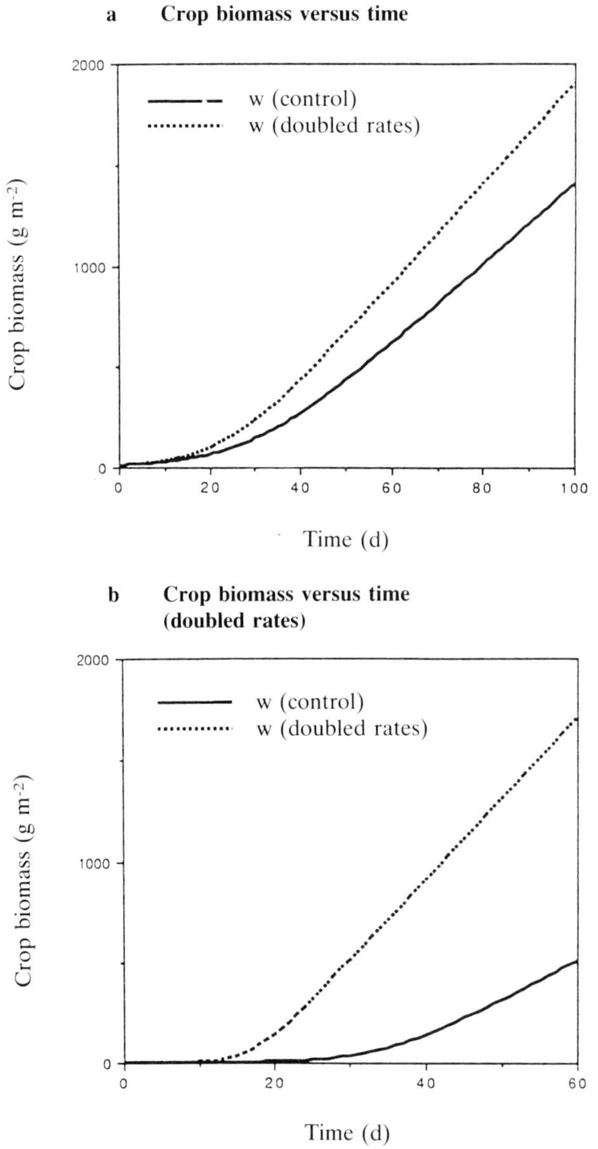

Figure 6.1 Calculated crop biomass according to the expolinear equation. The higher curve was calculated for increased growth parameters c_m and r_m. (**a**) Growth parameters differed by a factor 1.3 (**b**) Growth parameters differed by a factor 2

according to the expolinear equation, if we assume that there is *no down-regulation at all* in any functional or morphological parameter that describes plant growth. First of all, let us assume that the CO_2 effect is expressed by an increased maximum linear growth rate c_m. Let us also assume that there is no morphological adaptation, i.e., that the partitioning coefficient p_l and leaf area per leaf dry weight s are unchanged. Clearly, this is a simplification of reality, but it helps to discern the effects of increased mutual shading from adaptation in functional and morphological parameters. In this simplified situation the potential relative growth rate r_m according to Equation 6 will be changed in direct proportion to c_m. The ratio c_m:r_m will not be affected, but the timing parameter t_b (Equation 2) in inversely related to r_m.

The behaviour of the expolinear growth equation in these two situations can now be easily evaluated and compared, using altered values for c_m and r_m. When both c_m and r_m are increased by 30%, which is a typical value for the effect of CO_2-enrichment, graphs such as shown in *Figure 6.1a* are obtained for the time courses of dry biomass. The biomasses attained in the end differ by a number that is slightly greater than the stimulus factor 1.3 on growth rate, because the value of t_b ('lost time'; Monteith, 1981b) has also been reduced under these improved growth circumstances.

The time course of relative growth rate shows a peculiar behaviour (*Figure 6.2a*). Its initial value is about a factor 1.3 higher than in the control, but it soon declines to a value quite close to the control, and it even falls below it. This kind of pattern, if observed in experimental data, might be interpreted as down-regulation of the growth processes, attributable to all sorts of internal adaptations that return the photosynthetic rate to the value of the control situation.

In fact, the primary explanation is more mechanistic. In order to highlight this effect, the maximum growth rate c_m and potential relative growth r_m were multiplied by a factor of 2 instead of 1.3, so that the results became more contrasting (*Figure 6.1b*, *6.2b*). *Absolute* growth rate was much larger at any time during the growth period, but *relative* growth rate soon declined to a value that was even lower than in the control run. The explanation becomes quite clear, when absolute and relative growth rate are plotted not *versus* time, but *versus* biomass (*Figure 6.3*). In this representation the multiplication factor of 2 is retained. In *Figure 6.3a* absolute growth rate is given every 5 days, showing that the distance between each point in time is much larger in the doubled-rate run, than in the control run. The relationship between absolute growth rate c and biomass w is given by Equations 3 and 4, and by using $L = p_l\, s\, w$:

$$c = c_m \{1 - \exp(-k\, p_l\, s\, w)\} \qquad [\mathrm{g\ m^{-2}\ d^{-1}}] \qquad (7)$$

The factors $k\, p_l\, s$ are not affected by doubling c_m, and so the graphs differ only by the multiplication factor that operates on c_m, a scaling factor in the vertical direction. The scaling factor in the horizontal direction $(k\, p_l\, s)^{-1}$ remains the same. But, relative growth rate can also be read from this graph. It is the slope of a straight line from the origin to the point of the graph at the moment considered. Initially these slopes differ by a factor of two, but for the doubled-rate situation the curve levels off much sooner, because the scaling biomass $(k\, p_l\, s)^{-1}$ is reached much earlier. In simpler terms, the occurrence of plant shading is so strongly accelerated that the reduction of relative growth rate sets in much earlier. This effect dominates the time course of relative growth rate, even though the absolute

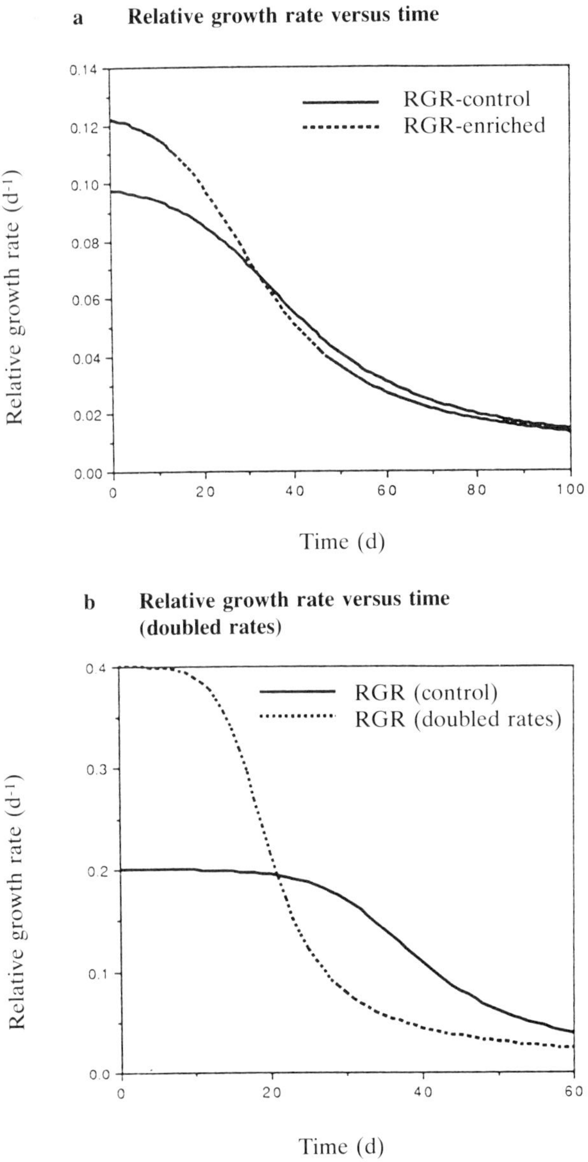

Figure 6.2 Time courses of actual relative growth rate for the curves as in Figure 6.1. The initial difference in relative growth rate soon disappears. (**a**) Growth parameters differed by a factor 1.3 (**b**) Growth parameters differed by a factor 2

growth rate remains much higher at any moment. Biomass can also be written as initial biomass times exp($\int r \, dt$).

Therefore, the largest ratio of biomasses occurs when the integrals of r dt differ most, e.g. at the moment of crossing over of the two lines in *Figure 6.2*, here at time 21 (*Figure 6.2b*). After that moment, the ratio of the biomasses starts to decline again. In practice, the graph of RGR versus time does not offer a convenient way to assess the effect on final biomass. A barely visible difference in integrated relative growth rate may have a large effect on final biomass. The time plot of absolute growth rate is a better measure for this purpose. The best method to see if *functional* adaptation occurs, is plotting growth rate *versus* biomass, or leaf area, rather than *versus* time (*Figure 6.3b*).

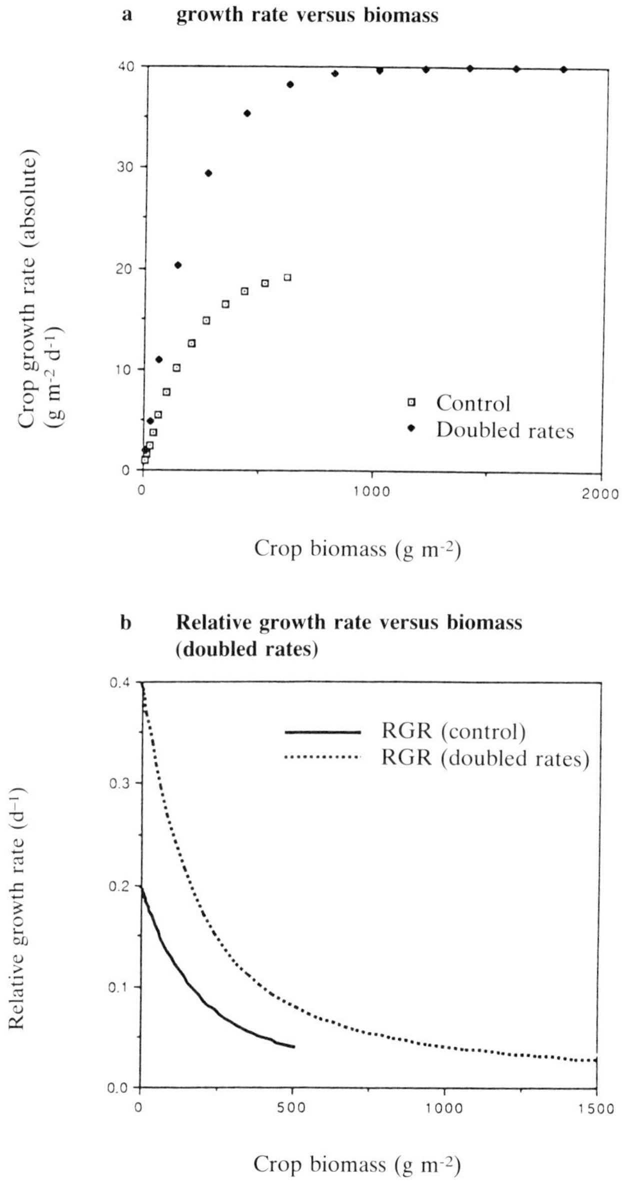

Figure 6.3 Absolute growth rate (**a**) and relative growth rate (**b**) as a function of biomass, for the control and doubled parameter situation.

The expolinear equation for nitrogen limited situations

Canopy photosynthesis clearly increases as a function of leaf area index, due to increasing light interception. Equation 4 expresses the relation between crop growth rate c and fraction of intercepted radiation f as a simple linear relation. However, the relation between gross canopy photosynthesis and intercepted fraction is not simply a straight line because lower leaves operate at lower light levels and tend to exhibit a higher light use efficiency (when photosynthetic parameters are uniform). Gross canopy photosynthesis thus increases more rapidly than fraction f of intercepted radiation.

Complications tend to straighten out this upward curved relation. Respiration increases when leaf area index increases, and this will certainly restrict *net* photosynthesis. Because a large portion of respiration is proportional to growth itself, the relative effects on gross and net photosynthesis are less different than one would initially think.

Another important factor is adaptation of leaf photosynthetic capacity to lower light levels deep in the canopy. This adaptation mainly affects maximum photosynthetic capacity ('A_{max}') (DeJong and Doyle, 1985), and is mediated by a lower leaf nitrogen content in the shade. Photosynthetic capacity is strongly related to leaf nitrogen (Van Keulen and Seligman, 1987; Leuning, Cromer and Rance, 1991a), but light use efficiency is not. Therefore the scaling factor for light intensity with respect to photosynthesis declines together with average light level, and mean light use efficiency is rather invariant with canopy depth. This effect tends to linearize the relationship of canopy photosynthesis with fraction of light intercepted (Leuning, Wang and Cromer, 1991b), and indeed even with canopy nitrogen (Field, 1991). For high leaf area indices it also means an improvement of nitrogen use efficiency, as the lower leaves require less nitrogen.

The linear growth phase of the original expolinear equation never ceases. A supposedly constant partitioning of assimilates to new leaf area causes the leaf area index to grow indefinitely as well. This behaviour, although unrealistic, does not always pose a problem. For instance, beyond a leaf area index of 5, the fraction of radiation intercepted is almost unity anyway and it hardly matters to the computed rate of growth whether leaf area index increases any further.

However, for calculation of nitrogen expenditure the situation is different. Green leaf area is particularly costly in terms of nitrogen demand, and when there is a limited amount of nitrogen available, this will pose an upper limit to the formation of leaf area index. A rather crude way to impose this upper limit in a simple model is to truncate any further increase in leaf area index as soon as the upper limit imposed by shortage of nitrogen (or of another requirement such as water) is reached. This method will however cause a discontinuity in the growth rate of leaf area, and also in the calculated uptake rate of nitrogen flowing to the leaves.

A more realistic growth pattern must allow for a gradual reduction in the formation rate of leaf area, while it approaches a maximum possible leaf area. Such a growth curve can be obtained by a modification of the expolinear growth equation. The maximum growth rate c_m is now further constrained by a factor f_m between 0 and 1 and which is independent of t or w:

$$w = \frac{c_m f_m}{r_m} \ln\{1 + \exp(r_m (t - t_b))\} \qquad [\mathrm{g\ m^{-2}}] \qquad (8)$$

This factor f_m allows for the model assumption that light interception will not exceed a value of $1 - \exp(-k\ L_m)$. The shape of this modified growth curve is still expolinear, but the difference with the original expression is that the linear growth phase proceeds with a reduced, 'apparent', linear growth rate $c_{m,app}$, which is given by the product of the potential growth rate c_m and the maximum interception fraction f_m. This modification has no effect on the early exponential phase, since the maximum relative growth r_m is the same.

Now we should consider what this modification means in terms of the growth pattern of leaf area index. First Equations 3 and 4 are combined to give:

$$\frac{dw}{dt} = c_m \{1 - \exp(-k\ L)\} \qquad [\mathrm{g\ m^{-2}\ d^{-1}}] \qquad (9)$$

This expression for dry matter growth rate is equated to the first derivative of the modified expolinear growth equation:

$$\frac{dw}{dt} = c_m \{1 - \exp(-k\ L)\} = \frac{c_m f_m \exp\{r_m (t - t_b)\}}{1 + \exp\{r_m (t - t_b)\}} \qquad (10)$$

This combination can give us leaf area index L as an explicit function of time t and of its maximum value L_m, defined by $f_m = 1 - \exp(-k\ L_m)$:

$$L = \frac{1}{\mathrm{k}} \ln\left[\frac{1 + \exp\{r_m (t - t_b)\}}{1 + \exp\{\mathrm{r}_m (t - t_b) - k\ L_m\}}\right] \qquad [\mathrm{m^2\ m^{-2}}] \qquad (11)$$

From this equation we see that $L \rightarrow L_m$ when $t >> t_b$. The equation for the rate of leaf area formation (Equation 5) is now modified to:

$$\frac{dL}{dt} = p_1 s\, c \frac{1 - \exp\{-k(L_m - L)\}}{1 - \exp\{-k\ L_m\}} \qquad [\mathrm{m^2\ m^{-2}\ d^{-1}}] \qquad (12)$$

This multiplication factor allows for a reduction that increases until the maximum leaf area index L_m is reached.

When this methodology is applied, an S-shaped growth curve of leaf area index appears. It is remarkable that the expolinear equation still stands up in this generalization. By permitting levelling off of leaf area growth a considerable theoretical objection against the expolinear equation is removed.

Nitrogen uptake rate of the foliage can be calculated as the rate of increase of the amount of nitrogen in the foliage. When all leaves have uniform nitrogen content, the uptake rate is simply the derivative of the S-shaped growth curve of leaf area index multiplied by leaf nitrogen content. This growth pattern generates

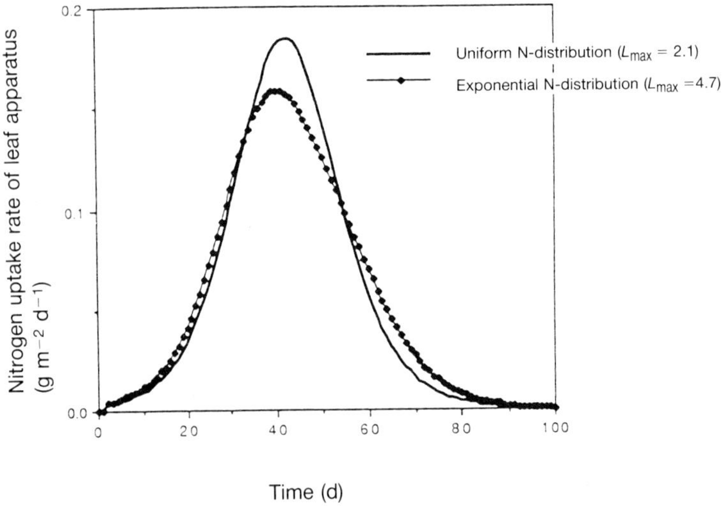

Figure 6.4 Nitrogen uptake rate to sustain leaf area growth. The total amount of nitrogen available for leaf area growth was set at 5.5 g m^{-2} (ground area). For the uniform distribution the maximum LAI was 2.1, but the optimized distribution an LAI of 5 could be formed

a bell shaped curve of nitrogen uptake rate of the green leaf mass (*Figure 6.4*). When leaf nitrogen content declines with canopy depth, the pattern of nitrogen uptake rate is still bell-shaped, but with an earlier and lower peak. In both nitrogen distribution modes, the bell shape applies qualitatively.

When there is only a limited amount of nitrogen available, this limitation will constrain leaf area growth, usually forcing it to a growth curve very much like the bell-shape presented before. For a given amount of nitrogen as a constraint, we can conversely infer the leaf area growth pattern that is possible on basis of this limited amount. As a typical situation, total amount of nitrogen available for uptake by the leaves was set at 5.5 g m^{-2} of ground area. When all leaves have a uniform nitrogen content of 2.6 g m^{-2} of leaf area, a total leaf area index can be formed of 2.1 , resulting in a maximum fraction of intercepted radiation of about 0.77. At the peak of leaf area formation, the nitrogen uptake rate reached about 0.18 gN m^{-2} d^{-1} and the mean duration of the uptake period was about 30 days.

The next step was to model the situation in which the nitrogen content was adapted to the light profile (Werger and Hirose, 1991). Only the topmost leaves have then the same content as in the first case, and deeper down in the canopy the leaf nitrogen content declines in such a way that the A_{max} profile follows the light profile. As a result a larger leaf area index could be formed (4.7 instead of 2.1), more light could be intercepted, and a higher assimilation rate per unit ground area could be maintained. Using Equation 12, by day 100 the N-optimized canopy had formed a biomass of 1250 g m^{-2}, whereas the N-uniform canopy had only formed 1000 g m^{-2}. This example shows the potential benefit of optimizing the distribution of nitrogen over the leaf canopy. For a given amount of available nitrogen, the difference between the two ways of distributing the amount of nitrogen over the leaves is only slight in terms of the time course of nitrogen uptake rate (*Figure 6.4*).

Discussion

In this paper no comparison with actual experimental data has been given. The ideas and graphs are generalizations based on literature data that provided the background information. The phenomena described in this paper can certainly be modelled by more complex, comprehensive models, and indeed they should be. Concentrating on one simple model structure, this paper intends to highlight a few interesting processes, which are perhaps less clearly seen when a more complex model is used. Monteith (1990) has recognized the importance of simple models throughout his work, and this paper was written in this tradition. In the real world, the complexity of processes may blur the clean effects suggested here, but not totally. The author preferred to concentrate on a few main lines of thought, rather than to get lost in a synthesizing complex model. Yet complex models are needed to bring together the many effects that simultaneously occur in the real world. It would be a mistake to abandon complex growth models for fear of chaos. The most fruitful way is to explore both methodologies and to see how they can be used to complement each other.

Acknowledgements

The author wishes to thank Dr R. Leuning for inspiring discussions and for his help when formulating the manuscript.

References

Amthor, J.S. (1991) Respiration in a future, higher-CO_2 world. *Plant, Cell and Environment*, **14**, 13–20

Besford, R.T. and Hand, D.W. (1989) The effects of CO_2 enrichment and nitrogen oxides on some Calvin cycle enzymes and nitrite reductase in glasshouse lettuce. *Journal of Experimental Botany*, **40**, 329–336

Besford, R.T., Ludwig, L.J. and Withers, A.C. (1990) The greenhouse effect: acclimation of tomato plants growing in high CO_2, photosynthesis and ribulose-1,5–biphosphate carboxylase protein. *Journal of Experimental Botany*, **41**, 925–931

Bowes, G. (1991) Growth at elevated CO_2: photosynthetic responses mediated through Rubisco. *Plant, Cell and Environment*, **14**, 795–806

DeJong, T.M. and Doyle, J.F. (1985) Seasonal relationships between leaf nitrogen content (photosynthetic capacity) and leaf canopy light exposure in peach (*Prunus persicata*). *Plant, Cell, and Environment*, **8**, 701–706

Field, C.B. (1991). Ecological scaling of carbon gain to stress and resource availability. In *Response of Plants to Multiple Stresses*. pp. 35–65 Edited by H.A. Mooney *et al*. San Diego: Academic Press

Goudriaan, J. and Monteith J.L., (1990) A mathematical function for crop growth based on light interception and leaf area expansion. *Annals of Botany*, **66**, 695–701

Keulen, H. van, and Seligman, N. G., (1987) Simulation of water use, nitrogen nutrition and growth of a spring wheat crop. *Simulation Monographs*. Wageningen: Pudoc

Kimball, B.A. (1983) Carbon dioxide and agricultural yield: An assemblage and analysis of 430 prior observations. *Agronomy Journal*, **75**, 779–788

Lawlor, D.W. and Mitchell, R.A.C. (1991) The effects of increasing CO_2 on crop photosynthesis and productivity: a review of field studies. *Plant, Cell and Environment*, **14**, 807–818

Lemon, E.R. (ed) (1983) CO_2 and plants. *AAAS Selected Symposium* No 84, Boulder CO: Westview Press

Leuning, R., Cromer, R.N., and Rance, R. (1991a) Spatial distributions of foliar nitrogen and phosphorus in crowns of *Eucalyptus grandis*. *Oecologia*, **88**, 504–510

Leuning, R., Wang, Y.P. and Cromer, R.N. (1991b) Model simulations of spatial distributions and daily totals of photosynthesis in *Eucalyptus grandis* canopies. *Oecologia*, **88**, 494–503

Monteith, J.L. (1981a) Climate variation and growth of crops. *Quarterly Journal Royal Meteorological Society*, **107** (454), 749–774

Monteith, J.L. (1981b). Does light limit crop production? In *Physiological Processes Limiting Plant Productivity*, Edited by C.B. Johnson. London: Butterworths

Monteith, J.L. (1990) Conservative behaviour in the response of crops to water and light. In *Simulation and Systems Management in Crop Protection*, pp 3–16 Edited by Rabbinge *et al*. Wageningen: Pudoc

Poorter, H. (1993) Interspecific variation in the growth response of plants to an elevated ambient CO_2 concentration. *Vegetatio*, **104/105**, 77–97

Rabbinge, R. and De Wit, C.T. (1989) Systems, models and simulation. In *Simulation and Systems Management in Crop Protection*, pp 3–15 Edited by R. Rabbinge *et al*. Wageningen: Pudoc

Spitters, C.J.T. (1990) On descriptive and mechanistic models for inter-plant competition, with particular reference to crop-weed interaction. In *Theoretical Production Ecology: Reflections and Prospects*, pp 217–236 Edited by R. Rabbinge *et al*. Wageningen: Pudoc

Waggoner, P.E. (1990) Defoliation, disease and growth. In *Theoretical Production Ecology: Reflections and Prospects*. pp 149–180 Edited by R. Rabbinge *et al*. Wageningen: Pudoc

Werger, M.J.A. and T. Hirose (1991) Leaf nitrogen distribution and whole canopy photosynthetic carbon gain in herbaceous stands. *Vegetatio*, **97**, 11–20

7

OPTIMAL ALLOCATION OF ASSIMILATE IN RELATION TO RESOURCE LIMITATION

H.G. JONES and J.R. LYNN
Horticulture Research International, Wellesbourne, Warwick CV35 9EF, UK

Introduction

Optimal growth requires effective capture of a range of resources including light, water and nutrients. Any one of these may be in especially short supply in a particular environment, so plants need mechanisms that ensure a balanced uptake of resources to minimize the limitation to growth imposed by the scarcest resource(s). In general this requires greater 'effort' (usually in terms of allocation of a greater proportion of assimilate) to be devoted to the capture of the most limiting resource. For example the typical response to limited availability of any resource required for plant growth (CO_2, nutrients, water, light) is an increased allocation to structures involved in uptake of that resource or an increased efficiency of uptake (e.g. Field, Merino and Mooney, 1983; Fredeen, Gamon and Field, 1991).

One might expect that the success of a plant in a particular environment would depend on the extent to which the allocation of carbohydrate and nutrients between root and shoot are modulated so as to achieve an appropriate balance between the capture of different resources such as light, water or nutrients, making the best use of available carbohydrate (or other materials). Achievement of such an optimal combination of characters involves co-ordination of different types of plant process (such as biosynthesis, metabolism, transport, development, etc.) in all the different parts of the plant from the leaves to the roots. This chapter considers both the ways in which the plant might detect the need for an altered pattern of carbohydrate allocation, and the different ways in which the necessary co-ordination is achieved.

Optimal carbohydrate allocation

ALLOCATION OF CARBOHYDRATE TO RESOURCE CAPTURE AS OPPOSED TO YIELD PRODUCTION

For many situations, success can be related to reproduction and the production of the maximum number of surviving progeny. It is of interest to consider what might be the optimal relative allocation of carbohydrate during growth to increasing

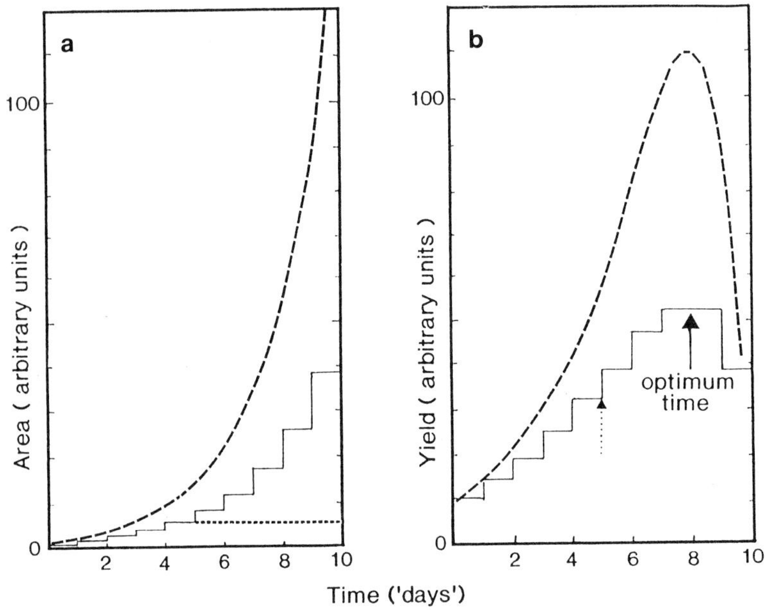

Figure 7.1 An illustration of how (**a**) leaf area and (**b**) yield of grain for a cereal might depend on the time of switch from vegetative growth to reproductive growth, showing the optimum time for the switch to occur (after Jones, 1992). The dashed line is for a continuous growth model, while the solid line represents a simplified discontinuous model where growth occurs in a number of finite 'daily' increments (in this case for a season of only ten 'days'). The dotted line shows the time course of leaf area (**a**) and the final yield for a switch at day 5 (**b**)

resource capture (i.e. allocation to leaf or root growth) as opposed to allocation to the production of reproductive structures such as flowers and seeds or organs of perennation. For situations where there is a finite growing period, such as for annual plants which have to reproduce before winter, it has been shown that maximal production of final yield or stored assimilate is achieved by dividing the total growing period into an initial phase where all effort is devoted to acquiring resource (i.e. all carbohydrate is allocated to extending root or leaf as necessary), followed by a final phase where all newly synthesized carbohydrate is allocated to reproduction (e.g. Cohen, 1971; Paltridge and Denholm, 1974). That is, a determinate growth habit appears optimal (*Figure 7.1*). Where, however, the length of the available growing season is uncertain, or where the availability of resources (such as rainfall) is uncertain, this conclusion no longer holds and the switch to reproductive growth occurs over an extended period. In this case, if the growing season does end prematurely, at least some seed is produced (see e.g. Jones, 1981). Plants having an indeterminate flowering habit behave in this way.

BALANCE BETWEEN CAPTURE OF DIFFERENT RESOURCES

Although one might intuitively assume that optimal growth in any environment would be achieved when resource capture is balanced such that no one resource is limiting, it is worth considering this statement a little further. In particular it is necessary to consider not only the relative limitation to growth imposed by the scarcity of any one resource, but the relative 'costs' of acquiring that resource.

Resource 'Limitations'

If one takes light as an example of an important resource, it is clear that the availability of the resource depends both on environmental factors (such as irradiance and its spatial and temporal distribution) and on plant factors (such as the leaf area and its distribution). In addition, growth also depends on factors that affect the utilisation of the resource once captured (such as quantum yield of photosynthesis). Similarly the amount of water available depends both on the concentration of water in the soil (environmental) and the volume of soil explored by the roots (plant). The term 'resource limitation' is sometimes used to refer to the combined effect of both environmental and plant factors affecting availability, though some authors restrict the use of resource limitation to describe the environmental factors, and refer to the plant factors as determining the 'capture limitation'. In this chapter the term resource limitation will be used in the more restricted sense and an attempt made to define it quantitatively.

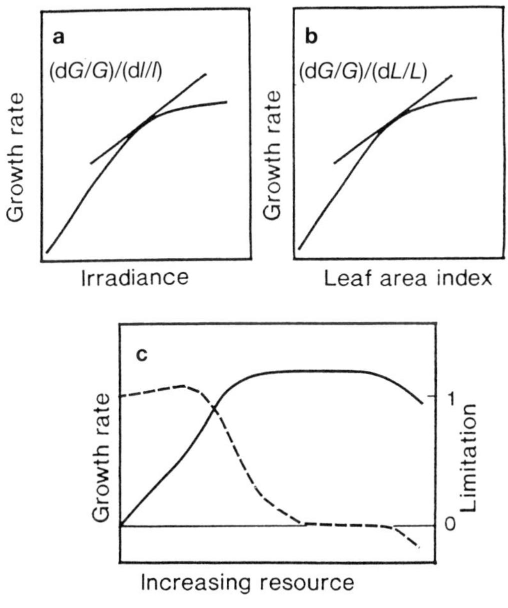

Figure 7.2 (**a**) A hypothetical relationship between growth rate (G) and resource availability (in this case for light, I), showing the calculation of the relative resource limitation ($\ell_i = (dG/G)/(dI/I)$); (**b**) hypothetical relationship between growth rate and a plant factor (leaf area index, L) determining resource capture, showing the relative capture limitation ($\ell_c = (dG/G)/(dL/L)$); (**c**) variation of growth rate (———) and ℓ_i (– – – –) with increasing resource availability, showing that ℓ_i may range from less than zero to more than unity

A hypothetical response curve relating growth rate (G) to a change in availability of any resource, R_i, is illustrated in *Figure 7.2a*. The sensitivity of growth to changes in R_i can be written

$$\text{Sensitivity} = \mathrm{d}G/\mathrm{d}R_i \tag{1}$$

The greater the sensitivity of growth to the change of a resource, the more that resource can be considered as limiting growth. Because the 'sensitivity' or 'limitation' defined by Equation (1) is dependent on the units used for G and R it is more convenient to normalize this equation to obtain a relative resource limitation, ℓ, given by:

$$\ell_i = (\mathrm{d}G/G)/(\mathrm{d}R_i/R_i) \tag{2}$$

Equally it is possible to define an analogous relative capture limitation (ℓ_c) in terms of one or more of the plant characters that can be changed to improve capture (*Figure 7.2b*). For example, for leaf area index (L), one can write:

$$\ell_c = (\mathrm{d}G/G)/(\mathrm{d}L/L) \tag{3}$$

In either case the relative limitation will normally vary between 0, where growth is unaffected by a change in availability of resource, and 1 where growth is proportional to the amount of resource. A calculated limitation <0 implies that resource availability is supraoptimal, with increases actually inhibiting growth, while values >1 indicate that relative growth rate increases disproportionately in relation to changes in resource supply (*Figure 7.2c*).

An analogous approach has been used to define the relative importance of stomatal and non-stomatal processes in limiting photosynthesis, with the gas-phase limitation (ℓ_g) being defined as $(\mathrm{d}A/A)/(\mathrm{d}g_s/g_s)$, where A is photosynthesis rate and g_s is the stomatal conductance (Jones, 1973; 1985). Particular advantages of this definition of limitation are that it is precisely defined and that it does not involve any extrapolation to unrealistic conditions (compare the definition proposed by Farquhar and Sharkey, 1982). A related approach was developed independently for the analysis of metabolic control (Kacser and Burns, 1973; Heinrich and Rapoport, 1974); this could also provide a basis for the analysis of the relative control exercised by component processes over the rate of any overall process such as growth. The main theorems of metabolic control analysis, however, apply only to closed systems such as a single metabolic pathway or network. In such a system all the individual flux control coefficients (C_i, defined in a similar way to Equation (2) as the relative change in flux through the system divided by the relative change in activity of a particular enzyme in the pathway) sum to unity. For growth, a number of separate external resources or effectors may be controlling the rate (*Figure 7.3*); in such situations the 'control' coefficients calculated as above but describing growth responses to changes of resource availability, are not constrained to sum to unity. It is best to distinguish these coefficients from true flux control coefficients by calling them response coefficients (Kacser and Porteous, 1987) or equivalently by using the term relative limitations (ℓ_i as in Equation 2). For example, it is possible for growth rate to be proportional simultaneously to changes in several different resources, such as irradiance and nitrogen supply, with both calculated relative resource limitations equalling unity at the same time.

a

→ E1 → E2 → E3 → ...... →

$$(dJ/J)/(dE1/E1) + (dJ/J)/(dE2/E2) + \ldots = 1$$

b

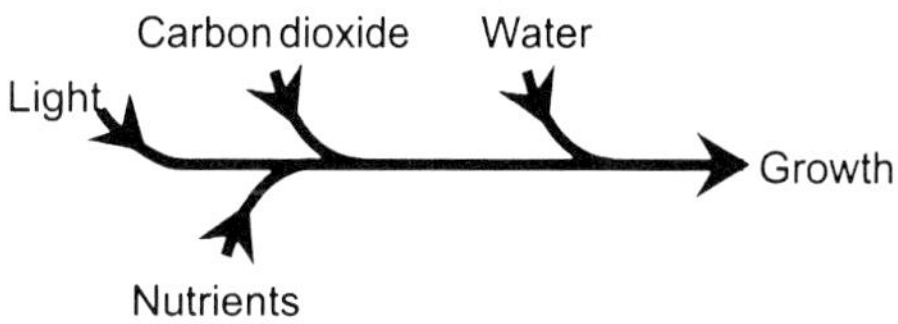

$$(dG/G)/(dC/C) + (dG/G)/(dI/I) + \ldots > 1$$

Figure 7.3 A diagram illustrating the difference between (**a**) flux control coefficients (in a closed system such as a single metabolic pathway with intermediate enzymes *E*1, *E*2, *E*3, etc.); and (**b**) response coefficients in the control of growth

This approach enables one to determine the sensitivity of growth to a change in the availability of different resources, giving relative resource limitations. A relative limitation calculated in this way would be a logical quantitative definition of the narrow sense (environmental) resource limitation. It is implicit from what has been stated above, however, that the availability of resource is only one of the factors that must be considered. Although one might imagine that the optimum pattern of allocation of carbohydrate to capture of different resources would lead to an equalization of limitations, this is not the case. For example, where the capture of two different resources depends on the same capture mechanism (e.g. water and phosphorus are both absorbed by the roots), the relative resource limitations could be very different in a well adapted plant where one resource might be in surplus ($\ell_i = 0$) and the other restricting growth ($\ell_i \simeq 1$). More importantly, however, it is necessary to consider the relative costs of acquiring the different resources and not solely to consider limitations calculated in terms of the effect of changing the *environmental* component of the availability of resources when one attempts to derive an optimal allocation of carbohydrate.

One needs to determine the response of growth to a change in carbohydrate (for example) allocation that will modify the resource capture process. The optimal pattern of carbohydrate allocation is the one that maximizes growth, and any deviation from the optimum leads to decreased net growth. When calculating the environmental limitations it is easy to consider instantaneous values, but in order to determine optimal carbohydrate allocation it is necessary to consider longer term consequences. For example, where one considers the effect of altered allocation between root and shoot of newly synthesized carbohydrate, the instantaneous growth rate (in terms of total dry weight) is unaffected by the proportion that goes into leaf growth as opposed to root growth, it is only over the longer term that the consequences of changes in allocation become apparent with increased allocation to roots reducing the potential for photosynthesis but increasing the potential for water uptake in the future. This point is discussed further below.

Modelling to determine optimal carbohydrate allocation

There have been a number of attempts to model the allocation of assimilate between roots and shoots to provide a framework for understanding the principles that determine optimal allocation of carbohydrate. In a particularly interesting analysis, Givnish (1986) investigated the optimal allocation, when water is limiting, of carbohydrate between the formation of roots and leaves in relation to stomatal conductance. He concluded that for optimal growth the ratio of the carbohydrate allocated to leaves and to roots should equal the ratio of the mesophyll (r_m) and stomatal (r_s) resistances for photosynthesis. That is:

$$r_m / r_s = f / (1 - f) \qquad (4)$$

where f is the fractional allocation of carbohydrate to the leaves. Interestingly this Equation was found to hold whatever the detailed dependence of the photosynthetic parameters on leaf water potential.

Similarly, there have been many attempts to consider the optimal partitioning of nitrogen (e.g Cowan, 1986: Thornley and Johnson, 1991). In such models partitioning of nitrogen between root and shoot is often achieved by the introduction of a specific partitioning function (see Thornley and Johnson, 1991), though Ågren and Ingestad (1987) have attempted to model the partitioning of nitrogen on a functional basis without the need to introduce any arbitrary partitioning function. They showed that the observed root:shoot partitioning can be modelled on the basis of a balance between nitrogen productivity (growth rate per unit total plant nitrogen) and photosynthesis.

Although there are many models that consider different resources individually, or in pairs, there have been relatively few attempts to develop unifying principles applicable to a number of resources. A particular problem is that most models are based on instantaneous assimilation rates, but as pointed out above, maximization of instantaneous growth does not necessarily lead to maximal long-term production. Nevertheless, we believe that it is possible to choose an appropriate function that can be optimized instantaneously to maximize long-term growth (at least under constant environmental conditions). As an initial hypothesis it is suggested that with optimal allocation of carbohydrate to different resource capture mechanisms (root growth, leaf area expansion, photosynthetic enzyme activity, etc.)

$$\partial^2 G'/(\partial t.\partial r) = \partial^2 G'/(\partial t.\partial l) = \partial^2 G'/(\partial t.\partial s) \; .\; .\; .\; . = 0 \qquad (5)$$

where r, l, s, etc. are the fractions of newly synthesized carbohydrate that are allocated to roots, leaves, stem, and so on, and given that the r, l, s, etc. maximize dG′/dt. In Equation 5 G′ is an appropriate measure of 'effective growth' that takes account of the future resource capture ability of that growth.

The proof that an optimization based on instantaneous optimisation of distribution of new carbohydrate to different organs can give the long-term optimal pattern of allocation follows. The assumption is made that the instantaneously optimal distribution of the total effective dry matter about the plant (i.e. what proportion of the total is in leaves, what in roots and so on) changes 'fairly slowly' over time. Note that in a more complex model this would not be true; for example if we allowed rainfall to occur at (random) times during the growth of the plant, then, assuming no shortage of nutrients, after a rainfall had occurred the instantaneously optimal distribution would require more leaves, and less roots than the distribution which was optimal before the rainfall.

Under the above assumption, where 'fairly slowly' is defined as being at a rate no faster than that which can be provided for in the distribution of new effective dry matter about the plant, we suppose that at $t=0$ the plant has an instantaneously optimal distribution of dry matter. Suppose further that the plant follows some arbitrary policy of resource distribution $p(t)$, and let $q(t)$ be the candidate optimal policy. We further define $H(t)$ as the rate of growth obtainable under our candidate optimal policy. Defining t to be the first time

p differs from q, by the definition of q, $\frac{dG}{dt} < \frac{dH}{dt}$ at time t. Hence, at time

$t+\epsilon$, for sufficiently small $\epsilon>0$, $G(t+\epsilon)<H(t+\epsilon)$, and the total effective dry mass of the plant is less than it would have been under policy q.

Now, we assume (for a contradiction), that policy p is not worse than policy q. Then there must exist some time $t'>t$ such that the total effective dry mass of the plant at time t' is the same under policies p and q. Hence there must exist some time t'' with $t<t''<t'$ and such that $G(t'')>H(t'')$, but for all times between t and t' the total effective dry mass of the plant is less than what it would have been under q, and so the distribution of effective dry mass at time t'' must be instantaneously better under policy p than under policy q. However, because of the assumption the plant following policy q has maintained an instantaneously optimal distribution of resources, and so the distribution cannot be better under policy p, thus contradicting the assumption that p is no worse than q.

Policy p, however, was arbitrary, and hence policy q is optimal. In the above argument it is possible to define growth rates in terms of total dry matter or any other relevant measure of production.

If we do allow random changes in the environment, leading to 'fast' changes in the instantaneously optimal effective dry matter distribution, then we would like to apply the policy of changing resource allocation in advance of the random event. This would allow the resource distribution after the random event to be closer to being instantaneously optimal for the new situation. This attractive policy is, however, inadmissable as it requires prediction of the future (i.e. when the random event will occur). The only intuitively reasonable admissible policy which accounts for, for example, randomly occurring rainfall, is to grow less root than q suggests, so that following a rainfall the plant does not have too much 'waste' root. The amount by which this policy differs from q should not be too great however, since as we have seen above, it is causing losses all the time before the rainfall occurs, and so policy q should not be too far off optimal even in this case. Strategies for an unpredictable rainfall environment have been discussed by Jones (1981).

Sensing the need for altered assimilate allocation

Given that plants respond to changing resource availability by altering partitioning, it is of interest to know how the need for changed partitioning is sensed and how the changes in partitioning are controlled. It is often not clear how far such adaptation is a biological example of Le Chatelier's Principle ('If a constraint is applied to a system in equilibrium, the system tends to adjust itself so as to oppose the constraint and restore equilibrium') with the constraints being resource limitations; or how far more positive control mechanisms are involved. Optimal growth and allocation of assimilate requires the existence of mechanisms by which the plant can determine which resources are most limiting and modify growth accordingly so as to restore the processes of resource capture to balance so that no one resource is limiting.

It is usually assumed that the lack of a resource is directly sensed with the resultant signal causing an appropriate modification of carbohydrate allocation. An example of such direct sensing is the synthesis of abscisic acid in roots as the root water potential falls, with this synthesis directly enhancing root growth (see Trewavas and Jones, 1991). Alternatively there might be no direct link between the control mechanism that gives optimal allocation and the allocation process. As an example of this indirect regulation, consider the time of the switch from vegetative to reproductive development. The control of the reproductive transition is often directly under daylength or vernalization control with no regard for the actual physiological status of the plant. Plants adapted to a particular locality tend to evolve so that they have a photoperiodic sensitivity that ensures that the switch occurs at the optimal time. There is no functional link between the system that controls the time of the switch and the physiological status of the plant, so that when the plants are transferred to a different environment, the timing of the switch may no longer be optimal. In some cases there may be overrides that modify such environmentally programmed responses; for example, high temperatures can override the long-day requirement for flowering in wheat (Evans, Wardlaw and Fischer, 1975).

It is also worth considering the extent to which the control of partitioning might be regulated solely by responses to local conditions without the need to invoke long-distance signalling. Optimal resource capture in this case would require a mechanism that stimulates root growth as either the nutrient concentration in the roots or the water availability (root water potential) decrease. Similarly leaf growth or shoot extension would need to be stimulated at low light, as would synthesis of photosynthetic enzymes, and stomatal closure would need to be activated as leaf water potential declines. The fact that some resource capture may involve complex interactions with other organisms must also be born in mind when considering the sensing and control mechanisms. For example mycorrhizal associations can be particularly important, especially in relation to phosphate uptake in nutrient deficient soils. The growth of the fungal partner appears to be to some extent under the control of the plant, with mycorrhizal growth being stimulated (perhaps allowed?) only when the plant has a need for the additional phosphate. The control of the fungal partner is probably exercised through the supply of carbohydrate made available, so the control may be similar to that involved in the stimulation of root growth when water or nutrients are limiting.

Signalling and the control of allocation

Co-ordination of developmental responses to optimize resource capture requires the existence of long-distance signalling systems. It is not clear, however, to what extent the interactions between different parts of plants can be explained simply as direct consequences of the supply of substrates or water from one organ to another, or whether more subtle regulatory controls involving hormones or other physiological signals are involved (Jackson, 1990). In animals, co-ordination is achieved at a number of levels which include both conscious activity under control by the brain, unconscious co-ordination involving the autonomous nervous system and the hormonal system. Plants, of course, lack obvious analogues of the nervous systems. Nevertheless it is worth considering the various possible mechanisms of long-distance signalling in plants, and evidence for their operation in the control of carbohydrate and nutrient allocation to optimize growth.

It is commonly assumed that co-ordination in plants is achieved by means of plant growth regulators such as auxin, abscisic acid, cytokinins and gibberellins, but the evidence for such signalling in the regulation of growth and resource capture in unstressed plants is patchy. Trewavas and Cleland (1983) have argued that these plant growth substances may be involved in growth regulation not through changes in concentration, but through changes in sensitivity in the responding cells, and that attempts to explain co-ordination in terms of rather simplistic growth regulator hypotheses are probably misleading. Nevertheless, changes in growth regulator concentration provide the only obvious way in which growth regulators could be involved in long-distance signalling. Some of the possible mechanisms involved in long-distance signalling are outlined below.

POSSIBLE TYPES OF SIGNALLING MECHANISM

A number of types of long distance signal have been suggested as candidates in the control of plant development in remote parts of plants, and as means of co-ordinating developmental responses such as those involved in optimizing resource capture or resource utilization. The chemical signals have been classified by Jackson (1990) into:

1. Positive messages, where the response depends on the increased supply of some physiologically active substance from the remote organ. The increased production of abscisic acid by roots in drying soil and the consequential stomatal closure in leaves is an example of this type of regulation.
2. Negative messages, which represent a decreased supply of some essential factor. The decreased supply of cytokinins from roots to shoots as soil dries is an example.
3. Accumulative or indirect messages which occur where there is a build-up of a compound in an organ because export is restricted, or because the ability of the organ to metabolize the substance is impaired. As an example of this type of message, the build-up of carbohydrate in leaves as transport is inhibited could lead to feedback inhibition of photosynthesis.

Although these signals are usually conceived of as analogue signals, it is possible that the signalling mechanisms that operate within plants are digital in nature. For example if the response to a hormone or other elicitor were to require a threshold concentration to be exceeded rather than being proportional to concentration, this would be a digital signal. In general analogue signals are notoriously unreliable, while digital signals are effectively immune to noise and to attenuation by distance.

It is also possible that the overall integration and control of plant development and resource capture may well involve a degree of redundancy with alternative mechanisms for control if one system is disabled for any reason. Computer systems based on fuzzy logic might be an appropriate engineering analogy.

In addition to chemical signals there are a number of other possibilities. In particular, physical signalling mechanisms such as those based on electrical signals (Davies, 1987) or turgor pressure (Malone and Stanković, 1991) are possible candidates in certain situations, while Due (1989) has suggested that many plant responses might be controlled by the frequency of the signalling mechanism. Both electrical and pressure signals have been implicated in the induction by insect attack of proteinase inhibitor synthesis in remote parts of plants (Green and Ryan, 1972), but there is no clear evidence that such regulation is involved in control of carbohydrate allocation to optimize resource capture.

EVIDENCE FOR ROOT TO SHOOT SIGNALLING IN DROUGHT ADAPTATION

A system that has been extensively studied in relation to the capture and use of a particular resource (water) is that involved in the control of plant responses to soil drying and especially the control of stomatal aperture. The control of water loss is dependent on the availability of water; although for many years it has been assumed that stomata respond to *leaf* water status, there is increasing evidence that a mechanism based on leaf water status is not adequate to explain the control of stomatal behaviour (see Davies and Zhang, 1991 for a recent review). The various lines of evidence include:

1. Stomatal conductance in experiments with *Nerium oleander* and *Helianthus annuus* was found to be more closely related to soil water status than to leaf water status when the latter was manipulated by changes in whole plant transpiration (Gollan, Turner and Schulz, 1985).
2. In other experiments stomatal conductance was shown to decline with soil water status even if leaf turgor was maintained by pressurization of the root system (Gollan, Passioura and Munns, 1986).
3. A number of split-root experiments have been conducted (see Davies and Zhang, 1991), where leaf conductance (and leaf growth) was found to be inhibited by drying a small proportion of the root zone, without any evidence of an effect on leaf water status.
4. Another interesting, though circumstantial, argument is based on the fact that the short-term variability of leaf water potential tends to be much greater than the difference between droughted and well watered plants (see Jones, 1990). It is hard to envisage a control system that would respond to the small treatment differences and not to the larger short-term fluctuations in leaf water status.
5. A recent analysis of the information required for optimal control of stomata (Jones and Sutherland, 1991) concluded that leaf water potential alone could not provide the necessary information for optimal stomatal control, while root water potential coupled with a measure of evaporative demand of the atmosphere could.

All these separate lines of evidence indicate that long-distance signalling must be involved as at least part of the mechanism controlling stomatal behaviour in response to water deficits. There is also good evidence that shoot development in water-stressed plants is also controlled by similar long-distance signalling, possibly by abscisic acid transported in the xylem sap (Gowing, Davies and Jones, 1990; Davies and Zhang, 1991).

Conclusion

Optimal growth requires effective capture of a range of resources including light, water and nutrients, any one of which may be in especially short supply. In general greater 'effort' (usually in terms of allocation of a greater proportion of assimilate) needs to be devoted to the capture of the most limiting resource. This chapter has attempted to define what is meant by resource limitation and to discuss optimal partitioning in relation to the nature of the mechanisms that enable a plant to achieve optimal form and assimilate partitioning, both in terms of the sensing and signalling mechanisms involved, and in terms of the resulting control of development.

Acknowledgement

This work was supported by the Agriculture and Food Research Council and by STD2 programme of the European Community (Project TS2.0121.M(H)).

References

Ågren, G.I. and Ingestad, T. (1987) Root:shoot ratio as a balance between nitrogen productivity and photosynthesis. *Plant, Cell and Environment*, **10**, 579–586

Cohen, D. (1971) Maximising final yield when growth is limited by time or by limiting resources. *Journal of Theoretical Biology*, **33**, 299–307

Cowan, I.R. (1986) Economics of carbon fixation in higher plants. In *On the Economy of Plant Form and Function*, pp 133–170 Edited by T.J. Givnish. Cambridge: Cambridge University Press

Davies, E. (1987) Action potentials as multifunctional signals in plants: a unifying hypothesis to explain apparently disparate wound responses. *Plant, Cell and Environment*, **10**, 623–631

Davies, W.J. and Zhang, J. (1991) Root signals and the regulation of growth and development of plants in drying soil. *Annual Review of Plant Physiology and Molecular Biology*, **42**, 55–76

Due, G. (1989) Frequency as a property of physiological signals in plants. *Plant, Cell and Environment*, **12**, 145–149

Evans, L.T., Wardlaw, I.F. and Fischer, R.A. (1975) Wheat. In *Crop Physiology*, pp 101–149 Edited by L.T. Evans. Cambridge: Cambridge University Press

Farquhar, G.D. and Sharkey, T.D. (1982) Stomatal conductance and photosynthesis. *Annual Review of Plant Physiology*, **33**, 317–345

Field, C.B., Merino, J. and Mooney, H.A. (1983) Compromises between water-use efficiency and nitrogen-use efficiency in five species of Californian evergreens. *Oecologia*, **60**, 384–389

Fredeen, A.L., Gamon, J.A. and Field, C.B. (1991) Responses of photosynthesis and carbohydrate-partitioning to limitations in nitrogen and water availability in field-grown sunflower. *Plant, Cell and Environment*, **14**, 963–970

Givnish, T.J. (1986) Optimal stomatal conductance, allocation of energy between leaves and roots, and the marginal cost of transpiration. In *On the Economy of Plant Form and Function* pp 171–213 Edited by T.J. Givnish, Cambridge: Cambridge University Press

Gollan, T., Turner, N.C. and Schulze, E.-D. (1985) The responses of stomata and leaf gas exchange to vapour pressure deficits and soil water content. III. In the schlerophyllous woody species *Nerium oleander*. *Oecologia*, **65**, 356–362

Gollan, T., Passioura, J. and Munns, R. (1986) Soil water status affects the stomatal conductance of fully turgid wheat and sunflower leaves. *Australian Journal of Plant Physiology*, **13**, 459–464

Gowing, D.J.C., Davies, W.J. and Jones, H.G. (1990) A positive root-sourced signal as an indicator of soil drying in apple, *Malus* × *domestica* Borkh. *Journal of Experimental Botany*, **41**, 1535–1540

Green, T.R. and Ryan, C.A. (1972) Wound induced proteinase inhibitor in plant leaves: a possible defence mechanism against insects. *Science*, **175**, 776–777

Heinrich, R. and Rapoport, T.A. (1974) A linear steady-state treatment of enzymatic chains. *European Journal of Biochemistry*, **42**, 107–120

Jackson, M.B. (1990) Communication between the roots and shoots of flooded plants. In *Importance of Root to Shoot Communication in the Responses to Environmental Stress* pp 115–133 Edited by W.J. Davies and B. Jeffcoat. Bristol: British Society for Plant Growth Regulation

Jones, H.G. (1973) Limiting factors in photosynthesis. *New Phytologist*, **72**, 1089–1094

Jones, H.G. (1981) The use of stochastic modelling to study the influence of stomatal behaviour on yield-climate relationships. In *Mathematics and Plant Physiology* pp 231–244 Edited by D.A. Rose and D.A. Charles-Edwards. London: Academic Press

Jones, H.G. (1985) Partitioning stomatal and non-stomatal limitations to photosynthesis. *Plant, Cell and Environment*, **8**, 95–104

Jones, H.G. (1990) Physiological aspects of the control of water status in horticultural crops. *HortScience*, **25**, 19–26

Jones. H.G. (1992) *Plants and Microclimate: A Quantitative Approach to Environmental Plant Physiology*, (2nd Edn). Cambridge: Cambridge University Press

Jones, H.G. and Sutherland, R.A. (1991) Stomatal control of xylem embolism. *Plant, Cell and Environment* **6**, 607–612

Kacser, H. and Burns, J.A. (1973) The control of flux. *Symposia for the Society of Experimental Biology*, **27**, 65–107

Kacser, H. and Porteous, J.W. (1987) Control of metabolism: what do we have to measure? *Trends in Biochemical Science*, **12**, 5, 7, 9, 11, 13–14

Malone, M. and Stanković, B. (1991) Surface potentials and hydraulic signals in wheat leaves following localized wounding by heat. *Plant, Cell and Environment*, **14**, 431–436

Paltridge, G.W. and Denholm, J.V. (1974) Plant yield and the switch from vegetative to reproductive growth. *Journal of Theoretical Biology*, **44**, 23–34

Thornley, J.H.M. and Johnson, I.R. (1991) *Plant and Crop Modelling: A Mathematical Approach to Plant and Crop Physiology*. Oxford: Oxford University Press

Trewavas, A.J. and Cleland, R.E. (1983) Is plant development regulated by changes in the concentration of growth substances or by changes in the sensitivity to growth substances? *Trends in Biochemical Sciences*, **8**, 354–357

Trewavas, A.J. and Jones, H.G. (1991) An assessment of the role of ABA in plant development. In *Abscisic Acid: Physiology and Biochemistry* pp 169–188 Edited by W.J. Davies and H.G. Jones. Oxford: Bios Scientific Publishers

8

PERSPECTIVES IN MODELLING RESOURCE CAPTURE BY CROPS

S.N. AZAM-ALI, N.M.J. CROUT and R.G. BRADLEY
Tropical Crops Research Unit, University of Nottingham, Sutton Bonington Campus, LE12 5RD, UK

Introduction

Our incomplete grasp of biology inevitably means that even the most precise descriptions of organic behaviour can only be approximations of reality: at best, they depend on the depth and extent of our knowledge. Traditional biology relies heavily on the collection of observational data without the development of principles to explain the underlying processes involved. As a result, the experience gained by individuals has often been used to make predictions based on judgement rather than through general quantitative theories.

Recently, it has become fashionable to develop *models* of crops that aim to organize observational data into more general rules. A crop model is a mathematical statement of how that crop behaves. Because this statement is mathematical, it has to be explicit at every stage. The need for an equation forces assumptions to be made and the model provides the means of testing those assumptions. If modelled predictions are not an accurate description of reality we are forced to accept that the system in question is not fully understood.

The purpose of this chapter is to review the types of crop models that exist and, in particular, to consider how they may help in our understanding of the way in which crop systems capture and use resources.

Why model resource capture?

Irrespective of their composition, the life of all green plants depends on their capture and use of external resources, principally solar radiation, carbon dioxide, water and nutrients. These resources are finite in space and time and have to be shared between individual plants for a crop community to progress towards maturity or to produce an economic yield.

The behaviour of a crop *stand* can be described or modelled in terms of its capture and use of these resources individually or collectively. Such models can be used to assess the state of a crop at any specific time and/or to predict its state at some future stage, often at the point when it is harvested. Crop 'yield' models have a number of practical applications and Steiner (1987) identified seven of these as follows:

- reduce the need for site-specific long-term field experiments,
- interpret climatological records in terms of production potential and climatic limitations,
- evaluate expected returns to soil and crop management practices,
- evaluate risks associated with management practices,
- communicate research results between locations,
- enhance understanding of biological and physical systems and their interactions,
- conceptualize multidisciplinary activities.

Of these, the first four are probably of most interest to individuals and agencies involved in devising farming strategies for different locations and circumstances. Despite their inadequacies, model predictions have a role to play in guiding planners and researchers before field experimentation. However, in many cases, models are appropriate as *relative* rather than *absolute* tools – in all probability models can be justifiably used to indicate whether a given management option is likely to increase yield but the magnitude of the increase may be difficult to estimate accurately.

It is Steiner's last three points that identify important but underestimated uses of crop models. First, they have an important role to play in the synthesis of ideas and data. Second, they identify weak areas in our understanding of processes and systems. Finally, they may encourage scientists from disparate backgrounds to co-operate in worthwhile research.

How can we model resource capture?

Once it has been established that modelling may have a role to play in the understanding of crop behaviour, there remains the question of which type of model should be selected. Models of resource capture can broadly be divided into two major categories:

- empirical and semi-empirical (index),
- simulation or mechanistic.

Within these, two features can be identified. The model, or parts of it, may be:

- dynamic or static,
- deterministic or stochastic.

EMPIRICAL AND SEMI-EMPIRICAL (INDEX) MODELS

There is a general divide between so-called mechanistic, or simulation, models and empirical models. The latter are primarily based on relationships derived from regression analysis and usually require large amounts of data for their development. The disadvantage of this type of model is that it cannot be applied outside the region for which it was developed. It is simply a mathematical description of the data, adding nothing to our understanding. Despite these limitations this type of simple model should not be rejected out of hand. For the site at which it was developed, an empirical model is likely to perform better than a simulation model in terms of the accuracy of prediction. This type of model can therefore be a very useful tool, *provided* it is not used outside its region of calibration.

Between empirical models, based entirely on regression analysis, and mechanistic simulation models, lies a large group of useful practical models which can be classified as 'index' models. In these, some kind of index is defined whose value is related to parameters that will mechanistically influence crop yield etc. The index is then often empirically related to the quantity that we are interested in, generally yield. The success of this type of model relies upon several factors, in particular:

- The choice of an index appropriate to the conditions of the region where it is to be applied. For instance, the choice of a soil moisture index would be less appropriate in, say, the humid tropics where water is not limiting.
- As with the empirical approach, a large data set is required for the region concerned in order to calibrate the model, which cannot be applied safely beyond that region. However, because it is based on a more-or-less mechanistic index the model should be reliable in abnormal seasons.

In summary, index models do not simulate crop processes but empirically relate crop performance to an appropriate index which varies with the limiting environmental factors.

SIMULATION OR MECHANISTIC MODELS

With this type, there is an attempt to base the structure of the model on real processes that take place in the system concerned. In other words, such models attempt description based upon *understanding*. France and Thornley (1984) regard these categories as based on the level in the organizational hierarchy of a system at which the model is implemented. They suggest that every system can be considered at an almost infinite number of levels. The simplest empirical model considers just the first level i.e. the whole system. In this case, there is no consideration of the system subcomponents.

However, a model could be based at a more fundamental level in the organizational hierarchy of the system. For instance, we could consider leaf areas, calculate intercepted light, and thereby estimate photosynthesis, and partition assimilates between the various plant parts. This could be done day-by-day allowing plant development to continue until we choose to 'harvest' our modelled crops. The model could also contain mechanisms whereby available soil water could restrict growth. In practice, each of these subprocesses in the model would be considered empirically. The rate of photosynthesis in relation to temperature and radiation intensity may be calculated from functions derived from experimental observations. The approach to plant development would probably take the form of a thermal sum, again simply based upon experimental observations. Therefore each of the subcomponents is an empirical description of that particular process but they are combined in a mechanistic fashion.

Of course, a model could be developed at a still more basic level and in such cases one might consider the kinetics of the chemical processes which constitute photosynthesis. The parameters in the model would then be empirically derived from experimental observations of photosynthetic reactions, rather than simple observations of gross assimilate produced versus light intensity etc.

DYNAMIC AND STATIC MODELS

Put simply, a static model is one which does not contain time as a variable. Therefore, such a model neglects any time-dependent aspect of the system in question. Of course, everything does change with time and therefore any static model must be an approximation, although if the system changes slowly the approximation may be a good and useful one. A map of the world is a good example of a very useful static model. In reality, continents move, mountains rise and fall, rivers dry up and so on but these processes are slow and therefore a static map can still serve a useful purpose for short-lived human beings.

In contrast, a dynamic model explicitly includes time as a variable and provides a means of modelling systems that alter over the timescale of interest. Clearly, most crop models are dynamic in that quantities such as mass, leaf area and yield change during the life of a crop. However, a crop model may contain submodels that are static. For instance, a light interception model might allow for variation in irradiance during the day; others might simply assume an average daily value. The former is a dynamic model, the latter are static.

DETERMINISTIC AND STOCHASTIC MODELS

Most crop yield models currently in use are deterministic, that is, they make *definite* predictions of yield etc. There is no attempt to estimate the level of uncertainty in the predictions. In practice, there is always some uncertainty surrounding the predictions, as indicated by the poor level of agreement often obtained between models and observations! Such uncertainty has three sources:

- Uncertainties in input data, e.g. meteorological data, soil characteristics.
- Uncertainties in model parameters, e.g. approximate values of crop parameters for processes such as respiration or leaf expansion rates.
- Systematic uncertainties introduced by the simplifications and approximations inherent in the actual structure of the model, e.g. the effect of assuming uniform irradiance during the day.

It is possible to examine the influences of the first two sources of uncertainty on the predictions of a model, and this is basically what stochastic models attempt to do. Although the means by which this is done are conceptually straightforward, in practice it can be difficult and time-consuming.

All models have some kind of input data or parameters. For instance, the Johnson and Thornley (1983) photosynthesis model has a parameter 'PM20' i.e. the light saturated value of photosynthetic rate at 20°C. For grass, they give a value of 1.5×10^{-6} $kgCO_2$ m^{-2} s^{-1}. This figure has some uncertainty associated with it. For the sake of argument, let us assume an uncertainty of 10% which produces a range of values for PM20 with most probably a normal distribution. In order to estimate the effect of this distribution of possible values on the model results, we could run the model many times on each occasion randomly selecting a value for PM20 from within its normal distribution range. This would produce a distribution of model results with the main prediction based on the actual estimate of PM20 with confidence limits derived from the *distribution* of results obtained by running the model repeatedly. This enables a range of possible crop weights to be predicted given the underlying uncertainty in the input parameter PM20. Of course, the basic approach can be extended to include any number of parameters or input data, although difficulties arise if the parameters are not strictly independent.

Stochastic models have two significant benefits.

- They enable uncertainties on predictions to be calculated. This is essential if model predictions are to be used for planning purposes. A planner is more likely to act on a yield prediction of 1.3 ± 0.1 than one of 1.3 ± 0.8. However, if the planner has a prediction of 1.3 with no uncertainty estimate, its value in terms of decision making is greatly limited.
- Such models enable key parameters which strongly affect model results to be identified. This can then lead either to further experimental work to improve the knowledge of those parameters or efforts to 'desensitize' the model to uncertainties in those parameters. In this way stochastic models can serve as useful tools for planning research.

Model structure and development

This chapter is not aimed primarily at those intending to actually develop working models but, rather, at those who may wish to use a model for some practical purpose. Nevertheless, it may be worthwhile to highlight some of the considerations that should apply before and during the development of any model. Such considerations have been outlined most lucidly by Norman (1989) and his observations are paraphrased below.

- The model should have an objective. This rule not only applies to modelling but is especially relevant in that context! How the objective of a model influences the type of model required has already been discussed. The temptation to develop a model for its own sake should be avoided. A major consideration is whether the model will actually add to knowledge or understanding.
- Model inputs, such as climate data and physiological responses, should be based on values from the relevant discipline. Inputs unique to the model should be avoided.
- Models should be tested against independent measurements. The temptation to adjust parameters to improve agreement should be avoided.
- A structured approach should be adopted to the definition of the system being modelled. If large computer programs are needed, it is important to use a structured programming language (e.g. PASCAL, C, or possibly FORTRAN). However, the need for specific programming skills may soon be reduced as modelling environment software becomes available. Such, systems may transform modelling in a manner similar to the way in which primary data manipulation has been transformed by the advent of 'spreadsheet' programs.
- Consider how model documentation is to be produced and at what level of complexity it will be aimed. This is crucial if the model is to be widely used.

Obviously this list is not exhaustive and Norman himself makes many other valid points. Some of the more important ones have been selected in this chapter in an attempt to emphasize the importance of clear planning and definition in any significant modelling exercise.

Modelling resource capture

Models are made up of several components that describe the partitioning of captured resources between various organs and their loss through respiration, senescence etc. Some model components may depend on relationships for phenological development that are based on temperature and/or photoperiod. However, at the heart of any model will be a 'growth engine' which provides some means of calculating the amount of assimilate that plants will produce i.e. some estimate of *photosynthesis*. The process of photosynthesis is extremely complex but ultimately involves the reduction of CO_2 to form organic molecules such as glucose using absorbed radiant energy. The waveband in which radiant energy is available for photosynthesis is that within the range 400–700 nm. This *photosynthetically active radiation* (PAR) is about 50% of solar radiation (Szeicz, 1974). Glucose molecules produced via photosynthesis either provide the building blocks for virtually all the organic constituents of plants or are respired to provide energy for metabolic processes. The basis of how photosynthesis can be estimated in crop models is now considered.

The solar engine

Photosynthesis can be modelled in several ways, for instance by using light response curves for individual leaves or, more spectacularly, by attempting to integrate the activity of individual cells. As an alternative to these complex approaches, it simply could be assumed that there is an approximately linear relation between the total dry matter that a crop accumulates and the amount of light that it intercepts over the same period. Irrespective of their complexity, both simple and complex 'solar' engines characterize the *potential* photosynthetic productivity. The accuracy of model predictions still depends on how successfully they quantify the detrimental effects of biological and physical factors on the final product.

It is worth reconsidering the basis for light-driven growth engines in terms of the efficiencies of individual processes. For example, the link between crop yield and accumulated intercepted radiation can be examined to see what factors influence model predictions.

Borrowing from the approach taken by Monteith (1972), the productivity of an ecosystem can be defined in terms of the efficiencies with which the solar resource is captured and converted to biomass i.e. the overall efficiency, ϵ, of the system can be described by;

$$\epsilon = \epsilon_g \epsilon_a \epsilon_p \ \epsilon_i \ \epsilon_s \epsilon_q \epsilon_r \epsilon_h \tag{1}$$

where, ϵ_g is a geometrical ratio i.e. the ratio of solar energy flux to solar constant, ϵ_a is the atmospheric transparency, ϵ_p is the spectral ratio i.e. PAR as a fraction of total radiation, ϵ_i is the interception efficiency of foliage, ϵ_s is the conversion efficiency of intercepted radiation, ϵ_q is the photochemical efficiency i.e. the energy value of carbohydrates, proteins and lipids, ϵ_r is the respiration efficiency and ϵ_h is the yield efficiency.

AVAILABILITY OF RESOURCE (ϵ_g ϵ_a ϵ_p)

These three factors influence the amount of PAR that is received above a crop canopy. In practice, they have little influence on the reliability of model predictions. Nevertheless, it should be noted that there is both geographical and seasonal variation in their values and errors associated in their measurement.

CAPTURE OF RESOURCE (ϵ_i)

For a crop to produce dry matter its leaves must intercept radiation and absorb CO_2. The size and duration of crop foliage determine the rate and duration of dry matter accumulation. The size of the intercepting surface depends on the green leaf area index (L) of a crop which can be expressed as:

$$L = N_p \times N_s \times A_s \tag{2}$$

where N_p is the number of plants per unit of ground area, N_s is the number of leaves per plant and A_s is the mean area of leaves per plant.

The amount of light that penetrates the canopy and strikes the ground depends both on L and on the angular arrangement of individual leaves. In order to describe the pattern of light penetration through a crop canopy, it is convenient to imagine a crop as consisting of a number of horizontal layers each of $L = 1$. If radiation is measured at a number of levels down the crop profile, each corresponding to unit L, then the measured irradiance at any level is a function of the angular arrangement of leaves above that level. In these circumstances, the relationship for light extinction can be described by the Monsi–Saeki equation

$$\frac{I}{I_o} = e^{-kL} \tag{3}$$

where I_0 is the irradiance above the crop canopy, I is the irradiance at a level within the canopy below a leaf area index of L and k is an extinction coefficient for radiation. However, it should be noted that the Monsi–Saeki equation assumes that the canopy is a homogeneous medium whose leaves are randomly distributed and, in this case, there is no effect of row structure on the pattern of light transmission through the canopy.

In these (ideal) circumstances, light transmission obeys Beer's law of exponential decay. Strictly, for attenuation to be exponential, the leaves should be black i.e. opaque to radiation. However, Goudriaan (1977) has shown Beer's law to be a good approximation in many real canopies and, in these circumstances, a plot of $\ln(I/I_0)$ against L gives a straight line with a gradient k which depends on the architecture of the crop.

If the radiation transmitted through the canopy follows the Monsi–Saeki equation and the extinction coefficient is known, the fraction, f, of radiation intercepted by a crop can be calculated from a knowledge of L, i.e.

$$f = 1 - \exp(-kL) \tag{4}$$

Although the Monsi–Saeki equation has been widely used to develop models of canopy architecture, there are often problems with its application in many field conditions. A major limitation of the model is that it assumes a random arrangement of leaves within the crop canopy. This may be a reasonable approximation for crops that completely cover the ground for most of the season but it is less valid where there is a distinct clumping of foliage. Nilson (1971) used a Markov chain model to theoretically describe the penetration of radiation through non-randomly distributed foliage. In this analysis he defined a parameter which was 1 for a perfectly random canopy, >1 for canopies with a regular row structure and <1 for canopies with clumped foliage. Non-random distribution of foliage is particularly evident in tropical environments where crops are either planted in wide rows or where drought, or some other stress, prevents crops achieving significant ground cover until late in the season, if at all. The value of k varies with the size of a crop canopy and, in practice, there may be considerable variation in the mean value of k for a particular species as indicated in *Table 8.1*.

Table 8.1 THE RANGE OF SOME REPORTED VALUES FOR THE EXTINCTION COEFFICIENT, k, FOR TEN MAJOR CROPS OF THE WORLD

Crop	k
Wheat	0.40–0.70
Rice	0.43–0.86
Maize	0.56–0.78
Barley	0.48–0.69
Sorghum	0.43–0.60
Millet	0.30–0.59
Cassava	0.58
Soyabean	0.45–0.96
Groundnut	0.40–0.66
Sugar beet	0.61–0.85

CONVERSION OF RESOURCE

Conversion efficiency of solar radiation (ϵ_s)

To calculate the productivity of a stand, we need to integrate the photosynthetic rate of individual leaves with respect to space and time. The rate of leaf photosynthesis is limited by the rate at which it can assimilate CO_2 from the atmosphere and reduce it to carbohydrate. The theoretical minimum energy required to reduce a CO_2 molecule is about 9.5 quanta of PAR (Penning de Vries *et al.*, 1989). However, the lowest values observed in C_3 plants, in the absence of photorespiration, is about 13 quanta per molecule (Farquhar and von Caemmerer, 1982). (The difference is largely because some light is absorbed by non-photosynthetic pigments.) Nevertheless, leaf photosynthesis is initially limited by the amount of radiant energy incident on the leaf surface. Eventually, as the incident energy increases, the concentration of CO_2 in the air around the leaves limits the rate of photosynthesis. Thus, the spatial distribution of leaves within a crop canopy markedly affects whether canopy photosynthesis is light- or CO_2-limited.

For uniform crop stands, the spatial integration of photosynthesis can again be described by the Monsi–Saeki model and the irradiance that each leaf intercepts can be modelled in terms of the mean orientation of leaves in relation to incident radiation. Individual leaf photosynthesis therefore proceeds at a rate that depends on irradiance and the relation between photosynthetic rate and irradiance for each element of foliage.

The temporal integration of photosynthesis requires a knowledge of crop behaviour over the whole growing season. Experimentally, the relationship between dry matter accumulation and intercepted radiation has been shown to be approximately linear for many crops, at least during the vegetative stage of growth. Thus, the rate of dry matter production per day $\delta W/\delta t$ can be expressed as:

$$\frac{\delta W}{\delta t} = \epsilon_s \, fS_0 \tag{5}$$

where, fS_0 (MJ m^{-2}) is the amount of light intercepted by the crop during the day (i.e. the product of the fractional interception, f, and the daily total of incident radiation, S_0) and ϵ_s (g MJ^{-1}) is the efficiency with which the crop uses that light to produce dry matter.

The total biomass accumulated over a growing season can be expressed as:

$$W = \epsilon_s \int fS_0 \delta t \tag{6}$$

As long as its value is conservative, a knowledge of ϵ_s therefore allows us to model the photosynthetic productivity of a crop in terms of captured solar radiation. However, there are a number of circumstances in which ϵ_s is lower than expected for a particular species or appears to decrease during the life of a crop. There are at least three types of departure from the linear relation between total dry matter and accumulated intercepted radiation. The slope of the relation may:

1. *level off during reproductive growth*. This well-documented phenomenon occurs in many determinate crops, such as cereals, and may result from a number of causes. For example, Gallagher and Biscoe (1978) found that the value of ϵ_s for PAR in barley and wheat crops fell by 30% after ear emergence. They ascribed this decline to both a reduced photosynthesis by ageing leaves and an increase in respiration per unit of assimilation after anthesis. However, it should be noted that in many species the continued production of new leaves delays the impact that senescing older leaves have on the value of ϵ_s until leaf production stops.

   Because temperature determines the duration of the reproductive phase in determinate crops which are not limited by stresses, such as drought, the proportion of time that crops intercept radiation when ϵ_s is not constant will depend on seasonal weather.
2. *break down during the season*. This may occur as a result of drought, pests and/or diseases. In each of these cases, ϵ_s may decline at any time after stress is imposed on the plant system. Drought reduces ϵ_s by decreasing photosynthetic capacity and leaf conductance. As a result, the rate of dry matter production may decrease or eventually stop even though the crop canopy continues to intercept radiation.

Pests and diseases can have a complex influence on ϵ_s. They may directly affect canopy structure by removing or destroying active leaf area so that a greater proportion of the intercepting surfaces of the crop are composed of non-photosynthetic elements e.g. stems, flowers, grain etc. Alternatively, pests and diseases can affect canopy function and photosynthesis by disrupting leaf tissue and/or by removing assimilates that have already been fixed by the crop.

3. *be reduced in gradient throughout the season*. The value of ϵ_s is sometimes less than that expected for a particular species throughout the season. This may occur as a result of nutrient stress or because of an increase in the saturation deficit, D (kPa), of the air surrounding the canopy. In both cases, photosynthetic rates may be reduced throughout the life of the crop resulting in a lower slope for the relation between dry matter and intercepted radiation.

There have been few attempts to quantify the effect of nutrient supply on the value of ϵ_s. However, Muchow and Davis (1988) found a reduction in ϵ_s for maize and sorghum growing in low nutrient soils in Australia and Muchow (1992) showed a similar effect in kenaf where the value of ϵ_s for a crop supplied with 240 kg ha^{-1} declined from 1.20 g MJ^{-1} to 0.94 g MJ^{-1} in a crop which was not supplied with nitrogen. Squire (1990) described a similar reduction in ϵ_s in relation to the leaf potassium content of oil palm fronds growing in Malaysia. He also noted that for the sorghum, maize and oil palm examples given above, the decline in ϵ_s was greater than the reduction in the fraction (f) of radiation intercepted by the crops and that measured changes in f and ϵ_s are seldom consistent between sites, even for one genotype.

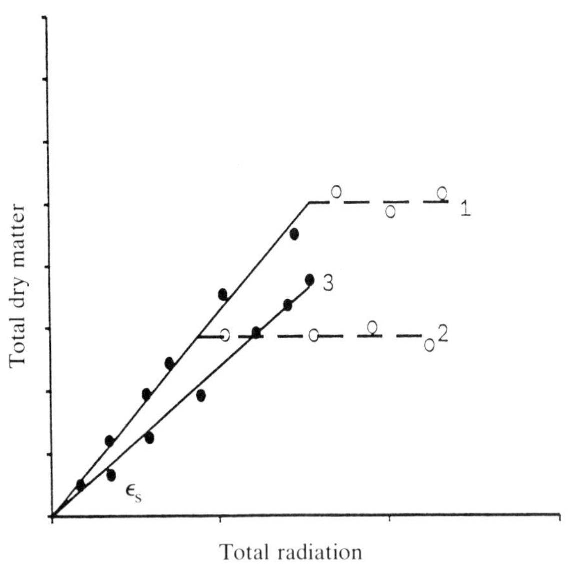

Figure 8.1 Schematic examples of how the relation between crop dry weight and accumulated intercepted radiation can break down: (1) late in the season during reproductive growth; (2) at any stage in the season when stress limits dry matter production e.g. drought; (3) Throughout the season when stress limits dry matter production e.g. nutrient shortage

The sensitivity of ϵ_s to saturation deficit, even under non-stressed conditions has previously been demonstrated for field crops of sorghum and maize (Stockle and Kiniry, 1990) and for sorghum growing in controlled-environment glasshouses (Hamdi, Harris and Clarke, 1987). Stockle and Kiniry (1990) reported that the average daily mean value of D accounted for 76% of the variability in ϵ_s for sorghum and 50% of the variability in ϵ_s for maize for crops that were reported as having no water, nutrient or low temperature stresses.

The three types of departure from the linear relation between dry matter and radiation are schematically illustrated in *Figure 8.1* and *Table 8.2* shows the range in some reported values of ϵ_s under apparently non-limiting and limiting conditions.

Table 8.2 VALUES OF THE CONVERSION COEFFICIENT (g MJ^{-1}) FOR INTERCEPTED RADIATION IN SOME MAJOR CROPS OF THE WORLD

Crop	ϵ_s	ϵ_s'
C_4 species		
Maize	2.43	1.30
Sorghum	2.69	1.20
Millet	2.62	0.57
C_3 species		
Wheat	1.50	0.80
Rice	2.05	1.00
Barley	1.20	1.10
Potato	1.84	1.00
Cassava		0.90
Sweet potato	1.55	
Soyabean	1.30	0.23
Groundnut	1.47	0.47
Sugar beet	1.80	1.00

The *maximum* value (ϵ_s) was derived from experiments under apparently optimal conditions. The *minimum* value (ϵ_s') was derived from experiments where shortage of water and/or nutrients appear to have reduced the conversion efficiency

Photochemical efficiency (ϵ_q)

The process of plant growth consists of the conversion of glucose to other organic compounds, the translocation of the glucose to the site of growth and, in legumes, the cost of nitrogen reduction. In terms of relative energy contents, plant tissue is composed of varying proportions of five distinct biochemical groups; nitrogenous compounds (especially proteins), carbohydrates (cellulose, hemicellulose, starch), lipids (fats, fatty acids, oils), lignin and organic acids. The weight ratio of these biochemical products to substrate varies between 0.35 to 1.0 g g^{-1} (Penning de Vries *et al.*, 1989). This is because the synthesis of products that are rich in lipids and proteins requires considerably more energy per unit dry weight than does the synthesis of carbohydrates. *Table 8.3* summarizes the energy content of the different plant compounds in relation to their biochemical composition. From this, it can be seen that the lipid fraction of seeds contains about 2.19 times as much

Table 8.3 SOME CHARACTERISTICS OF THE FIVE MAJOR GROUPS OF PLANT COMPONENTS AND MINERALS (FROM PENNING DE VRIES *ET AL*., 1989)

	Heat of combustion (kJ g^{-1})	*Nitrogen content* (g g^{-1})	*Carbon content* (g g^{-1})
Carbohydrates	17.3	0.00	0.45
Proteins	22.7	0.15	0.53
Lipids	37.7	0.00	0.77
Lignins	29.9	0.00	0.69
Organic acids	13.9	0.00	0.38
Minerals (K, Ca, P, S)	0.0	0.00	0.00

energy as an equivalent weight of carbohydrate. Therefore, calculations of ϵ_s based on total dry weight need to be adjusted to account for the energy equivalents of the lipid and protein fractions of the dry weight. Clearly, the value of ϵ_s must decrease during the grain filling of crops which produce seeds that are high in oil or protein (Muchow *et al*., 1982; Kiniry *et al*., 1989).

One aspect that is poorly understood is whether stresses, such as soil moisture deficits, affect the relative biochemical composition and therefore the energy content of plant material. Limited evidence suggests that drought does not substantially alter the composition of seeds. For example, Brough and Azam-Ali (1992) examined the yield and composition of seeds harvested from bambara groundnut crops growing in controlled-environment glasshouses at the University of Nottingham Tropical Crops Research Unit (TCRU). Five soil moisture treatments were imposed ranging from no irrigation after crop establishment to 300 mm applied uniformly throughout the season. Daily mean temperature in each glasshouse was maintained at 28°C and D was kept below a maximum of 2.0 kPa. At final harvest, seed yields ranged from 0.04 t ha^{-1} in the driest treatment to 3.20 t ha^{-1} in the irrigated control (Chavula, 1991). Despite this 80–fold variation in seed yield the relative proportions of carbohydrates, lipids and proteins were unaltered across the irrigation treatments.

Respiration efficiency (ϵ_r)

Respiration is an important component in the carbon budget of crops because it is responsible for the efflux of CO_2 from plant cells. It can be considered at two levels; that which occurs as a result of the *growth* of crops and that which is required for their *maintenance*. It is generally assumed that, at any given temperature, respiration continues in the light at a comparable rate to that in the dark. However, during the life of a crop, the relative contributions of the growth and maintenance components of respiration change with the age and weight of the crop. Maintenance respiration becomes more important later in the life of a crop because a larger amount of biomass requiring maintenance exists. It is conventional to assume that the maintenance respiration of a crop is proportional to its total dry weight. Based on several studies, Amthor (1989) estimated that over the lifetime of a crop total respiration is divided roughly equally between growth and maintenance processes, with the latter increasing from about 20% to more than 90% during the life of a crop.

One method of quantifying the significance of respiration to the total carbon balance of a crop is to express growth efficiency (GE) as the ratio of the increase in biomass to the sum of the increase in biomass and respiration over the same period. Amthor (1989) estimated that the growth efficiency of several field crops ranged between about 0.5 and 0.6 i.e. about half the carbon assimilated in photosynthesis is eventually lost in respiration.

The process of *photorespiration* is induced in C_3 plants by the presence of oxygen. The relative importance of photorespiration increases with temperature resulting in a reduction in the initial efficiency of light use of individual leaves. In C_4 plants the initial light use efficiency at low temperatures may be less than that of C_3 plants because of the energy costs involved in suppressing photorespiration. Penning de Vries *et al.* (1989) summarized studies that investigated the effects of temperature on the value of the initial light use efficiency (kgCO_2 ha^{-1} h^{-1} per J m^{-2} s^{-1} of absorbed PAR) for a number of C_3 and C_4 species. They showed that the initial value declined from between 0.4 and 0.5 at 0°C to virtually 0 at 50°C.

Accurate measurements of respiration are difficult to make because the photosynthetic rate, temperature, size and age of a crop vary throughout its life and affect the relative amounts of photosynthesis and respiration. Therefore, the exact influence of respiration on ϵ_s is extremely difficult to assess. Nevertheless, in theory the value of ϵ_s should decrease as crop dry weight increases during the season because of an increase in maintenance respiration (McCree and Silsbury, 1978). In temperate regions, the seasonal increase in temperature should also cause a reduction in ϵ_s. However, Kiniry *et al.* (1989) working with five species in the USA showed no effects of changing temperatures on the value of ϵ_s. They concluded that energy loss due to respiration appeared to be proportional to plant growth.

Yield efficiency (ϵ_h)

For many determinate crops, the ratio of reproductive (or economic) yield to total dry weight, i.e. the harvest index, h, is conservative. This is because the reproductive weight of individual plants, g, is closely related to the total dry weight, w, of each plant above a minimum weight, w_o, i.e. the minimum amount of vegetative 'infrastructure' necessary before reproductive growth can commence. Therefore,

$$g = h(w - w_0) \tag{7}$$

In these circumstances, the total reproductive yield, G, per unit ground area is given by

$$G = P_g = h(W - Pw_0) \tag{8}$$

where P is the plant population and W is the total dry weight per unit area of ground.

In fact, few models use a fixed value of *h per se* to determine reproductive yield. Rather, they sequentially partition new growth to reproductive and vegetative structures during the life cycle of the crop. Nevertheless, in many circumstances the allocation of assimilate to the reproductive or economic components are not

Table 8.4 REPORTED VALUES OF HARVEST INDICES FOR SOME MAJOR CROPS OF THE WORLD

Crop	h	h′
Wheat*	0.50	0.22
Rice	0.50	0.25
Maize	0.45	0.30
Barley	0.52	0.35
Sorghum	0.40	0.20
Oat	0.50	0.23
Millet	0.45	0.20
Rye	0.29	
Potato†	0.65	0.55
Cassava‡	0.77	0.30
Sweet potato§	0.74	0.50
Soyabean	0.59	0.25
Groundnut	0.48	0.16
Sugar cane¶	0.30	0.20
Sugar beet‖	0.45	0.06

The *maximum* harvest index (h) is presented for crops growing under apparently optimal conditions. The comparable *minimum* harvest index (h') is shown for crops whose growth appears to have been limited by adverse environmental factors.
* refers to bread and durum wheat; † fresh tuber at 32.5% dry weight; ‡ (fresh tuber at 35.0% dry weight; § fresh tuber at 30.0% dry weight; ¶sugar at 10–12% of fresh cane; ‖ sugar at 16% of beet dry weight

conservative (see *Table 8.4*) and estimates of yield based on such an assumption may be prone to substantial error. Again, this is particularly the case for crops that are grown in marginal areas often on a stored water profile. Here, the vegetative phase may continue more-or-less as normal as long as there is adequate water but drought will become an increasingly important factor during grain filling. This will lead to premature senescence of leaves and a reduction in crop photosynthetic potential. The net effect will be a crop with a reasonable vegetative growth but poor final yield i.e. a lower value of h.

The water engine

To assimilate CO_2 and produce dry matter, green plants *transpire*. The link between the amount of CO_2 that enters the crop and the amount of H_2O that leaves it means that, at least in principle, seasonal transpiration can be used to estimate the productivity of a crop.

Although calculating the transpirational component of evaporation is fraught with difficulties, this approach has potential advantages in rainfed tropical environments where it is the shortage of water rather than solar radiation which determines crop productivity. Further, in many circumstances, stomatal closure means that little CO_2 enters or H_2O exits from leaves even when they are intercepting solar radiation. Therefore, when water stress restricts crop growth, a knowledge of the potential conversion efficiency of radiation may be of little predictive value and, often, it may be more appropriate to consider the relation between crop productivity and transpiration.

Stanhill (1986) noted that almost 300 years ago Woodward quantitatively related plant growth to transpiration by calculating 'the proportion of the encrease [*sic*] of the plant to the expense of the water' (Woodward, 1699). Since then, there have been many studies of the link between the dry matter produced by a crop and its transpiration over the same time. Despite more recent investigations, such as Tanner and Sinclair (1983), de Wit's (1958) monograph is probably still the most notable review of the subject. In terms of crop modelling, Monteith (1990) provides a theoretical framework for using transpiration as a basis for calculating crop productivity.

There is accumulating evidence that, for any particular crop and season, the relation between total dry weight and transpiration is linear with a slope, ϵ_w (g kg^{-1}), known as the *dry matter:transpired water ratio* (Squire, 1990). Unlike ϵ_s, the value of ϵ_w does not seem to be seriously affected by plant nutrition (de Wit, 1958; Wong, Cowan and Farquhar, 1979; Tanner and Sinclair, 1983). However, when the same crop is grown at different sites or in different seasons, the calculated value of ϵ_w often appears to vary between experiments. There are three principal causes of this apparent variation in ϵ_w:

1. *The contribution of roots to total productivity*:
When crops are well supplied with water, the contribution of roots to the total dry weight is usually small and can often be ignored in calculations of ϵ_w. However, where crops are grown on limited soil moisture, root weight may be a substantial component of total biomass and root:total weight ratios of between 0.3 and 0.5 have been reported for a number of crops including sorghum (Myers, 1980), barley (Gregory, Shepherd and Cooper, 1984), millet (Azam-Ali, Gregory and Monteith, 1984) and groundnut (Nageswara Rao *et al.*, 1989). A further complication which needs to be taken into account is that root:total weight ratios are strongly time-dependent with proportionately less assimilate allocated to roots late in the season. Nevertheless, despite their importance, because measurements of root weight are tedious and time-consuming most calculations of ϵ_w are based on the above-ground component of crop weight. Clearly, the inconsistency of using total (i.e. including roots) or only above-ground dry weight in calculations of ϵ_w may lead to marked variations in published values both between and within experiments where the fractional weight of roots changes during the season and if experimental treatments include differences in soil moisture. For example, for three populations of groundnut growing on stored soil water in India, Azam-Ali *et al.* (1989) showed that the calculated value of ϵ_w increased from 1.57 g kg^{-1} to 3.00 g kg^{-1} when the contribution of roots was included in measurements of crop dry weight.

   Even when root weights are included in calculations of ϵ_w, measured values are prone to substantial errors because of difficulties in extracting roots from soil, representative sampling of the soil volume and, often, the poor replication of sampling in space and time compared with measurements of above-ground components of crop weight.
2. *The component of total evaporation lost from the soil surface*:(E_s)
When crops completely cover the ground for a large part of the season, evaporation, E_s, directly from the surface of the soil may be a small fraction of total evaporation and can effectively be ignored in calculations. However, this is rarely the case in tropical field environments in which (1) there is usually

a large proportion of bare ground because crops never achieve full cover and/or (2) the soil surface is frequently rewetted by relatively small amounts of rainfall which may be insufficient to penetrate more than the uppermost soil layer. In these circumstances, E_s may dominate estimates of total evaporation. For example, Cooper, Keatinge and Hughes (1987) showed that for unfertilized barley crops growing in a Mediterranean environment, E_s was 75% of total evaporation.

Water lost from the soil surface is of little practical benefit to growing crops (although exposed wet soil can influence transpiration by reducing the value of D above the canopy). Where E_s is a significant component of total evaporation it must therefore be taken into account when calculating crop yield responses to rainfall or irrigation. Physical measurements of E_s are laborious and indirect methods often rely on estimates of the proportion of ground covered by the crop – which is itself a difficult factor to quantify. However, the omission of E_s from calculations may lead to large variations in published estimates of crop water use.

3. *The influence of atmospheric saturation deficit*: (D)
Because the air in the sub-stomatal cavities of leaves is saturated at the leaf temperature, the driving force for evaporation is the difference between the saturation vapour concentration in the leaf and that in the air. As a consequence, the value of ϵ_w is inversely related to the saturation deficit (D) of the atmosphere during the season (Tanner and Sinclair, 1983). It is not the absolute atmospheric moisture content but the difference between the saturation vapour pressure and actual vapour pressure of the air, that is important to crop growth. When leaf and air temperatures are the same, the driving force for evaporation is the true saturation deficit of the air, and in these circumstances D sets the potential rate for transpiration. Although transpiration at any time is closely linked to the current saturation deficit, the seasonal mean value of D is often an acceptable approximation because changes in D during the season are usually small compared with differences between seasons (Monteith, 1986).

The theoretical influences of root weight, surface evaporation and saturation deficit on calculated values of ϵ_w are shown in *Figure 8.2*. When root weight and E_s are included in the calculations, they improve the general applicability of ϵ_w. Further, when D is included as a normalizing factor (Monteith, 1990), the value $\epsilon_w D$ (g kg^{-1} kPa^{-1}) seems to remain conservative throughout the season (Squire, 1990) and, unlike ϵ_w, remains fairly constant from site-to-site and year-to-year. Therefore $\epsilon_w D$ may be a more useful index of crop productivity than either ϵ_s or ϵ_w, particularly when the supply and/or demand for water is limiting.

For any particular crop, the value $\epsilon_w D$ can be treated as a quasi-constant in terms of crop modelling and yield prediction. Strictly, the correct terminology for $\epsilon_w D$ is '*the product of the dry matter: transpiration ratio and the mean seasonal saturation deficit*'. However, a less clumsy surrogate is the '*transpiration equivalent*', Ω_w, of total dry matter.

There are virtually no estimates or reports of Ω_w for the major crops of the world. It is therefore difficult to assess the relative merits of light- or water-driven estimates of crop productivity for use in crop models. However, as an example, the potential for using Ω_w as a basis for model predictions is illustrated for two crops in *Table 8.5*. The table is based on a summary presented by Squire

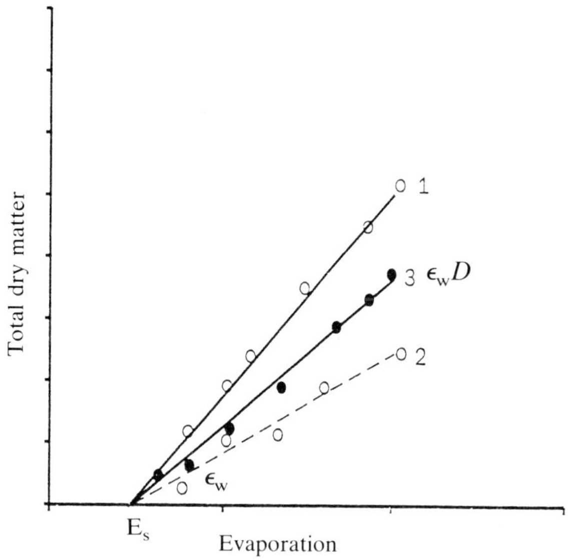

Figure 8.2 Schematic examples of how total crop dry matter including roots is linearly related to transpiration: 1) When the atmosphere is wet; 2) When the atmosphere is dry; 3) When the dry matter:transpired water ratio (ϵ_w) is multiplied by the mean saturation deficit throughout the season (D). E_s indicates that soil surface evaporation needs to be deducted from total evaporation to provide linear fits between dry matter and transpiration

(1990) for different stands of pearl millet and groundnut growing at field sites in India and Africa and in controlled-environment glasshouses at the University of Nottingham. For both species, the use of Ω_w is clearly more reliable than ϵ_w as a basis for calculating crop productivity across a wide range of atmospheric and soil conditions. For millet, the coefficient of variation reduces from 41% for ϵ_w to 10% for Ω_w and, for groundnut, comparable values are 49% and 17%.

Further, it is worth pointing out that similar estimates based on intercepted radiation alone (ϵ_s) are prone to substantial variation – presumably because of the greater influence of drought on crop growth than on intercepted radiation.

The use of D as a normalizing factor for ϵ_s i.e. by calculating a '*radiation equivalent*' $\Omega_s = \epsilon_s D$ allows us to compare estimates of crop productivity based on Ω_w and Ω_s. Again, including the influence of D in radiation-based estimates of productivity improves the reliability of predictions. However, despite the considerable improvement in predictions based on Ω_s versus ϵ_s, it is still evident that, at least for the examples shown in *Table 8.5*, predictions of crop productivity based on Ω_w still appear to be more reliable than those based on Ω_s.

Despite their potential use in predicting crop productivity, estimates of transpiration are far more difficult to obtain than comparable estimates of the combined evaporation, E, from crop (E_p) and soil (E_s) surfaces. Nevertheless, if E_s can be measured or estimated, then a knowledge of the total evaporation, λE, at any site can be used to predict the total dry matter, W, of a crop, i.e.

Table 8.5 COMPARISONS OF CROP DRY MATTER EXCLUDING ROOTS (W) TRANSPIRATION (ΣE), SATURATION DEFICIT (D), DRY MATTER: TRANSPIRED WATER RATIO (ϵ_w), TRANSPIRATION EQUIVALENT (Ω_w) ACCUMULATED INTERCEPTED RADIATION (ΣS_i), CONVERSION COEFFICIENT FOR INTERCEPTED RADIATION (ϵ_s), AND RADIATION EQUIVALENT (Ω_s) FOR STANDS OF PEARL MILLET AND GROUNDNUT GROWING IN EXPERIMENTAL GLASSHOUSES AT THE UNIVERSITY OF NOTTINGHAM AND IN THE FIELD IN CENTRAL INDIA OR WEST AFRICA. FURTHER DETAILS OF THE EXPERIMENTS AND THE ORIGINAL VERSION OF THIS TABLE APPEAR IN SQUIRE (1990)

Crop and region	W ($g\ m^{-2}$)	ΣE_t (kg)	D (kPa)	ϵ_w ($g\ kg^{-1}$)	Ω_w ($g\ kg^{-1}\ kPa^{-1}$)	ΣS_i ($MJ\ m^{-2}$)	ϵ_s ($g\ MJ^{-1}$)	Ω_s ($g\ MJ^{-1}\ kPa^{-1}$)
Pearl millet								
Nottingham	1440	220	1.4	6.55	9.16	550	2.62	3.67
India	660	150	2.4	4.40	10.56	450	1.47	3.52
India	320	70	2.3	4.57	10.51	300	1.07	2.45
Africa	170	80	4.0	2.13	8.50	297	0.57	2.29
Mean	**648**	**130**	**2.53**	**4.41**	**9.68**	**399**	**1.43**	**2.98**
CV%				**41**	**10**		**61**	**24**
Groundnut								
Nottingham	270	52	1.0	5.19	5.19	191	1.41	1.41
Nottingham	250	76	1.4	3.29	4.61	175	1.43	2.00
Nottingham	200	82	1.6	2.44	3.90	136	1.47	2.35
Nottingham	110	72	2.0	1.53	3.06	130	0.85	1.69
India	220	110	2.1	2.00	4.20	420	0.52	1.10
India	420	220	2.5	1.91	4.77	900	0.47	1.17
Mean	**245**	**102**	**1.77**	**2.73**	**4.29**	**325**	**1.02**	**1.62**
CV%				**49**	**17**		**46**	**30**

$$W = \frac{(\Omega_w)(\lambda E - E_s)}{D} \qquad (9)$$

and from equation (6)

$$W = \epsilon_s \int f S_0 \mathrm{d}t = \frac{(\Omega_w)(\lambda E - E_s)}{D} \qquad (10)$$

Where the reproductive fraction of total dry matter can be reasonably predicted from a knowledge of the total productivity, grain yield, G, can either be calculated as a function of solar radiation when water is not limiting, i.e.

$$G = h(\epsilon_s \int f S_0 \mathrm{d}t) \qquad (11)$$

or transpiration when water is limiting, i.e.

$$G = h\left(\frac{(\Omega_w)(\lambda E - E_s)}{D}\right) \qquad (12)$$

In this brief discussion of light-driven and water-driven methods of calculating crop productivity, it is important to note that it is the method by which they predict *total* biomass that is relevant. Irrespective of how well the water engine may or may not estimate productivity, it still suffers from the same limitations as the solar engine in terms of the subsequent influence of factors such as respiration, photochemical composition and the fraction allocated to harvestable yield.

Synthesis

Clearly, there are special circumstances in which either the supply of water or the interception of solar radiation alone dominate the process of dry matter production by crops. However, in most cases crops are continuously responding to short-term changes in both solar radiation and water supply with neither factor being optimal. Consequently, in order to successfully predict the performance of crops throughout their growth and across the widest possible range of circumstances, robust crop models should take account of both light and water-driven productivity.

One such model which adopts this approach is the PARCH (**P**redicting **A**rable **R**esource **C**apture in **H**ostile Environments) system which has recently been developed at Nottingham (Bradley and Crout, 1992). The model converts daily potential water uptake by roots to carbohydrate using Ω_w. Intercepted radiation is similarly converted using the appropriate value of ϵ_s. This provides two possible estimates of the daily increment of crop carbohydrate. The lesser of the two is used in the calculation of daily dry matter production and the difference between the two is used as an index of crop water stress and represents plant water status. This index is used where translocation of stored assimilates forms a component in the calculation of grain yield. A preliminary version of the PARCH model has recently been produced at Nottingham for sorghum based on soil, climate and crop data supplied from experiments in Africa, India (Monteith, Huda and Midya, 1989) and elsewhere.

In 1990, the five TCRU glasshouses at the University of Nottingham were used as an experimental 'testbed' for the PARCH model. Sorghum crops were grown across soil moisture conditions which ranged from a 'control' which was irrigated to field capacity each week to a 'droughted' treatment which received no irrigation

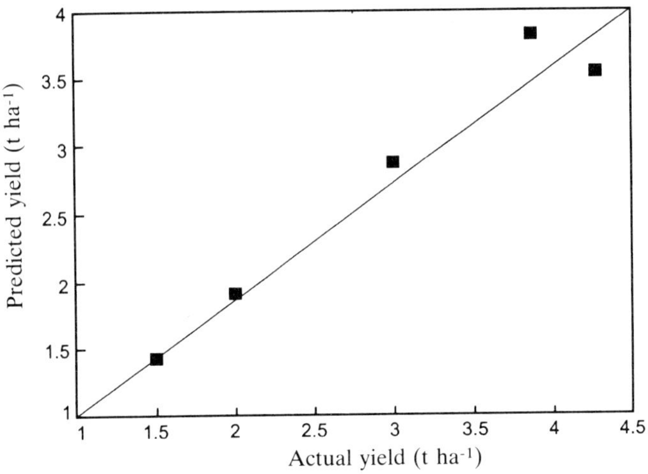

Figure 8.3 The use of the PARCH model to predict grain yield of sorghum and actual yields obtained from sorghum crops growing in controlled-environment glasshouses at the University of Nottingham UK in 1990

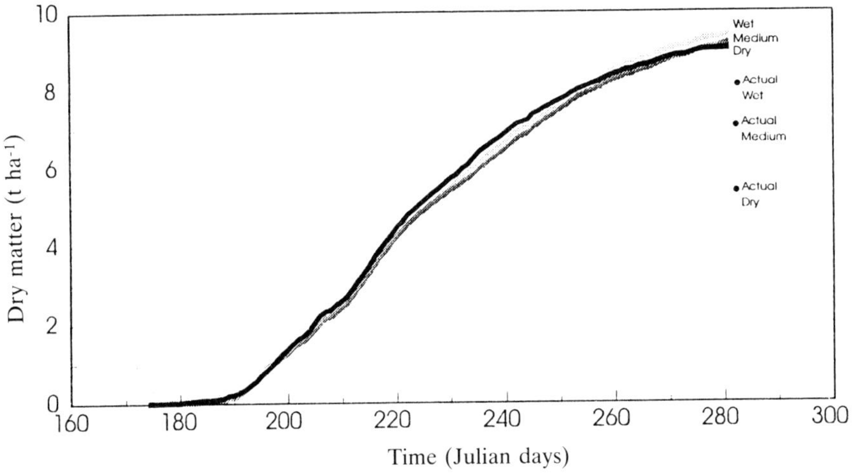

Figure 8.4 Seasonal predictions from the PARCH model for total dry matter of wet, intermediate and dry irrigation treatments and actual grain yields of sorghum crops growing in controlled-environment glasshouses at the University of Nottingham UK in 1990. Predicted yields were obtained after the water dependent functions of the model had been disabled

after crop establishment. The model successfully predicted most aspects of crop growth and soil water use and *Figure 8.3* shows the correlation between predicted and actual crop yields for all five soil moisture regimes. *Figures 8.4* and *8.5* illustrate the implications of basing predictions of crop yield exclusively on the capture of a single resource for three of the five sorghum stands presented in *Figure 8.3*. In *Figure 8.4*, the PARCH model was run with the water-dependent components disabled. In this case, yields were calculated purely as a function of intercepted radiation. It is evident that predicted yield was relatively close to that actually achieved in only one treatment i.e. that which was least short of water. In both other cases, predictions substantially exceeded the achieved yields. Similarly, in *Figure 8.5* the radiation-dependent components of the model were switched off and predictions were based entirely on water use. Here, the correlation between predicted and actual yield is closest where water was the most limiting factor.

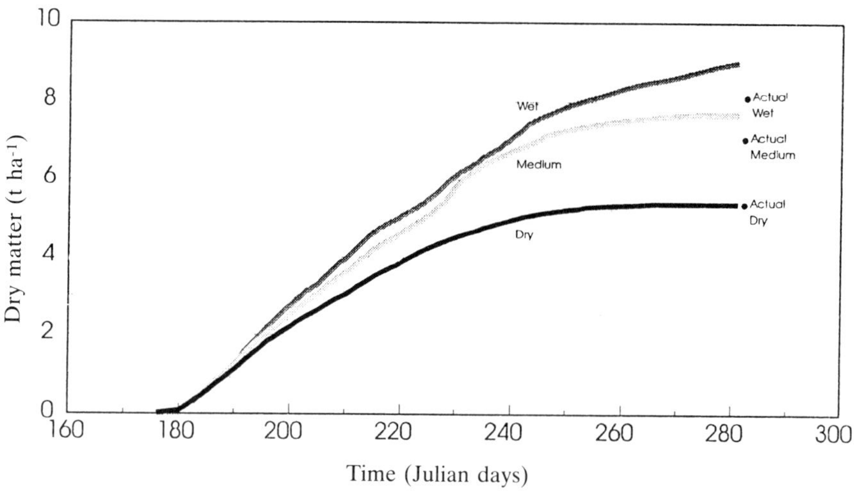

Figure 8.5 Seasonal predictions from the PARCH model for total dry matter of wet, intermediate and dry irrigation treatments and actual grain yields of sorghum crops growing in controlled-environment glasshouses at the University of Nottingham UK in 1990. Predicted yields were obtained after the light-dependent functions of the model had been disabled

Where next?

Earlier reference was made to the practical applications of crop models listed by Steiner (1987). From this brief review, it is evident that there is still much to be done in terms of understanding the interactions between biological and physical processes in crop systems and in our attempts to conceptualize across disciplines. However, the greatest need for progress appears to be in the development of models that successfully predict the performance of crops in suboptimal or low input environments where crops are growing closest to their margins of existence. Penning de Vries *et al.* (1989) noted that in developing countries there is less agricultural research per crop, per hectare and per person than that in rich countries. The lack of crop-specific research information makes the task of integrating findings and making practical recommendations to farmers even more difficult. It seems clear that it is in these environmentally and informationally *resource*-poor regions that models of *resource* capture are most appropriate. At present, the development of such models is especially limited by our ignorance of below-ground interactions between plant roots, soil water and nutrients and our incomplete knowledge of how heterogeneous crop canopies intercept and use solar radiation.

In terms of resource capture by crop canopies, the following areas can be identified as priorities for experimental and modelling effort:

- velocity and activity of root extraction fronts;
- soil water storage, transport and release characteristics;
- techniques to measure transpiration e.g. using stem-flow methods, remote-sensing;
- quantifying the precise effect of saturation deficit on transpiration and the value of Ω_w for different species;
- the pattern of light transmission and photosynthetic activity in crops that do not cover the ground for all or most of the season.

References

Amthor, J.S. (1989) *Respiration and Crop Productivity*. New York: Springer Verlag, 215 pp

Azam-Ali, S.N., Gregory, P.J. and Monteith, J.L. (1984) Effects of planting density on water use and productivity of pearl millet (*Pennisetum typhoides*) grown on stored water. I. Growth of shoots and roots. *Experimental Agriculture*, **20**, 203–214

Azam-Ali, S.N., Simmonds, L.P., Nageswara Rao, R.C. and Williams, J.H. (1989) Population, growth and water use of groundnut maintained on stored water. III. Dry matter, water use and light interception. *Experimental Agriculture*, **25**, 77–86

Bradley, R.G. and Crout, N.M. (1992) *The PARCH model – A Users Guide*. University of Nottingham Tropical Crops Research Unit, Sutton Bonington Campus LE12 5RD

Brough, S.H. and Azam-Ali, S.N. (1992) The effect of soil moisture on the proximate composition of bambara groundnut (*Vigna subterranea* (L) verdc.). *Journal of the Science of Food and Agriculture* **60** (2), 197–204

Chavula, K.M. (1991) *Growth, light interception and yield of bambara groundnut (*Vigna subterranea *(L) verdc.) in response to soil moisture*. M.Sc. Thesis, University of Nottingham

Cooper, P.J.M., Keatinge, J.D.H. and Hughes, G. (1987) Crop evapotranspiration – a technique for calculation of its components by field measurements. *Field Crops Research*, **7**, 299–312

Farquhar, G.D. and von Caemmerer, S. (1982) *Modelling of Photosynthetic Response to Environmental Conditions*. Encyclopedia Plant Physiology, New Series 12b. Berlin: Springer Verlag, pp 549–587

France, J. and Thornley, J.H.M. (1984) *Mathematical Models in Agriculture. The Quantitative Approach to Problems in Agriculture and Related Sciences*. London: Butterworth

Gallagher, J.N. and Biscoe, P.V. (1978) Radiation absorption, growth and yield of cereals. *Journal of Agricultural Science* (Cambridge), **91**, 47–60.

Goudriaan, J. (1977) *Crop Micrometeorology: A Simulation Study*. Wageningen: Pudoc

Gregory, P.J., Shepherd, K.D. and Cooper, P.J.M. (1984) Effects of fertilizer on root growth and water use of barley in northern Syria. *Journal of Agricultural Science (Cambridge)*, **103**, 429–438

Hamdi, Q.A., Harris, D. and Clark, J.A. (1987) Saturation deficit, canopy formation and function in *Sorghum bicolor* (L.). *Journal of Experimental Botany*, **38**, 1272–1283

Johnson, I.R. and Thornley, J.H.M. (1983) Vegetative crop growth model incorporating leaf area expansion and senescence, and applied to grass. *Plant Cell and Environment*, **6** 721–729

Kiniry, J.R., Jones, C.A., O'Toole, J.C., Blanchet, R., Cabelguenne, M. and Spanel, D.A. (1989) Radiation-use efficiency in biomass accumulation prior to grain-filling for five grain-crop species. *Field Crops Research*, **20**, 51–64

McCree, K.J. and Silsbury, J.H. (1978) Growth and maintenance requirements of subterranean clover. *Crop Science*, **19**, 13–18

Monteith, J.L. (1972) Solar radiation and productivity in tropical ecosystems. *Journal of Applied Ecology,* **9**, 747–766

Monteith, J.L. (1986) How do crops manipulate water supply and demand? *Philosophical Transactions of the Royal Society of London* A, **316**, 245–259

Monteith, J.L. (1990) Conservative behaviour in the response of crops to water and light. In *Theoretical Production Ecology: Reflections and Prospects*, pp 3–14 Edited by R. Rabbinge, J. Goudriaan, H. van Keulen, F.W.T. Penning de Vries and H.H. van Laar, Wageningen: Pudoc

Monteith, J.L., Huda, A.K.S. and Midya, D. (1989) RESCAP: A resource capture model for sorghum and pearl millet. In *Modelling the growth and development of sorghum and pearl millet* Edited by S.M. Virmani, H.L.S. Tandon and G. Alagarswamy pp 30–34, ICRISAT Research Bulletin 12, Patancheru, India

Muchow, R.C. (1992) Effect of water and nitrogen supply on radiation interception and biomass accumulation of kenaf (*Hibiscus cannabinus*) in a semi-arid tropical environment. *Field Crops Research* **28**, 281–293

Muchow, R.C. and Davis, R. (1988) Effect of nitrogen supply on comparative productivity of maize and sorghum in a semi-arid tropical environment. II. Radiation interception and biomass accumulation. *Field Crops Research* **18**, 17–30

Myers, R.J.K. (1980) The root system of a grain sorghum crop. *Field Crops Research* **3**, 53–64

Nageswara Rao, R.C., Simmonds, L.P., Azam-Ali, S.N. and Williams J.H. (1989) Population, growth and water use of groundnut maintained on stored water. I. Growth of roots and shoots. *Experimental Agriculture*, **25**, 51–61

Nilson, T. (1971) A theoretical analysis of the frequency of gaps in plant stands. *Agricultural Meteorology* **8**, 25–38

Norman, J.M. (1989) Modelling plant processes. In *Plant Canopies; Their Growth, Form and Function*, pp 161–175 Edited by G. Russell, B. Marshall and P.G. Jarvis, Cambridge: Cambridge University Press

Penning de Vries, F.W.T., Jansen, D.M., ten Berge, H.F.M. and Bakema, A. (1989) *Simulation of Ecophysiological Processes of Growth in Several Annual Crops*. Wageningen: Pudoc 271 pp

Stanhill, G. (1986) Water use efficiency. *Advances in Agronomy*, **39**, 53–83
Steiner, J.L. (1987) Modelling agroclimatic systems: guidelines and perspectives. Soil, crop and water management systems for the Sudano-Sahelian zone. *Proceedings of an international workshop*, 7–11 January 1987. pp 341–351 Patancheru, India: ICRISAT
Stockle, O.S. and Kiniry, J.R. (1990) Variability in crop radiation-use efficiency associated with vapor-pressure deficit. *Field Crops Research* **25**, 171–181
Squire, G.R. (1990) *The Physiology of Tropical Crop Production*. Wallingford, Oxon: CABI
Szeicz, G. (1974) Solar radiation in canopies. *Journal of Applied Ecology*, **11**, 1117–1156
Tanner, C.B. and Sinclair, T.R. (1983) Efficient water use in crop production: research or research? In *Limitations to efficient water use in crop production*. Edited by H.M. Taylor, W.R. Jordan and T.R. Sinclair. Madison Wisconsin: ASA, CSSA, SSSA, 538 pp
Wit, C.T. de, (1958) *Transpiration and Crop Yields*. Agricultural Research Reports 64.6, Wageningen: Pudoc 88 pp
Wong, S.C., Cowan, I.R. and Farquhar, G.D. (1979) Stomatal conductance correlates with photosynthetic capacity. *Nature*, **282**, 424–426
Woodward, L. (1699) Some thoughts and experiments concerning vegetation. *Philosophical Transactions of the Royal Society of London*, **21**, 192–227

9

THE APPLICATION OF THE CONCEPTS OF RESOURCE CAPTURE TO THE EFFECT OF PEST INCIDENCE ON CROPS

H.F. van EMDEN and P. HADLEY
Department of Horticulture, School of Plant Sciences, University of Reading, UK

Introduction

The concept of resource capture by crops is now well established among crop physiologists. There is an increasing body of literature which shows that, for a diverse range of plant species, a simple relationship exists between crop yield (both in terms of total dry matter produced and the harvestable fraction) and the total solar radiation intercepted by the crop canopy thus:

$$W = \int RUE.IR\,dt \qquad (1)$$

where W is the total dry matter produced (or total yield) (g m^{-2}), RUE is the efficiency of conversion of radiation into total dry matter (or yield) (g MJ^{-1}) and IR is the amount of intercepted solar radiation by the crop canopy (MJ m^{-2}). Thus, there is now a well defined approach to describing variation in crop yield through effects on intercepted radiation and radiation use efficiency. It is surprising, though, that the equation or any similar approaches have been used only rarely by plant pathologists and virtually never by entomologists in studying the effects of pests and disease on crop yield.

For entomologists, the concept embodied by the equation is well established in the sense that reference has frequently (e.g. van Emden, 1978) been made to the reasons why defoliation by insects has a very different effect on yield from manual defoliation. Also in the insect literature of the biological control of weeds, it has been recommended (albeit in different words) that the most effective biological control agent is likely to be the insect which reduces RUE maximally per unit decrease in IR (Harris, 1971; Winder and van Emden, 1980). What has been lacking is the quantification of the relative magnitude of the two components in individual examples.

Removal of leaf area (% defoliation) by chewing insects has frequently been measured, as has reduction in leaf area following feeding by sucking insects. Such measurements do not, however, accurately reflect changes in IR, and thus observed increases in leaf area ratios in plants showing reduced growth (quoted for example by van Emden, 1990 in relation to damage by aphids) do not necessarily imply RUE must have decreased. Effects of insects on leaf number, on leaf curling (e.g. by aphids on the flag leaf of cereals) and the distribution of damage on individual leaves will all effect *IR* through effects on canopy structure rather than via the photosynthesis leaf area duration component, which is what has usually been measured.

It is likely, therefore, that entomologists have placed too much emphasis on yield loss that cannot be explained by reductions in photosynthesis leaf area duration. Nevertheless, there is no doubt that yield losses resulting from reduced *RUE* must be important when plants are damaged by insects. Assimilates may be withdrawn by sap-sucking insects, or lost through the increased plant respiration caused by the foreign proteins in insect saliva (Kloft, 1960). The whole spectrum of effects of insect saliva in plants, from tissue necrosis to hypertrophy, has been reviewed by Miles (1968, 1990).

It is therefore not surprising that relations established by entomologists between pest burdens and crop yield have not always been clear, since the effect of a particular pest burden may be different at different stages in the development of a crop, or defoliation has not been incorporated into the severity assessment. Therefore, there is considerable scope for the concepts of resource capture to be applied to assessments of the impact of pest attack on yields.

The purpose of this chapter is to review the available literature where the concepts of resource capture have been applied to pest attack. In addition, through re-analysis of previously published data on which resource capture concepts have not been applied, the potential for applying such concepts to studies of pest attack is illustrated.

Effects of pests on crops

Boote *et al.* (1983) categorize pest effects on plant growth into seven groups: tissue consumers, leaf senescence accelerators, stand reducers, light stealers, photosynthetic rate reducers, assimilate sappers and turgor reducers. However, Johnson (1987) argues that, when considering resource capture concepts, pest effects can be linked into two larger groups: those that affect primarily the interception of radiation (tissue consumers, leaf senescence accelerators, stand reducers and light stealers) and those that primarily effect radiation use efficiency (photosynthetic rate reducers, assimilate sappers and turgor reducers).

THEORY

Both entomologists and plant pathologists usually examine the effect of the severity of the pest or disease attack on yield. However, such relationships are often poor because this approach does not take into account the effect of the timing of the attack during the development of the crop, nor does it consider effects on yield determinants, such as the efficiency of plant photosynthesis organs and/or radiation interception.

A theoretical basis for considering both pest and disease in relation to resource capture is given by Waggoner and Berger (1987) as follows:

At any instant of time, (from Equation 1):

$$dW/dt = RUE.IR \qquad (2)$$

and (assuming that the transmission of radiation through the canopy follows Beer's Law):

$$IR = I.(1\text{–}\exp^{(-KL)}) \tag{3}$$

where I is total radiation (MJ m^{-2} d^{-1}), K is the light extinction coefficient and L is the leaf area index.

Thus, the rate of crop growth is (assuming RUE remains constant) linearly related to IR. Intercepted radiation increases curvilinearly with increasing leaf area index (L) until almost all available radiation is absorbed by the crop (*Figure 9.1*).

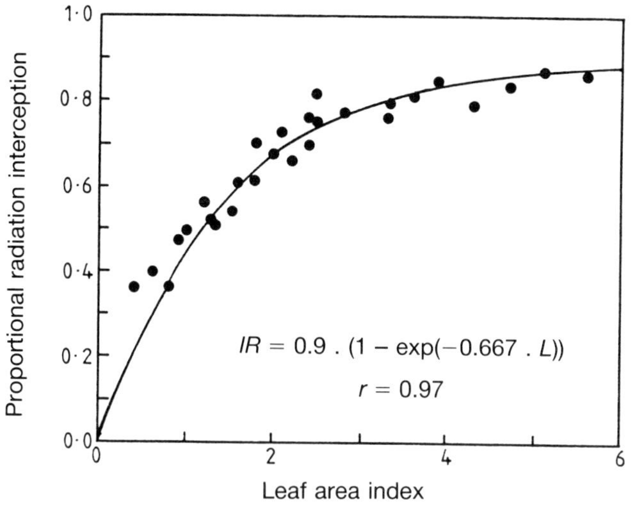

Figure 9.1 The relationship between leaf area index and the proportion of solar radiation intercepted in potato. Data are from Allen and Scott (1980)

For a crop subject to a pest attack that only results in leaf removal from the crop, then, if the area of leaves removed by the pest is (L_R), it is possible to calculate the amount of energy absorbed by the crop subject to pest attack (IR_P):

$$IR_P = I.(1 - e^{(-K(L_H - L_R))}) \tag{4}$$

where L_H is the healthy leaf area index.

Thus it is possible to define the radiation interception loss through pest attack as $IR_H - IR_P$. Therefore:

and
$$W_H = RUE. \int IR_H \mathrm{d}t$$
$$W_P = RUE. \int IR_P \mathrm{d}t \tag{5}$$

where W_H and W_P are the dry matter accumulation of the healthy and pest infested crop respectively, and:

$$W_L = RUE. \int (IR_H - IR_P)\, \mathrm{d}t \tag{6}$$

where W_L is the dry matter lost through the pest attack.

However, as indicated earlier, pests can also affect radiation use efficiency (RUE), so that the pest may not only reduce the interception of radiation by direct effects on the canopy through damage to the assimilatory area, but also by indirect effects on canopy expansion through a reduction in radiation use efficiency. Whichever is the case, it is more appropriate to consider the effect of pests affecting radiation use efficiency as also affecting radiation interception by the canopy so that the reduction in dry matter accumulation can be resolved thus:

$$W_L = W_H - W_P = RUE_H . \int IR_H \, dt - RUE_P . \int IR_P \, dt \tag{7}$$

where RUE_H and RUE_p are the radiation use efficiencies of healthy and attacked crops respectively.

Effects of manual defoliation on crop growth

In assessing the damage caused by pests, it is important to be able to measure the yield losses they cause. These can be measured by comparing crops subjected

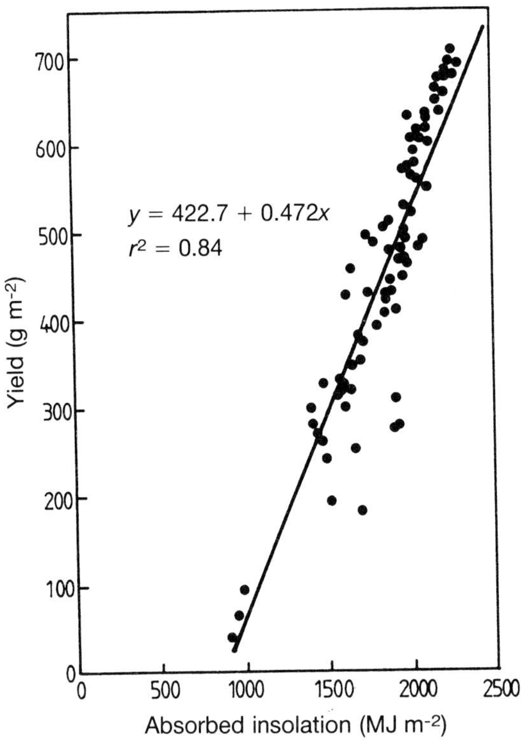

Figure 9.2 The healthy leaf area absorption HAA and pod yield of 78 crops of peanut cultivar 'Florunner' grown over a period of 14 years by five growers. Reproduced from Waggonner and Berger (1987)

to pest infestation with those where pests are excluded by using cages or by controlling them chemically. The difficulty with this approach is the problem of being able to obtain a reliable and repeatable infestation to enable accurate yield assessments to be made. To avoid this, crops have been artificially defoliated to simulate the effects of defoliating insects on crop yield.

Scott and Jaggard (1985) have recently reviewed manual defoliation data published by a number of workers studying yield loss in sugar beet. Generally these defoliation experiments have produced somewhat confusing and conflicting results on final yield of sugar beet. This is largely a result of differences in methodology but, even with standard methodology, inconsistencies are likely from year to year as a result of differing radiation receipts and temperatures as well as differences in water and nutrient status.

However, using published data for peanuts, Waggoner and Berger (1987) showed that there were good relationships between yield and intercepted radiation when crops were manually defoliated (*Figure 9.2*). Johnson *et al.* (1986, 1987) have developed a simple potato growth model that simulates crop loss in relation to pest attack. The model describes the growth of potato and uses solar radiation and temperature as the main driving variables. When older leaf tissue is removed

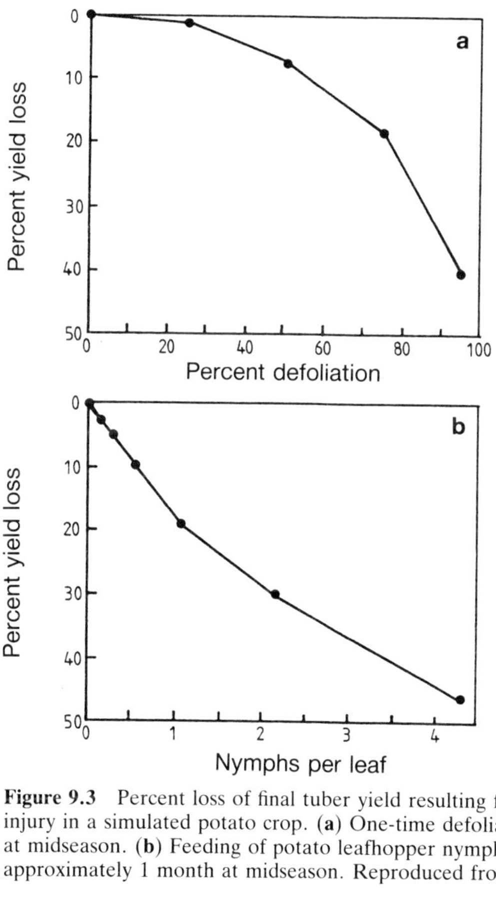

Figure 9.3 Percent loss of final tuber yield resulting from two kinds of injury in a simulated potato crop. (**a**) One-time defoliation of older tissue at midseason. (**b**) Feeding of potato leafhopper nymphs over a period of approximately 1 month at midseason. Reproduced from Johnson (1987)

to simulate leaf removal by pests, in mid-season the effect is a reduction in final tuber yield largely as a function of reduced light interception (*Figure 9.3a*). There is a curvilinear reduction in tuber yield with increasing defoliation, largely because a small loss of leaves at the base of the canopy has little effect on radiation interception, but interception loss gets larger as the rate of defoliation increases. Reference to the relationship between leaf area index and the proportion of radiation intercepted by a crop shows why this should be the case (*Figure 9.1*). Since the relationship is asymptotic, small changes in leaf area index have little effect on a fully grown potato plant until the leaf area index has been reduced by approximately 50%. Below this, progressively greater reductions in leaf area index have an increasing effect on the reduction in radiation interception.

Modelling plant–insect interrelations

A number of models of insect damage to plants exist. However, although the modelling of loss of leaf area is often considered in some detail, the concept of resource capture as a determinant of dry matter and yield accumulation is rarely used. Rudd (1980) described a simulation of insect damage on soybean. The basis of the model is an empirical relationship between dry matter production versus leaf area index derived from data of Shibles and Weber (1965), such that there are two relationships between dry matter production rate and L above and below a L value of 3.835. Simple partitioning functions then allocate dry matter to leaf dry weight, stem dry weight and later pod growth. This relationship was then used to simulate the effects of defoliation on crop yield. Interestingly, although there is no mention of light interception, the equation used to describe dry weight accumulation once the canopy has closed is:

$$\frac{dW}{dt} = (1 - \exp(.91435\,L).\,PTSMAX \qquad (8)$$

where w is total dry weight and *PTSMAX* is dry matter production rate. The L function therefore represents a Beer's Law extinction curve and *PTSMAX* would be equivalent to *RUE*.

A validation of a model for estimating leaf feeding insect damage in soybean has recently been published by Herbert *et al.* (1992). Yield is predicted from relationships between rate of dry weight accumulation and leaf area as described by Rudd (1980) above. The model calculates the leaf area consumed by the several defoliating species attacking the crop, based on published data for each species relative to temperature-dependent development and consumption rates. This involves estimation of leaf consumption based on species abundance and size distribution. A component for leaf area increase after insect defoliation is also included in the model. When compared with real data, the model appeared to overestimate net total leaf area loss, although the overall relationship between observed and predicted values was good (*Figure 9.4*). This suggests a simple scaling error associated with the model. Yield was also accurately predicted based on the leaf area after insect feeding. The above model represents a considerable advance over empirical approaches to insect–yield estimation, and it would clearly be only a small step to develop a model based on resource capture such that yield is related not only to L but also to actual measurements of crop light interception and *RUE*.

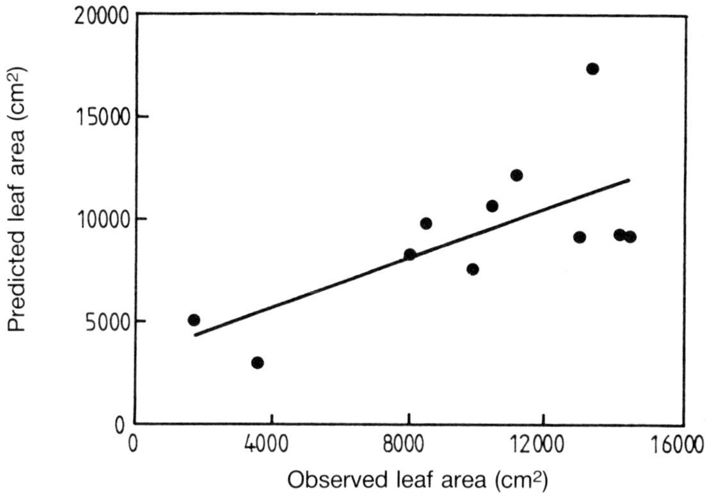

Figure 9.4 Relationship between the measured total leaf area lost (total leaf area at the onset of insect feeding less the total leaf area after the insect feeding period) and the predicted leaf area consumption from a model of insect feeding. Reproduced from Herbert *et al.* (1992)

The above example relates to a study involving a simulation of insect pest damage. The present authors have not been able to find data where the true effects of defoliator insects have been examined. However, there are data for the spotted tentiform leaf miner (*Phyllonorycter blancardella* (F.)) which has

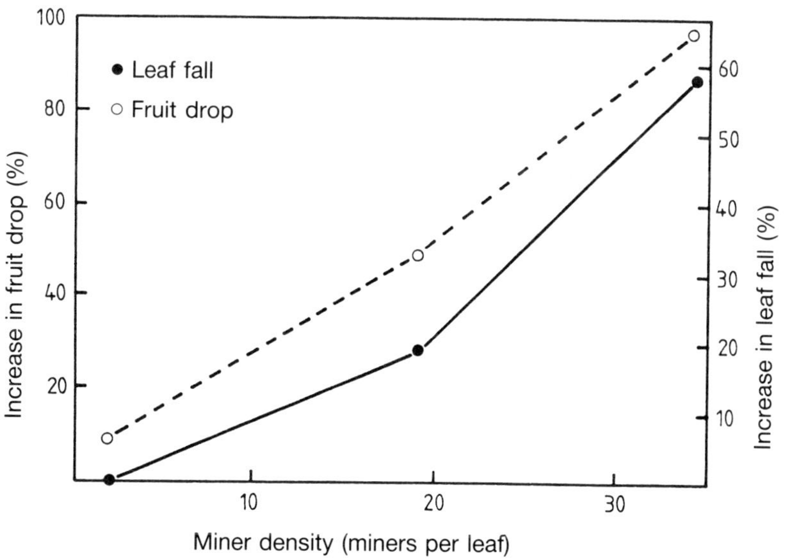

Figure 9.5 Effects of increasing numbers of spotted tentiform leaf miners on leaf and fruit fall in apple. Data from E.A.C. Hagley (cited by Proctor *et al.*, 1982)

become a widespread pest of apple (*Malus domestica* Borsch) almost everywhere it is commonly grown. It damages leaves and thereby reduces their photosynthetic activity, but it may be that the effect of external feeders is different. Circumstantial evidence points to the fact that these leaf miners reduce yield through their effects on leaf area. Work of E.A.C. Hagley (cited by Proctor *et al.* (1982)) indicates a more or less linear decline in yield with increasing densities of miners per leaf (*Figure 9.5*). This tends to suggest that leaf miner damage is generally mechanical and yield decline results from such attacks as a function of a decline in intercepted radiation. Proctor *et al.* (1982) examined the effects of spotted tentiform leaf miner on apple leaf photosynthesis. Leaf mining reduced photosynthesis so that at 20 miners per leaf an average of 32.9% of the leaf area was injured, yet photosynthesis was reduced by only 23.2% (*Figure 9.6*). Photochemical efficiency, measured as the initial slope of the light response curve, was not affected by miners. However, mesophyll conductance and chlorophyll concentration were reduced and internal CO_2 concentration was increased. The implication therefore is that photosynthetic rate is reduced as a function of the mechanical damage to the leaf with no evidence for an effect on the efficiency of photosynthesis.

The effect of defoliation may not, however, be quite as simple, particularly for perennial species. Heichel and Turner (1983) studied the photosynthetic activity of red maple (*Acer rubrum* L.) and red oak (*Quercus rubra* L.) in response to simulated insect defoliation. They noted a 30 to 60% enhancement of photosynthesis of residual primary foliage remaining on trees that had been defoliated by 50 and 75% and this was paralleled by similar changes in leaf conductance to water vapour exchange (*Figure 9.7*).

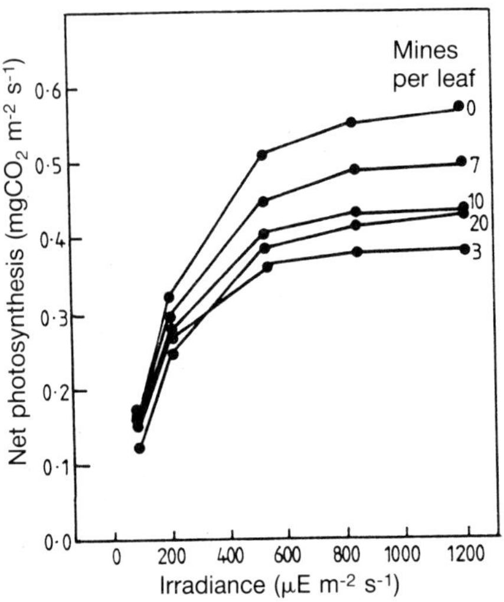

Figure 9.6 Light-response curves of net photosynthesis for undamaged and spotted tentiform leaf miner damaged apple leaves containing 3, 7, 10 and 20 mines per leaf. Each symbol represents an average of measurements on six replicate leaves. Reproduced from Proctor *et al.* (1982)

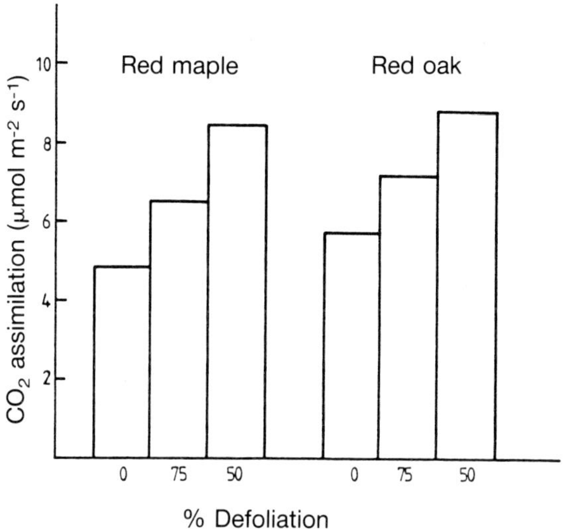

Figure 9.7 Effect of defoliation on carbon dioxide assimilation in red maple (*Acer rubrum* L.) and red oak (*Quercus rubra* L.). Data from Heichel and Turner (1983)

Pest attack in relation to radiation use efficiency

Effects of pest attack on radiation use efficiency may be very large and, for sap sucking insects such as aphids, this is not simply limited to the removal of assimilates. Aphid damage is not only caused by assimilate removal but also by wounding, saliva injection which, among other effects, raises plant respiration rates (Kloft, 1960) and excretion of honeydew, sometimes sufficient to occlude the stomata. Other effects of honeydew are to encourage the growth of sooty moulds which effects radiation interception and, with heavy honeydew deposits, plasmolysis of the cells in the underlying leaf may occur. Aphids can also act as vectors for plant pathogens, particularly viruses, which themselves affect plant metabolism. That aphid attack reduces *RUE* was shown by Dixon (1971); an infestation of sycamore by sycamore aphid (*Drepanosiphum platanoides* Schr.) caused a reduction in the growth of annual rings many times greater then could be accounted for by the observed loss in leaf area. A recent example is given by van Emden (1990) where the effect of cabbage aphid (*Brevicoryne brassicae* L.) on the growth of Brussels sprouts was examined. Dry matter loss was great at low densities while the rate of increase in yield loss was little changed at higher rates of infestation (*Figure 9.8*). These changes were also accompanied by significant reduction in partitioning of dry matter to the roots and an approximately 20% reduction in stomatal number. Both these latter features have important implications for both nutrient and water relations of pest infested crops; however, this appears to have received little attention.

Johnson (1987) studied the yield losses of potato as a result of the feeding of potato leafhopper (*Empoasca fabae* F.). As with the data of van Emden (1990), yield loss was greatest at low densities and the rate of response diminished at higher insect densities (*Figure 9.3b*). While feeding, the potato leafhopper injects

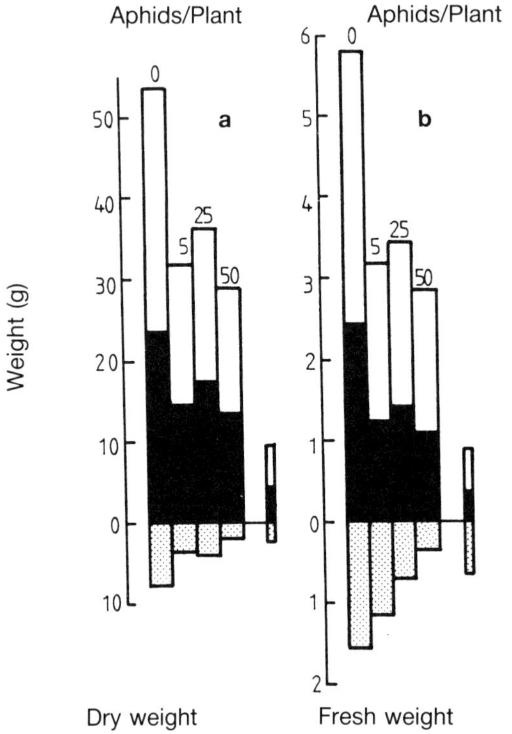

Figure 9.8 The distribution of (**a**) dry weight and (**b**) fresh weight of Brussels sprout plants receiving different initial infestations of *Brevicoryne brassicae* (means of nine plants). White area, leaves; black area, stem; stippled area, roots; narrow columns, least significant difference (P=0.05). Reproduced from van Emden (1990)

a toxic substance into the leaf tissue that significantly reduces photosynthesis. This has a major effect on radiation use efficiency of the crop. Assuming that radiation interception is unaffected, then this suggests that photosynthetic efficiency is curvilinearly reduced by increasing pest density.

Effects of aphid infestation on photosynthetic activity in tolerant and intolerant crop genotypes

A more detailed study of the effects of aphids on photosynthetic activity on wheat plants grown in controlled environment conditions (Ryan *et al.*, 1990) involved infesting single leaves of both non-tolerant and tolerant wheat genotypes with adult greenbugs for 5 days. After this, aphids and offspring were counted and weighed and honeydew was collected and its dry weight recorded. Photosynthetic leaf characteristics were then measured. Following the 5-day period, infestation levels were not different between the non-tolerant and tolerant genotypes, and honeydew dry weight was also not different, suggesting that both tolerant and susceptible plants received equal feeding pressure. Photosynthetic activity and

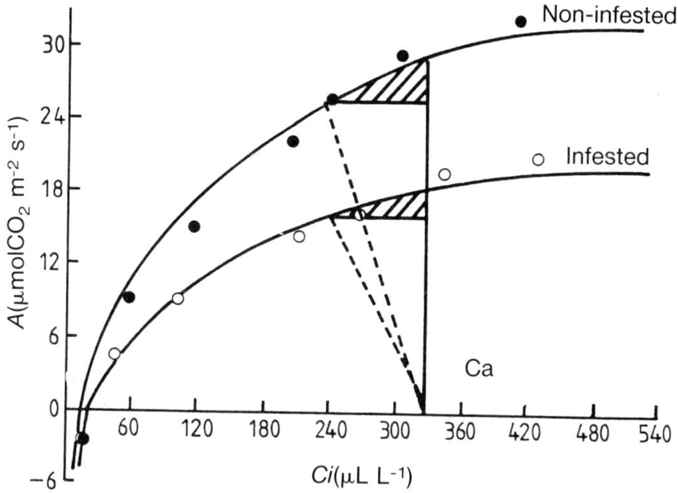

Figure 9.9 Carbon dioxide assimilation (A) plotted against internal CO_2 concentration (Ci) for leaves of greenbug infested and non-infested non-tolerant wheat (cv. Sturdy). The slanted dashed lines describe the effects of gaseous diffusion of CO_2 from the atmosphere to the intercellular spaces for each treatment. The vertical solid line, as it crosses the A versus Ci curve gives the rate of A when there is no stomatal barrier to CO_2 diffusion into the leaf. The asterisk marks indicate the A and Ci values, measured at ambient CO_2 concentrations. Reproduced from Ryan *et al.* (1990)

chlorophyll content of infected leaves of the non-tolerant genotype showed a stepwise reduction with increasing levels of infestation (up to 75% lower than controls) (*Figure 9.9*). However, no such effects could be detected in the tolerant genotype. More detailed analysis of the internal CO_2 concentration (Ci) curves indicated that a reduction in stomatal conductance was not the reason for low photosynthetic activity in the infected susceptible leaves. When Ci was calculated for these leaves, it remained constant with different levels of aphid infestation. If changes in stomatal conductance had been implicated, a reduction in Ci with increasing level of infestation would have been anticipated. However, C_E levels (photosynthetic efficiency) were significantly reduced in aphid infected leaves implying effects on the amount or activity of Rubisco (von Caemmerer and Farquar, 1981).

Thus aphids clearly affect the biochemistry of photosynthesis, whereas leaf miners (see earlier) appear to have a more mechanical effect. This correlates well with their effect at the crop level in that aphids affect yield through effects on radiation use efficiency while miners reduce yield by decreasing interception. Moreover, the data described above presented by Ryan *et al.* (1990) have important implications in relation to crop tolerance to aphid infestation. The tolerance mechanism does not appear to be associated with a reduction in feeding activity but is instead associated with the ability of the plants to maintain the integrity of their photosynthetic machinery when subjected to aphid infestation.

Estimating yield loss from multiple pest infestation

Loss of potential crop yield is often brought about by multiple pest and disease attacks. Johnson (1987) provides an interesting example of applying the resource capture concept to the effects of potato leafhopper, early blight (*Alternaria solani* (Ell. & Mart.) Jones & Grout) and *Verticillium dahliae* Kleb. on the yield of potato. Their experiment included a factorial contribution of multiple levels of all three factors. When the relationship between yield and effective leaf area was analysed separately for each pest, the slopes of the regression of yield on leaf area were dependent on the specific pest (*Figure 9.10*). For early blight and *Verticillium* wilt, which reduced yield through their effect on intercepted radiation by prematurely senescing older leaves, both relations had similar slopes. However, the average slope obtained from regressing yield and leaf area across potato leafhopper treatments was approximately three times greater as a result of a reduction in *RUE* by the effects of the potato leafhopper on green leaf tissue.

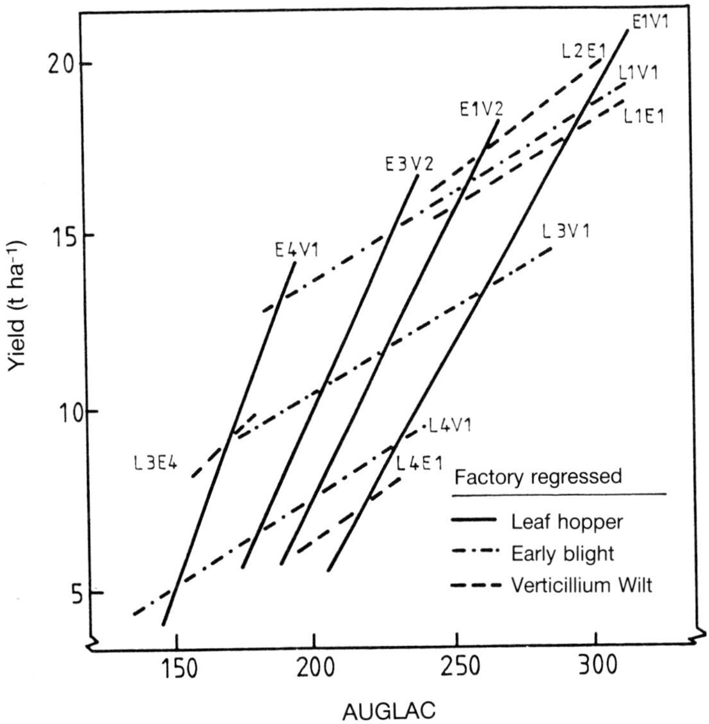

Figure 9.10 Results of the regression of total yield on area under the proportion of green leaf area remaining curve (AUGLAC) for Russet Burbank potato in a factorial experiment involving multiple infestations of early blight (E), *Verticillium* wilt (V) and potato leafhopper (L). Regressions were done separately on each pest factor. A subset was defined as all plots of a particular pest factor that were related by having similar infestations of the other two pests. Labels in the figure indicate the relative infestation levels of the other two pests for each subset. Low pest infestation levels = 1. Reproduced from Johnson *et al.* (1987)

Partitioning crop yield reduction between pest effects on radiation use efficiency and radiation interception

It is clear that pest infestations can vary from those in which the species concerned primarily affect the interception of radiation, to those that mainly affect *RUE* through their effect on photosynthesis. However, these effects are generally not wholly of one type or the other, and it is more appropriate to consider most pests as affecting both yield determinants. Taking this one step further, it should be possible to partition yield loss due to a particular pest (or group of pests) into that caused by a reduction in interception of radiation and that caused by a reduction in *RUE*. This could then be presented as a ratio or a proportion.

The information necessary to attribute yield losses to a reduction in *RUE* and intercepted radiation would include:

1. measurements of *L* of healthy crops and pest infested crops;
2. measurements of radiation interception by the healthy and infested crop;
3. measurement of dry matter accumulation/yield accumulation in both healthy and infested crops.

Measurements performed on healthy crops should enable the construction of the relation between interception of radiation and *L* and between rate of yield/dry matter accumulation and radiation interception. Measurements performed on the infested crop will enable the loss in *L* to be calculated, and from this, the reduction in radiation interception to be estimated. The yield reduction on the basis of a reduction in radiation interception can then be estimated and compared with actual yield reduction. Where the actual yield reduction exceeds that calculated on the basis of a reduction in radiation interception, the difference can be attributed to a reduction in *RUE*. As a result, the partitioning of yield reduction into its two components can be determined.

Unfortunately, such data are not available. However, as an example, the following analysis is drawn from a number of sources. Wratten (1975) describes the effects of field caged, postanthesis populations of the aphids *Sitobion avenae* (F.) and *Metopolophium dirhodum* (Walk.) on grain yield of winter wheat. *S. avenae* and *M. dirhodum* reduced overall grain weight by 14% and 7% respectively. A feature of the infestations was premature leaf senescence, which occurred approximately one week earlier than in healthy plants in both cases (*Figure 9.11*). It is possible to estimate a yield depression on the basis of the earlier senescence by estimating the reduction in radiation interception as a result of loss of effective leaf area and estimating its conversion into grain yield. Assuming that average radiation receipts in July are 17 MJ m^{-2} d^{-1}, 25% reflection from the crop, full interception of remaining incident radiation by the crop and a radiation conversion efficiency of 1.4 gDW MJ^{-1} (Monteith and Scott, 1982), the estimated loss in dry weight occurring as a result of a 7 day earlier onset of senescence would be 125 g m^{-2}. With the further assumptions that harvest index was 0.40 and average crop yield was approximately 7 t ha^{-1} (700 g m^{-2}), this is equivalent to a yield reduction of 7.1%. This suggests that the yield depression of 7% for wheat infested by *M. dirhodum* can be completely accounted for by its effect on leaf senescence whereas only 50% of the yield depression of wheat infested with *S. avenae* can be accounted for similarly; half of the loss in yield is determined by a fall in *RUE*. The data presented by Wratten (1975) indicate that, although there was no marked difference in numbers between both aphid species, there were

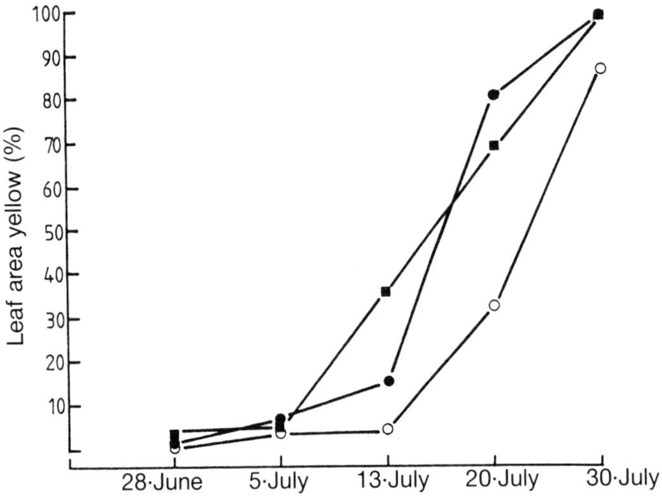

Figure 9.11 Wheat flag leaf senescence of control plants, ○; *Sitobion avenae* infested, ●; *Metopolophium dirhodum* infested plants, ■. Reproduced from Wratten (1975)

twice as many *S. avenae* on the flag leaf and ear as *M. dirhodum*. Since this is the main site for carbohydrate supply to the developing grains, it is not surprising then that *S. avenae* had a more significant effect on *RUE*. Clearly, this cannot be considered as an accurate analysis; nevertheless, it does illustrate the usefulness of such the approach.

Data presented earlier from Johnson *et al.* (1987) are used as a second example (*Figure 9.10*). The damage caused by leafhoppers on potato can be partitioned into direct effects on photosynthetic efficiency and indirect effects through hopperburn in reducing the size of the assimilatory surface. Rather than relating yield loss to intercepted radiation, Johnson *et al.* (1986) use a term called 'area under the proportion of green leaf remaining curve' (*AUGLAC*) which is, in effect, the integral of leaf area through development. Although radiation interception was not measured, yield appeared to be linearly related to *AUGLAC*. In addition, since Burstall and Harris (1982) have shown good relationships between percentage ground cover in potatoes and intercepted radiation, it seems reasonable to assume that *AUGLAC* is reasonably correlated with radiation interception and can be substituted for it in this case. Making the same assumption that Johnson (1987) makes, when discussing the same data, that early blight and *Verticillium* wilt reduce radiation interception, the slope of yield against *AUGLAC* for crops infected with these diseases can be assumed to represent crops in which yield is reduced as a function of a decline in radiation interception. The slope of the yield against *AUGLAC* for leafhopper infested crops is much higher than that caused by the two diseases. If the effect of leafhopper infestation on yield were purely an effect of a reduction in *RUE* without any effect on leaf area, the line relating yield and *AUGLAC* across leafhopper treatments should have been vertical. This is not the case and the slope of the relationship lies somewhere between the two, suggesting that yield reduction by leafhoppers is a function of both a reduction in *RUE* and a reduction in effective leaf area. The partitioning of yield reduction between these factors can be estimated from the slopes of the lines as the ratio of the angles that each line subtends from the vertical. Thus, 57% of the yield reduction is caused by a reduction in *RUE*, whereas 43% can be associated with the leaf area effects of hopperburn.

Conclusion

The concept of resource capture provides an excellent framework for the analysis of the effects of pest attack on reducing yield in crops. It appears that the effect of pest species on crops can largely be resolved into infestations that affect yield by reducing the assimilatory surface and so *L*, or those where the effect is to reduce photosynthetic activity and hence *RUE*. Conventionally, entomologists have considered simple relationships between pest density and yield and have largely ignored how pests may affect the relationships between yield and environment. However, models which have attempted to quantify the effects of pests on crop yield through changes in intercepted radiation and *RUE* appear to provide a sound basis for studying pest crop interaction. The data available to test such models are, at present, scant. Little attention has been paid, so far, to whether defoliating insects may also change canopy structure such that light extinction coefficients are changed as well as *L*. The possible effects of pests on the partitioning of assimilates within the plant, nutrient uptake and water relations has recieved scant attention. Considerable scope exists for crop physiologists to work with entomologists on pest–crop interactions. Radiation use efficiency provides an appropriate conceptual basis for quantifying the damage caused by pests on plants and provides a particularly useful tool for use in crop-pest management in two main ways:

First, and perhaps most important, a critical concept in the development of improved insect control in the last 30 years has been that of intervening with pesticide only in response to a real threat, with a consequent reduction of prophylactic treatments. Farmers and growers increasingly treat only in relation to an advised 'economic threshold' of pest density. Economic thresholds have proved quite difficult to quantify, because the number of pests per plant is only part of the picture, even in the absence of any biological control restraint on pest increase. Many of the difficulties of establishing thresholds appropriate for different stages of plant growth or different cultural regimes would be alleviated if the plant–pest interaction could be modelled in resource capture terms.

Second, the capacity of some cultivars of a crop to tolerate pest damage better than others is well known to plant breeders and is a characteristic well worth combining with types of plant resistance that only reduce pest numbers following colonization (antibiosis). Especially with fast breeding insects (where the insect burden changes during the course of the experiment), it has proved extremely hard to develop an experimental protocol to identify tolerance in cultivars differing in antibiosis. Resource capture analysis seems an obvious way forward.

Finally, how to choose between candidate herbivores as potential biological control agents for weed problems on the basis of how far the different agents would debilitate plants has been a debate in an important area of applied entomology. The debate has tended to generate rather more heat than light. As indicated earlier, the resource capture concept could provide the basic principle that has been sought so actively.

References

Allen, E.J. and Scott, R.K. (1980) An analysis of growth of the potato crop. *Journal of Agricultural Science (Camb.)*, **94**, 583–606

Boote, K.J., Jones, J.W., Mishoe, J.W. and Berger, R.D. (1983) Coupling pests to crop growth simulators to predict yield reductions. *Phytopathology*, **73**, 1581–1587

Burstall, L. and Harris, P.M. (1982) The estimation of percentage light interception from leaf area index and percentage ground cover in potatoes. *Journal of Agricultural Science (Camb.)*, **100**, 241–244

von Caemmerer, S. and Farquar, G.D. (1981) Some relationships between the biochemistry of photosynthesis and the gas exchange of leaves. *Planta*, **153**, 376–387

Dixon, A.F.G. (1971) The role of aphids in wood formation. 1. The effect of the sycamore aphid, *Drepanosiphum platanoides* (Schr.) (Aphididae), on the growth of the sycamore, *Acer pseudoplatanus* (L.). *Journal of Applied Ecology*, **8**, 165–179

van Emden, H.F. (1978) Ecological information necessary for pest management in grain legumes. In *Pests of Grain Legumes: Ecology and Control*, pp 297–307 Edited by S.R. Singh, H.F. van Emden and T.A. Taylor. London: Academic Press

van Emden, H.F. (1990) The effect of *Brevicoryne brassicae* on leaf area, dry matter distribution and amino acids of the Brussels sprout plant. *Annals of Applied Biology*, **116**, 199–204

Harris, P. (1971) Current approaches to biological control of weeds. In *Biological Control Programmes against Insects and Weeds in Canada*, pp 67–76. Farnham Royal: Commonwealth Agricultural Bureaux

Heichel, G.H. and Turner, N.C. (1983) CO_2 assimilation of primary and regrowth foliage of red maple (*Acer rubrum* L.) and red oak (*Quercus rubra* L.): response to defoliation. *Oecologia (Berlin)*, **57**, 14–19

Herbert Jr, D.A., Mack, T.P., Backmann, P.A. and Rodriguez-Kabana, R. (1992) Validation of a model for estimating leaf-feeding by insects in soybean. *Crop Protection*, **11**, 27–34

Johnson, K.B. (1987) Defoliation, disease and growth: a reply. *Phytopathology*, **77**, 1495–1497

Johnson, K.B., Johnson, S.B. and Teng, P.S. (1986) Development of a simple potato growth model for use in crop-pest management. *Agricultural Systems*, **19**, 189–200

Johnson, K.B., Teng, P.S. and Radcliffe, E.B. (1987) Coupling feeding effects of the potato leafhopper, *Empoasca fabae*, nymphs to a model of potato growth. *Environmental Entomology*, **16**, 250–258

Kloft, W. (1960) Wechselwirkungen zwischen pflanzensaugenden Insekten und den von ihnen besogenen Pflanzengeweben. *Zeitschrift für angewandte Entomologie*, **45**, 337–381; **46**, 42–70

Miles, P. (1968) Insect secretions in plants. *Annual Review of Phytopathology*, **6**, 137–164

Miles, P. (1990) Aphid salivary secretions and their involvement in plant toxicoses. In *Aphid–Plant Genotype Interactions*, pp 131–147 Edited by R.K. Campbell and R.D. Eikenbary. Amsterdam: Elsevier

Monteith, J.L. and Scott, R.K. (1982) Weather and yield variation of crops. In *Food, Nutrition and Climate*, pp 127–153 Edited by K. Blaxter and L. Fowden. London: Applied Science Publishers

Proctor, J.T.A., Bodnar, J.M., Blackburn, W.J. and Watson, R.L. (1982) Analysis of the effects of the spotted tentiform leafminer (*Phyllonorycter blancardella*) on the photosynthetic characteristics of apple leaves. *Canadian Journal of Botany*, **60**, 2734–2740

Rudd, W.G. (1980) Simulation of insect damage to soybeans. In *Proceedings of the 2nd World Soybean Research Conference*, 26–29 March, 1979, North Carolina State University, Raleigh, NC. pp 547–555 Edited by F.T. Corbin. Boulder, Colorado: Westview Press

Ryan, J.D., Morgham, P.E., Richardson, R.C., Johnson, R.C., Mort, A.J. and Eikenbary, R.D. (1990) Greenbug and wheat. A model system for the study of phytotoxic Homoptera. In *Aphid-Plant Genotype Interactions*. pp 171–186 Edited by R.K. Campbell and R.D. Eikenbary Amsterdam: Elsevier

Scott, R.K. and Jaggard, K.W (1985) The effects of pests and diseases on growth and yield of sugar beet. *Proceedings of the 48th Winter Congress of the International Institute for Sugar Beet Research*, pp 153–169. Brussels: International Institute for Sugar Beet Research

Shibbles, R.M. and Weber, C.R. (1965) Leaf area, solar radiation interruption and dry matter production in soybeans. *Crop Science*, **5**, 575–577

Waggoner, P.E. and Berger, R.D. (1987) Defoliation, disease, and growth. *Phytopathology*, **77**, 393–398

Winder, J.A. and van Emden, H.F. (1980) Selection of effective biological control agents from artificial defoliation/insect cage experiments. *Proceedings 5th International Symposium on the Biological Control of Weeds*, Brisbane, 1980, pp 425–439 Edited by E.S. Der Fosse. Canberra: CSIRO

Wratten, S.D. (1975) The nature of the effects of the aphids *Sitobion avenae* and *Metopolophium dirhodum* on the growth of wheat. *Annals of Applied Biology*, **79**, 27–34

10

DISEASE AND RESOURCE CAPTURE

A.C. MADEIRA
Departamento de Ciências do Ambiente, Instituto Superior de Agronomia, Lisboa, Portugal

J.A. CLARK
Department of Physiology and Environmental Science, University of Nottingham, UK

Introduction

The depression of crop yields by disease is common, and is of major concern in agriculture. Diseases can reduce yield and/or quality, for example in terms of flavour and appearance of edible products. Although most plant diseases are localized and losses moderate, epidemics can be severe in species susceptible to particular diseases. *Phytophthora* late blight of potato, *Helminthosporium* brown spot of rice and *Puccinia* stem rust of wheat are diseases which have caused not only economic losses but famine. However, the extent of crop loss due to disease is difficult to assess accurately. *Table 10.1* shows estimated world production (FAO, 1992) and losses due to diseases, insects and weeds in several important arable crops (Cramer, 1967). Globally, 10–20% of crop production is estimated to be lost due to disease, 30–70% of total losses, despite much research on chemical and other preventative measures. Even these figures may underestimate the impact of disease as they exclude post-harvest losses and damage. There is

TABLE 10.1 ESTIMATED WORLD CROP PRODUCTION AND PRE-HARVEST LOSSES (FAO, 1992) AND PERCENTAGE LOST TO DISEASES, AND TO DISEASES, INSECTS AND WEEKS ACCORDING TO CRAMER (1967)

	Actual production	*Estimated losses to disease*	*Estimated total losses*	*Potential production*	*Crop lost to disease* (%)	*Total crop losses* (%)
	(millions of tonnes)					
Cereals	1881	264	991	2872	9.2	34.5
Potatoes	259	82	122	377	21.8	32.3
Other root crops	583	171	440	1023	16.7	43.0
Sugar cane	1073	458	1311	2384	19.2	55.0
Sugar beet	299	41	97	396	10.4	24.5
Vegetables	445	62	171	616	10.1	27.7
Legumes	60	10	30	90	11.3	33.3

therefore an urgent need to improve both understanding of the ways that diseases affect crops and the effectiveness of control methods.

For a disease to occur, develop and become epidemic three elements are needed: a susceptible host plant, a virulent pathogen, and persistently favourable environmental conditions for the growth, reproduction and spread of the pathogen. The effects of pathogens depend not only on their aggressiveness, but also on the subsequent duration and severity of infection, which are affected by environmental factors such as the availability of water from rainfall, fog, dew or groundwater, temperature, wind and solar radiation (e.g. Waggoner, 1965; Colhoun, 1973; Leach and Anderson, 1982; Fitt and McCartney, 1986; Royle and Butler, 1986).

Pathogens may affect plants either:

1. directly, by using assimilates (biotrophy) or structural tissues killed in advance of pathogen spread (necrotrophy); and by changing rates of metabolic and growth processes; or
2. indirectly, by reducing light, CO_2, water and nutrient inputs. Changes in photosynthesis, translocation, plant water balance and leaf senescence may reduce assimilates for plant growth and yield.

This review is concerned mainly with the effects of fungal diseases on plants and, secondly, the consequences for resource capture. The former area is much better understood than the latter, in part because field experiments on diseased crops are difficult to manage and are unpopular with farm managers!

Responses to disease: plant level

LEAVES

Changes in the photosynthetic rate of plants infected by pathogens (fungi, bacteria and viruses) have often been reported. In foliar diseases, visible necroses decrease the size and/or duration of the photosynthetic leaf area. In the late stages of infection, photosynthesis may also be reduced by a depression of chloroplast efficiency or number, or by inhibition of the enzyme systems associated with photosynthesis (Buchanan *et al.*, 1981; Ahmad, Farrar and Whitbread, 1983; Hutcheson and Buchanan, 1983; Ayres, 1984; Scholes and Farrar, 1985).

However, photosynthesis may be unchanged or even stimulated by biotrophic fungal pathogens, when chlorophyll is retained in 'green islands' at infected sites and/or in uninfected regions of infected leaves or plants (Livne, 1964; Ayres, 1976; Habeshaw, 1984; Scholes and Farrar, 1985). At later stages of infection, high rates of photosynthesis in the uninfected tissue of infected leaves are probably the result of changes in photosynthate transport between healthy and infected tissue or between the leaf and the rest of the plant (Habeshaw, 1984). For example, Williams and Ayres (1981) and Paul and Ayres (1984) found that photosynthesis was stimulated (and photorespiration inhibited) in healthy leaves of barley infected by powdery mildew and of rust-infected groundsel plants. Such compensatory activity may protect the yield of lightly infected plants and may be critical for the survival of those heavily infected. According to Whipps and Lewis (1981) biotriphic fungi (e.g. powdery mildews and rusts) have greater effects on both photosynthesis and translocation in infected plants than nectrotrophs. However, necrotrophic fungi, bacteria and viruses can all cause retention of assimilates at the infection sites, and have significant effects on their hosts (Coffey, Marshall and Whitbread, 1970; Rowe and Reid, 1979).

In infected plants, control of gas exchange between the leaf and the environment is modified by changes in stomatal and mesophyll conductances and by damage to the leaf cuticle. Foliar pathogens can affect stomatal movements both directly by damage or stimulation of stomata, and indirectly by disrupting leaf structure and by competing with host cells for water and nutrients. Infection by foliar pathogens can stimulate stomata to remain open even in the dark, as for example in late blight (*Phytophthora infestans*) of potato (Farrell, Preece and Wren, 1969) and leaf blotch (*Rynchosporium secalis*) of barley (Ayres, 1972). The opening increases transpiration, but this may be small in cool moist conditions if the effects are confined to colonized tissues.

Stomatal closure in infected leaves (or failure to open in light) reduced transpiration of rusted (*Uromyces phaseoli*) bean leaves (Duniway and Durbin, 1971) and net photosynthesis during the later stages of mildew infection (*Erysiphe pisi*) in pea leaves (Ayres, 1976). Although both rusts and mildews inhibit stomatal opening, the rupture of cells in the host cuticle and epidermis by sporulating rust fungi provides an uncontrolled path for water loss. Rust infection may therefore increase transpiration and cause wilting, and consequently reduce leaf expansion and biomass accumulation. In contrast, powdery mildew fungi grow mainly on the surface of their hosts, exerting less control on water loss. For example, Ayres (1981a) and Williams and Ayres (1981) suggested that powdery mildew infection may even reduce transpiration from pea and barley plants and decrease competition for water, which may explain the stimulation of net photosynthesis

TABLE 10.2 MEASURED AND CALCULATED COMPONENTS OF PHOTOSYNTHESIS ($mgCO_2$ m^{-2} s^{-1}), ESTIMATED INTERNAL CO_2 CONCENTRATION (c_i) AND CONDUCTANCES IN HEALTHY LEAVES (H) AND DISEASED LEAVES (D) INFECTED BY RUST FUNGI (AFTER OWERA, FARRER AND WHITBREAD, 1981). P_n, NET PHOTOSYNTHETIC RATE; R_l, CO_2 EFFLUX IN THE LIGHT; R_d, DARK RESPIRATION; P_g, GROSS PHOTOSYNTHETIC RATE; g_a, ATMOSPHERIC CONDUCTANCE; g_s, STOMATAL CONDUCTANCE; g_m, MESOPHYLL CONDUCTANCE

	Puccinia hordei *(Barley)*		Uromyces phaseoli *(French bean)*		Puccinia poarum *(Wheat)*	
	H	*D*	*H*	*D*	*H*	*D*
P_n (mg m^{-2} s^{-1})	0.340	0.249	0.078	0.051	0.318	0.246
R_l (mg m^{-2} s^{-1})	0.069	0.117	0.072	0.087	0.049	0.051
R_d (mg m^{-2} s^{-1})	0.023	0.050	–	–	–	–
c_i (mg m^{-3})	95	176	95	156	108	128
P_g (mg m^{-2} s^{-1})	0.409	0.419	0.150	0.138	0.367	0.297
$g_a + g_s$ (10^{-3} m s^{-1})	4.8+1.3	4.8+2.1	1.5	1.1	0.64	0.52
g_m (10^{-3} m s^{-1})	4.3	4.1	1.6	0.88	3.4	2.3

in uninfected young leaves. However, there is contradictory evidence in the literature: Gordon and Duniway (1982a) found that sugar beet leaves infected by the powdery mildew *Erysiphe polygoni* lost control of transpiration and had lower water potentials than healthy leaves, especially under water stress induced by the failure of stomata to close in the dark.

In physical terms, both leaf conductance and the concentration of CO_2 in intracellular spaces within the leaf (c_i) determine the net flux of CO_2 from air to leaf. As rates of respiration are usually greater in infected than in healthy leaves, the respiratory production of CO_2 increases c_i, as reported by Owera, Farrar and Whitbread (1981). It is clear that net photosynthesis was depressed by disease, even when gross photosynthesis increased (*Table 10.2*). Increased c_i may be associated with reductions in net photosynthesis, even when leaf damage due to fungal attack raises the conductance to CO_2 (Farrar and Lewis, 1987). McGrath and Pennypacker (1990) found increased c_i in wheat leaves infected by stem rust and leaf rust, but associated with decreased leaf conductance, and thus a decrease in stomatal water loss.

Marked reductions in net photosynthesis and transpiration, measured by Duniway and Slatyer (1971) in tomato leaves infected by vascular wilt (*Fusarium oxysporum*), were ascribed to reductions both in stomatal conductance and mesophyll CO_2 conductance. Gordon and Duniway (1982 b, c) also related reduced rates of photosynthesis in mildewed sugar beet leaves to decreased stomatal and mesophyll conductances to CO_2 (*Figure 10.1*). Photosynthetic rates were reduced mainly by decreases in the mesophyll conductance, which were the result of metabolic changes in the chloroplasts. Reductions in net photosynthesis and stomatal conductance were also observed in field beans infected by *Ascochyta fabae* or *Ascochyta* blight (Madeira, 1988) and *Botrytis fabae* (Malthus and Madeira, 1992; Malthus and Madeira, 1993), as shown in *Table 10.3*, but were not accompanied by significant changes in the chlorophyll *a* content of leaves.

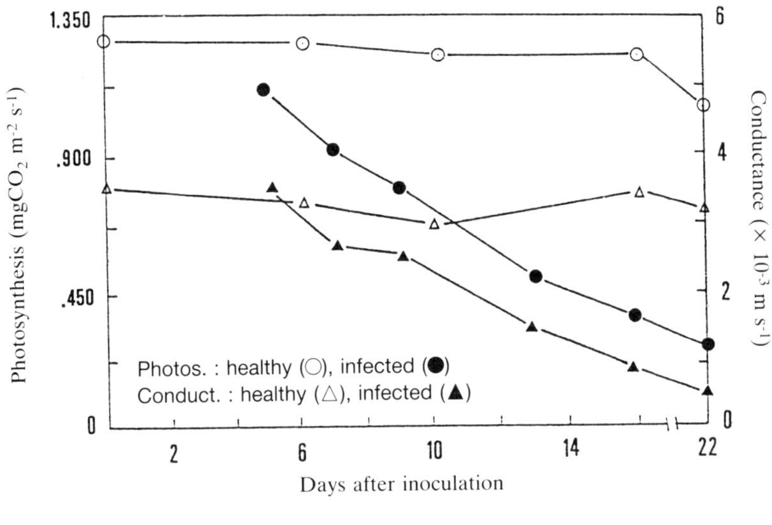

Figure 10.1 Net photosynthesis and mesophyll conductance for healthy and powdery mildew infected sugar beet leaves (redrawn from Gordon and Duniway, 1982c)

TABLE 10.3 MEASURED NET PHOTOSYNTHETIC RATE (P_n, $mgCO_2$ m^{-2} s^{-1}) AND STOMATAL CONDUCTANCE TO WATER VAPOUR AND CALCULATED TOTAL CONDUCTANCE TO CO_2 TRANSPORT OF HEALTHY (H) AND DISEASED (D – WITH % DISEASED AREA). (FROM MADEIRA, 1988; MALTHUS AND MADEIRA, 1993).

	Ascochyta fabae		Botrytis fabae		
	H	*D (>10%)*	*H*	*D (<10%)*	*D (>10%)*
P_n (mg m^{-2} s^{-1})	0.635	0.304 **	0.250	0.158 *	0.110 ***
g_s (10^{-3} m s^{-1})	11.3	6.9 *	2.8	2.2	1.7 **
g_t (10^{-3} m s^{-1})	6.4	4.1 *	1.7	1.3	1.0 *

* $P<0.05$; ** $P<0.01$; *** $P<0.001$

However, the mechanisms by which foliar fungi affect photosynthesis are uncertain, and Ayres (1981b) suggested that altered stomatal behaviour is probably a consequence rather than a cause of reductions in photosynthesis.

Increased respiration is a general response of plant tissues to pathogen invasion, which leads to a more rapid use of reserves than in healthy tissue (Daly, 1976; Hutcheson and Buchanan, 1983; Ayres, 1984; Smedegaard-Petersen, 1984). Usually, this increase in respiratory activity appears shortly after infection, but before the appearance of visible symptoms, and continues to rise during the multiplication and sporulation of the pathogen. Subsequently, respiration may decline to normal or even lower levels, or may continue to rise, depending on the host–pathogen combination. In plant varieties resistant to pathogens, respiration may increase more rapidly in the early stages of infection than in susceptible varieties, as stored energy is used for defence mechanisms activated by pathogen attack. According to Smedegaard-Petersen and Tolstrup (1985), this response of resistant plants to infection may explain observed yield reductions even in apparently healthy crops: for example, the yield of resistant barley plants was decreased after inoculation with *Erysiphe graminis*, although no disease symptoms were visible during the growing season.

Changes in the metabolism of diseased plants related to increased respiration may be due to increased membrane permeability of mitochondria, to increased activity of enzymes involved in respiration, or to activation of the less efficient pentose phosphate oxidative pathway. Kosuge and Kimpel (1981) suggested that the increased respiration in diseased plants may be associated with the production of secondary metabolites, such as isoflavonoids and lignin, implying that respiration is tightly coupled to biosynthesis of stress metabolites by the pentose pathway. Compounds produced in response to attack by pathogens include phytoalexins with fungitoxic effects, which may determine the resistance of the plants to disease (Friend, 1981).

ROOT AND VASCULAR SYSTEMS

There have been many reports of the impact of root diseases on plants, but their consequences for yield are poorly quantified. Damage to cereals by root-infecting fungi has been reviewed by Hornby And Fitt (1981). For some root diseases (for example, take-all, root rot) an indirect indication of root dysfunction may be provided by decreases in leaf area, plant height, shoot dry weight and grain yield. Yellowing, a consequence of reduced chlorophyll content in leaves, is a common shoot symptom of root diseases and is also a general symptom of nitrogen shortage. However, other root diseases may have little initial effect on shoot growth (Fitt and Hornby, 1978).

Many plant pathogens interfere with the transport of water and solutes by affecting directly the extraction of water from the soil, by increasing the resistance to water transport in the xylem, by disrupting stomatal control of transpiration or by a combination of effects. Even without visible damage, pathogens may reduce the capacity of cells to retain solutes by affecting cell membrane permeability, and thus cause or exacerbate water stress. Pathogens such as the damping-off fungi, the root rots and some bacteria, may destroy roots before symptoms are visible in the shoots. Pathogens may also affect the size of root system, measured in terms of root density, root depth and number of functional roots and again decrease the water and nutrient uptake by the roots. Some other fungal and bacterial pathogens, that cause damping-off, stem rots, cankers or vascular browning, may affect the xylem vessels and in severe cases kill young plants. The vascular system may be destroyed or blocked either by the pathogen itself (as mycelium, spores or bacterial cells) or by substances produced by the pathogen (for example, toxins or enzymes), or by the plant in response to infection (for example, gels and tyloses in the vessels). Disturbance of the vascular system reduces the transport of water through the xylem and metabolites in the phloem. The water flow rate in infected stems has been reported to be inversely proportional to the number of vessels blocked and directly proportional to the radius of the vessels affected by the pathogen and/or by the infection products (Ayres, 1981a). Vascular wilts caused by *Fusarium* and *Verticillium* spp. (Dimond, 1970) and bacteria such as *Pseudomonas solanacearum* (Nelson and Dickey, 1970) may destroy the xylem in roots and stems. Foliar pathogens, for example rusts and powdery mildews, may also affect the host root system and root distribution in soil.

The severity of attack by a particular disease affecting plant roots and stems depends on soil moisture and nutrient contents. Micro-organisms are adapted to the full range of soil moisture. For example, *Pythium* damping-off of seedlings and *Phytophthora* root and crown rot are most severe in very wet or waterlogged soil conditions, which may be interpreted as an effect of poor soil aeration favouring the pathogen. Soil-borne fungi causing root rots (*Phytophthora* sp.), root and stem rots (*Rhizoctonia* and *Sclerotium* sp.) and bacteria such as *Erwinia* and *Pseudomonas* spp. produce their worse effects when the soil is wet but not flooded. Vascular wilts also are characteristically more severe in wet than in dry soils, although symptoms of wilting develop most rapidly under hot, dry weather. In contrast, other fungi grow best in dry soil conditions and attack plants stressed by insufficient water, e.g., the fungal diseases causing root rot, stalk rot and stem canker (Schoeneweis, 1986). Soil water deficits seem to increase the severity of stalk and root rots caused by non-aggressive fungi, such as *Diplodia*, *Fusarium*, *Gibberela* and *Macrophomina* (Jordan, Clark and Seetharama, 1984). Duniway (1982) and Schoeneweis (1986) suggested that the effects of flooding or water stress on disease development may involve a change in host susceptibility and predisposition to disease as well as an effect on the growth and development of the pathogen, since processes in plants, such as hormonal balance and cell-wall and protein synthesis, are significantly affected by water stress. Water stress is also known to promote premature or accelerated plant senescence, which may facilitate infection. For example, symptoms of stalk rot may be caused by early season water stress, which by accelerating the death of parenchyma cells may predispose roots and stalk to infection (Schneider and Pendery, 1983).

Water stress has been reported to facilitate the growth of some foliar necrotrophic pathogens. For example, the leaf blight of Kentucky bluegrass produced by *Helminthosporium sativum* increases significantly in water stressed plants and sporulation is higher on wilted than on turgid leaves. In contrast, the growth and development of some foliar biotrophic fungi (for example, *Erysiphe pisi* in pea leaves) are inhibited by drought and salinity stress (Ayres and Paul, 1986).

The susceptibility of plants to diseases is also influenced by the availability and/or form and balance of nutrients in the soil (Huber and Watson, 1974; Cunfer, Toughton and Johnson, 1980; Engelhard, 1989). For example, high nitrogen levels in soil are associated with severe foot rot (eye spot) caused by *Cercosporella* sp. in wheat and foliar chocolate spot (*B. fabae*) in broad bean, when the nitrogen is applied as ammonium, while wheat stripe rust (*Puccinia* sp.) and potato and tomato wilt (*Verticillium* sp.) are associated with nitrate. Water, pH, other nutrients and temperature influence the uptake and use of specific forms of nitrogen both by plants and pathogens (Huber and Watson, 1974).

According to Huber (1989) the uptake, transport and distribution of soil nutrients by plants can also be altered by many pathogens, such as fungi and bacteria causing root rot, vascular wilts, leaf spots and blights and viruses, with symptoms similar to those produced by nutrient deficiency or excess. Pathogens can also induce nutrient deficiencies indirectly by immobilizing Mn and Fe in the soil. For example, *Gaeumannomyces graminis, Fusarium oxysporum, Verticillium dahliae* and *Streptomyces scabie* oxidize Mn^{3+} to the non available Mn^{4+} form.

Responses to disease: stand level

CANOPY AREA AND DURATION

Many methods for the assessment of disease severity are based on subjective scoring of visible symptoms, which can be poorly related to the impact of disease on resource capture. In contrast, leaf area is an objective measure, directly related to the ability of a canopy to capture light energy. In healthy growing crops, rates of leaf production and expansion that determine the development of the leaf canopy depend on factors that include temperature, water and nitrogen supply and plant density. The rate of leaf senescence, which determines leaf area duration and photosynthesis and translocation patterns, also depends on environmental factors (Thomas and Stoddart, 1980). Diseases influence the development of leaf area and subsequent canopy senescence through their effects on assimilate production and use.

Foliar pathogens therefore depress dry matter production when they reduce canopy leaf area and duration. For example, Simkin and Wheeler (1974) and Jenkyn (1976) found that leaf area duration in barley attacked by mildew (*Erysiphe graminis*) was less than in plots protected with fungicide, as a result of more rapid senescence and grain yield was also affected. However, it is not clear from these and similar experiments whether the differences between infected and treated plants are caused primarily by the infection accelerating senescence or by the fungicide delaying senescence. Trials including an untreated control have shown that prophylactic treatments can increase the persistence of green foliage, as noted in cereal crops by Jenkins, Melville and Jemmett (1972) and in field bean by Madeira (1988). Increased leaf duration may reflect the inhibition of saprophytic leaf surface fungi by the fungicide and a consequent delay in leaf senescence, as suggested by Dickinson (1973) and Hislop (1976), or be caused by a direct growth-regulator effect of the fungicide on leaves. Such effects may explain some reports of yield increases associated with fungicide treatments that are not attributable to the control of disease. However, according to Lorenz and Cothren (1989) measurements of leaf gas exchange in wheat showed no effect of fungicide treatments on yield in the absence of disease; the yield benefits of fungicide were exclusively related to control of foliar pathogens in wheat.

In wheat and barley crops, most of the carbohydrate stored in the grain is derived from photosynthate produced after anthesis, mainly by the ear, flag leaf and upper stem. Lower leaves are less important for grain filling and yield, but may support root activity. In soybean crops also, seed yield is related mainly to photosynthesis during the pod filling period (Hawkins,1982). Consequently, yield losses caused by pathogen infection depend both on the stage of plant development when infection occurs and on the site of infection. For example, Mignucci and Boyer (1979) suggested that infection of poorly illuminated soybean leaves by the powdery mildew *Microsphaera diffusa* reduces yield much less than when favourable conditions for early infection permit rapid invasion of upper leaves of the canopy. Carver and Griffiths (1981) found that mildew reduced green leaf area and root weight in barley. When mildew was controlled, leaf area duration up to anthesis was strongly correlated with grain yield, the amount of soluble carbohydrate in the culm at anthesis and the amount of carbohydrate lost between anthesis and harvest.

According to Ben-Kalio and Clarke (1979), powdery mildew infection in groundsel, caused by the biotrophic fungus *Erysiphe fisheri*, reduced both leaf area and dry matter production only when levels of infection were high (more than 75% of aerial surfaces covered). However, low levels of mildew (less than 10%) had no effect on growth as increased leaf expansion compensated for the increased senescence of older leaves. In barley, the loss in grain yield due to net blotch, caused by the necrotrophic fungus *Drechslera teres*, was related to the severity of the disease, the growth stage when the pathogen was inoculated and the subsequent decrease in green leaf area (Deadman and Cooke, 1987). Similarly, good correlations between the severity of bacterial infection and yield in spring wheat and barley were found by Shane, Baumer and Teng (1987).

When foliar damage is visible, observations of yield depression due to disease are commonplace, although the causes are not always clear. For example, according to Williams (1975) decreases in the yield of broad beans infected by *Botrytis fabae* (chocolate spot), measured in terms of decreased number of pods per plant, could not be explained by the measured reduction in leaf area: leaf infection and resulting decreases in leaf area were compensated by increased photosynthetic efficiency of the remaining leaves, maintaining dry matter production. Williams suggested that a number of mechanisms may reduce pod set in faba bean infected by *B. fabae* and hence limit pod number and the harvested yield. These included reductions in assimilate supply due to disease, direct infection of flowers and young pods and/or changes in the metabolism of the plants due, for example, to ethylene production by the fungus. Ishag (1973) had previously concluded that the number of faba bean pods per plant is controlled by the availability of assimilates. However, in Williams' (1975) experiment, yield losses induced by *B. fabae* infection could not be explained entirely by limitation of assimilate supply, as the loss in leaf area began only after pod set. Studies of the effects of defoliation and infection by *B. fabae* (Griffiths and Amin, 1977) showed that yield was unaffected by treatments applied to leaves below the flowering nodes. In contrast, removal of leaves or infection induced at flowering nodes and above reduced the yield by decreasing the number and size of seeds, but did not affect the number of pods. They concluded that faba beans are exceptionally vulnerable to attack by *B. fabae* at the stage of flowering and early pod development, but at later stages of crop development, infection is unlikely to reduce yield much.

Similar results were found in the field by Madeira, Clark and Rossall (1988) and Madeira (1988) for faba bean infected by *Ascochyta fabae*, a necrotrophic fungus which produces leaf and pod spot disease. The reduced dry matter production (*Figure 10.2*) of treatments in which the disease was induced deliberately was consistent with observed decreases in green leaf area, leaf area duration and light interception. The corresponding time courses of leaf area are shown in *Figure 10.3*. Leaf area development can also be affected by fungi which invade the vascular system (Harrison and Isaac, 1968). For example, the effects of *Verticillium* spp. on potato include stunted growth if infection occurs during the growth phase reducing maximum leaf area, and premature defoliation of mature crops reducing leaf area duration.

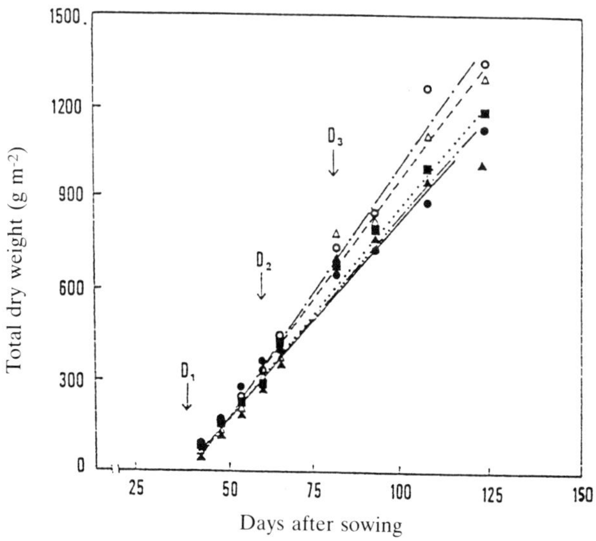

Figure 10.2 Above ground total dry weight per unit ground area (g m^{-2}) *v*.time for healthy and *Ascochyta fabae* infected field bean crops. The regression equations for the treatment data are: Fungicide (△), $y = 16.0x - 719$; Control (○), $y = 17.1x - 778$; Disease 1 (●), $y = 13.6x - 578$; Disease 2 (▲), $y = 13.2x - 553$; Disease 3 (■), $y = 14.1x - 611$. The arrows show inoculation dates

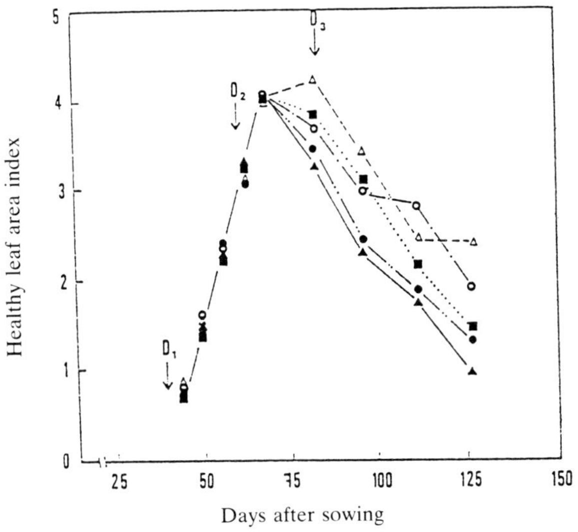

Figure 10.3 Progress of green leaf area index for healthy and *Ascochyta fabae* infected field bean crops. Symbols as in Figure 10.2

LIGHT INTERCEPTION

Light interception is the aspect of resource capture by plants most amenable to analysis for both healthy and diseased plants because it depends on canopy size and therefore integrates the impact of the environment, including pests and diseases, on the canopy.

The effects of disease and other stresses on the ability of plants to produce photosynthate, measured as dry matter production (W, g m^{-2}), can be considered in terms of those which affect the fraction f of light intercepted by foliage at a particular time during the growing season, those which influence the efficiency of conversion of intercepted solar energy to energy stored in carbohydrates (ϵ, g MJ^{-1}), and those which affect canopy duration (Monteith, 1977 and Chapter 1). If I (MJ m^{-2} d^{-1}) is the incident solar energy, then:

$$W = \int \epsilon f I \, dt \qquad (1)$$

Several studies have suggested that stresses due to both environmental conditions and disease influence dry matter production and yield mainly through reductions in the size and duration of the leaf canopy and consequent effects on light interception (Biscoe and Gallagher, 1977; Gallagher and Biscoe, 1978; Haverkort and Bicamumpaka, 1986; Waggoner and Berger, 1987; Russell, Jarvis and Monteith, 1988). As suggested by Haverkort and Bicamumpaka (1986) and Hughes (1988), the relative insensitivity of ϵ to disease may be important in the use of crop-loss assessment models to study the effects of disease, because it permits the use of fixed conversion coefficients. However, Green, Hebblethwaite and Ison (1985) have shown that in field bean crops ϵ may be reduced by water stress, while Hughes (1988) suggested that similar effects should be expected due to

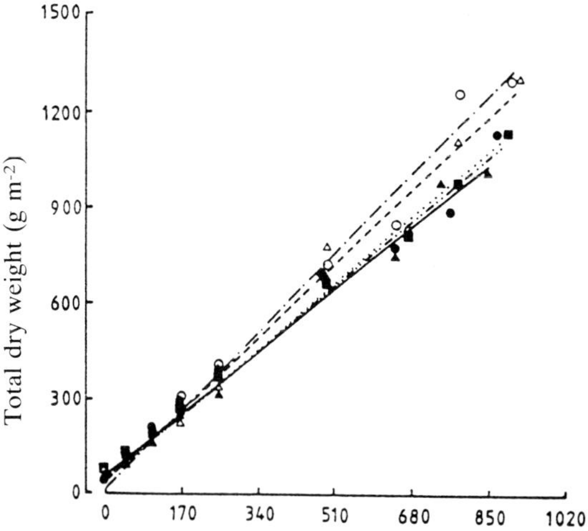

Figure 10.4 Total dry weight (g m^{-2}) *v*. accumulated intercepted radiation (MJ m^{-2}) for healthy and *Ascochyta fabae* infected field bean crops. The regression equations for the treatment data are: Fungicide (△), $y = 1.34x + 24$; Control (○), $y = 1.45x + 14$; Disease 1 (●), $y = 1.19x + 44$; Disease 2 (▲), $y = 1.19x + 49$; Disease 3 (■), $y = 1.21x + 38$. The arrows show inoculation dates

disease. Reductions in ϵ of over 20% have been observed in field bean subjected to *Ascochyta fabae* infection (see *Figure 10.4*, Madeira, 1988). It is likely that photosynthetic efficiency is also depressed by diseases in other crops, but the principal mechanism reducing yield remains light interception, which can itself be a consequence of reduced efficiency if a crop is infected during leaf expansion.

Beer's Law is usually invoked to relate the fraction f of light intercepted by a canopy to its leaf area index L and to an extinction coefficient K, determined principally by canopy geometry (more complex treatments, discussed by Campbell in Chapter 3 are not justified in this content). The relation:

$$f = 1 - \exp(-KL) \tag{2}$$

allows reductions in f due to disease to be estimated from reductions in leaf area, provided all leaves remain green and of equal efficiency. Changes in canopy duration due to premature senescence can also be accommodated by reducing the period of integration in Equation 1.

Changes in canopy geometry due to disease, such as *Fusarium* wilt, reduce canopy light interception when the leaf laminae become more vertical, decreasing K. The effects of water stress due to disease may be indistinguishable in the short term from those due to water shortage *per se*, and both transpiration and photosynthesis are usually depressed when wilting caused by vascular fungi reduces light interception, even when stomatal conductance is unchanged.

Changes in leaf colour due to disease may result in changes both in the transmission of light by the canopy and of ϵ. When leaves become chlorotic they reflect and transmit more PAR than healthy leaves, for example, when sugar beet is attacked by beet yellows virus. Chlorotic leaves may therefore contribute more to *measured* ground cover than to light interception and photosynthesis (if ϵ is reduced). At the other extreme, necrotic areas of leaves infected by fungi and bacteria intercept more incident solar radiation than healthy leaves, but without contributing to photosynthesis. Since measurements of light interception in field crops include the light intercepted by diseased tissue, exaggerated estimates of light interception by healthy leaf area in diseased canopies will underestimate ϵ. It is therefore surprising that Waggoner and Berger (1987) and Hughes (1988) were unable to find examples of depressed values of ϵ in the literature.

The problem of estimating light interception by healthy leaves may be approached in two ways. First, if the proportion of leaf area x affected by disease is small, f may be estimated using the Beer's Law equation (Equation 2), but with the healthy leaf area index $L_H = L(1 - x)$ substituted for total leaf area index, i.e.:

$$f_1 = 1 - \exp(-KL_H) \tag{3}$$

The value of f_1 is less than the radiation interception estimated from solarimeter data, and must be estimated from measurements of L_H obtained from crop harvests and values of K established for healthy stands of the same crops. Since light is also intercepted by diseased tissue, the above equation overestimates the light available for photosynthesis and Waggoner (personal communication) has pointed out that the errors of this approximation become increasingly significant with increases in the proportion of diseased leaf area.

Waggoner and Berger (1987) suggested two alternative forms of Equation 2 to cater for diseased leaves. The equation:

$$f_2 = (1 - x)[1 - \exp(-KL)] \tag{4}$$

is the rigorous form applicable to most field measurements. To incorporate defoliation d as well as the effects of disease symptoms on remaining leaves they propose:

$$f_2 = (1 - x)[1 - \exp(-KL_i(1 - d))] \tag{5}$$

where L_i is the leaf area index before attack.

Equation 5 is useful for modelling the effects of defoliation on crops due to pests and disease, but not for interpreting field measurements, since L_i $(1-d)$ is the leaf area measured from a crop harvest. Analysis of the light interception of diseased crops therefore requires the measurement of both total and healthy leaf areas, for use in Equation 4. However, observation suggests that it is rare to encounter large proportions of diseased leaf area, since badly diseased leaves and necrotic lesions within leaves are lost over periods short compared with the leaf duration. For example, Madeira (1988) found in a field experiment that the proportion of diseased leaf area in faba bean inoculated with *Ascochyta fabae* remained below 15%, except in the final stages of senescence, despite large changes in canopy area. In these conditions, the results of using Equations 3 and 4 were not significantly different. Calculations show that for uniformly diseased crops with $L = 2$ and the fraction of necrotic leaf area $x = 0.1$, Equation 3 will overestimated f by about 5%, increasing to over 30% for $x = 0.5$. The corresponding errors in the estimation of ϵ from crop data will be approximately of equal proportion but opposite in sign. Equation 4 should therefore be employed for the analysis of light interception by crops with more than about 10% of diseased leaf area. The corollary is also true: late season estimates of light interception and ϵ in diseased and senescent crops will be inaccurate even when based on measurements by such simple and apparently reliable devices as tube solarimeters, unless allowance is made for interception by non-photosynthesizing surfaces.

The above theory assumes that the disease is uniformly distributed in the canopy. In real crops there is the additional complication that severity varies with position. For example, beet yellows virus is transmitted by aphid vectors and the date of infection varies according to the suitability of the weather for aphid migration. Infection from an initial site follows translocation patterns in the plants: fully expanded leaves which are acting only as assimilate sources remain uninfected, but leaves expanding at the date of infection become badly damaged and chlorotic, while leaves initiated after infection carry the virus but are green (Peters, 1988). Thus, within an infected plant, the diseased leaves are distributed in a characteristic pattern rather than at random. Many diseases attack mainly older leaves, usually at the bottom of the canopy, while others attack only young expanding leaves. Each pattern of attack produces a different disease distribution, which should be accounted for in a rigorous treatment of light interception.

The following calculation illustrates the possible consequences of systematic differences of infection with height in the canopy. The example assumes a mature canopy intercepting 95% of incident light. Uniform infection of 0.15 of the canopy area by necrotic lesions would result in reduction of light interception by healthy leaves from 95 to 81%. If, overall, the diseased proportion is the same, but the top half of the canopy is disease free while the lower half has $x = 0.30$, healthy leaves will intercept 90% of the incident light, only 5% less than the healthy canopy. However, if the disease is concentrated in the upper leaves the figure becomes 72%, 23% less than the healthy canopy (*Table 10.4*).

The influence of layering has been considered in work on the inverse problem to that of light interception – the spectral reflectance of crops and the use of the resulting signal in remote sensing. For example, the SAIL model has been used to simulate the effect of chlorosis distribution in sugar beet on the spectral signature

TABLE 10.4 FRACTIONAL LIGHT INTERCEPTION (*f*) IN A DISEASED CROP WITH DISEASE INCIDENCE DIFFERENCES BETWEEN LAYERS. LIGHT INTERCEPTION ESTIMATED USING EQUATION 4 ASSUMING 95% LIGHT INTERCEPTION FOR $L = 4.5$. DISEASED FRACTION $= x$. SEE TEXT FOR DETAILS

Leaf area index	*Canopy top 1/2* 2.25	*Canopy lower 1/2* 2.25	*Total* 4.5
	f *Top*	f *Lower*	f *Total*
$x=0$ No disease	78	17	95
$x=0.15$ Uniform disease	66	14	81
$x=0.30$ Top 1/2 Only	55	17	72
$x=0.30$ Lower 1/2 Only	78	12	90

of diseased crops (Eastwood and Steven, 1991), and could also be adapted to estimate their light interception.

A further characteristic of disease attacks is that they are usually patchy, aerial photograph showing the spread of diseases from initial foci (Blakeman, 1990). Experimental assessments of the effects of disease on resource capture by crops usually assume uniform infection – but in the real world some parts of the canopy may be unaffected while others may be killed. Quantitative assessments of the effects of spatial distribution of disease on yield are rare.

Exceptionally, Ferrandino (1989) examined the spatial and temporal variation of pest and disease damage and their consequences for yield. Ferrandino was able to construct a model of disease spread from point foci which gave a good match to field observations, in terms both of the time course of defoliation (*Figure 10.5*), and dry matter production. This study produced the interesting conclusion that yield losses are likely to be greatest in a patchy epidemic – 'characterized by steep spatial gradients' – and less severe when infection is more uniform. In contrast, a subsequent theoretical assessment of the effects of spatial gradients of disease severity by Hughes (1990) suggests that the effects of patchiness on losses may depend on the shape of the loss/severity relation, and that losses may be either less or more severe in a patchy epidemic than in a uniformly diseased crop.

It is simplistic to use Beer's law to predict light interception in a canopy with patchy disease because factors other than light interception may influence yield. However, the exercise can at least indicate the likely effects of disease on dry matter production. When lesions are retained, applying Equation 4 to the canopy considered in *Table 10.4* predicts that the mean light interception for uniform disease should be the *same* as that of a patchy crop in which half is healthy and

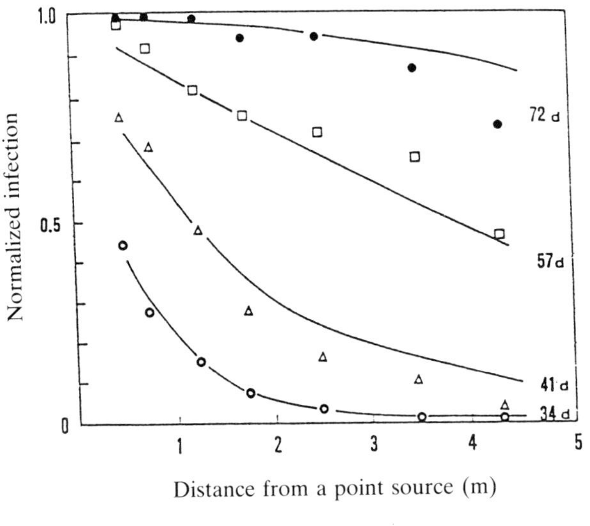

Figure 10.5 Normalized infection as a function of distance from a point source at 34, 41, 57 and 72 days after emergence – potato plants were inoculated with *Phytophthora infestans* 27 days after emergence (redrawn from Ferrandino, 1989)

half is diseased. For example, the light interception of a uniformly diseased crop with lesion area $x = 0.15$ is 81%, which is the average of 95 and 66% for patches with no disease and lesion area $x = 0.30$, respectively. Disease pattern should therefore have little effect on light interception and dry matter production when lesions are retained and when disease has little effect on ϵ. When a disease causes defoliation, Equation 5 suggests that Ferrandino's (1989) conclusion is correct. For the canopy considered in *Table 10.4* (initial $L = 4.5$, $K = 0.67$), the value of light interception calculated for uniform loss of 0.15 of leaf area is about 1% less than that for a crop with the same mean defoliation, but with half healthy and half losing 0.30 of the initial leaf area. For a mean defoliation of 0.25 the difference in f increases to 3% (of incident light intensity) and to 30% when half the canopy area has been lost. Rectangular variations in disease incidence can occur only in models, but the above calculations suggest that the effects of disease pattern on productivity may depend on the relation between the development of lesions and defoliation, and probably on the host–disease combination and conditions of infection. Similar calculations can be made for a pattern of infection at a particular time, but analyses of the effects of disease on crop yields also need to consider the consequences of the dynamics of pathogen spread for light interception and dry matter production over the life of the crop (Ferrandino, 1989), and the effects of disease on harvest index.

The consequences of patchiness in diseased crops are far from fully understood and warrant more attention, not just because most diseases develop in patches but also because patchy crops result in inefficient nutrient utilization (Biscoe, personal communication). Detection of disease at an early stage could enable control measures to be concentrated on diseased areas, minimizing the environmental load of toxic chemicals as well as their cost to the farmer (Steven *et al.*, 1990).

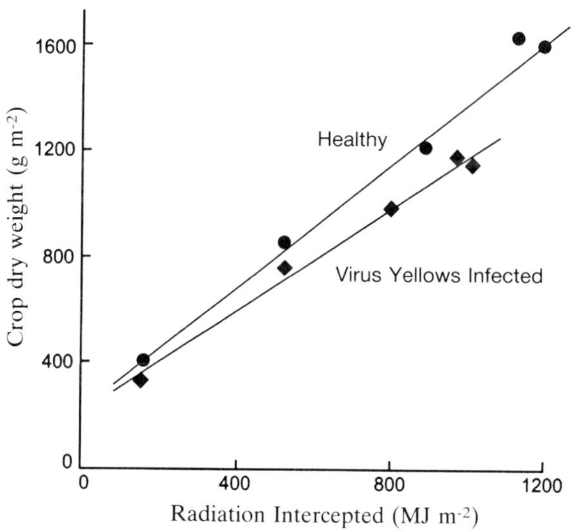

Figure 10.6 Dependence of total dry matter production on accumulated intercepted radiation for healthy and virus infected sugar beet (Scott and Jaggard, 1985). Data points correspond to harvests at approximately monthly intervals. The values for dry matter yield of energy are: Healthy, 1.28 g MJ^{-1}; Virus infected, 0.98 g MJ^{-1}

Little reference has been made to virus diseases. In terms of crop physiology, their effects are analogous to those of fungi and bacteria. However, since direct control is ineffective, the only defences against virus diseases of crops are those of plant resistance, attack on the vectors, and hygiene. As with fungal diseases, measurements of the effects of plant viruses on crop resource capture are difficult and rare. We are indebted to Keith Jaggard for evidence that sugar beet yellows virus alters the relation between dry matter production and light interception, as shown in *Figure 10.6* (Scott and Jaggard, 1985). Production of dry matter and light interception by diseased stands were significantly lower than those of healthy crops and ϵ was also smaller.

Conclusions

Our current understanding of the processes by which diseases affect crop yields is far from perfect, although the main mechanisms are clear. At the biochemical and cellular level, diseases may depress photosynthesis, enhance respiration and interfer with translocation. At the canopy level, it is clear that both fungal and viral infections, like environmental stresses, can depress yields both by decreasing the efficiency of light use (by about 20 to 30%) and by reducing canopy light interception. In consequence, measures which maintain a healthy canopy will maximize resource capture, as is true for other stresses.

Mathematical models have been developed to simulate and predict a number of plant diseases and potential epidemics, based on knowledge of the interacting factors that affect disease development. For example, weather data have been used in day-degree models to predict stripe rust (*Puccinia striiformis*) in winter wheat (Coakley and Line, 1981; Coakley, Line and Boyd, 1983) and potato early blight caused by *Alternaria solani* (Franc, Harrison and Lahman, 1988). The impacts of

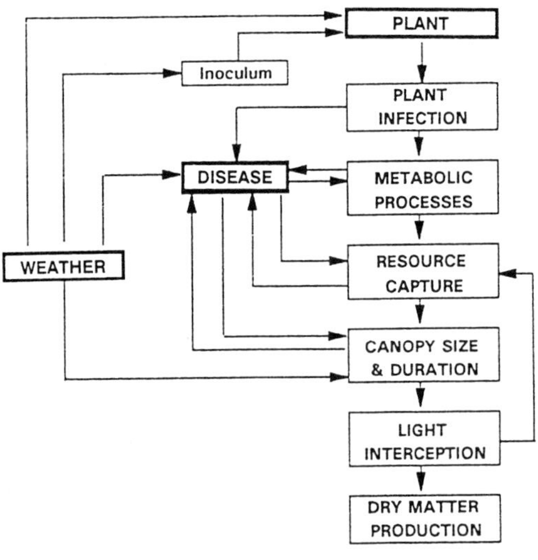

Figure 10.7 Suggested minimum elements for a model of the interaction between crop, disease and environment

plant disease on yield and quality have been related to critical stages in disease development in 'critical-point' or 'multiple-point' linear regression models. Non-linear models or integrated measures of disease progress, such as the area under the disease progress curve, time of onset or progress rate have also been used in modelling the effects of disease (James, 1974; Madden, 1983). However, Shaw and Royle (1987) and Rouse (1988) have suggested that such yield loss-models should take account of non-disease variables that constrain yield and interact with plant growth and disease development throughout the growing season. The principal aim of such models has been to exploit our understanding of crop diseases in order to identify the most effective period for chemical treatment and to minimize the number of applications for pathogen control. Such models cover few of the complex interactions which determine the effects of a particular pathogen on the yield of its host and might best be extended by coupling to existing crop yield models. Even the simplest model of the interacting effects of disease and weather on crop growth and yield should include the elements in *Figure 10.7.* Careful testing of models is required to ensure realistic predictions of crop growth and yield over a representative range of conditions. A major limitation to progress is the lack of published field data suitable for validation.

References

Ahmad, I., Farrar, J.F. and Whitbread, R. (1983) Photosynthesis and chloroplast functioning in leaves of barley infected with brown rust. *Physiological Plant Pathology*, **23**, 411–419

Ayres, P.G. (1972) Abnormal behaviour of stomata in barley leaves infected with *Rhynchosporium secalis* (Oudem.) J.J. Davis. *Journal of Experimental Botany*, **23**, 683–691

Ayres, P.G. (1976) Patterns of stomatal behaviour, transpiration and CO_2 exchange in pea following infection by powdery mildew (*Erysiphe pisi*). *Journal of Experimental Botany*, **27**, 1196–1205

Ayres, P.G. (1981a) Effects of disease on plant water relation. In *Effects of Disease on the Physiology of the Growing Plant*, pp 131–148. Edited by P.G. Ayres. Cambridge: Cambridge University Press

Ayres, P.G. (1981b) Responses of stomata to pathogenic organisms. In *Stomatal Physiology*, pp 205–221. Edited by P.G. Jarvis and T.A. Mansfield. Cambridge: Cambridge University Press

Ayres, P.G. (1984) The interaction between environmental stress injury and biotic disease physiology. *Annual Review of Phytopathology*, **22**, 53–75

Ayres, P.G. and Paul, N.D. (1986) Foliar pathogens alter the water relations of their hosts with consequences for both host and pathogen. In *Water, Fungi and Plants*, pp 267–285. Edited by P.G. Ayres and L. Boddy. Cambridge: Cambridge University Press

Ben-Kalio, V.D. and Clarke, D.D. (1979) Studies on tolerance in wild plants: effects of *Erysiphe fisheri* on the growth and development of *Senecio vulgaris*. *Physiological Plant Pathology*, **14**, 203–211

Biscoe, P.V. and Gallagher, J.N. (1977) Weather, dry matter production and yield. In *Environmental Effects on Crop Physiology*, pp 75–100. Edited by J.J. Landsberg and C.V. Cutting. London: Academic Press

Blakeman, R.H. (1990) The identification of crop disease and stress by aerial photograph. In *Applications of Remote Sensing in Agriculture*, pp 229–254. Edited by H.D. Steven and J.A. Clark. London: Butterworths

Buchanan, B.B., Hutcheson, S.W., Magyarosy, A.C. and Montalbini, P. (1981) Photosynthesis in healthy and diseased plants. In *Effects of Disease on the Physiology of the Growing Plant*, pp 13–28. Edited by P.G. Ayres. Cambridge: Cambridge University Press

Carver, T.L.W. and Griffiths, E. (1981) Relationship between powdery mildew infection, green leaf area and grain yield of barley. *Annals of Applied Biology*, **99**, 255–266

Coakley, S.M. and Line, R.F. (1981) Quantitative relationships between climatic variables and stripe rust epidemics of winter wheat. *Phytopathology*, **71**, 461–467

Coakley, S.M., Line, R.F. and Boyd, W.S. (1983) Regional models for predicting stripe rust on winter wheat in the Pacific Northwest. *Phytopathology*, **73**, 1382–1385

Coffey, M.D., Marshall, C. and Whitbread, R. (1970) The translocation of ^{14}C-labelled assimilates in tomato plants infected with *Alternaria solani* (Ell. and Mart.) Jones and Grout. *Annals of Botany*, **34**, 605–615

Colhoun, J. (1973) Effects of environmental factors in plant disease. *Annual Review of Phytopathology*, **11**, 343–364

Cramer, H.H. (1967) *Plant Protection and World Crop Production*. Leverkusen: Bayer Pflanzenschutz

Cunfer, B.M., Toughton, J.T. and Johnson, J.W. (1980) Effects of phosphorus and potassium fertilization on *Septoria* glume blotch of wheat. *Phytopathology*, **70**, 1196–1199

Daly, J.M. (1976) The carbon balance of diseased plants: changes in respiration, photosynthesis and translocation. In *Encyclopedia of Plant Physiology*, New Series, vol. 4, *Physiological Plant Pathology*, pp 450–476. Edited by R. Heitefuss and P.H. Williams. Berlin: Springer-Verlag

Deadman, M.L. and Cooke, B.M. (1987) Effects of net blotch on growth and yield of spring barley. *Annals of Applied Biology*, **110**, 33–42

Dickinson, C.H. (1973) Interactions of fungicides and leaf saprophytes. *Pesticide Science*, **4**, 563–574

Dimond, A.E. (1970) Biophysics and biochemistry of the vascular wilt syndrome. *Annual Review of Phytopathology*, **8**, 301–322

Duniway, J.M. and Durbin, R.D. (1971) Some effects of *Uromyces phaseoli* on the transpiration rate and stomatal response of bean leaves. *Phytopathology*, **61**, 114–119

Duniway, J.M. (1982) Role of biometeorology in integrated pest management: soil-plant-water relations and disease. In *Biometeorology in Integrated Pest Management*, pp 307–326. Edited by J.L. Hatfield and I.J. Thomason. London: Academic Press Inc.

Duniway, J.M. and Slatyer, R.O. (1971) Gas exchange studies on the transpiration and photosynthesis of tomato leaves affected by *Fusarium oxysporum* f.sp. *lycopersici*. *Phytopathology*, **61**, 1137–1381

Eastwood, J.A. and Steven, M.D. (1991) The effect of the distribution of chlorotic material on the reflectance of mixed canopies. *Proceedings of the 5th International Colloquium – Physical Measurements and Signatures in Remote Sensing*, Courchevel, France (14–18 Jan., 1991). Paris: ESA

Engelhard, A.W. (ed.) (1989) *Soil Borne Plant Pathogens: Management of Diseases with Macro- and Microelements*. St Paul: The American Phytopathological Society, 217 pp.

FAO (1992) *Quarterly Bulletin of Statistics*, **5**

Farrar, J.F. and Lewis, D.H. (1987) Nutrient relations in biotrophic infections. In *Fungal Infection of Plants*, pp 92–132. Edited by G.F. Pegg and P.G. Ayres. Cambridge: Cambridge University Press

Farrell, G.M., Preece, T.F. and Wren, M.J. (1969) Effects of infection by *Phytophthora infestans* (Mont.) de Bary on stomata of potato leaves. *Annals of Applied Biology*, **63**, 265–275

Ferrandino, F.J. (1989) Spatial and temporal variation of a defoliating plant disease and reduction in yield. *Agricultural and Forest Meteorology*, **47**, 273–289

Fitt, B.D.L. and Hornby, D. (1978) Effects of root-infecting fungi on wheat transport processes and growth. *Physiological Plant Pathology*, **13**, 335–346

Fitt, J.M. and McCartney, H.A. (1986) Spore dispersal in splash droplets. In *Water, Fungi and Plants*, pp 87–104. Edited by P.G. Ayres and L. Boddy. Cambridge: Cambridge University Press

Franc, G.D., Harrison, M.D. and Lahman, L.K. (1988) A simple day-degree model for initiating chemical control of potato early blight in Colorado. *Plant Disease*, **72**, 851–854

Friend, J. (1981) Alteration in secondary metabolism. In *Effects of Disease on the Physiology of the Growing Plant*, pp 179–200. Edited by P.G. Ayres. Cambridge: Cambridge University Press

Gallagher, J.N. and Biscoe, P.V. (1978) Radiation absorption, growth and yield in cereals. *Journal of Agricultural Science*, **91**, 47–60

Gordon, T.R. and Duniway, J.M. (1982a) Stomatal behaviour and water relations in sugar beet leaves infected by *Erysiphe polygoni*. *Phytopathology*, **72**, 723–726

Gordon, T.R. and Duniway, J.M. (1982b) Photosynthesis in powdery mildewed sugar beet leaves. *Phytopathology*, **72**, 718–723

Gordon, T.R. and Duniway, J.M. (1982c) Effects of powdery mildew infection on the efficiency of CO_2 fixation and light utilization by sugar beet leaves. *Plant Physiology*, **69**, 139–142

Green, C.F., Hebblethwaite, P.D. and Ison, D.A. (1985) A quantitative analysis of varietal and moisture status effects on the growth of *Vicia faba* in relation to radiation absorbtion. *Annals of Applied Biology*, **106**, 143–155

Griffiths, E. and Amin, S.M. (1977) Effects of *Botrytis fabae* infection and mechanical defoliation on seed yield of field beans (*Vicia faba*). *Annals of Applied Biology*, **86**, 359–367

Habeshaw, D. (1984) Effects of pathogens on photosynthesis. In *Plant Diseases: Infection, Damage and Loss*, pp 63–72. Edited by R.K.S. Wood and G.J. Jellis. Oxford: Blackwell Scientific Publications

Harrison, J.A.C. and Isaac, I. (1968) Leaf-area development in King Edward potato plants inoculated with *Verticillium alboatrum* and *V. dahliae*. *Annals of Applied Biology*, **61**, 217–230

Haverkort, A.J. and Bicamumpaka, M. (1986) Correlation between intercepted radiation and yield of potato crops infected by *Phytophthora infestans* in Central Africa. *Netherlands Journal of Plant Pathology*, **92**, 239–247

Hawkins, A.F. (1982) Light interception, photosynthesis and crop productivity. *Outlook on Agriculture*, **11**, 104–113

Hislop, E.C. (1976) Some effects of fungicides and other agrochemicals on the microbiology of the aerial surfaces of plants. In *Microbiology of Aerial Plant Surfaces*, pp 41–74. Edited by C.H. Dickinson and T.F. Preece. London : Academic Press
Hornby, D. and Fitt, B.D.L. (1981) Effects of root-infecting fungi on structure and function of cereal roots. In *Effects of Disease on the Physiology of the Growing Plant*, pp 101–130. Edited P.G. Ayres. Cambridge: Cambridge University Press
Huber, D.M. (1989) Introduction. In *Soil Borne Plant Pathogens: Management of Diseases with Macro- and Microelements*, pp 1–8. Edited by A.W. Engelhard. St Paul: The American Phytopathological Society
Huber, D.M. and Watson, R.D. (1974) Nitrogen form and plant disease. *Annual Review of Phytopathology*, **12**, 139–165
Hughes, G. (1988) Models of crop growth. *Nature*, **332**, 16
Hughes, G. (1990) Characterizing crop responses to patchy pathogen attack. *Plant Pathology*, **39**, 2–4
Hutcheson, S.W. and Buchanan, B.B. (1983) Bioenergetic and metabolic disturbances in diseased plants. In *Biochemical Plant Pathology*, pp 327–345. Edited by J.A. Callow. Chichester: John Wiley & Sons, Ltd.
Ishag, H.M. (1973) Physiology of seed yield in field bean. I. Yield and yield components. *Journal of Agricultural Science, Cambridge*, **80**, 181–189
Kosuge, T. and Kimpel, J.A. (1981) Energy use and metabolic regulation in plant-pathogen interactions. In *Effects of Disease on the Physiology of the Growing Plant*, pp 29–45. Edited by P.G. Ayres. Cambridge: Cambridge University Press
James, W.C. (1974) Assessment of plant diseases and losses. *Annual Review of Phytopathology*, **12**, 27–48
Jenkyn, J.F. (1976) Effects of mildew (*Erysiphe graminis*) on green leaf area of Zephyr spring barley, 1973. *Annals of Applied Biology*, **82**, 485–488
Jenkins, J.E.E., Melville, S.C. and Jemmett, J.L. (1972) The effects of fungicides on leaf diseases and on yield in spring barley in South-West England. *Plant Pathology*, **21**, 49–58
Jordan, W.R., Clark, R.B. and Seetharama, N. (1984) The role of edaphic factors in disease development. In *Sorghum Root and Stalk Rots: A Critical Review*, pp 81–96. Proceedings Consultive Group Discussion on Research Needs and Strategies for Control of Sorghum Root and Stalk Rot Diseases, Bellagio, Italy. Hyderabad: ICRISAT
Leach, C.M. and Anderson, A.J. (1982) Radiation quality and plant diseases. In *Biometeorology in Integrated Pest Management*, pp 267–306. Edited by J.L. Hatfield and I.J. Thomason. London: Academic Press
Livne, A. (1964) Photosynthesis in healthy and rust-affected plants. *Plant Physiology*, **39**, 614–621
Lorenz, E.J. and Cothren, J.T. (1989) Photosynthesis and yield of wheat (*Triticum aestivum*) treated with fungicides in a disease-free environment. *Plant Disease*, **73**, 25–27
Madeira, A.C.A. (1988) *Growth, Light Interception and Disease in Field Bean*. Ph.D. Thesis, University of Nottingham.
Madeira, A.C., Clark, J.A. and Rossall, S. (1988) Growth, light interception and disease in field bean (*Vicia faba*): the effect of late infection by *Ascochyta fabae*. *Annals of Applied Biology*, **112**, 585–595
Madden, L.V. (1983) Measuring and modelling crop losses at the field level. *Phytopathology*, **73**, 1591–1596

Malthus, T.J. and Madeira, A.C. (1992) Spectral reflectance of field bean leaves infected by *Botrytis fabae*. *Journal of Experimental Botany*, **43** (Supplement), 46

Malthus, T.J. and Madeira, A.C. (1993) High resolution spectroradiometer: spectral reflectance of bean leaves infected by *Botrytis fabae*. *Remote Sensing of Environment*, **45**, 107–116

McGrath, M.T. and Pennypacker, S.P. (1990) Alteration of physiological processes in wheat flag leaves caused by stem rust and leaf rust. *Phytopathology*, **80**, 677–686

Mignucci, J.S. and Boyer, J.S. (1979) Inhibition of photosynthesis and transpiration in soybean infected by *Microsphaera diffusa*. *Phytopathology*, **69**, 227–230

Monteith, J.L. (1977) Climate and the efficiency of crop production in Britain. *Philosophical Transactions of the Royal Society of London B* **281**, 277–294

Nelson, P.E. and Dickey, R.S. (1970) Histopathology of plants infected with vascular bacterial pathogens. *Annual Review of Phytopathology*, **8**, 259–280

Owera, S.P.A., Farrar, J.F. and Whitbread, R. (1981) Growth and photosynthesis in barley infected with brown rust. *Physiological Plant Pathology*, **18**, 79–90

Paul, N.D. and Ayres, P.G. (1984) Effects of rust and post-infection drought on photosynthesis, growth and water relations in groundsel. *Plant Pathology*, **33**, 561–569

Peters, D. (1988) Effects of virus yellows on the plant. In *Virus Yellows Monograph*, pp 19–26. I.I.R.B. Pests and Disease Study Group.

Rouse, D.I. (1988) Use of crop growth-models to predict the effects of disease. *Annual Review of Phytopathology*, **26**, 183–201

Rowe, J. and Reid, J. (1979) Some aspects of carbon relations in the barley-*Helminthosporium teres* complex. II. The effects of infection upon net accumulation of carbon in the tissues. *Canadian Journal of Botany*, **57**, 208–214

Royle, D.J. and Butler, D.R. (1986) Epidemiological significance of liquid water in crop canopies and its role in disease forecasting. In *Water, Fungi and Plants*, pp 139–156. Edited by P.G. Ayres and L. Boddy. Cambridge: Cambridge University Press

Russell, G., Jarvis, P.G. and Monteith, J.L. (1988) Absorption of radiation by canopies and stand growth. In *Plant Canopies: Their Growth, Form and Function*, pp 21–40. Edited by G. Russell, P.G. Jarvis and M. Marshall. Cambridge: Cambridge University Press

Schneider, R.W. and Pendery, W.E. (1983) Stalk rot of corn: mechanism of predisposition by an early season water stress. *Phytopathology*, **73**, 863–871

Schoeneweiss, D.F. (1986) Water-stress predisposition to disease – an overview. In *Water, Fungi and Plants*, pp 157–174. Edited by P.G. Ayres and L. Boddy. Cambridge: Cambridge University Press

Scholes, J.D. and Farrar, J.F. (1985) Photosynthesis and chloroplast functioning within individual pustules of *Uromyces muscari* on bluebell leaves. *Physiological Plant Pathology*, **27**, 387–400

Scott, R.K. and Jaggard, K.W. (1985) The effect of pests and diseases on growth and yield of sugar beet. In *Proceedings of the 48th Winter Congress of the IIRB*, pp 153–169

Shane, W.W., Baumer, J.S. & Teng, P.S. (1987) Crop losses caused by *Xanthomonas* streak on spring wheat and barley. *Plant Disease*, **71**, 927–930

Shaw, M.W. and Royle, D.J. (1987) A new method for disease-loss studies used to estimate effects of *Pyrenophora teres* and *Rhynchosporium secalis* on winter barley yields. *Annals of Applied Biology*, **110**, 247–262

Simkin, M.B. and Wheeler, B.E.J. (1974) Effects of dual infections of *Puccinia hordei* and *Erysiphe graminis* on barley, cv. Zephyr. *Annals of Applied Biology*, **78**, 237–250

Smedegaard-Petersen, V. (1984) The role of respiration and energy generation in diseased and disease-resistant plants. In *Plant Diseases: Infection, Damage and Loss*, pp 73–85. Edited by R.K.S. Wood and G.J. Jellis. Oxford: Blackwell Scientific Publications

Smedegaard-Petersen, V. and Tolstrup, K. (1985) The limiting effect of disease resistance on yield. *Annual Review of Phytopathology*, **23**, 475–490

Steven, M.D., Malthus, T.J.; Demetriades-Shah, T.H.; Danson, F.M. and Clark, J.A. (1990) High spectral resolution indices for crop stress. In *Applications of Remote Sensing in Agriculture*, pp 209–227. Edited by M.D. Steven and J.A. Clark. London: Butterworths

Thomas, H and Stoddart, J.L. (1980) Leaf senescence. *Annual Review of Plant Physiology*, **31**, 83–111

Waggoner, P.E. (1965) Microclimate and plant disease. *Annual Review of Phytopathology*, **3**, 103–126

Waggoner, P.E. and Berger, R.D. (1987) Defoliation, disease and growth. Phytopathology, **77**, 393–398

Whipps, J.J. and Lewis. D.D. (1981) Patterns of translocation, storage and interconversion of carbohydrates. In *Effects of Disease on the Physiology of the Growing Plant*, pp 47–83. Edited by P.G. Ayres. Cambridge: Cambridge University Press

Williams, G.M. and Ayres, P.G. (1981) Effects of powdery mildew and water stress on CO_2 exchange in uninfected leaves of barley. *Plant Physiology*, **68**, 527–530

Williams, P.F. (1975) Growth of broad beans infected by *Botrytis fabae*. *Journal of Horticultural Science*, **50**, 415–424

11

AIR POLLUTION – WHAT'S IN IT FOR CROPS?

M. H. UNSWORTH
Center for Analysis of Environmental Change, Weniger Hall 283, Oregon State University, Corvallis, Oregon 97331–6511, USA

J. J. COLLS
Department of Physiology and Environmental Science, University of Nottingham, Sutton Bonington Campus, Loughborough, Leics, LE12 5RD, UK

Introduction

Pollute: *to defile, to befoul, to contaminate*: by its dictionary definition, air pollution implies an undesirable situation, damaging to living and non-living systems. The early history of research on air pollution and crops supports this interpretation, but more recent research has revealed that human influences on air quality may not always be detrimental to agriculture. Hence the choice of title for this paper, which is intended to review benefits as well as disbenefits of air pollution.

The earliest studies of air pollution and crops focused on either phytotoxicity observed around major pollution sources such as smelters, power plants and pulp mills, or animal toxicity arising, for example, from fluorine or heavy metals deposited on grazing land. When foliage is destroyed, as it is by high concentrations of many gaseous pollutants, the implications for resource capture by crops are rather obvious. But more recently, with the application of Clean Air Acts in many countries, and with improved pollution control by industry, cases of gross damage to crops are rare, and attention has turned to the much more subtle effects of exposure to low concentrations of pollutants. In such cases, especially with sulphur- or nitrogen-containing pollutants which contain essential elements for crop growth, it is clear that air pollution is not always damaging. We will review some recent work on air pollution and crops from which new insights have developed into how pollution influences the capture of resources such as light, water, carbon and other essential elements, and we will give examples of growth stimulation as well as growth inhibition. We confine our attention to effects of sulphur dioxide, nitrogen oxides, ammonia, and tropospheric ozone, and use a number of case studies to indicate where there are strengths and weaknesses in our current understanding of effects on resource capture. For broader reviews of effects of air pollution on crops, readers are referred to Heck, Taylor and Tingey (1988), and Unsworth and Ormrod (1982).

Radiation capture

Analysis of crop growth in terms of interception of radiation and dry matter production is a strong theme in many papers in this proceedings. In spite of the power of this form of analysis (e.g. Monteith 1977 and Chapter 1 of this volume), it has seldom been applied to analysis of air pollution studies. In the past, this has been partly because of the operational difficulty of making destructive harvests from field exposure chambers occupying only a small area, but recent advances in non-destructive methods of estimating plant biomass should allow the method to be applied more often in future. The following case study summarizes an investigation of radiation interception and the growth of field-grown soybeans exposed to ozone (Unsworth, Lesser and Heagle, 1984), and draws out some general points concerning the influence of air pollution on radiation capture.

The aim of the study was to investigate the relationship between intercepted radiation and growth, and to define the influence of O_3 on the relationship. In particular, early season growth was analysed in terms of:

$$M = \epsilon_m \int_{t_1}^{t_2} f S_o \, dt \qquad (1)$$

where M is the increase in dry matter (g m^{-2}) in the time interval t_2-t_1, S_o is the incident radiation (W m^{-2}), f is the fraction of radiation intercepted by the canopy and ϵ_m is the dry matter – radiation conversion coefficient (g J^{-1}). The analysis sought to quantify how O_3 stress influenced ϵ_m and f.

Soybeans (*Glycine max* (L.) Merr. 'Davis') were grown in the field in North Carolina in open-top chambers (Heagle, Body and Heck, 1973) which were ventilated with either charcoal-filtered air, or non-filtered air to which about 20, 40, or 60 nl l^{-1} ozone (O_3) was added for 7 h day^{-1} to simulate photochemical air pollution. (More detail about this approach is given in a later section.) Intercepted radiation was measured throughout the season in one chamber of each treatment and 14 harvests were made throughout the season to determine leaf area, above-ground biomass and its components.

Figure 11.1a shows the seasonal changes in f in the lowest and highest O_3 treatments (seasonal mean O_3 concentrations were 19.8 and 104 nl l^{-1}, respectively). Values of f for the intermediate O_3 concentrations fitted consistently within the envelope in increasing order of O_3 concentration. There are three phases apparent: rapid canopy expansion up to about day 45, when O_3 did not appear to affect interception; a maximum radiation interception phase, in which O_3 influenced the duration of the phase and the maximum interception f_m; and a senescence phase, in which the rate of decline of f appeared to be relatively independent of O_3 concentration. More detailed analysis revealed that f_m declined linearly with increasing O_3 concentration C,

$$f_m = 0.915 - 6.03 \times 10^{-4} C$$
where C is expressed in nl l^{-1}. (2)

The duration of the maximum interception phase decreased from about 72 d in the lowest O_3 treatment to about 55 d in the highest. *Figure. 11.1b* shows in more detail the variation of f with time towards the end of the season for all five chamber treatments and for open (ambient air) plots. The onset of the senescence phase is difficult to define precisely, but its timing appears to be increasingly early for the three highest treatments (seasonal mean O_3 concentrations of 61.4, 76.9 and 104.0 nl l^{-1} respectively).

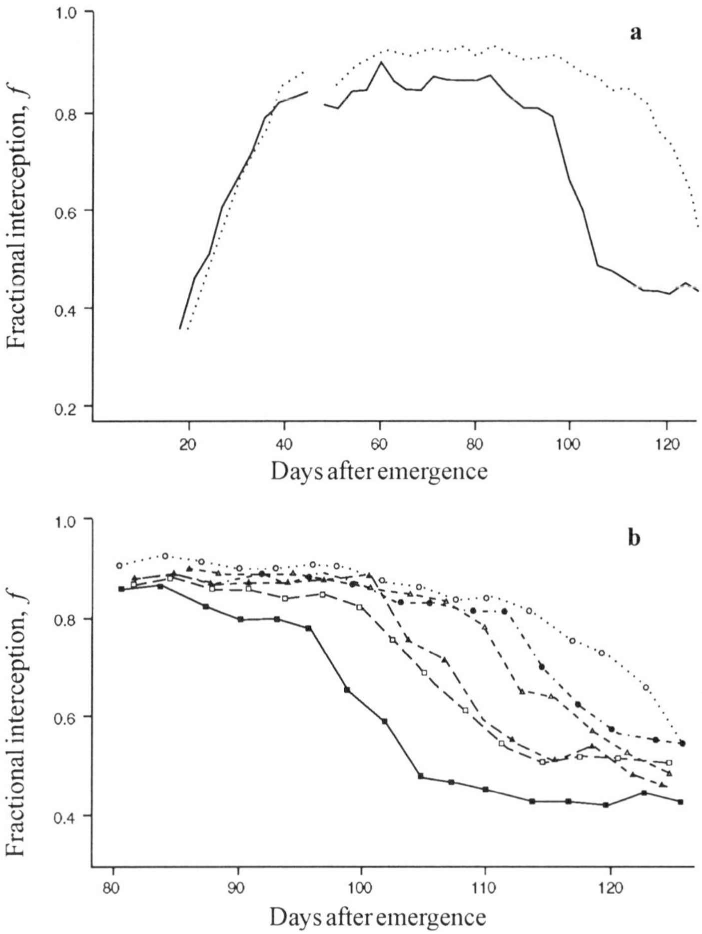

Figure 11.1 (**a**) The fraction of total solar radiation intercepted by canopies of soybeans in open-top field chambers ventilated with charcoal filtered air (····) or with non-filtered air to which 60 nl l-1 O_3 was added (———). (**b**) Decline in *f* after day 80 in soybeans exposed to six O_3 treatments in open-top chambers and in the field. Seasonal mean O_3 concentrations were 20 (○····○), 43 (●–·–●), 61 (▲ – – – ▲), 77 (□———□) and 104 (■———■) in the chambers, and 48 (△ – – △) in the field. (from Unsworth *et al.* 1984)

Thus seasonal radiation interception of this crop was influenced by O_3 in two ways: maximum interception was decreased, and canopy duration was decreased. As a consequence, the seasonal mean fractional radiation interception decreased by about 5% (0.684 to 0.649) between the lowest and highest O_3 treatments.

During early growth (up to 63 days after emergence (DAE)), the relation between intercepted radiation and dry matter production was linear, as reported by many others. *Figure 11.2* shows the variation of ϵ_m with ozone concentration (see Equation 1), and indicates that the coefficient was significantly decreased

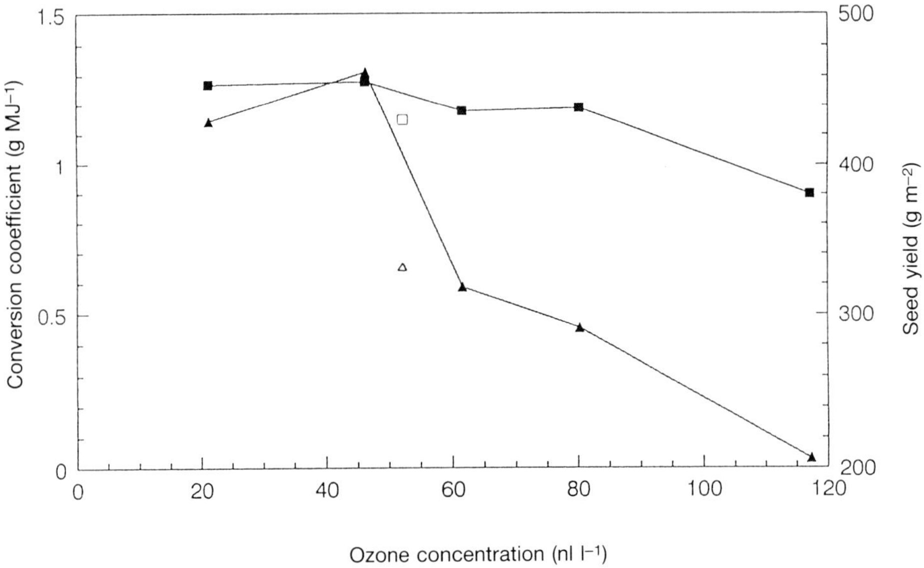

Figure 11.2 Variation with O_3 concentration of the dry matter-radiation conversion coefficient ϵ_m (■) and the bean yield (▲) for soybeans grown in open-top chambers. The symbols □ and △ show the corresponding values from open plots. (redrawn from Unsworth *et al.* 1984)

only in the highest treatment, although there was a trend of reduction in the intermediate values.

So, was the influence of O_3 on radiation interception an important factor in reducing the final yield of the soybean crop? The simple answer is – no. It is clear from the final yield values in *Figure 11.2* that the small losses in radiation interception could not account for the large losses in yield which were observed in all but the lowest two O_3 treatments. Detailed analysis of the components of yield indicated that ozone also decreased the pod number per plant, and reduced the duration of seed-filling, and that changes in these two features accounted for most of the yield loss. Increased pod abscission is a well-known response when there is competition with other organs for limited supplies of assimilate, so it is possible that the reduction in pod number was a secondary response to ozone, with the primary responses being reductions in carbon assimilation and/or diversion of assimilate to leaf repair or adaptation.

Although the influence of ozone on radiation interception did not appear to be of major importance in this study, this response may be more important for perennial crops, forests and natural systems where early senescence of foliage could either disturb the production of metabolites important for over-wintering, or, if occurring over a series of years, could reduce the standing biomass of sensitive plants in ways that would decrease competitive ability. We have recently shown in open-top chamber studies with field-grown apple and ash trees, that early leaf abscission is strongly related to ozone exposure, confirming a wide body of evidence from pot experiments and uncontrolled observations in the field (Wiltshire, Wright and Unsworth, 1992).

Carbon capture

CARBON ASSIMILATION

It has been known for many years that exposure of crop plants to sulphur dioxide (SO_2) and O_3 reduces net photosynthesis P_n. In many cases this response is also associated with stomatal closure, but it is only relatively recently that experiments have been completed to investigate whether P_n was reduced because of stomatal closure or because of damage to the photosynthetic process.

Kropff (1987) used the analysis of Jones (1985) to analyse relationships between assimilation rates A and internal CO_2 concentration c_i in leaves of *Vicia faba* exposed to SO_2. He concluded that his observations of rapid decreases in P_n were accounted for by an increase in mesophyll resistance, probably because sulphur metabolites were competing with CO_2 for the enzyme ribulosbisphosphate (RubP) carboxylase/oxygenase. He observed a slow increase in stomatal resistance, and interpreted this as a secondary response as a result of feed-back between P_n and c_i.

Farage *et al.* (1991) studied the photosynthetic systems of wheat leaves after short-term exposure to O_3, analysing light-response and A/c_i curves after short (4–16 h) exposures to rather high concentrations, 200 and 400 nl l^{-1} O_3. At both concentrations, the rate of light-saturated photosynthesis A_{sat} decreased in close correlation with decreasing stomatal conductance g_s, but, at the lower

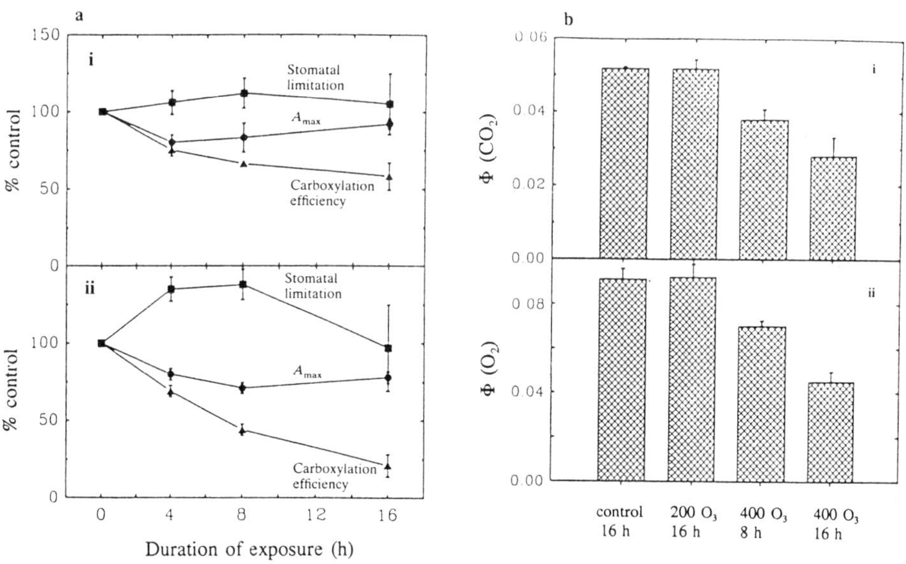

Figure 11.3 (**a**) Changes (relative to controls in clean air) in the CO_2-saturated assimilation rate A_{max}, the carboxylation efficiency, and stomatal limitation of wheat leaves exposed to (**i**) 200 nl l^{-1} O_3 or (**ii**) 400 nl l^{-1} O_3. Stomatal limitation was calculated from the method of Farquhar and Sharkey (1982).
(**b**) Quantum yields Φ for CO_2 fixation (**i**) and O_2 evolution (**ii**) in wheat leaves exposed to 200 or 400 nl l^{-1} O_3 compared with controls in clean air. (from Farage *et al.* 1991)

concentration, A_{sat} and g_s stabilized after 8 h, suggesting that leaves had the capacity to detoxify the O_3 absorbed at this concentration. Since the ratio of c_i to external CO_2 concentration c_a did not decrease in these experiments, Farage *et al.* concluded that the decrease in g_s was a result of the loss of photosynthetic capacity, rather than the reverse. *Figure 11.3* shows Farage *et al.*'s results from analysis of A/c_i curves. Ozone significantly decreased the initial slope of the A/c_i curve, which is a measure of primary carboxylase efficiency. On the other hand, the effect of O_3 on the CO_2-saturated rate of photosynthesis, A_{max}, which is related to the maximum rate of RubP regeneration and hence to electron transport, was much less. Analysis of light-limited photosynthesis showed that quantum yields Φ for CO_2 fixation and O_2 evolution were not affected by 200 nl l^{-1}; Φ for both cases declined with increasing duration of exposure (*Figure 11.3b*). The cause of these declines was not clear, but did not seem to be associated with patchiness of stomatal closure. This elegant piece of work should encourage further research by physiologists, especially to clarify the mechanisms of action of O_3 on carbon assimilation in the light-limited conditions that are common in field crops.

DRY MATTER PRODUCTION AND CROP YIELD

Turning to the other end of the carbon-capture story, there has been a great deal of work investigating effects of air pollution exposure on growth, development and yield of crops. For brevity, we will summarize here only a few examples that have provided insight into controlling factors.

In the mid 1970s, a series of studies began in England to attempt to explain why high-yielding ryegrass cultivars performed disappointingly when planted in areas of Northern Britain where there were relatively high concentrations of SO_2. Work by Bell, Roberts and their colleagues (reviewed in Bell, 1985 and Roberts, 1984) led eventually to sufficient reliable data to construct dose response relationships (*Figure 11.4a*). Although it is clear that ryegrass yield is significantly depressed at SO_2 concentrations above 100 nl l^{-1}, there are both positive and negative yield responses at lower concentrations. It seems likely that some of the scatter arises from interactions with temperature, which influences both growth rate and the ability of cells to detoxify the absorbed sulphur metabolites. Other reasons for the scatter observed in *Figure 11.4a* may be the role of sulphur as a plant nutrient, or in controlling leaf pathogens. *Figure 11.4b* illustrates the large seasonal variation in the response of dry matter production of grasses to SO_2, and includes results from an exposure to a mixture of SO_2 and NO_2, which often occur at similar concentrations in polluted atmospheres. There is considerable evidence to show that responses of many species to $SO_2 + NO_2$ are larger than to SO_2 alone (Mansfield and McCune, 1988).

Two major co-ordinated programmes in America and Europe have investigated the responses of agricultural crops to air pollutants. Both programmes have focused on the photochemical pollutant, ozone, associated mainly with vehicle emissions, although nitrogen and sulphur oxides have also been studied. The National Crop Loss Assessment Network (NCLAN) operated in the USA during a core period from 1980 to 1985. At five regional centres in the principal agricultural areas, experiments on the economically-important crop species were carried out according to agreed protocols, and the results were analysed by a common statistical procedure. Open-top chambers (OTCs) were used to achieve near-field

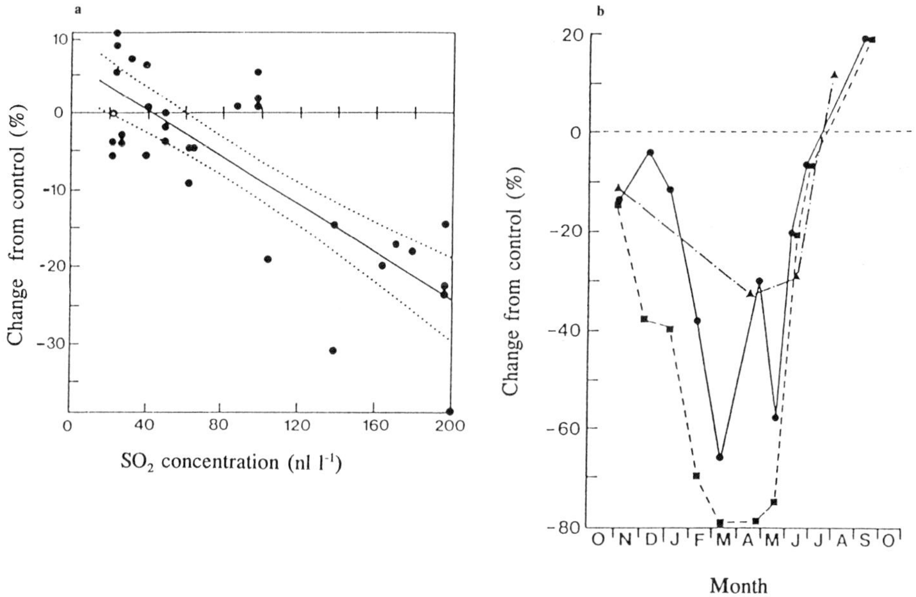

Figure 11.4 (**a**) Changes in yield (relative to controls in clean air) of *Lolium perenne* exposed to SO_2 in chambers for more than 20 days. The linear regression is statistically significant at $P < 0.001$ and the dotted lines show the 95% confidence limits of the regression.
(**b**) Changes in yield (relative to controls in clean air) of grasses grown in various months in chambers containing SO_2 or $SO_2 + NO_2$. Key: ▲ *Lolium perenne*, 50 nl l^{-1} SO_2; ● *Poa pratensis*, 68 nl l^{-1} SO_2; ■ *Poa pratensis*, 68 nl l^{-1} SO_2 + 68 nl l^{-1} NO_2. (from Roberts 1984)

growing conditions for the crops while retaining control over the air quality. Ten crops, which together make up about 85% of US acreage, were studied: corn, soybean, wheat, hay, tobacco, sorghum, cotton, barley, groundnut and bean. Results from the NCLAN programme were reviewed in the volume edited by Heck, Taylor and Tingey (1988).

The NCLAN results are a good example of the range of responses which must be reconciled if they are to be used to predict, for example, national economic yield losses due to air pollutants. In many cases, individual experiments provided convincingly smooth changes in yield with increasing ozone concentration. However, repeating the experiments from one season to another, or allowing natural variations in a variable such as soil water status, sometimes led to markedly different exposure–response functions.

In 1985, the Commission of the European Communities (CEC) established a network of sites in Europe to address the question of the effects of air pollutants on European agricultural crops. Eleven experimental sites in eight countries were involved. A common experimental protocol was implemented, although the degree of uniformity of the NCLAN design was neither attempted nor attained. Two sets of experiments were conducted, one on the responses to existing ambient air pollutants, and another on exposure–response functions over a range of concentrations of a specific pollutant (usually O_3) or mixture of pollutants.

The accepted methodology for the first of these sets was to grow the crop of interest in OTCs, some of which were ventilated with air that had been passed through a filter to remove pollutants (CF), and others ventilated with unfiltered 'ambient' air (NF). Various measurements of plant development, growth, and yield were then compared between the treatments. Filters containing activated charcoal made from coconut shells are efficient at removing ozone, less so at removing sulphur dioxide or nitrogen dioxide, and ineffective on nitrogen monoxide. The latter can, however, be removed by first passing the air through 'Purafil', which oxidizes NO to NO_2, so that it can be filtered by the charcoal. In this way, typically 60–70% of the ambient air pollutants were removed from filtered chambers, as an average over the growing season.

The second set of experiments was directed mainly at the responses to ozone of four crops: spring wheat, dwarf green bean, oats and barley. The crops were grown in OTCs as described for the filtered air experiments described above. Ozone was added to the air flowing into the chambers; since ambient ozone has a strong diurnal cycle with a peak in the early afternoon, ozone additions were

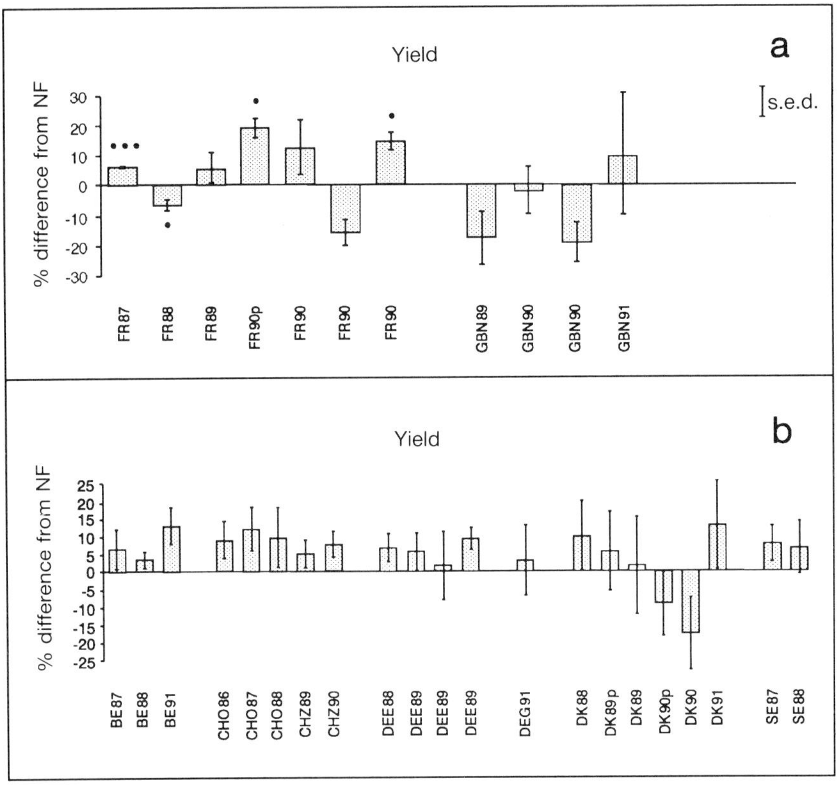

Figure 11.5 Percentage difference between crop yields in non-filtered air (NF) and charcoal-filtered air in studies in the EOTC network from 1987–1991. (**a**) *Phaseolus vulgaris* (green bean) (**b**) *Triticum aestivum* (spring wheat). Key to sites: Fr – France; GBN – U.K.; BE – Belgium; CHO, CHZ – Switzerland; DEE, DEG – Germany; DK – Denmark; SE – Sweden
* $P < 0.05$; *** $P < 0.001$. (from Unsworth and Geissler, in press)

normally made in synchrony with this cycle, and expressed as the average over the 7-h period of addition. The number of concentration steps varied from one to five between different experiments. As with the NCLAN, a wide range of results was found.

Figure 11.5 (from Unsworth and Geissler, in press) shows examples of CF/NF comparisons of yields for beans (*Phaseolus vulgaris*) and wheat (*Triticum aestivium*) at several sites in the European (EOTC) network. For beans, there was no consistent evidence that removing pollutants altered yield significantly. Wheat was studied at a wider range of sites, and *Figure 11.5* shows that, of the 21 studies, CF yields were larger than NF on 19 occasions. Although individual studies did not reveal statistically significant differences between CF and NF yields, analysis of the pooled data across years, sites and cultivars demonstrated a statistically significant benefit of air filtration (Lesser and Geissler, in press). It is not possible, of course, from this design to determine which pollutant or mix of pollutants caused the yield depression.

Figure 11.6 is an example of exposure–response relationships for wheat from the EOTC network. When expressed as proportional yield (relative to yield at 20 nl l^{-1} O_3), there was excellent agreement between results obtained in the field in Sweden and at two sites in Switzerland (Skärby *et al.*, in press).

Skärby *et al.* also found that plants grown in pots in OTCs were less sensitive to O_3 than plants in field soil. This may be an indication of different water relations, as discussed later. On the other hand, there was good agreement between pot and field studies of responses of bean yields to O_3; identical cultivars grown in Pau (SW France) and Sutton Bonington (Central England) agreed well across sites, seasons and cultivation media (*Figure 11.7*, from Colls *et al.*, in press) and this is strong confirmation that the O_3 exposure-response relation is robust, in spite of the very different climates of these two sites.

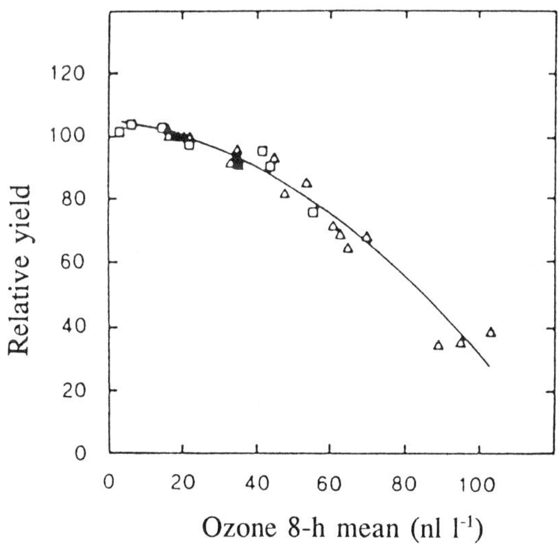

Figure 11.6 Proportional yield of spring wheat (*Triticum aestivum*) in response to seasonal mean O_3 concentration in Sweden (□, 1987, 1988) and Switzerland (△, 1986, 1987, 1988, 1989 and 1990). Yield at 20 nl l^{-1} O_3 was used as a reference. (from Skärby *et al.*, in press)

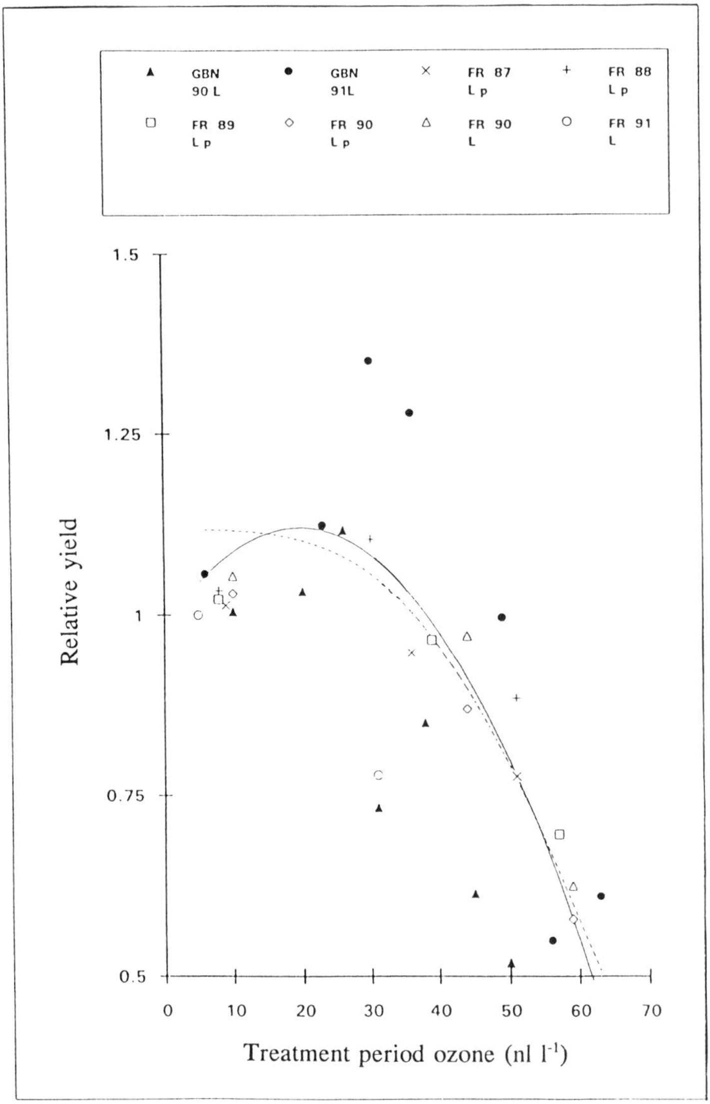

Figure 11.7 Yield of *Phaseolus vulgaris* cv. Lit at Nottingham, England and Pau, France, relative to the yield at 5 nl l^{-1} O_3. L = Lit grown directly in soil, Lp = Lit grown in pots. Country codes as for Figure 11.5. ——— and - - - - - are least squares regression fits of quadratic and Weibull models, respectively (from Colls *et al.*, in press)

Analysis of EOTC data made it clear that, in searching for consistent patterns of response across years and sites, various approaches to presenting the data must be tried. For example, when quadratic equations were fitted to all of the absolute yields of beans determined at Sutton Bonington, the responses of *Figure 11.8a* resulted. However, when these data were all expressed as a proportion of the extrapolated yield at an ozone concentration of 5 nl l^{-1}, a similar approach to *Figure 11.6*, it became clear that a common pattern existed (*Figure 11.8b*). As an indication that responses to air pollution stress are not always clear-cut, *Figure 11.8a* also shows that exposure of the bean cultivar Nerina to low concentrations of O_3 gave a stimulation in yield that was due to an increase in the number of pods per plant. At higher concentrations the yield fell, due to reductions in both number of seeds per pod and individual seed weight.

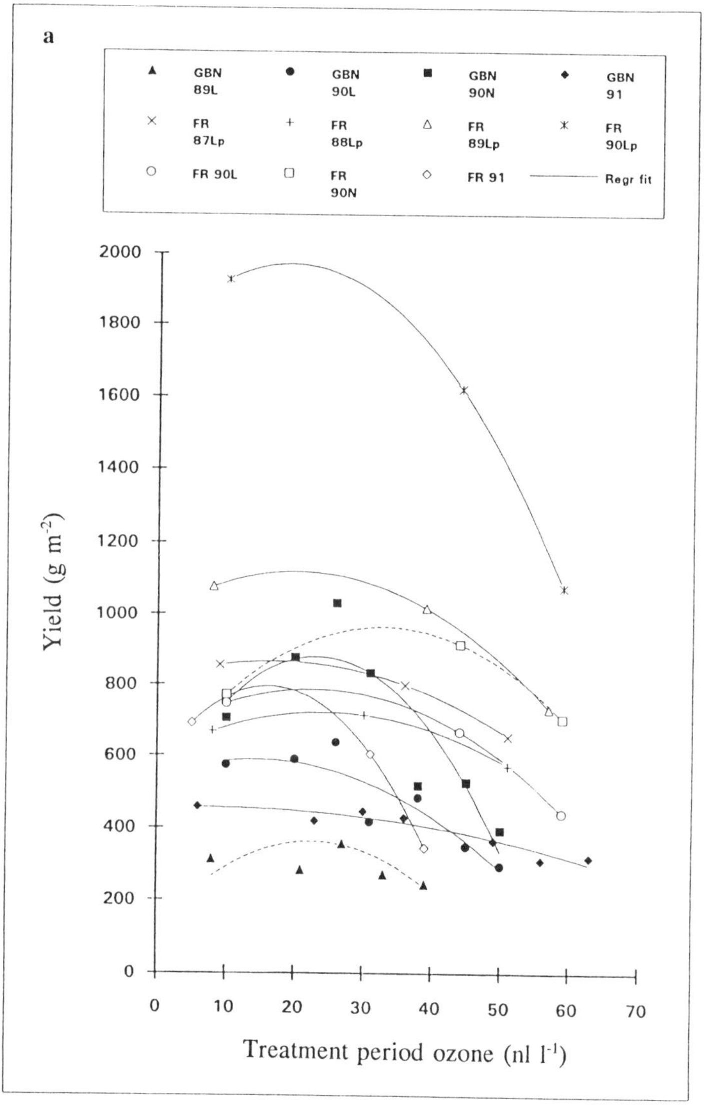

Figure 11.8a Absolute yields of *Phaseolus vulgaris* cvs Lit (L) and Nerina (N) at Nottingham, England and Pau, France. Country codes as for Figure 11.5. Quadratic curves have been fitted to data sets (from Colls *et al.*, in press)

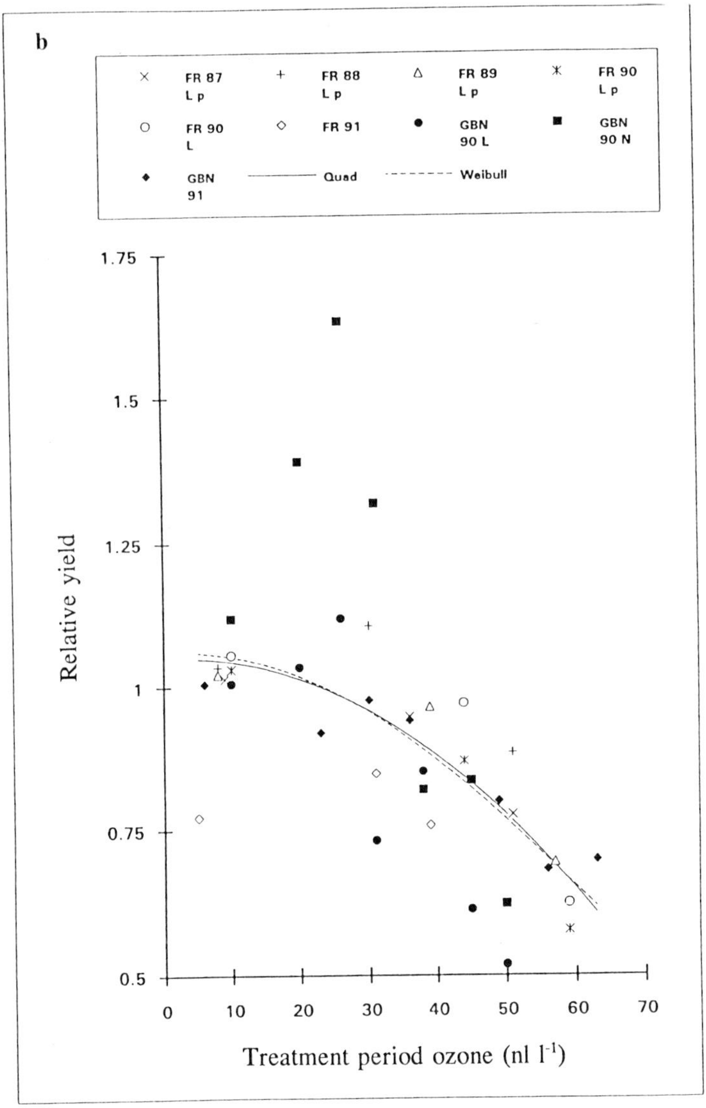

Figure 11.8b Variation of relative yield with ozone concentration; yields are expressed as a proportion of those predicted at an ozone concentration of 5 nl l-1 from the curves fitted to the data sets of Figure 8a (from Colls *et al.*, in press)

Responses of other plant parameters to ozone

The influence of O_3 on yield is the out-turn of earlier plant responses that may involve changes in leaf area development and duration, root growth, dry matter allocation, or alteration to components of yield. We give here some examples of these precursive responses and draw attention to their implications for resource capture.

We discussed earlier the influence of O_3 on radiation interception by soybean, and some consequences for yield. Sanders *et al.* (1990) made more detailed studies of leaf area development and duration and of flowering and pod development in *Vicia faba* growing in CF and NF chambers in the EOTC experiment. The crop, which was a determinate cultivar, attained maximum height and shoot number about 60 days after emergence, at the end of the flowering growth phase. In the last 40 days of this phase, upper leaves in the CF treatment expanded to areas which were 4–20% higher than comparable leaves in the NF treatment (*Figure 11.9*). These differences were sufficient to alter the radiation interception of the crop, so that the accumulated intercepted radiation was 9% higher in the CF treatment by 60 DAE. Over the period 29–49 DAE the net assimilation rate and the relative and absolute growth rates were all larger in the CF treatment than in the NF.

Flowering began about 46 DAE, and by 60 DAE plants in the CF treatment had 11% more flowers than in the NF. This difference was maintained during early pod development (up to 90 DAE) and was caused by retention of more pods on the upper racemes of the CF plants. As the pods dried, however, more pods were abscised in the CF treatment, and by harvest, there were no significant differences in pod number or seed yield between treatments.

One of the difficulties in interpreting the results of Sanders *et al.* lies in the complexity of the air pollution exposure over the season, with the NF treatment

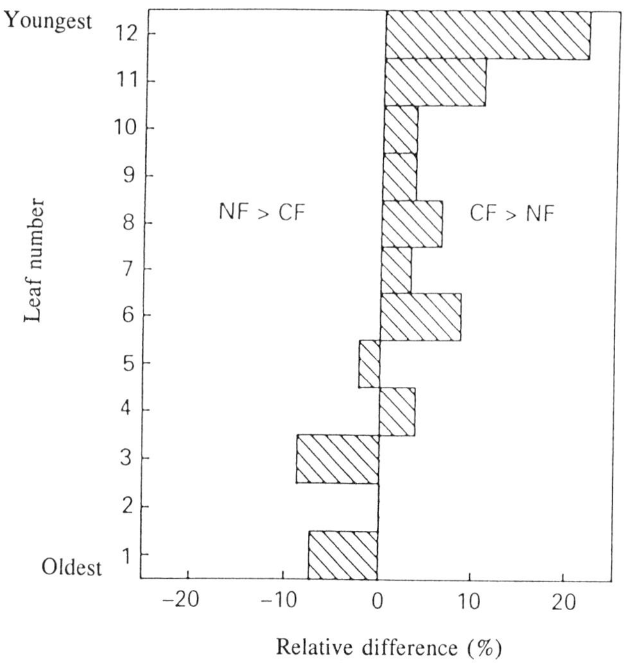

Figure 11.9 Relative differences between the areas of leaves on plants of *Vicia faba* cv. Ticol grown in open-top chambers ventilated with either charcoal-filtered air (CF) or non-filtered air (NF) at Nottingham, England. (from Sanders *et al.*, 1990)

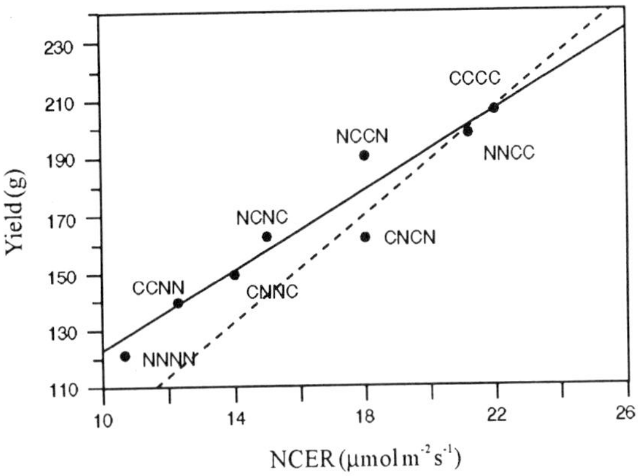

Figure 11.10 Relationship between seed yield of 'Davis' soybean grown at Raleigh, N Carolina, USA and mean net carbon exchange rate (NCER) during the fourth quarter of the growing season. Symbols beside each point refer to the sequential treatments given in open-top chambers in each quarter of the growing season: C = charcoal-filtered air; N = non-filtered air with approximately 100 nl l^{-1} O_3 added for 7 h d^{-1} (from Miller *et al.* 1991)

being exposed at various growth stages to episodes of O_3, SO_2 and NO_2, sometimes individually, sometimes in mixtures. Miller *et al.* (1991) avoided some of this complexity in their study of net carbon exchange of soybeans in response to ozone at different stages of growth, by using controlled additions of O_3 at about 100 nl l^{-1} applied to selected OTCs in different quarters of the growing season. They measured net carbon exchange rate (NCER) of leaflets near the top of the canopy. *Figure 11.10* shows the relationship between final seed yield and NCER for treatments applied in the fourth quarter. This quarter provided the strongest relation between NCER and yield ($r^2 = 0.919$), although statistically significant relationships were also found for the third quarter. This was probably for two reasons: it is well established that a large fraction of the carbon harvested in soybean seeds is assimilated after anthesis, and the peak contribution to seeds is in mid- to late podfill (Yamagata, Kouchi and Yoneyama 1987; Wells *et al.* 1982); and NCER measurements in the fourth quarter incorporate not only effects during that treatment period but also effects carried over from previous quarters. The latter explanation is likely to be important, because ozone causes irreversible injury to leaves at all growth stages at which exposures occur.

Exposure dynamics and crop yields

Although the concentration of air pollutants in the atmosphere is never constant, convention in air pollution research relates crop yields to (usually) seasonal mean concentrations. Clearly this is unlikely to be a robust approach in environments where the occurrence of a particular pollutant is strongly episodic with greatly varying concentrations. Ideally, responses should be related to some measure of the pollutant absorbed by the plant, but this is an impractical quantity to use for air quality regulation, and so better forms of exposure indices have been sought. Tingey, Hogsett and their colleagues (Lee *et al.*, 1988; Tingey *et al.*, 1991) have developed the concept of O_3 dose (atmospheric concentration above a threshold × time) as an improved predictor of crop yield, and this index seems likely to be increasingly used in air quality regulation to protect crops and natural ecosystems.

Many of the quantitative determinations of crop responses to elevated sulphur dioxide concentrations have been undertaken at constant elevations above ambient. However, SO_2 is characteristically emitted from point sources such as electricity generating stations, and hence the exposure of plants in the field is strongly influenced by meteorological factors such as wind direction. The concentration fluctuates in time, peaking when the source is upwind and falling to much lower values when the wind direction changes. In recognition of these factors, an experiment on winter barley was designed at Nottingham specifically to investigate the effect of intermittency of exposure on crop responses to SO_2. The experiment used a free-air field-release system for SO_2 (Colls, Geissler and Baker, 1992) which avoids chamber effects. Three field release plots were used to generate cycles of exposure which were repeated throughout the growing season. One plot was exposed to a high concentration (200 nl l^{-1}) of SO_2 for 1 day, followed by 5 days with no release. Within this 6-day period, the two other plots received 68 nl l^{-1} for 3 days, and 34 nl l^{-1} for 6 days respectively. By the end of the 6-day cycle, all three plots had been exposed to approximately the same dose, 200 nl l^{-1} days. The cycle then started again, and continued in this way through the growing season; 24 cycles were completed before harvest.

Although previous work with the field release system had clearly demonstrated yield suppression when barley crops were exposed *continuously* to SO_2, none of the treatments in the 'intermittent' experiment showed changes in either dry weight increase during the season or in the final grain yield when compared with control plots, demonstrating that even persistent exposure to such high SO_2 concentrations may not affect crops provided that there are 'recovery' periods.

Interactions with water

There are two distinct aspects of crop water relations that interact with air pollution responses: first, air pollutants may alter the ability of crops to capture water through their root systems and transpire it through foliage; second, drought may alter the sensitivity of crops to air pollutants.

There is increasing evidence that several air pollutants can cause a major reallocation of dry matter away from the roots to the foliage (e.g., Jones and Mansfield, 1982 (SO_2); Ashenden and Williams, 1980 (SO_2+NO_2); Okano *et al.*, 1984 (O_3)). Mansfield and others have suggested that this response is likely to predispose plants to drought, by limiting their capability to capture water at a rate required by the above-ground demand. A more precise test of this hypothesis involves assessing the length or area of absorbing roots in relation to foliage area, as only a fraction of the root system is active in water uptake. Recent work by Taylor and Davies (1990) has suggested that (for beech seedlings) air pollution may decrease both active root length and root diameter.

Laboratory and field experiments have shown that exposure to O_3 reduces stomatal conductance, and consequently affects water use (e.g., Tingey and Hogsett, 1985). Exposure to moderate concentrations of SO_2 elicits a similar response, but low concentrations of SO_2 may increase conductance in some species (see Black, 1985 for review).

Short-term stomatal responses may, however, paint a misleading picture, because crop water use depends on diurnal and seasonal patterns of stomatal control. Mansfield and McCune (1988) have shown that leaves exposed to SO_2 + NO_2, or O_3, often exhibit inhibited stomatal closure and sluggish responses when faced with severe water stress, and clearly this has implications for seasonal water use. Recently at Nottingham we have used stem-flow gauges to study the water use of ash saplings exposed in the field in open-top chambers, either to charcoal-filtered air (CF) or to O_3 episodes. *Figure 11.11* shows that water use per tree declined more rapidly with time in the O_3 treatments than in the CF treatments, probably as a result of a combination of stomatal responses and earlier senescence (J.J.J. Wiltshire, personal communication). Quantitative field studies of crop water use in response to air pollutants are clearly needed.

Turning to the second aspect of crop water relations in relation to air pollution, does drought protect crops from air pollution injury? It is well-documented that water-stress reduces foliar injury if crops in the greenhouse or the field are exposed to acute doses of ozone (reviewed, e.g., by Moser *et al.*, 1988). But it is less clear what the interactions of drought and chronic O_3 exposure are on yield, and

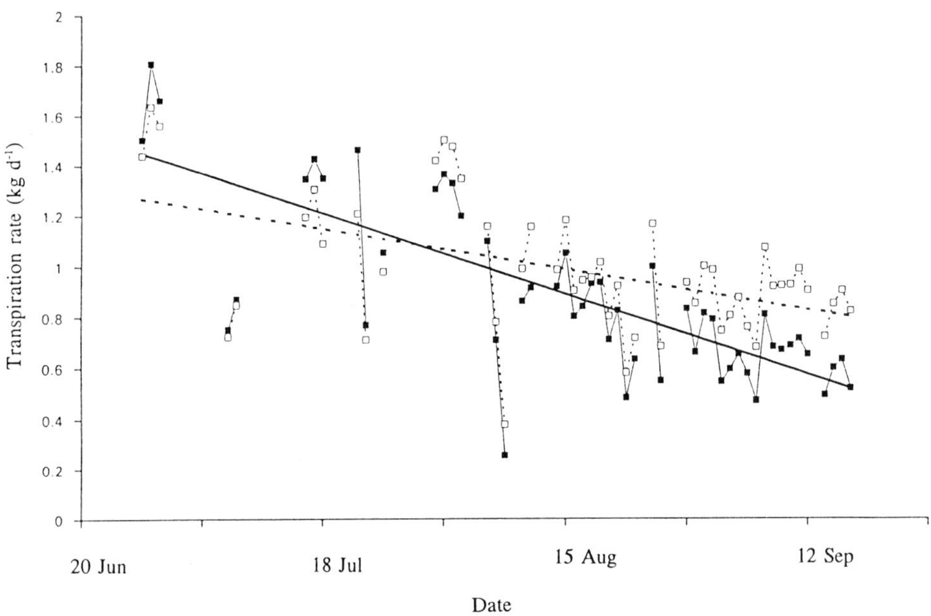

Figure 11.11 Mean daily transpiration per tree for young ash trees growing at Nottingham, England in 1992 in open-top chambers with (□) or without (■)O_3 episodes. Episodes lasted for 1–4 days at an O_3 concentration for 8 h d^{-1} of 150 nl l^{-1}. There were nine episode days before the measurements started, and 14 during the measurement period. Points have been joined within the intermittent periods of transpiration measurement. □ – – – – □ with O_3. ■ ———— ■ without O_3. The continuous lines ----- and ———— are the regressions to the high and low O_3 treatments, respectively (J.J.J. Wiltshire, personal communication)

there is very little work on this subject for other pollutants. Although controlled environment studies (e.g., Amundson *et al.*, 1986) have shown that soybean growth was less impaired by O_3 when plants experienced drought stress, Heggestad and Lesser (1990) found variable protection by drought in field studies with soybean. Temple, Taylor and Benoit, (1985), also in an NCLAN study, found that in one year the growth and yield of field-grown, drought-stressed cotton was less impaired by O_3 than were growth and yield of well-watered controls; but in a second year no protection was detected. The EOTC network also produced equivocal results from field experiments. It is likely that the protective effect of drought for O_3 injury also depends on evaporative demand, and hence on solar radiation, saturation deficit and wind speed. Quantifying the interaction between effects of soil water deficit and O_3 exposure on yield is an important future research topic, because the atmospheric conditions which generate large concentrations of photochemical O_3 are also those which create large evaporative demands, and practical issues such as irrigation policy require the results of carefully designed experiments on this theme.

Air pollutants as nutrients

Sulphur and nitrogen are essential nutrients for all plants. There is no absolute requirement for these chemicals to be obtained from the soil or soil water, and atmospheric sources may be important in some environments. Indeed, species such as mosses and lichens may rely almost entirely on atmospheric deposition for their nutrition. Other semi-natural ecosystems with low nitrogen budgets may be particularly sensitive to aerial deposition. Critical loads of N to natural vegetation on low fertility sites (i.e., inputs above which changes to ecosystem structure and function are likely) are probably between 5 and 35 kgN ha^{-1} yr^{-1}. Deposition of pollutants may be in precipitation (wet deposition), or by the impaction of cloud droplets, or as gas (dry deposition) from the air. Each of these mechanisms takes on a different relative significance according to local factors such as the prevailing meteorology, physical situation, plant structure and nutrient status. In Britain, and probably in much of continental Europe, deposition of N by these various pathways probably ranges between 15 and 50 kgN ha^{-1} yr^{-1}, and so exceeds the critical loads for some sensitive ecosystems. Agricultural crops are normally supplied with adequate S and N to maximize yield, yet there is evidence that they can respond to aerial deposition. Many chamber experiments on *Poa pratensis*, for example, have shown that low concentrations of S and/or N in the air can stimulate yield before the toxic effects at higher concentrations become apparent (Mansfield and McCune, 1988). Further, results from a field release experiment in which winter barley was exposed to SO_2 (added to air with a very low background SO_2 content) showed that yield was stimulated at concentrations less than 40 p.p.b. and suppressed at higher levels (McLeod and Baker, 1988).

Nitrogen from the atmosphere may be wet-deposited as nitrate or dry-deposited as HNO_3, NO, NO_2 or NH_3. The deposition of gaseous NH_3 is being increasingly recognized as significant. The main sources of NH_3 are agricultural operations such as fertilizer application to crops and intensive rearing of livestock. Although concentrations in the air tend to be low compared with other air pollutants, (1–10 nl l^{-1} compared with 10–100 nl l^{-1} for O_3), the deposition rate can be large because of the strong affinity of surfaces for NH_3. Hence the input of nitrogen from NH_3 can be high. In most environments the input is probably not more than 10–20 kgN ha^{-1} yr^{-1}, but at some sites, such as in Holland near intensive pig farms, it can exceed 50–100 kgN ha^{-1} yr^{-1}, and is enough to change the local ecosystem from nitrogen-poor heathland to grassland. Similar adjustments due to aerial deposition may be occurring in the UK; although the fluxes are lower and the changes less well studied, the implications of this impact of agricultural practice on the natural environment are a cause for concern. The joint presence of NH_3 and SO_2 in the air affects the deposition of both gases. Evidence from a SO_2 field release experiment at Liphook, UK, (McLeod *et al.* 1990) and from field measurements over moorland suggest that the presence of moderate concentrations of NH_3 may significantly increase the deposition of SO_2. This is believed to be because the deposition of ammonia raises the pH of the leaf surfaces, effectively increasing the SO_2 concentration gradient and hence the flux of SO_2 across the leaf boundary layer.

Carbon dioxide

No account of the implications of air pollution for resource capture could be complete without consideration of the increasing concentration of the most globally pervasive pollutant, carbon dioxide, but space does not permit any detailed discussion here. Whereas the 'conventional' air pollutants are usually local or, at worst, regional in spatial distribution, CO_2 concentrations are rising globally at an average rate of 1.8 μl l^{-1} yr^{-1}, have climbed from a pre-industrial value of 280 μl l^{-1} to a current 360 μl l^{-1}, and will certainly continue to rise throughout the next century even allowing for reasonable action by the major emitting nations. The potential consequences for agriculture are not yet understood fully (for review see Kimball *et al.*, 1990). If the only effect was the direct result of higher CO_2 concentration *per se*, there would almost certainly be a net benefit to agriculture. In many farming areas in which there are no nutrient or water limitations, production of dry matter is likely to increase. However, the full potential of this stimulation will not be realized if, for example, the increased canopy area causes water stress that cannot be ameliorated. Hence the effect of CO_2 on water use efficiency will be critical in marginal supply regions and may influence choice of species and cultivars. Interactions between increased concentrations of CO_2, and of air pollutants such as O_3, which are also likely to increase in developed regions, are not understood, and field studies are being planned in Europe and the USA to investigate these and other aspects of future climate.

Concluding remarks

Although resource capture is such a strong integrating theme in modern crop science, this review indicates that the concept has not been used much to analyse air pollution experiments. This may be partly because current methods for exposing crops to air pollutants do not lend themselves easily to harvesting the above- and, more importantly, below-ground components of crops which are often the starting point for resource capture analyses. But it is clear that there is a great need for better understanding of how air pollutants affect the ability of field crops to capture water from the soil and carbon from the air. This work also needs to be undertaken in systems that can simulate the higher CO_2 concentrations that will exist in conjunction with air pollution in the future. The most pervasive air pollutant globally is CO_2, and many of the exposure methods developed for exposing field crops to SO_2, O_3, etc., are readily adaptable for CO_2 research and should be used to address important questions of crop productivity and resource capture in the future.

References

Amundson, R.G., Raba, R.M., Schoettle, A.W. and Reich, P.B. (1986) Response of soybean to low concentrations of ozone: II. Effects on growth, biomass allocation and flowering. *Journal of Environmental Quality*, **15**, 161–167

Ashenden, T.W. and Williams, I.A.D. (1980) Growth reductions in *Lolium multiflorum* Lam. and *Phleum pratense* L. as a result of SO_2 and NO_2 pollution. *Environmental Pollution*, A, **21**, 131–139

Bell, J.N.B. (1985) SO_2 effects on the productivity of grass species. In *Sulfur Dioxide and Vegetation: Physiology, Ecology and Policy Issues* pp 209–226 Edited by W.E. Winner, H.A. Mooney and R.A. Goldstein, Stanford, California: Stanford University Press

Black, V.J. (1985) SO_2 effects on stomatal behavior. In *Sulfur dioxide and Vegetation* pp 96–117 Edited by W.E. Winner, H.A. Mooney and R.A. Goldstein, Stanford, California: Stanford University Press

Colls, J.J., Geissler, P.A. and Baker, C.K. (1992) Use of a field release system to distinguish the effects of dose and concentration of sulphur dioxide on winter barley. *Agriculture, Ecosystems and Environment*, **38**, 3–10

Colls, J.J., Sanders, G.E., Geissler, P.A., Bonte, J. and Galaup, S. (in press). The responses of beans exposed to air pollution in open-top chambers. In *Air Pollution and Crop Responses in Europe*. Edited by L. de Temmerman and P. Mathy. Brussels: CEC

Farage, P.K., Long, S.P., Lechner, E.G. and Baker, N.R. (1991) The sequence of change within the photosynthetic apparatus of wheat following short-term exposure to ozone. *Plant Physiology*, **95**, 529–535

Farquhar, G.D. and Sharkey, T.D. (1982) Stomatal conductance and photosynthesis. *Annual Review of Plant Physiology*, **33**, 317–345.

Heagle, A.S., Body, E.E. and Heck, W.W. (1973) An open-top field chamber to assess the impact of air pollution on plants. *Journal of Environmental Quality*, **2**, 365–368

Heck, W.W., Taylor, O.C. and Tingey, D.T. (Eds) (1988) *Assessment of crop loss from air pollutants*, London: Elsevier Applied Science

Heggestad, H.E. and Lesser, V.M. (1990) Effects of ozone, sulfur dioxide, soil water deficit and cultivar on yields of soybean. *Journal of Environmental Quality*, **19**, 488–495

Jones, H.G. (1985) Partitioning stomatal and non-stomatal limitations to photosynthesis. *Plant, Cell and Environment*, **8**, 95–104

Jones, T. and Mansfield, T.A. (1982) Studies on dry matter partitioning and distribution of ^{14}C-labelled assimilates in plants of *Phleum praiense* exposed to SO_2 pollution. *Environmental Pollution*, A, **28**, 199–207

Kimball, B.A., Rosenberg, N.J. and Allen, L.H. (Eds.) (1990) *Impact of carbon dioxide, trace gases and climate change on global agriculture*, Madison, Wisconsin: American Society of Agronomy Special Publication Number 53

Kropff, M.J. (1987) Physiological effects of sulphur dioxide. 1. The effect of SO_2 on photosynthesis and stomatal regulation of *Vicia faba* L. *Plant, Cell and Environment*, **10**, 753–760

Lee, E.H., Tingey, D.T. and Hogsett, W.E. (1988) Evaluation of ozone exposure indices in exposure-response modelling. *Environmental Pollution*, **53**, 43–62

Lesser, V.M. and Geissler, P. (in press) Statistical methods for crop dose-response studies in Europe and the United States.

McLeod, A.R. and Baker, C.K. (1988) The use of open field systems to assess yield response to gaseous pollutants. In *Assessment of Crop Loss from Air Pollutants* pp 81–210 Edited by W.W. Heck, O.C. Taylor and D.T. Tingey, London: Elsevier Applied Science

McLeod, A.R., Holland, M.R., Shaw, P.J.A., Sutherland, P.M., Darrall, N.M. and Skeffington, R.A. (1990) Enhancement of nitrogen deposition to forest trees exposed to SO_2. *Nature*, **347**, 277–279

Mansfield, T.A. and McCune, D.C. (1988) Problems of crop loss assessment when there is exposure to two or more gaseous pollutants. In *Assessment of Crop Loss from Air Pollutants* pp. 317–344 Edited by W.W. Heck, O.C. Taylor and D.T. Tingey, London: Elsevier Applied Science

Miller, J.E., Pursley, W.A., Vozzo, S.F. and Heagle, A.S. (1991) Response of net carbon exchange rate of soybean to ozone at different stages of growth and its relation to yield. *Journal of Environmental Quality*, **20**, 571–575

Monteith, J.L. (1977) Climate and the efficiency of crop production in Britain. *Philosophical Transactions, Royal Society of London*, Series B, **28**1, 277–294

Moser, T.J., Tingey, D.T., Rodecap, K.D., Rossi, D.J. and Clark, C.S. (1988) Drought stress applied during the reproductive phase reduced ozone-induced effects in bush bean. In *Assessment of Crop Loss from Air Pollutants* pp 345–364 Edited by W.W. Heck, O.C. Taylor and D.T. Tingey, London: Elsevier Applied Science

Okano, K., Ito, O., Takeba, G., Shimizu, A. and Totsuka, T. (1984) Effects of NO_2 and O_3 alone and in combination on kidney bean plants. 5. ^{13}C-assimilate partitioning as affected by NO_2 and/or O_3. *National Institute of Environmental Studies, Japan, Research Reports*, **66**, 49–57

Roberts, T.M. (1984) Long term effects of sulphur dioxide on crops: an analysis of dose-response functions. *Philosophical Transactions, Royal Society of London*, Series B, **305**, 299–316

Sanders, G.E., Turnbull, N.D., Clark, A.G. and Colls, J.J. (1990) The growth and development of *Vicia faba* L. in filtered and unfiltered open-top chambers. *New Phytologist*, **116**, 67–78

Skärby, L., Selldén, G., Mortensen, L., Bender, J., Jones, M., de Temmerman, L., Wenzel, A. and Fuhrer, J. (in press) Responses of cereals exposed to air pollutants in open-top chambers. In *Air Pollution and Crop Responses in Europe*. Edited by L. de Temmerman and P. Mathy. Brussels: CEC

Taylor, G. and Davies, W.J. (1990) Root growth in *Phagus sylvatica*: impact of air quality and drought at a site in southern Britain. *New Phytologist*, **116**, 457–464

Temple, P.J., Taylor, O.C. and Benoit, L.F. (1985) Responses of cotton yield to ozone as mediated by soil moisture and evapotranspiration. *Journal of Environmental Quality*, **14**, 55–60

Tingey, D.T. and Hogsett, W.E. (1985) Water stress reduces ozone injury *via* a stomatal mechanism. *Plant Physiology*, **77**, 944–947

Tingey, D.T., Hogsett, W.E., Lea, E.H., Herstrom, A.A. and Azevedo, S.H. (1991) An evaluation of various alternative ambient ozone standards based on crop yield loss data. In *Transactions: Tropospheric Ozone and the Environment* pp 272–288 Edited by R.L. Berglund, D.R. Lawson and D.J. McKee Pittsburgh: Air And Waste Management Association

Unsworth, M.H. and Ormrod, D.P. (Eds) (1982) *Effects of gaseous air pollution in agriculture and horticulture*, London: Butterworths Scientific

Unsworth, M.H., Lesser, V.M. and Heagle, A.S. (1984) Radiation interception and the growth of soybeans exposed to ozone in open-top field chambers. *Journal of Applied Ecology*, **21**, 1059–1079

Unsworth, M.H. and Geissler, P. (in press) Results and achievements of the European Open-top Chamber Network. In *Air Pollution and Crop Responses in Europe*. Edited by L. de Temmerman and P. Mathy. Brussels: CEC

Wells, R., Schultze, L.L., Ashley, D.A., Boerma, H.R. and Brown, R.H. (1982) Cultivar differences in canopy apparent photosynthesis and their relationship to seed yield in soybeans. *Crop Science*, **22**, 886–890

Wiltshire, J.J.J., Wright, C.J. and Unsworth, M.H. (1992) The effects of ozone episodes on autumn leaf fall in apple. *Forest Ecology and Management*, **51**, 115–120

Yamagata, M., Kouchi, H. and Yoneyama, T. (1987) Partitioning and utilization of photosynthate produced at different growth stages after anthesis in soybean: analysis by long-term ^{13}C-labelling experiments. *Journal of Experimental Botany*, **38**, 1247–1259.

12

CAN RESOURCE CAPTURE PRINCIPLES ASSIST PLANT BREEDERS OR ARE THEY TOO THEORETICAL?

J. HAMBLIN
ICARDA, PO Box 5466, Aleppo, Syria

Introduction

This chapter concentrates on breeding for cereal grain yield in variable and unpredictable rain-fed environments. This is because yield is a major objective in nearly all breeding programmes and the author's background of working in mediterranean climates has resulted in an interest in rain-fed environments. The discussion will concentrate on cereals (although not exclusively) as these have more data available than other plant families. Although this may seem restrictive, it matches the author's experience, and many of the problems faced by breeders in this situation have implications for other crops, breeding objectives and environments.

Farmers, agricultural scientists, industry and governments all have an interest in crop production and the efficient use of farm resources. Models are used frequently to estimate growth and production. These models can operate at many levels, for example from the molecular level through organelles, cells, tissues, organs, plants, crops, farms, regions, nations and across several nations (Stockle, 1989). The type of model used depends on the questions being asked.

Frequently but not exclusively, the problems that plant breeders face occur at the organizational level of the plant and crop. Often physiologists are interested in processes several levels below this; for example at the membrane, cell or tissue level (Passioura, 1981). Agronomists and crop physiologists tend to operate at the same level of complexity as plant breeders; farmers are more interested in economic interactions at the farm level. Government and business interests in crop models usually range from regional to international levels. If the level of operation of a model is not clearly recognized by different groups of specialists trying to use it, this can cause problems in developing effective collaboration between breeders and others interested in modelling plant or crop production (Passioura, 1981).

Plant breeders are usually highly pragmatic people with clear-cut breeding objectives who are dedicated to producing better plant varieties. Often these objectives are under complex, polygenic control. Increased and more stable yields are among the most frequent of these objectives. These characters frequently have large interactions with the environment, so that a high expression of the character in one environment does not necessarily mean a correspondingly high expression in another. Their heritability is low, making selection difficult even within an environment.

Plant breeders are also eternal optimists and look for simply inherited characters closely associated with grain yield,which have a higher heritability and a lower genotype by environment interaction than does yield itself. For physiologists this is a happy hunting ground for collaboration. The literature is full of lost causes of this type, for example the short-lived enthusiasm for the relationship between nitrogen reductase and yield or between proline and drought tolerance. Currently physiologists are being given a run for their money by the molecular biologists in trying to develop universal panaceas for plant breeders.

If a simply inherited character is of universal importance to survival or production, then it should have been selected for by all breeding programmes selecting for high yield in a given set of environments. Apart from the obvious, such as plants should be green and have a development pattern that is appropriate to the environment, no universal or even regional plant characteristics have been identified (Ceccarelli, Acevedo and Grando, 1991). This suggests that breeders should not expect to find one. For crops that have to face a rough and tumble existence in the dense stands of farmers' fields, the idea that indirect selection for simple, easily measured characters will increase yield potential in a range of environments is a holy grail that is not likely to exist in real life. This may be less true in situations where yield is not limited by water, nutrition, weeds or neighbours as occurs for horticultural crops. On the other hand, characters that give a marked selection disadvantage are usually selected against by nature or plant breeders. One thinks of an inappropriate day-length or vernalization response and shattering heads in field crops. More extreme examples include variegated plants, which rarely occur in nature but which are common in gardens, and albino plants that only exist in the laboratories of physiologists and molecular biologists.

With this background of personal prejudices and some of the problems of using models and of the unlikely event that selection for complex characters in a variable environment is likely to be simple, perceptions of the potential of resource capture models for plant breeders will now be discussed.

Resource Capture Models

Resource capture models operate at the level of complexity (plants and crops) that is of predominant interest to plant breeders and are thus scaled correctly to be of interest. As an example, which is only a small sample of the crop models available (Stapper, 1984), two models will be considered together in outline; the Ceres–Cereal model of Ritchie and co-workers (Singh *et al.*, 1989) and the Rescap model of Monteith and co-workers (Monteith, Huda and Midya, 1989; Singh *et al.*, 1990). At the same time some of the problems that models pose for breeders will be discussed in terms of input information and what breeders feel about them. Later, consideration will be given to the problems that plant breeders face and what they do about them. This is followed by discussion of where these models could assist breeders.

The plant input data for these models may seem limited to a physiologist dealing with, at most, a few hundred plots, but to a breeder dealing with several thousand plots they look pretty extensive. Some of the non-plant data needed might also be considered difficult to obtain, particularly if trials are carried out at many sites, often far from base.

PHENOLOGY

Both models have plant development subroutines that drive the plant's life cycle from a knowledge of temperature and day-length and variety responses to these. This may be modified by the effects of moisture and nutrient stresses and cultivar-specific characteristics such as vernalization (Singh *et al.*, 1989; Singh *et al.*, 1990). Strictly phenology cannot be considered a resource capture characteristic, although it has obvious influences on resource capture through such factors as leaf area and yield development and on resource partitioning.

Plant anthesis and maturity are often under complex genetic control and several components have been recognized: these include temperature and day-length responses (Ellis *et al.*, 1989; Erskine *et al.*, 1990) as well as vernalization and length of the basic juvenile phase (Ellis *et al.*, 1989; Roberts *et al.*, 1988; Boyd *et al.*, 1991; Masle, Doussinalt and Sun, 1989).

For modelling purposes the response patterns of varieties to these different environmental signals must be known. Obtaining these data for many genotypes, particularly in selection programmes, is often beyond the resources of a plant breeder. Therefore breeders doubt whether it is worth collecting detailed information on many lines. In some situations simplified techniques that classify genotypes from data sets of a few sites can predict a variety's flowering behaviour over a wide range of sites (Perry, Siddique and Wallace, 1987; Erskine *et al.*, 1990). Because fewer numbers are involved it is easier to assess parental than progeny response. If their likely developmental response to environments can be predicted, one can assemble parents for use in a wide range of environments.

Breeders are still concerned about using these predictive models to eliminate germplasm. Even though a model may be highly predictive, an error of a few days, for example 3, which is a small error for a 120–day period from emergence to anthesis, may be large in terms of yield, particularly if grain filling occurs in

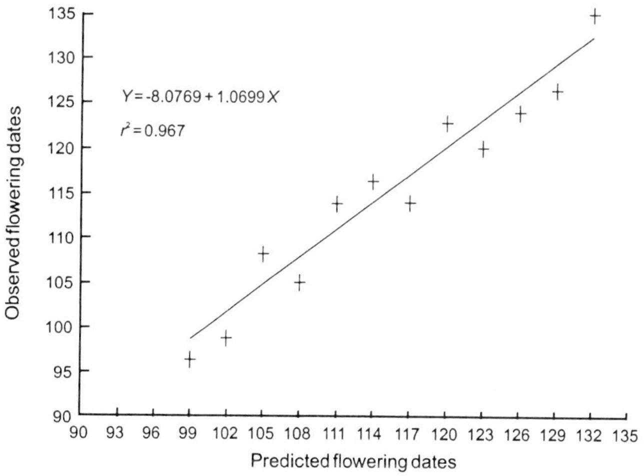

Figure 12.1 A plot of 12 hypothetical varieties differing by increments of 3 days in predicted flowering; the observed flowering dates differ from the predicted by ±3 days at random

a period of severe drought stress. *Figure 12.1* shows a hypothetical situation where the predicted flowering dates (on the *x* axis) for 12 genotypes varied by increments of 3 days, from 99 days to 132 days from planting. This approximates the actual range reported by Loss *et al.* (1989) for wheat. On the *y* axis is plotted a hypothetical set of observed values, where the observed date differs from the predicted by ±3 days at random. The correlation between observed and predicted days to flowering is 0.967, which also was the mean correlation of 10 different random runs (range 0.958–0.977). This would be considered an acceptable relationship between observed and predicted values for modelling purposes (Cao and Moss, 1989 a, b, c; Loss, Perry and Anderson, 1990; Singh *et al.*, 1989; Volk and Bugbee, 1991). When the deviation between observed and predicted flowering dates was increased to ±5 days the correlations were still high and varied from 0.885 to 0.931 (mean 0.901).

Using data from dryland experiments at ICARDA and from Australia it is possible to estimate the effect on yield potential of changing flowering date by one day. This has been done by regressing yield on anthesis date (Hamblin, 1971), on heading date (Oosterom, personal communication) and on an index of maturity (Perry and D'Antuono, 1989). In the case of Perry and D'Antuono (1989) only a single estimate of maturity was published and this has been used in all the regressions. If there is a relationship between maturity and yield then this procedure is likely to reduce the size of any correlation, as actual maturity dates for each trial are likely to give a better relationship than a single estimate. Also the maturity index showed a range of 34 units (unspecified), whereas the text states that the range in anthesis dates was about 14 days. Therefore the rate of yield loss per day is underestimated, probably by 50%. In the other two cases anthesis and heading date were estimated individually for each data set.

Using different genotypes to estimate the relationship between yield and development is open to the criticism that genotypes differ for other, undefined, characters as well as heading date. However, in one set of data (Hamblin, 1971) the material used was random sets of 100 lines drawn from a cross. It is likely that in this material the background genetic effects have been averaged out. van Oosterom, a Ph.D. student at ICARDA, used 36 lines which, although of different genetic origins, are likely to be of sufficient number to show real effects. Perry and D'Antuono (1989) used smaller numbers (16–19 entries) and here genetic effects may have been important. If an attempt had been made to generate these effects agronomically, for example by varying planting date, a similar criticism applies: namely that the treatments did not experience the same weather. Also the purpose is to show that good predictive models for phenology may still not be accurate enough to give reliable yield estimates for plant breeders and therefore the example is adequate. A range of values for the regression of flowering date on yield and their source is given in *Table 12.1*. The effect of a ±3 day error (6 day total) in predicting flowering date, especially in stress environments with low yield potential, is frequently much greater than the expected rate of genetic gain from selection. This was estimated at 0.57% per year by Perry and D'Antuono (1989) in Western Australia.

For plant breeders it is important to know what is the optimum flowering time in different environments. These may include different places, seasons or planting dates. Breeders also need to know whether there is more than one genetic route to this optimum. This information is of considerable use to plant breeders, particularly those working in programmes with a wide geographical or

Table 12.1 THE EFFECT OF DELAYING FLOWERING BY ONE DAY ON GRAIN YIELD, ESTIMATED FROM GENOTYPES WITH DIFFERING DAYS TO FLOWERING

Source	*Yield loss* (kg ha^{-1} d^{-1})	*Trial mean* yield (kg ha^{-1})	*% loss/day*	*loss/6 days* (kg ha^{-1})
Van Oosterom	54 ***	1330	4.1	324
(pers. comm.)	61 ***	1250	4.9	361
(36 entries)	79 ***	1340	5.9	474
	82 ***	2460	3.3	492
	82 ***	1470	5.6	492
	91 **	2520	3.6	546
	96 **	3310	2.9	576
Mean†	77.9	1954	3.99	467.4
Perry and D'Antuono, 1989;	19 +	1849	1.0	114
(maturity, Table 3;	46 ***	1731	2.7	276
(16 – 19 entries)	23 ***	1113	2.1	138
	38 ***	1393	2.7	228
	10 *	1296	0.8	60
	37 ***	1993	1.9	222
	44 ***	2031	2.2	264
	10 +	463	2.2	60
	34 ***	1257	2.7	204
	18 ***	700	2.6	108
	29 ***	1772	1.6	172
	22 ***	1340	1.6	132
	10 n.s.	981	1.0	60
	19 ***	1522	1.2	144
	8 +	1608	0.5	48
	18 **	1597	1.1	108
	23 ***	1459	1.6	138
	11 ***	460	2.4	66
	37 ***	1571	2.4	222
	13 ***	607	2.1	78
Mean†	24.1	1356	1.86	146.4
(from Table 3, 19 entries)	23***	1373	1.7	138
Hamblin 1971	5 n.s.	4240	0.1	30
(8 sets of 100	68 ***	3420	2.0	408
different entries).	82 ***	3520	2.3	492
	28 **	4300	0.7	174
	21 ***	1890	1.1	126
	24 *	4650	0.5	144
	31 *	4440	0.7	186
	30 ***	1790	1.7	180
Mean†	40.5	3430	1.18	243

n.s. = not significant, + = $P < 0.1$, * = $P < 0.05$, ** = $P < 0.01$, *** = $P < 0.001$.
(†) means include only significant values

environmental mandate. Small errors can have a significant impact on estimated yield; in the 35 cases considered, this varied from 0.1% to 5.6% per day. For the 33 cases where the regression of yield loss was significant, the average loss was 2.17% per day on a mean yield of 1923 kg ha^{-1} or 42 kg ha^{-1} d^{-1}.

Ceccarelli, Acevedo and Grando (1991) have considered this problem in a different context where they have suggested that different combinations of the same characters may produce equally high-yielding genotypes for stress environments. Recently, van Oosterom, (personal communication) has shown that, in the environment of northern Syria, there are at least two different development strategies that give high yield potential. A similar conclusion has been reached for the environment of Western Australia (Young, personal communication) where there are several genetic routes to ensure flowering at the optimum time for a given planting date and region, but so far plant breeders have exploited only a limited set of these options.

Phenology models are weak when genotypes require vernalization (Cao and Moss, 1991). Some 23% of the developing world's wheat production is from areas where varieties require some degree of vernalization (CGIAR/TAC, 1988). There is an urgent need for phenology models to be improved to better account for these genotypes.

LEAF AREA DEVELOPEMENT

In situations where water and nutrients are non-limiting the potential growth of a crop is initially related to the development of leaf area, the amount of photosynthetically active radiation it absorbs and the photosynthetic conversion ratio (Singh *et al.*, 1989; Singh *et al.*, 1990).

Studies of leaf area development and its effect on crop growth and yield go back a long way (Gregory, 1918) and in fact an earlier (1956) Nottingham Easter School was devoted to the growth of leaves. It included a chapter by Milthorpe (1956) on: 'The relative importance of different stages of leaf growth in determining the resultant area'. Here he stated: 'First, as we must relate (leaf) area to the environment in which the plant is growing we need to know which of the processes of leaf growth are most influenced by the environment . . .'. This seems an appropriate starting point for this discussion.

Recently much has been published on the rate of leaf appearance in cereals (e.g. Syme, 1974; Klepper, Belford and Rickman, 1984; Kirby, Appleyard and Fellowes, 1985; Baker *et al.*, 1986; Belford, Klepper and Rickman, 1987; Kirby and Perry, 1987; Cao and Moss, 1989 a, b, c, 1991; Masle, Doussinalt and Sun, 1989; Volk and Bugbee, 1991). This has been in part a response to the needs of modellers who have been interested in the relationship between environmental factors and the rate of leaf area development. This work clearly shows that, for a given genotype and environment, the rate of leaf appearance is approximately linear with thermal time. Now plant breeders are interested in variety effects and the range of variability between genotypes. *Table 12.2* gives the ratio of the variety with the fastest rate of leaf appearance divided by the variety with the slowest rate within an experiment, or if agronomic treatments have been applied, within a treatment. This shows the range of relative differences between varieties in the rate of leaf appearance independent of other factors.

The data in *Table 12.2* show that there is a considerable range in the rate of leaf appearance between varieties. This is at a compound rate, therefore even quite small differences in the rate per leaf can have a significant effect on leaf area development over the life span of a crop. Also, where more than one ratio is given in the Table for a particular reference, the ratios are for the rate of leaf appearance in different agronomic treatments. These also can have significant effects on rate of leaf appearance. There were frequently signs of interactions between treatments

Table 12.2 THE RATIO OF THE RATE OF LEAF APPEARENCE FOR THE VARIETIES WITH THE FASTEST AND SLOWEST RATES WITHIN AN EXPERIMENT OR AGRONOMIC TREATMENT

Source	*Ratio of the rate of leaf appearance for genotypes with the fastest and slowest rates*					
Kirby *et al.*,	1.18	1.12	1.11	1.20		Wheat 80/81 planting date
(1985)	1.41	1.21	1.07	1.24		Wheat 81/82 planting date
(Table 5)	1.16	1.25	1.26	1.28		Barley 80/81 planting date
	1.20	1.16	1.16	1.17		Barley 81/82 planting date
Kirby and	1.09	1.17	1.20	1.19	1.19	Wheat 83 planting date
Perry (1987)*	1.28	1.27	1.50	1.18	1.10	Wheat 84 planting date
(Figure 3)	1.15	1.12	1.44	1.27	1.15	Wheat 85 planting date
	1.14	1.32	1.32	1.61	1.64	
Baker *et al.*,			1.09			Wheat 82 Irrigated
(1986)*			1.26			Wheat 82 Unirrigated
(Figure 2)						
Syme 1974			1.23			Wheat 70 Field
(Table 2)			1.28			Wheat 71 Glasshouse
Cao and Moss	1.15	1.12	1.13	1.09		Wheat growth chamber
(1989a)*	1.11	1.07	1.06	1.05		temperature response
(Figure 2)	1.12	1.09	1.18	1.19		Barley growth chamber
	1.21	1.17	1.21	1.19		temperature response
Cao and Moss	1.09	1.10	1.12	1.09		Wheat growth chamber
(1989b)*	1.07	1.09	1.09	1.05		day-length response
(Figure 2)	1.14	1.17	1.15	1.23		Barley growth chamber
	1.15	1.11	1.11	1.10		day-length response

* Data from published figure, and therefore may be slightly inaccurate.

and varieties, but only in Kirby and Perry (1987) were ANOVAs presented to confirm this.

The above data show that although within a variety and treatment there is a high level of predictability about the rate of leaf appearance, this is not constant for varieties over treatments and the interactions may be considerable. Also the effects of winter and spring growth habit can be large and comparisons are only meaningful within and not between these groups. The message to plant breeders is similar to that reached for predicting anthesis. The models look pretty good, but genotypic differences and genotype by environment effects may be considerable.

If physiologists can provide convincing evidence about the benefits of selecting for rate of leaf appearance, it should be possible to get a reasonable estimate from total leaf number and time to flag leaf appearance. At the moment this is not a measurement made regularly by breeders.

Besides rate of leaf appearance, other factors influence the rate of leaf area development and photosynthetic efficiency. These include leaf size and disposition as well as rate of tiller production. The latter tends to be closely related to leaf appearance (Klepper, Belford and Rickman, 1984; Belford, Klepper and Rickman, 1987; Masle, Doussinalt and Sun, 1989). Plant breeders have on occasions shown some interest in leaves, for example, the leaf angle debate (Watson and Witts, 1959; Gardener, 1966; Tanaka, Kawana and Yamaguchi, 1966; Tanner *et al.*, 1966; Donald, 1968; Pendleton *et al.*, 1968; Hamblin and Donald, 1974). Breeders also have shown a considerable interest in early vigour, which almost certainly relates to leaf area and inclination. This influences ground cover, improves radiation interception and reduces soil evaporation giving a concomitant increase in transpiration (Asana, 1965; Acevedo, 1987; Turner and Nicolas, 1987; Acevedo *et al.*, 1991). Early vigour also may improve weed suppression (Pal *et al.*, 1960). Currently early growth is scored by breeders and is related to biomass at anthesis (Turner and Nicolas, 1987). However attempts are being made to measure early vigour more quantitatively using infra-red sensing technology (Portmann, personal communication).

Once beyond scoring for early vigour the researcher becomes involved in measuring crop growth rate, leaf area and other aspects of growth analysis. This approach stems from the work of Gregory (1918) and was reviewed by Watson in his address to this meeting in 1956. Some aspects of the approach are still subject to controversy among physiologists (Warren Wilson, 1981; Hardwick, 1984; Hunt *et al.*, 1984). Although these measurements have been recommended as breeding objectives (Lupton, 1966) they have not been taken up by plant breeders. This is because of the complexity of measurement and difficulties in showing differences between genotypes and relating these to yield. Even agronomists and physiologists have not explored the variation in leaf area development between genotypes with much enthusiasm. Some measurements of varietal differences in canopy development have been presented (Siddique *et al.*, 1989; Stapper and Fischer, 1990; Karimi and Siddique, 1991).

Breeders have not taken much interest in the photosynthetic conversion efficiency either. This is partly because of the problems of making valid comparisons between genotypes and partly because, within a species, leaf area development and deployment are more critical to growth than photosynthetic efficiency at the chloroplast level (Evans and Wardlaw, 1976).

The level of photosynthetically active radiation is primarily determined by latitude and season, factors over which breeders often have little or no control. However, the effect of altering both planting date and phenology is an area where a modelling approach would allow breeders to explore alternative strategies without investing a great deal of time or money in trying options that lead to reduced yields. The only case where reduced yields might be acceptable would be in multiple cropping. A reduction in growing period and yield may mean that a potential component of a crop cycle can fill a vacant time slot in the calendar which is currently not used for cropping. An example of this is the use of barley in Vietnam to allow two crops of rice and one of barley where before only two rice crops were grown (Vivar, 1990).

The above discussion on phenology and leaf area development shows that, although the subroutines used in the models are highly predictive, none the less the errors involved are high in relation to the rate of improvement of varieties in breeding programmes and on the whole the inputs required are beyond the resources of most breeding programmes. Their potential uses by plant breeders will be discussed later.

ROOT FACTORS AND WATER USE

The discussion of plant development and leaf growth was relatively extensive. Consideration of the root growth and water supply/use components of resource capture models will be less so partly because there is a scarcity of data and partly because the problems of measurement, particularly in the field, are horrendous. For plant breeders dealing with thousands of genotypes they are the stuff of nightmares.

The models have elegant feedback systems that grow roots in relation to light fixation, top growth, partitioning and water and nutrient availability. The partitioning of dry matter often interacts with water availability. Breeders appreciate the elegance of the models and the work involved in developing them, but they are not keen to get involved in root work, or if they do, then at one remove, using indirect measures to estimate root parameters.

For example, Turner and Nicolas (1987) following on the work of Blum, Mayer and Gozlan (1982) used an infrared thermometer to assess crop temperature postanthesis in dryland crops in a mediterranean environment. This is related to the plant's ability to extract water at this time; those that can extract more water will be cooler (Blum *et al.*, 1982). It is important that any measurements are not confounded by maturity effects. However, even with this caveat, during grain filling under water stress leaves are dying rapidly and a major problem is to decide what temperature is being measured. Is it leaf temperature, soil temperature or a combination of the two? Under rain-fed conditions this technique has problems (Turner and Nicolas, 1987).

Other ways of assessing a genotype's ability to regulate its water use in relation to the season have been suggested. Increasing the hydraulic conductivity of the roots by selecting for decreased xylem vessel size in seminal roots has been advocated by Passioura (1972). There is some evidence that this approach does spare water early in the season for use later (Richards and Passioura, 1981 a, b) but to date breeding programmes are not actively involved in using this approach. On free-draining soils using water while it is available is more important than sparing water (Turner and Nicolas, 1987). An alternative has been to increase water extraction, and here osmoregulation has been proposed (Morgan, 1983) and there is some evidence for its potential (Morgan and Condon, 1986). Increased rooting depth and higher rates of root penetration may be important in mediterranean climates on deep, free-draining soils (Passioura, 1983), but will be of much less importance on heavier soils where rooting depth is shallow and determined by the depth of the wetting front (Brown, Gregory and Wahbi 1987).

Just as breeders have not been overenthusiastic about canopy characteristics, so they have problems in assessing root and water use characteristics and how to utilize them in a breeding programme. While it is quite easy to show differences between genera in rooting patterns (Hamblin and Hamblin, 1985), it is more difficult to measure significant differences between varieties within a species (Hamblin and Hamblin, 1985; Bishop, 1986; Wahbi and Gregory, 1989). Also there is concern as to what these differences might mean in mediterranean, rain-fed environments.

A research area that is currently causing great interest among plant breeders is the use of ^{13}C analysis as it relates to water-use efficiency (Farquhar, O'Leary and Berry, 1982). It is still early days, but if the technique realizes its early promise it may become a regular screening tool for breeders (Farquhar and Richards, 1984; Anon, 1992).

The above brief discussion shows that although plant breeders are interested in root/soil water characteristics, they are not enthusiastic about approaches that require direct measurements on root systems. There is however, considerable interest among plant breeders in indirect ways of assessing root characteristics. It is important to remember that varieties are expected to perform well on a range of soil types and that these will influence root growth. The optimum pattern is likely to vary between soil types and seasons and varietal effects on the plasticity of root responses to environmental differences are not well understood.

What plant breeders do when breeding for yield in stress environments

Having considered resource capture models from a plant breeder's perspective, the work of cereal breeders in highly stressed and variable environments will be discussed. This will allow some discussion in the next section of where resource capture models can assist plant breeders.

Breeders work with variation and this is of three main types: genetic, environmental and the interaction of genotype with the environment. Experienced breeders know their material and target environments well and as a general rule make their crosses with specific objectives in mind. This part of the work is relatively straightforward although it requires considerable thought, and, except at international centres, is commonly of a small scale, usually in the range of 10s to low 100s of crosses per year. At international centres with a large mandated area, the crosses made are usually in the high 100s or low 1000s.

The major problem, easily stated but difficult to accomplish, is to identify superior genotypes that in fluctuating environments give consistently higher yields. There is a great deal of genetic theory and biometrical methodology to assist the breeders, but the process still has a large organizational and number 'crunching' side to it.

The importance of the genotype by environment interaction (g × e) is well illustrated by the fact that the g × e component of the analysis of variance is often two to three times larger than the genetic component of variance (Byth, Eismann and De Lacy, 1976; Zobel, Wright and Gauch, 1988; Braun, Pfeiffer and Pollmer, 1992; Oosterom, personal communication). This means that varieties are performing differently, relative to each other, over the range of environments used. The question then becomes one of how wide a level of adaption should breeders aim at and in which environments is selection most efficient. Should we select in high-yielding, favoured environments and expect the lines selected there to perform equally well in stress environments, or should we select in stress environments which represent the type of environments that the varieties will have to grow and yield in? This problem is of central importance to plant breeding methodology, and is hotly debated among plant breeders (e.g. Brennan and Byth, 1979; Simmonds, 1979, 1981, 1991; McVetty and Evans, 1980; Rosielle and Hamblin, 1981; Jensen, 1988; Ceccarelli, 1989; Braun, Pfeiffer and Pollmer, 1992).

Theory on this point is agreed. If the correlation between yield in stress and non-stressed environments is not perfect, then selection is more efficient in the environment in which the genotypes are expected to perform (Simmonds, 1991). Also, selection in stress environments will lead to yield improvement in stress environments, but this may lead to a reduction in yield in non-stress environments (Rosielle and Hamblin, 1981).

Plant breeding programmes often are based on research stations and the environment there is frequently more favourable than the target environments. The success of the green revolution varieties depends on high inputs (Simmonds, 1991). Their success has been less in more marginal environments. Recently work at ICARDA on barley is producing clear evidence that when both are grown in stress environments, lines selected in stress conditions outyield lines selected in favourable conditions. The reverse also holds, namely that when grown in favourable environments, lines selected in favourable environments outyield lines selected in stress environments (Ceccarelli and Grando, 1991a, b). At the 6th International Barley Genetics Symposium the reasons for the intensity of the debate over selection environment were considered in a workshop organized by ICARDA. The consensus was that part of the problem in the debate hinges on the definition of low-yielding sites (Hamblin, 1992). For example for people working in irrigated situations, then stress environments may have a higher yields than the 'high-yielding' environments of people working in areas of high abiotic stress and

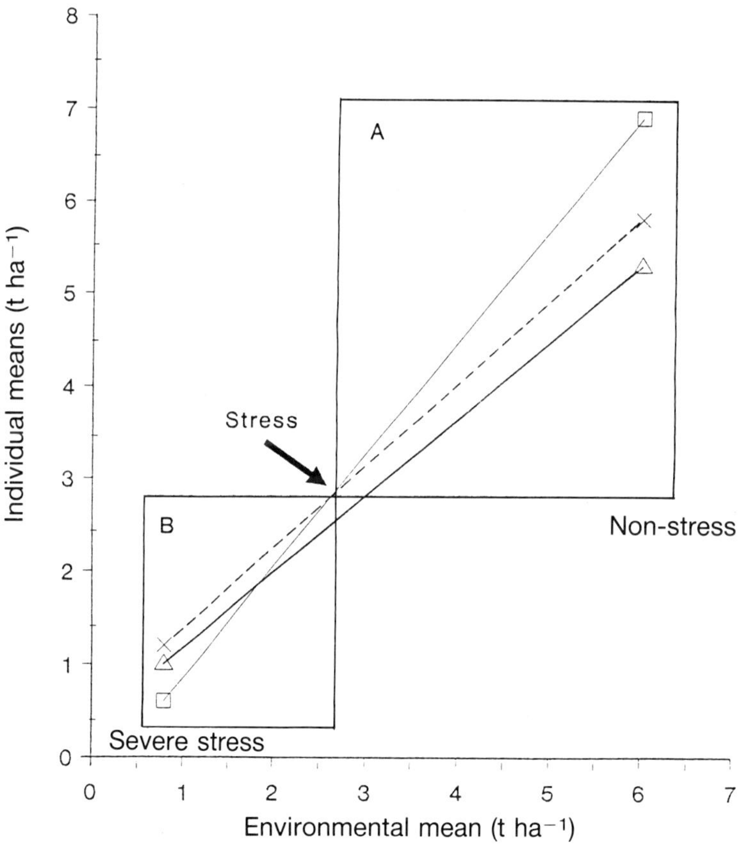

Figure 12.2 An example of different views of what constitutes a stress environment and its possible effect on individual variety assessment (modified from Ceccarelli, 1989) –□– b = 1.17 + 0.1; –△– b = 0.89 + 0.7; –×– b = 0.80 + 0.3

low yield potential such as those served in the ICARDA mandate. The argument (e.g. Srivastava, 1987; Ceccarelli, 1989) is well illustrated in *Figure 12.2* which shows that for yield levels above 3 t ha^{-1} ranking of varieties does not change, but below this level rankings can change dramatically.

The importance of defining stress environments is further illustrated by data from Syria. (The author is grateful to colleagues in the 'on-farm' testing programme at ICARDA and members of the Syrian National Agricultural Research programme, who have over the years accumulated these data.) The frequency of a particular site/year mean yield being obtained for barley grown in zones B (250–350 mm rainfall annually) and C (less than 250 mm) has been plotted. It is seen in *Figure 12.3* that the distribution is highly skewed to the low-yielding environments. Certain points stand out from this, the first being the low frequency of high-yielding environments. Second, the mean of varieties will be unduly affected by a few high-yielding site/year combinations. One 4 tonne site/year value has a four- to eight-fold greater influence on the mean of a variety than the 0.0–1.0 tonne site/year values. However, the 0.0–1.0 tonne site/years occur seven times more frequently than the site/years with more than 4 tonnes, which are rare. The mean

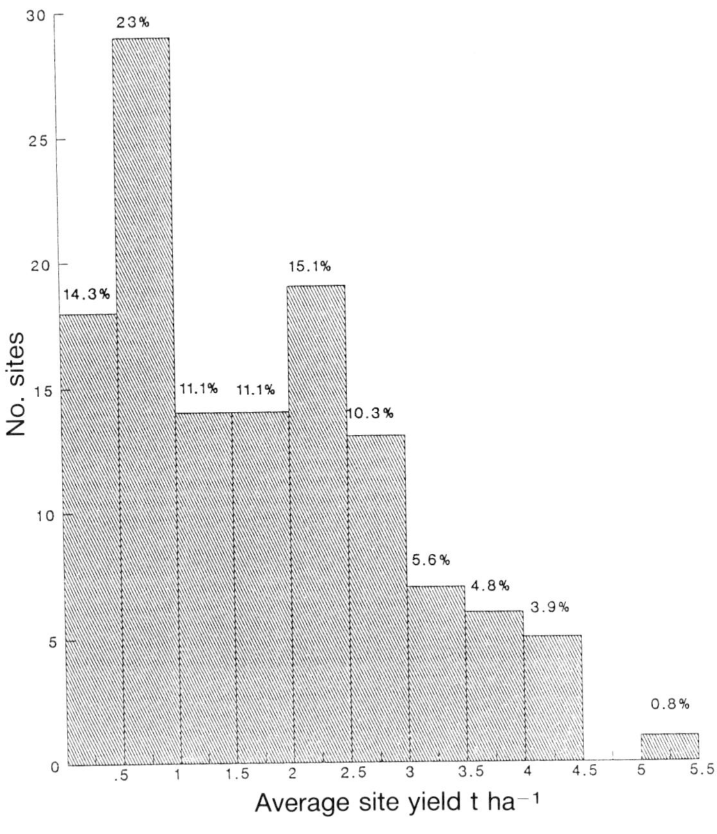

Figure 12.3 The frequency of different site mean yields for barley in the Syrian 'on farm' testing programme 1982–91 in zones B (250–350 mm) and C (less than 250 mm annual rainfall)

yield over environments is therefore biased upwards, in relation to the frequency with which these site/years occur. In the case illustrated here the mean is 1903 kg ha^{-1} whereas the median, a more appropriate measure, is less, 1715 kg ha^{-1}. The mean is about 11% higher than the median. This problem of weighting site yields has been discussed by Yau (1991), and is currently receiving considerable attention at ICARDA.

When wheat yields from zone A (between 350 and 650 mm) and irrigated areas are considered the pattern is different (*Figure 12.4*). Unfortunately direct comparison of wheat with wheat or barley with barley is not possible in these two contrasting groups of environments, stress and non-stress, as wheat trials are not grown in zone C and barley trials are not grown in zone A or under irrigation. The points made here are still valid; namely that the pattern of site/year yields is very different in stress and non-stress environments and that in the low-yielding environments there are major problems of scaling site/year yields to get a balance between frequencies and yield. This is less of a problem in non-stress environments where the distribution of yields is less skewed and their frequencies more equal. There is still a problem of the weighting given to high-yielding sites, but it is less pronounced in this situation. The mean yield only differed from the median by 5%.

When the frequency of high- and low-yielding environments is known, then theory (Rosielle and Hamblin, 1981) has some quite explicit recommendations. If the frequency of low-yielding sites is high (*Figure 12.3* situation), then it is better to select for high yields in stress environments (relative to the environmental

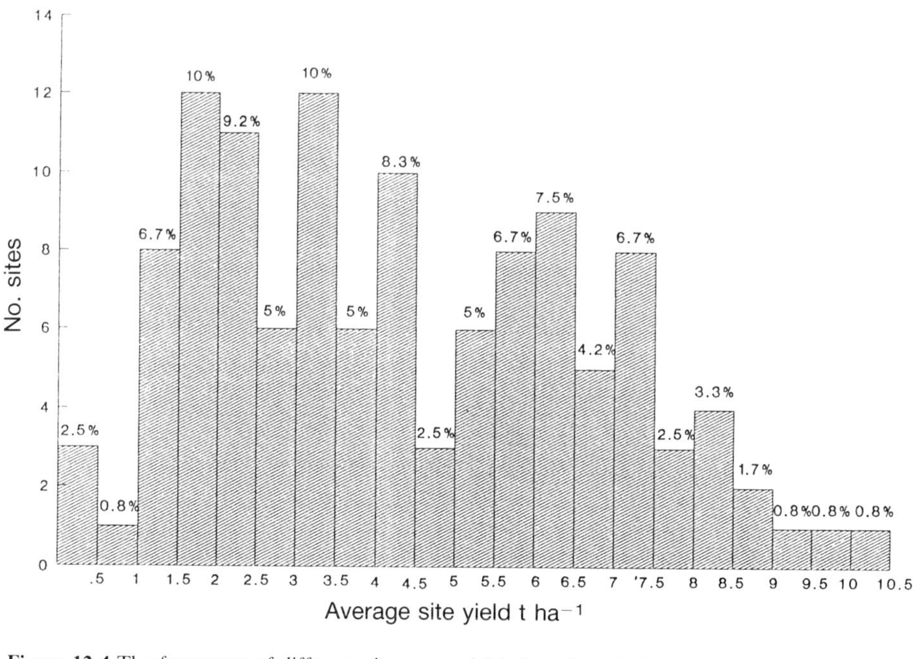

Figure 12.4 The frequency of different site mean yields for wheat in the Syrian 'on farm' testing programme 1982–91 in zone A (350–650 mm annual rainfall) and irrigated areas

potential) and forgo some yield potential in high-yielding environments as these are not frequent enough to compensate for reduced yield in the majority of low-yielding environments. However, when the frequency of high- and low-yielding environments is about the same (*Figure 12.4* situation) then one should select for a high mean yield and responsiveness to improved conditions.

Phenology and early leaf area development were identified earlier as two characters that relate to yield potential in stress environments. Using data from the spring barley and durum wheat programmes at ICARDA (the author is grateful to Drs Ceccarelli and Nachit for permission to use their data) a 'retrospective selection experiment' has been carried out. This allows an examination of material with the same development pattern and early leaf area growth to see what is the effect of selecting for both high and low yield in stress environments when the material is tested in both stress and non-stress environments.

For barley three data sets are considered (*Table 12.3*). They contained 332, 300 and 280 genotypes (the testing procedure has been reported by Ceccarelli and Grando, 1992). Each group (cohort) was grown for 3 years and in ten, nine and eight different site/year combinations respectively. The site/years were grouped into high- and low-yielding groups depending on whether they differed from the overall mean by more or less than one standard deviation (Ceccarelli and Grando, 1992). The data were sorted into groups by heading date on a daily basis and then within groups by growth vigour (an indication of early leaf area development). The data were then inspected to identify adjacent pairs of genotypes (or one apart)

Table 12.3 THE MEAN YIELDS (KG HA^{-1}) OF HIGH AND LOW YIELDING SELECTIONS IN STRESS ENVIRONMENTS AND THEIR CORRESPONDING AVERAGE YIELDS AND YIELDS IN NON-STRESS ENVIRONMENTS FOR 3 DIFFERENT COHORTS OF BARLEY GENOTYPES. (FOR DETAILS SEE CECCERELLI AND GRANDO, 1992)

	Cohort 1 *1986 – 88*	*Cohort 2* *1987 – 89*	*Cohort 3* *1988 – 90*
No. site/years	10	9	8
Total no. genotypes	332	300	280
No of genotypes selected	18 pairs	29 pairs	11 pairs
No of days with more than 30 genotypes headed	3	4	2
HSLYE kg ha^{-1}	1069	863	776
LSLYE kg ha^{-1}	491	436	382
SED	28	22	24
HSLYE in HYE kg ha^{-1}	4577	5270	2469
LSLYE in HYE kg ha^{-1}	5394	5522	3907
SED	105	85	148
HSLYE mean yield kg ha^{-1}	3053	2638	1737
LSLYE mean yield kg ha^{-1}	3456	2454	1979
SED	39	27	46
Early vigour HSLYE	2.74	2.51	2.59
Early vigour LSLYE	2.74	2.50	2.61

HSLYE, High selections, low yielding environments; LSLYE, Low selections, low yielding environments; LYE, Low yielding environments; HYE, High yielding environments; SED, Standard error of the difference for their growth vigour ranking (same heading date, very similar growth vigour) but that differed in yield in the low-yielding environments (selection in stress environments). The criterion for selection was that the difference between pairs of genotypes should be more than 50% of the mean yield in stress environments.

These were then compared for their yield in the high-yielding sites and for their average yield over all sites. Only heading dates on which more than 30 genotypes headed on that particular day were included for selection. The number of heading days in each cohort meeting this criterion varied from 2 to 4 (see *Table 12.3* for full details) and were so similar that the data over days was averaged. Also, as the comparison is between matched pairs of genotypes, the fact that the different pairs headed on slightly different days does not matter.

The data in *Table 12.3* show that:

1. Despite the selected genotype pairs having identical flowering dates and very similar early growth estimates, there were highly significant differences in mean yields of the groups and in their yields in the high- and low-yielding environments.
2. When tested in high-yielding environments the genotypes that were low yielding in stress environments were significantly higher yielding than the genotypes that were high yielding in stress environments.
3. This interaction has important implications for breeders working in stress environments.

The durum data are less extensive and only examine one cohort in one year and six environments. The material was separated by heading date into two groups, each having more than 30 entries per day; one group headed on days 102 and 103, the other on days 107, 108 and 109. The data within each day were ranked by growth vigour and high and low selections differing by 50% of the site mean yield were made at the lowest yielding site. The mean performance of these lines was compared at the highest yielding site and overall. The data are presented in *Table 12.4*.

The data in *Table 12.4* show a different response to barley, namely that there is no interaction in the performance of the two sets of selections at different

Table 12.4 THE MEAN YIELDS (KG HA^{-1}) OF LOW AND HIGH YIELDING SELECTIONS IN A STRESS ENVIRONMENT AND THEIR CORRESPONDING AVERAGE YIELDS AND YIELDS IN A NON-STRESS ENVIRONMENT OF ONE COHORT OF DURUM WHEAT GENOTYPES

	LYE	*HYE*	*Average*
Early heading group, days 102 and 103, 15 pairs.			
HSLYE	928	3097	1404
LSLYE	381	2918	1273
SED	18.5	426	101
Late heading group, days 107, 108 and 109, 20 pairs.			
HSLYE	791	3200	1352
LSLYE	265	2982	1124
SED	90	602	181

Abbrieviations as in Table 12.3. No of environments – 6; Total no. genotypes – 240 yield levels. The durum data are preliminary and therefore must be treated with more caution whereas the barley data, reported in *Table 12.3*, are consistent over cohorts, sites and years. However, if the durum results are confirmed, then there are challenges for crop physiologists and the modellers. Despite the selections being carefully paired for heading date and early vigour, the barley selections show an interaction with changed yield levels and the durum selections do not. From our current understanding of the growth of barley and durum wheat, can these results be predicted by the models? The author's initial reaction is that it would be unlikely, but this is an area worth further work as it is variations of this scale and type that are important to plant breeders.

Where resource capture models can assist plant breeders

There are several areas were resource capture models can help plant breeders: they are interrelated. The first is in examining phenological options. The second is to ask what/if questions about changing various components of growth and partitioning. The third is to assist in the choice of parents for hybridization.

If there are sufficient meteorological data available, one can use these models to assess whether the phenology of current varieties makes the best use of the growing season, and whether current phenology minimizes the risk of crop failures or low yields in the range of seasons that varieties are likely to encounter in one place. Having assessed the optimum development for different environments it should then be possible to examine the constancy of environmental signals and provide information to the breeders on possible genetic routes to this optimum by combining different degrees of vernalization, temperature and daylength response and inherent variation in the length of the basic vegetative phase. Based on this, breeders can then decide whether they will be able to select effectively for a particular set of developmental characteristics and also what parental material is most likely to give the desired recombinants. The risk associated with variation in different environmental signals on development and yield could be assessed against real seasonal data so that variability of different development strategies in variable environments can be assessed.

This area is one that could be quickly exploited by breeders and modellers. Different options could be assessed in real life to determine the effectiveness of this approach. However, the effect of small errors in development having large effects on potential yield must be remembered, particularly for low-yielding stress environments.

In the same vein, breeders could test the possibility of growing species in new areas by using crop growth models and climatic data. In this way the best crop options and the most suitable regions from which to obtain germplasm could be determined from a range of possible options.

The models could also be used to examine the potential of current varieties when planted over a range of sowing dates to determine if other development options would be more appropriate. One could assess if it is worthwhile developing varieties specifically for different sowing dates, or whether the advantages of doing so are not commensurate with the costs involved to the breeder and the farmer.

Another possibility is to explore how changes in patterns of growth and partitioning might affect yield in different environments (van Keulen and Seligman, 1987). For example: what are the effects of a faster rate of leaf development on water use through the season and the implications of this for grain yield and yield stability. Also how does the rate of root extension affect water and N availability and does this vary with soil type? There are obviously many questions of this type that could be explored initially by using models.

Using such questions one can carry out sensitivity analyses to see how much change in a particular character may be needed for the change to have a significant impact on production within and over seasons. This can be compared with the known level of variation for that character. The range of parental material needed to ensure genetic progress can be assessed. If genetic progress can be made, then the likely economic benefits also can be determined. This allows a best-bet assessment when attempts are made to modify plant characteristics.

One might be able to assess the potential of mixtures in reducing seasonal fluctuations in yield, using models to examine how a range of phenological options within a composite might maximize yield over a run of seasons. Having determined whether this was of benefit, one is still left with the question of how to ensure the survival of desirable components within such a mixture. The results may throw some light on how pure a variety needs to be from a biological, rather than a variety patenting, point of view.

Breeders would like to see some comparative runs of different models when the same questions are being addressed. If different models give similar results this would increase breeders' confidence in the output. If the answers were different breeders would probably have some choice phrases for the modellers!

To sum up, the challenge for the breeder working in stress environments is how to cope with highly fluctuating environments that cause large genotype by environment interactions and how to select most effectively for and in these environments. The challenge for the modeller is how and where can he help the breeders? Current models are capable of answering some of the broad questions of general interest, but are not yet able to answer the detailed questions that breeders cope with when producing varieties that must grow in randomly varying stress environments.

Acknowledgements

The author would like to thank colleagues at ICARDA for stimulating comments on various drafts of this chapter and George Kashour who carried out many of the calculations and who prepared the figures.

Footnote

After this manuscript was finalized the Agronomy Monograph no 31, '*Modelling plant and soil systems*' edited by Hanks and Ritchie (1991) was received, but too late to be included in this discussion.

References

Acevedo, E. (1987) Assessing crop and plant attributes for cereal improvement in water limited mediterranean environments. In *Drought Tolerance in Winter*

Cereals, pp 303–320 Edited by J.P. Srivastava, E. Porceddo, E. Acevedo and S. Varma. Chichester: John Wiley and Sons

Acevedo, E., Craufurd, P.Q., Austin, R.B. and Perez-Marco, P. (1991) Traits associated with high yield in barley in low-rainfall environments. *Journal of Agricultural Science (Cambridge)*, **116**, 23–36

Asana, R.R. (1965) Ideal and reality in crop plants. 1. Wheat. *Journal Indian Agricultural Research Institute Postgraduate School*, **3**, 63–68

Baker, J.T., Pinter Jr., P.J., Reginato, R.J. and Kanemasu, E.T. (1986) Effects of temperature on leaf appearance in spring and winter wheat cultivars. *Agronomy Journal*, **78**, 605–613

Belford, R.K., Klepper, B. and Rickman, R.W. (1987) Studies of intact shoot-root systems of field grown winter wheat. II. Root and shoot developmental patterns as related to nitrogen fertilizer. *Agronomy Journal*, **79**, 310–319

Bishop, A.M. (1986) *Growth and nutrient uptake of branching cultivars of* L. angustifolius. M.Sc. Thesis. Department of Soil Science and Plant Nutrition, University of Western Australia

Blum, A., Mayer, J. and Gozlan, G. (1982) Infrared thermal sensing of plant canopies as a screening technique for dehydration avoidance in wheat. *Field Crops Research*, **6**, 51–58

Boyd, W.J.R., Kaveeta, L., Potipibod, S. and Rossnagel, B. (1991) Variation in basic vegetative period and minimum leaf numbers. In *Barley Genetics VI, Proceedings 6th International Barley Genetics Symposium* Vol.1, pp 6–8 Edited by L. Munck. Copenhagen: Munksgaard

Braun, H.J, Pfeiffer W.H. and Pollmer, W.G (1992) Environments for selecting widely adapted spring wheats. *Crop Science*, **31**, 1420–1427

Brennan, P.S. and Byth, D.E. (1979) Genotype × environmental interactions for wheat yields and selection for widely adapted wheat genotypes. *Australian Journal of Agricultural Research*, **30**, 221–232

Brown, S.C., Gregory, P.J. and Wahbi, A. (1987) Root characteristics and water use in mediterranean environments. In *Drought Tolerance In Winter Cereals*, pp 275–283 Edited by J.P. Srivastava, E. Porceedu, E. Acevedo and S. Varma. Chichester: John Wiley and Sons

Byth, D.E., Eismann, R.L. and De Lacy, I.H. (1976) Two way pattern analysis of a large data set to evaluate genotypic adaption. *Heredity*, **37**, 215–230

Cao, W. and Moss, D.N. (1989a) Temperature effect on leaf emergence and phyllochron in wheat and barley. *Crop Science*, **29**, 1018–1021

Cao, W. and Moss, D.N. (1989b) Day-length effect on leaf emèrgence and phyllochron in wheat and barley. *Crop Science*, **29**, 1021–1025

Cao, W. and Moss, D.N. (1989c) Temperature and day-length interaction on phyllochron in wheat and barley. *Crop Science*, **29**, 1046–1048

Cao, W. and Moss, D.N. (1991) Vernalization and phyllochron in winter wheat. *Agronomy Journal*, **83**, 178–179

Ceccarelli, S. (1989) Wide adaptation: how wide? *Euphytica*, **40**, 197–205

Ceccarelli, S., Acevedo, E. and Grando, S. (1991) Breeding for yield stability in unpredictable environments: single traits, interaction between traits, and architecture of genotypes. *Euphytica*, **42**, 169–185

Ceccarelli, S. and Grando, S. (1991a) Selection environment and environmental sensitivity in barley. *Euphytica*, **57**, 157–167

Ceccarelli, S. and Grando, S. (1991b) Environment of selection and type of germplasm in barley breeding for low-yielding conditions. *Euphytica* in press

CGIAR/TAC (1988) *Report of the 3rd. external program review of CIMMYT* Mexico: CIMMYT

Donald, C.M. (1968) The breeding of crop ideotypes. *Euphytica*, **17**, 385–403

Ellis, R.H., Summerfield, R.J., Roberts, E.H. and Cooper, J.P. (1989) Environmental control of flowering in barley (*Hordeum vulgare*). III. Analysis of potential vernalization responses, and methods of screening germplasm for sensitivity to photoperiod and temperature. *Annals of Botany*, **63**, 687–704

Ellis, R.H., Roberts, E.H., Summerfield, R.J. and Cooper, J.P. (1989) Environmental control of flowering in barley (*Hordeum vulgare*). II. Rate of development as a function of temperature and photoperiod and its modification by low temperature vernalization. *Annals of Botany*, **62**, 145–158

Erskine, W., Ellis, R.H., Summerfield, R.J., Roberts, E.H. and Hussain, A. (1990) Characterization of responses to temperature and photoperiod for time to flowering in a world lentil collection. *Theoretical and Applied Genetics*, **80**, 193–199

Evans, L.T. and Wardlaw, I.F. (1976) Aspects of the comparative physiology of grain yield in cereals. *Advances in Agronomy*, **28**, 301–359

Farquhar, G.D. and Richards, R.A. (1984) Isotopic concentration of plant carbon correlates with water use efficiency of wheat genotypes. *Australian Journal of Plant Physiology*, **11**, 539–552

Farquhar, G.D., O'Leary, M.H. and Berry, J.A. (1982) On the relationship between carbon isotope discrimination and the intercellular carbon dioxide concentration in leaves. *Australian Journal of Plant Physiology*, **9**, 121–137

Gardener, C.J. (1966) *The physiological basis for yield differences in three high and three low yielding varieties of barley*. M.Sc. Thesis, University of Guelph, Ontario, Canada

Gregory, F.G. (1918) Physiological conditions in cucumber houses. *Third Annual Report, Experimental Research Station Cheshunt for 1917*, pp 19–28

Hamblin, A.P. and Hamblin, J. (1985) Root characteristics of temperate legume species and varieties on deep free draining entisols. *Australian Journal of Agricultural Research*, **36**, 63–72

Hamblin, J. (1971) *Single plant selection for yield*. Ph.D. Thesis. Department of Agronomy, University of Adelaide, South Australia

Hamblin, J. (1992) Workshop summary; Breeding for abiotic stress. In *Barley Genetics VI, Proceedings of 6th International Barley Genetics Symposum*, Vol.2 pp 909–910. Edited by L. Munck. Copenhagen: Munksgaard

Hamblin, J. and Donald, C.M. (1974) The relationship between plant form, competitive ability and grain yield in a barley cross. *Euphytica*, **23**, 535–542

Hardwick, R.C. (1984) Some recent developments in growth analysis – a review. *Annals of Botany*, **54**, 807–812

Hunt, R., Warren Wilson, J., Hand, D.W. and Sweeney, D.G. (1984) Integrated analysis of growth and light interception in winter lettuce. 1. Analytical methods and environmental influences. *Annals of Botany*, **54**, 743–757

Jensen, N.F. (1988) *Plant Breeding Methodology*. New York: John Wiley

Karimi, M.M. and Siddique, K.H.M. (1991) Crop growth and relative growth rates of old and modern wheat cultivars. *Australian Journal of Agricultural Research*, **42**, 13–20

Kirby, E.J.M., Appleyard, M. and Fellowes, G. (1985) Effect of sowing date and variety on main shoot leaf emergence and number of leaves of barley and wheat. *Agronomie*, **5**, 117–126

Kirby, E.J.M. and Perry, M.W. (1987) Leaf emergence rates of wheat in a mediterranean environment. *Australian Journal Agricultural Research*, **38**, 455–464

Klepper, B., Belford, R.K. and Rickman, R.W. (1984) Root and shoot development in winter wheat. *Agronomy Journal*, **76**, 117–122

van Keulen, H. and Seligman, N.G. (1987) *Simulation of Water Use, Nitrogen Nutrition and Growth of a Spring Wheat Crop*. Simulation Monographs. Wageningen: PUDOC

Loss, S.P., Perry, M.W. and Anderson, W.K. (1990) Flowering times of wheats in South-Western Australia: a modelling approach. *Australian Journal of Agricultural Research*, **41**, 213–223

Loss, S.P., Kirby, E.J.M., Siddique, K.H.M. and Perry, M.W. (1989) Grain growth and development of old and modern Australian wheats. *Field Crops Research*, **21**, 131–146

Lupton, F.G.H. (1966) The use of physiological characters in breeding for yield in wheat. *Acta Agriculturae Scandinavica. Supplementum*, **16**, 174–176

McVetty, P.B.E. and Evans, L.E. (1980) Breeding methodology in wheat. I. Determination of characters measured on spaced F2 plants for yield selection in spring wheat. *Crop Science*, **20**, 583–586

Masle, J., Doussinalt, G. and Sun, B. (1989) Response of wheat genotypes to temperature and photoperiod in natural conditions. *Crop Science*, **29**, 712–721

Milthorpe, F.L. (1956) The relative importance of the different stages of leaf growth in determining the resultant area. In *The Growth of Leaves*, pp 141–150 Edited by F.L. Milthorpe. Proceedings 3rd. Easter School in Agricultural Science, University of Nottingham. London: Butterworths

Monteith, J.L., Huda, A.K.S. and Midya, D. (1989) RESCAP: a resource capture model for sorghum and pearl millet. In *Modelling the growth and development of sorghum and pearl millet*, pp 30–34 Edited by S.M. Virmani, H.L.S. Tandon and G. Alagarswamy. Research Bulletin no 12. Patancheru A.P. 502 324, India: ICRISAT

Morgan, J.M. (1983) Osmoregulation as a selection criterion for drought tolerance in wheat. *Australian Journal of Agricultural Research*, **34**, 607–614

Morgan, J.M. and Condon, A.G. (1986) Water use, grain yield and osmoregulation in wheat. *Australian Journal of Plant Physiology*, **13**, 523–532

Pal, B.P., Upadhyaya, Y.M., Khan, H.R. and Ramanujam, S. (1960) A study of survival in a mixture of fourteen varieties of wheat. *Indian Journal of Genetics and Plant Breeding*, **20**, 102–112

Passioura, J.B. (1972) The effect of root geometry on the yield of wheat growing on stored water. *Australian Journal of Agricultural Research*, **23**, 745–752

Passioura, J.B. (1981) The interaction between the physiology and the breeding of wheat. In *Wheat Science – today and tomorrow*. pp 191–201 Edited by L.T. Evans and W.J. Peacock. Cambridge: Cambridge University Press

Passioura, J.B. (1983) Roots and drought resistance. *Agricultural Water Management*, **7**, 265–280

Pendleton, J.W., Smith, G.E., Winter, S.R. and Johnston, R.J. (1968) Field investigations of the relationships of leaf angle in corn (*Zea mays* L.) to grain yield and apparent photosynthesis. *Agronomy Journal*, **60**, 422–424

Perry, M.W. and D'Antuono, M.F. (1989) Yield improvement and associated characteristics of some Australian spring wheat cultivars introduced between 1860 and 1982. *Australian Journal of Agricultural Research*, **40**, 457–472

Perry, M.W., Siddique, K.H.M. and Wallace, J.F. (1987) Predicting phenological development for Australian wheats. *Australian Journal of Agricultural Research*, **38**, 809–819

Richards, R.A. and Passioura, J.B. (1981a) Seminal root morphology and water use of wheat. I. Environmental effects. *Crop Science*, **21**, 249–252

Richards, R.A. and Passioura, J.B. (1981b) Seminal root morphology and water use of wheat. II. Genetic variation. *Crop Science*, **21**, 252–255

Roberts, E.H., Summerfield, R.J., Cooper, J.P. and Ellis, R.H. (1988) Environmental control of flowering in barley (*Hordeum vulgare*). I. Photoperiod limits to long-day responses, photoperiod-insensitivity phases and effects of low-temperature and short-day vernalization. *Annals of Botany*, **62**, 127–144

Rosielle, A.A. and Hamblin, J. (1981) Theoretical aspects of selection for yield in stress and non-stress environments. *Crop Science*, **21**, 943–946

Siddique, K.H.M., Belford, R.K., Perry, M.W. and Tennant, D. (1989) Growth, development and light interception of old and modern wheat cultivars in a Mediterranean-type environment. *Australian Journal of Agricultural Research*, **40**, 473–487

Simmonds, N.W. (1979) *Principles of Crop Improvement*. London: Longman

Simmonds, N.W. (1981) Genotype (G) environment (E) and GE components of crop yields. *Experimental Agriculture*, **17**, 355–362

Simmonds, N.W. (1991) Selection for local adaption in a plant breeding program. *Theoretical and Applied Genetics*, **82**: 363–367

Singh, U., Godwin, D.C., Humphries, C.G. and Ritchie, J.T. (1989) A computer model to predict the growth and development of cereals. In *Proceedings 1989 Summer Computer Simulation Conference*. Austin Texas. pp 668–675 Edited by J.K. Clema

Singh, P., Saxena, N.P., Monteith, J.L. and Huda, A.K.S. (1990) Defining, modeling and managing water requirement of chickpea. ICRISAT 1990. *Chickpea in the nineties: Proceedings 2nd International Workshop on Chickpea Improvement*, 4–8 Dec 1989 ICRISAT Center, India. Patancheru, A.P. 502 324 India

Srivastava, J.P. (1987) Barley and wheat improvement for moisture limiting areas in West Asia and North Africa. In *Drought Tolerance in Winter Cereals*. Edited by J.P. Srivastava, E. Porceddo, E. Acevedo and S. Varma. Chichester: John Wiley

Stapper, M. (1984) *Simulations assessing the productivity of wheat maturity types in a mediterranean climate*. Ph.D. Thesis. University of New England, Armidale, N.S.W. Australia

Stapper, M. and Fischer, R.A. (1990) Genotype, sowing date and plant spacing influence on high yielding irrigated wheat. 1. Phasic development, canopy growth and spike production. *Australian Journal of Agricultural Research*, **41**, 997–1019

Stockle, C.O. (1989) Simulation models in agriculture: from cellular level to field scale. In *Proceedings 1989 Summer Computer Simulation Conference*, Austin Texas. pp 639–644 Edited by J.K.Clema

Syme, J.R. (1974) Leaf appearance rate and associated characters in some Mexican and Australian wheats. *Australian Journal of Agricultural Research*, **25**, 1–7

Tanaka, A., Kawana, K. and Yamaguchi, J. (1966) Photosynthesis, respiration, and plant type of the tropical rice plant. *Technical Bulletin, International Rice Research Institute*, No.3. Los Banos, The Phillipines

Tanner, J.W., Gardner, C.J., Stoskopf, N.C. and Reinsbergs, E. (1966) Some observations on upright-leaf-type small grains. *Canadian Journal of Plant Science*, **29**, 690

Turner, N.C. and Nicolas, M.E. (1987) Drought resistance of wheat for light textured soils in a mediterranean climate. In *Drought Tolerance in Winter Cereals*. pp 203–216 Edited by J.P. Srivastava, E. Porceedu, E. Acevedo and S. Varma. Chichester: John Wiley

Vivar, H. (1991) *Annual report cereal improvement program for 1990*. Aleppo, Syria: ICARDA

Volk, T. and Bugbee, B. (1991) Modelling light and temperature effects on leaf emergence in wheat and barley. *Crop Science*, **31**, 1218–1224

Wahbi, A. and Gregory, P.J. (1989) Genotypic differences in root and shoot growth of barley (*Hordeum vulgare*). II. Field studies of growth and water use of crops grown in northern Syria. *Experimental Agriculture*, **25**, 389–399

Warren Wilson, J. (1981) Analysis of growth, photosynthesis and light interception for single plants and swards. *Annals of Botany*, **48**, 507–512

Watson, D.J. and Witts, K.J. (1959) The net assimilation rate of wild and cultivated beets. *Annals of Botany (NS)*, **23**, 431–439

Yau, S.K. (1991) Variance of relative yield as an agronomic type of stability measure. *Proceedings 8th Meeting of Eucarpia section Biometrics on Plant Breeding*. pp 297–306 Edited by J. Pesek, M. Herman, and H. Hartmann. Brno: Eucarpia

Zobel, R.W., Wright, M.J. and Gauch Jr., H.G. (1988) Statistical analysis of a yield trial. *Agronomy Journal*, **80**, 88–393

13

COMPETITION FOR RESOURCE CAPTURE IN AGRICULTURAL CROPS

M.J. KROPFF
International Rice Research Institute, P.O. Box 933, 1099 Manila, The Philippines

J. GOUDRIAAN
Department of Theoretical Production Ecology, Wageningen Agricultural University, P.O. Box 430, 6700 AK Wageningen, The Netherlands

Introduction

Interplant competition for capture of the essential resources for plant growth i.e. light, water and nutrients, strongly affects the performance of natural, semi-natural and agricultural ecosystems. Ecologists have studied competition to understand the diversity and stability of plant communities, succession patterns of vegetation, and to help define management strategies for semi-natural ecosystems. Agroecologists have studied competitive phenomena to optimize plant densities of crops, to optimize intercropping systems and to quantify crop–weed interactions to improve weed management systems with minimum herbicide inputs. Similar approaches have been used to study interplant competition in natural and disturbed systems. However, because of the complex nature of interplant competition, it has taken a long time to develop generalized concepts and theories.

The first systematic approaches for studying competitive phenomena were developed in the 1960s. For monocultures, much of our current understanding is based on the work of a Japanese group (e.g. Shinozaki and Kira, 1956) whereas de Wit (1960) developed the first systematic approach to study competition in mixtures. He introduced an experimental design (the replacement series in which the mixing ratio varies, but total density remains constant) with a model to analyse the results. These approaches were based on a hyperbolic relationship between plant density and yield, and have been used extensively in agricultural and ecological sciences to study competition between plants, plant population dynamics, and component contributions of intercropping systems (see reviews by Trenbath, 1976; Harper, 1977; Radosevich and Holt, 1984; Grace and Tilman, 1990). Recently, several papers discussed the disadvantages and pitfalls of the replacement series approach, such as its total density dependence (cf. Connolly, 1986; Taylor and Aarssen, 1989). Only in the early 1980s, approaches were developed to describe competition over a range of population densities with varying mixing ratios and at a range of total densities, generally also based on the hyperbolic yield–density relationship (Suehiro and Ogawa, 1980; Wright, 1981; Spitters, 1983a, b; Cousens, 1985; Spitters, Kropff and de Groot, 1989). Similar approaches have been developed using the neighbourhood approach, in which the number of neighbours of an individual plant in a predefined area is related to the weight of the central plant (Silander and Pacala, 1985; Firbank and Watkinson,

1987; Pacala and Silander, 1987). However, a great deal of plant size frequency remains unexplained by the spatial arrangement of the plants alone (Firbank and Watkinson, 1990). All these approaches have a phenomenological character: the outcome of competition is described at a given moment, but no explanation is given of the process. The extrapolation domain of these descriptive approaches is often limited; they only account for the effect of plant density on the competition process. This results in a variation of model parameters from year to year and site to site (Firbank and Watkinson, 1990). Also, these descriptive approaches cannot help identify the mechanisms for evolution of species traits or for the structure and dynamics of populations, communities and ecosystems (Tilman, 1988), and they cannot help in explaining the mechanisms for variation in competition effects between seasons and sites (Kropff, 1988).

In the 1980s, more mechanistic approaches were developed to quantify interplant competition by taking into account the temporal dynamics of competition. Spitters and colleagues developed ecophysiological models for interplant competition, based on ecophysiological models for monoculture crops (Spitters and Aerts, 1983; Kropff *et al.*, 1984). They focussed on competition in weedy crop situations. Several other groups developed similar approaches to quantify crop–weed interactions (Graf *et al.*, 1990; Ryel *et al.*, 1990; Wilkerson *et al.*, 1990), competition between grassland species (Rimmington, 1984), and competition in row crops (Hodges and Evans, 1990). In ecology, mechanistic models for interplant competition have been developed with a focus on competition for nutrients and succession in more complex semi-natural plant communities (e.g. Tilman, 1988; Berendse, 1985). In these long-term studies, the approach is vastly different from the approach taken in the crop–weed interaction studies where instantaneous processes are taken into account.

The actual mechanisms of competition for resource capture by plants are complex. Plants are morphologically and physiologically extremely plastic in their response to their environment, making generalization of plant responses difficult. This chapter reviews currently used approaches to the quantification of effects of interplant competition in monoculture crops, intercrops and weedy crops. Available descriptive as well as mechanistic approaches will be discussed.

Approaches to describe the effect of competition on resource capture in agricultural crops

COMPETITION WITHIN MONOCULTURES

Competition in monocultures has been studied extensively by agronomists conducting density experiments to optimize harvestable yield and minimize seed inputs in the system. Montgomery (1912, cited by de Wit, 1960) reported the yield density response of oats, showing that yields increased with reduced spacing until a maximum yield level was reached. In the 1960s, a group of Japanese scientists identified the major effects of competition in monocultures and developed approaches to mathematically describe relationships that appeared to be consistent (Kira, Ogawa and Sakazaki, 1953; Shinozaki and Kira, 1956; Yoda *et al.*, 1963). They related plant size and plant survival (self-thinning) to competition effects as a result of differences in plant spacing. De Wit (1960) also discussed the issue of competition in monocultures, used the hyperbolic equation for the

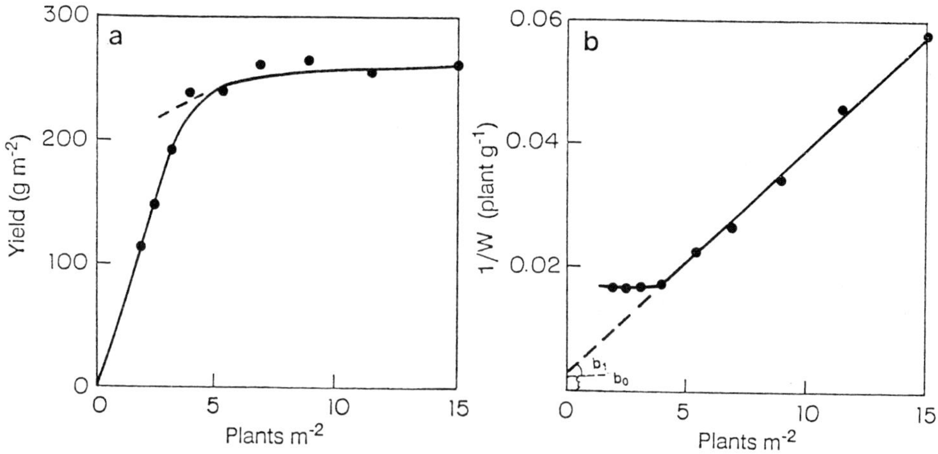

Figure 13.1 Density response of maize. Plots of (**a**) biomass per unit area and (**b**) the reciprocal of per-plant weight versus plant density. (After: Spitters, 1983a)

yield density response and discussed differences with the approach introduced by Shinozaki and Kira (1956). The most widely used approach to describe the density response is a rectangular hyperbola (e.g. Shinozaki and Kira, 1956; de Wit, 1960) (*Figure 13.1*):

$$Y = N \,/\, (b_0 + b_1 N) \tag{1a}$$

where Y is the yield of the crop in monoculture in g m^{-2}, N is the plant density of the crop in numbers per square metre, and b_0 is the intercept and b_1 is the slope of the relationship between N/Y and N. The reciprocal per plant weight then is a linear function of plant density:

$$1/w = N/Y = b_0 + b_1 N \tag{1b}$$

where w is the weight per plant (g plant^{-1}). The intercept b_0 is the reciprocal of the virtual biomass of an isolated plant. It deviates from the real biomass of and isolated plant, because at wide spacings, without interplant competition the biomass is density independent and does not decrease with density as the hyperbola suggests. This is illustrated in *Figure 13.1* where the relation between $1/w$ and density levels off at low densities. The parameter b_1 is the reciprocal of the maximum biomass per unit area e.g. the asymptote in *Figure 13.1a*.

This relationship can be used to describe the response of total dry matter production of crops to density reasonably well except for extremely low densities (Figure 13.2). For harvestable yield, however, a very narrow optimum density has been observed for several species. This is illustrated in *Figure 13.2* for some crops that respond differently (de Wit, van Laar and van Keulen, 1979). In cereals, seed yield and total dry matter production are constant over a wide range of densities in which the crop is able to form a closed canopy. These plants have the characteristics of a determinate growing point that terminates in an ear or panicle for every tiller that survives. Interplant competition determines

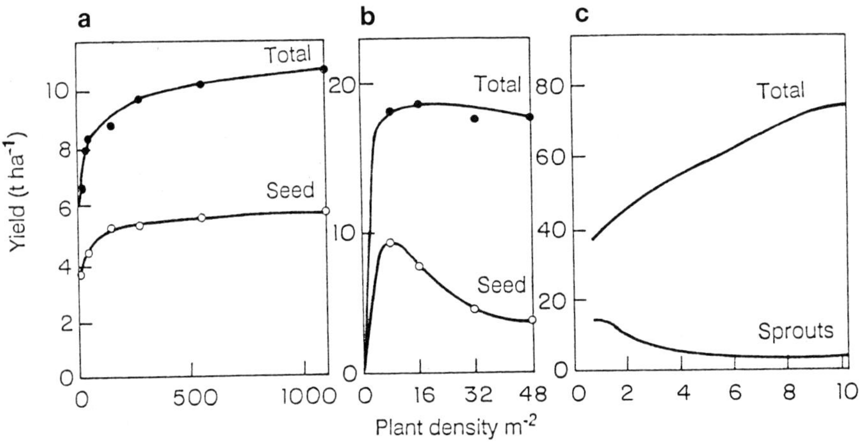

Figure 13.2 Relation between plant density and yield (total and harvestable) for barley (**a**), maize (**b**), and Brussels sprouts (**c**). (After: de Wit, van Laar and van Keulen, 1979)

tiller survival and not so much the size of the ears or panicles. Therefore, the number of tillers per unit of area in cereals is relatively constant over a wide range of densities (for wheat: Darwinkel, 1978; for rice: Yoshida, 1981). However, optimum spacing may differ between varieties. For rice it was found that early maturing varieties need a higher density than late maturing varieties to form a dense canopy (Yoshida, 1981).

In indeterminate plants such as Brussels sprouts, the growing point remains vegetative and the harvestable yield is formed by axillary leafy buds that have to reach an appropriate size. If interplant competition is too strong, the amount of resources captured by a single plant is too small to produce enough assimilates for the main and secondary growing points. For harvestable yield of Brussels sprouts, the optimum density is very low (almost free growing plants are needed), whereas total dry matter production per unit of area continues to increase with plant density, because the crop as a whole captures more resources. For maize, an intermediate situation has been observed. The cob emerges also from axillary buds, and when plant density is higher than the plant density needed to form a closed canopy for optimum resource capture, total dry matter production remains constant, but harvestable yield decreases (*Figure 13.2*). If the relationship between harvestable yield and density shows a maximum, an alternative model has to be used. Spitters (1983b) added a quadratic term to Equation 1:

$$1/w = b_0 + b_1N + b_2N^2 \qquad (2)$$

Watkinson (1980) introduced a power term to account for optimum shape of responses:

$$w = w_m (1 + aN)^{-b} \qquad (3)$$

where w_m is the weight of an isolated plant.

In plant monocultures, interplant competition may result in mortality of the least competitive plants. This phenomenon is known as self-thinning, which was first described by Yoda *et al.*, 1963. They found a close relationship between the number of plants per unit area that survived and the weight per plant. Their so-called 2/3 power law describes the dynamics of the system i.e. how density declines as mean weight per plant increases in a single stand in time. This relationship is intensively used in forestry, to determine thinning rates for tree stands, to optimize timber production. For a detailed discussion of these approaches refer to Firbank and Watkinson (1990).

Another aspect of competition in monoculture stands is the variation in individual plant size. Firbank and Watkinson (1990) showed evidence that the variability in individual plant weight increases with increasing density i.e. competition. Differences in individual plant weight and size are caused by differences in starting position of the individual plants (moment of germination, seed size) and genetic variation in growth parameters. In general, relative differences remain the same throughout the season.

COMPETITION IN INTERCROPPING SYSTEMS

Intercropping is intensively practiced in the tropics. These systems range from complex systems in which farmers fill up all space with crop plants that are needed for subsistence, to agroforestry systems, where tree species are combined with annual crop species. The most commonly used systems are: (1) mixed intercropping – growing two or more crops simultaneously, without a specific spatial arrangement; (2) row or strip intercropping – growing two or more crops simultaneously where one or more crops are planted in rows or strips; (3) relay intercropping – growing two or more crops simultaneously with only a partial overlap of the growing cycle.

The agronomic aspects of these systems have been thoroughly reviewed by Willey (1979a, b) and Vandermeer (1989). These systems can have several advantages over monoculture systems, of which the potential higher efficiency of resource use is the most important one studied. The advantage of intercropping over monoculture systems is generally expressed using the LER (Land Equivalent Ratio) (Willey, 1979a, b):

$$\mathrm{LER} = Y_{1,\mathrm{mix}}/Y_{1,\mathrm{mono}} + Y_{2,\mathrm{mix}}/Y_{2,\mathrm{mono}} \tag{4}$$

where $Y_{1,\mathrm{mono}}$ and $Y_{2,\mathrm{mono}}$ are the yields of the species in monoculture and $Y_{1,\mathrm{mix}}$ and $Y_{2,\mathrm{mix}}$ are the yields in mixture. When LER=1, the same yields can be obtained with monoculture, using the same area of land. When LER>1 a larger area of land is needed to produce the same yields in monocultures as in the mixture. Often LER values are reported that are much higher than 1 (Willey, 1979a, b). However, in these standard mixed cropping trials, the individual crops in mixture are grown at the same density as in the respective monocultures. In many situations where high values for LER are reported, the advantage could therefore also be obtained by growing the crop in monoculture at a higher density. The replacement series approach developed by de Wit (1960), is much more suitable to address questions related to the real yield advantage of intercropping over monoculture cropping. The relative total density is the same in mixtures and monocultures.

The yield of a species is expressed relative to its yield in monoculture. The sum of the relative yields is the RYT (Relative Yield Total). If RYT>1, there is a true advantage of mixed cropping. However, in the replacement series design, some pitfalls exist as well (Connolly, 1986).

An alternative approach was developed by Spitters (1983a, b). This approach starts from the same principles as the approach of de Wit (1960). The starting point in the derivation of the model is the response of crop yield to plant density, which can be described by a rectangular hyperbola (Equation 1). The effect of other species can be introduced in this equation in an additive way (Suehiro and Ogawa, 1980; Wright, 1981; Spitters, 1983a):

$$Y_{12} = N_1 / (b_{10} + b_{11}N_1 + b_{12}N_2) \quad (5a)$$

The reciprocal per plant weight then equals:

$$1/w_1 = b_{10} + b_{11}N_1 + b_{12}N_2 \quad (5b)$$

where Y_{12} is the yield of species 1 in a mixture with species 2, and N is the number of plants. The parameter b_{11} measures intraspecific competition between plants of species 1 and the parameter b_{12} measures interspecific competition effects between the species. A similar equation can be derived for species 2. *Figure 13.3* demonstrates the shape of such a relationship for groundnut and maize (Spitters, 1983a). This approach allows the analysis over a range of total densities and mixing ratios. The parameter values can be used to derive indices for the relative competitiveness of the species (b_{12}/b_{11}) and niche differentiation: If the ratio $(b_{11}/b_{12})/(b_{21}/b_{22})$ exceeds unity, there is niche differentiation and a RYT>1, indicating that the mixture as a total captures more resources than the respective monocultures. This can be the case when species have different

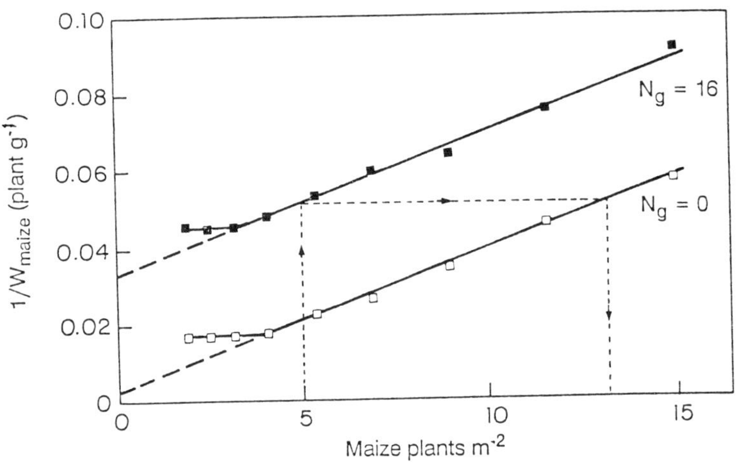

Figure 13.3 The effect of adding 16 groundnut plants (N_g =16) to a monocrop of maize (N_g=0) had the same effect on maize biomass as adding eight maize plants. (After: Spitters, 1983a)

rooting systems, exploiting different compartments of the soil or in the situation of legumes, intercropped with other crops that can use the N fixed by the Rhizobia. The parameters can be estimated using non-linear regression techniques on the logarithm of both sides of the equation because the yields are distributed log-normally.

For harvestable yield, Spitters (1983b) derived a parabolic equation based on Equation 2, to account for the effect of strong competition on harvestable yield.

COMPETITION IN WEEDY CROPS

The hyperbolic yield density equation for the description of yield loss in relation to weed density is the most widely used regression model to describe effects of competition at one point in time (Cousens, 1985; Spitters, van Keulen and van Kraalingen, 1989). Based on the general formulation of Spitters (1983a) (Equation 5), a one parameter hyperbolic function can be derived for the effect of weeds on crops (Spitters, Kropff and de Groot, 1989):

$$Y_L = a\, N_w \,/\, (1 + a\, N_w) \qquad (6)$$

where Y_L gives the yield loss relative to the weed-free crop, N_w is the weed density and a describes the yield loss caused by adding the first weed per square metre. In this one-parameter regression model, maximum fractional yield loss is 1 (or 100%) at high weed densities.

The approach described by Cousens (1985) involves an additional parameter which permits a maximum yield loss of less than 100% (m):

$$Y_L = a\, N_w \,/\, (1 + a/m\, N_w) \qquad (7)$$

These hyperbolic yield–density equations fit well to data from experiments where only the weed density is varied (Kropff *et al.*, 1984; Cousens, 1985; Weaver, Smits and Tan, 1987; Spitters, Kropff and de Groot, 1989; Kropff, Weaver and Smits, 1992). However, the parameters a and m may vary strongly among experiments due to the effect of other factors on competition processes. An example demonstrating the large difference in effects of *Echinochloa crus-galli* (L.) P. Beauv. on maize in two subsequent growing seasons is given in *Figure 13.4*. In this situation, the extra parameter for a limited maximum yield loss (Equation 7) was not needed to obtain a good fit of the model. In other situations, a very clear maximum yield loss smaller than 100% was observed at high weed densities (Weaver, Smits and Tan, 1987). Because variation in the yield loss–weed density function is often largely determined by differences in the period between crop and weed emergence, (Håkansson, 1983; Kropff *et al.*, Cousens *et al.*, 1987; Kropff, Weaver and Smits, 1992), precise prediction of yield loss on the basis of early observations should be based on both weed density and the period between crop and weed emergence. An additional variable in the hyperbolic yield–density equation to account for the effect of differences in the period between crop and weed emergence was introduced by Håkansson (1983) and Cousens *et al.* (1987). However, in practice, weeds often emerge in successive flushes, making it difficult to apply a descriptive model that accounts for the effect of both weed density and the relative time of weed emergence because every flush has to be regarded as

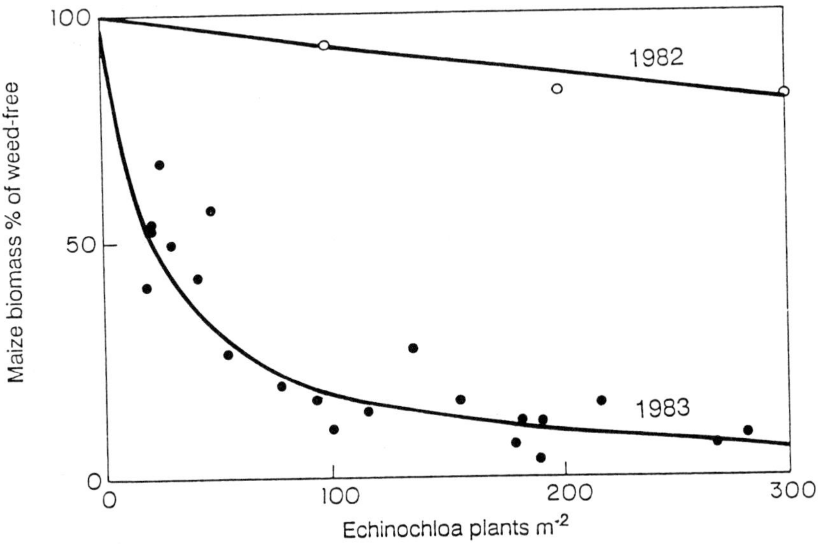

Figure 13.4 The effect of different densities of *Echinochloa crus-galli* L. on maize yield in 2 years at the same site in Wageningen, The Netherlands. (After: Kropff *et al.*, 1984)

a different weed species. Therefore alternative approaches are needed to predict yield loss by weeds for use in weed management systems.

Toward a mechanistic understanding of competition for resource capture in agricultural crops

The approaches previously discussed are all descriptive at the system level and do not explain competition effects at the process level. To improve the productivity and resource use efficiency of crop plants by varietal improvement and optimization of crop management in various competition situations, quantitative insight in the mechanisms of competition and the role of morphological and physiological plant traits is needed. Relationships between morphological and physiological characteristics of species and their competitive strength have been widely studied (cf. Pearcy, Tumosa and Williams 1981; Ampong-Nyarko and De Datta, 1989). Pearcy, Tumosa and Williams (1981) found a close relationship between competitive ability and photosynthetic responses of *Chenopodium album* L. (C_3 plant type) and redroot pigweed (*Amaranthus retroflexus* L., C_4 plant type). However, causal relationships should include processes related to resource capture and not only resource use efficiency. The complexity of relationships between morphological and physiological characteristics and competitive ability of plants in mixtures has been recognized by many researchers.

Ecophysiological models for interplant competition for the resources light, water and nutrients can be used as a tool to study these complex relationships. These models are based on the assumption that competition is a dynamic process which can be understood from the distribution of the limiting resources between the competing neighbouring plants and the efficiency with which each plant uses the resources captured. Such a mechanistic approach provides insight into the processes underlying competition effects observed in (field) experiments and may so be of help in searching for ways to manipulate competitive relations, like those between crop and weeds.

Most effort to simulate interplant competition using ecophysiological models has been on crop–weed interactions (Spitters and Aerts, 1983; Kropff *et al.*, 1984, Kropff, 1988; Graf *et al.*, 1990; Ryel *et al.*, 1990; Wilkerson *et al.*, 1990; Kropff and Spitters, 1992). A few attempts have been made to develop ecophysiological models for interplant competition in intercropping systems (e.g. Caldwell and Hansen, 1990). Because of the complexity of these systems, ecophysiological models may help to optimize the systems and to suggest optimum plant densities, planting dates, cultivar combinations, crop management practices, seeding rates, etc. Detailed data from field experiments, however, will be required to parameterize, calibrate and validate the models. Ecophysiological models for monoculture crops, generally do not simulate the growth of individual plants that compete for resource capture. A simple preliminary model for competition between individual plants was developed by R. Stokkers, M.J. Kropff, J. Goudriaan and J.A. den Dulk, Wageningen Agricultural University, unpublished), using a simulation model for interplant competition that is based on the assumption that the leaves are distributed homogeneously over an area. The model simulates competition for light for every individual plant by taking into account the leaf area and height of its neighbours. Preliminary results demonstrated the principle that relative differences between plants remain the same throughout the growing season. Hodges and Evans (1990) developed a model for the effect of row spacing on maize yield. They added to the CERES-Maize model (Jones and Kiniry, 1986) a light distribution model, only based on distribution of direct radiation. Their model accurately predicted effects of differences in spacing and plant density on yield.

An ecophysiological model that was developed to simulate crop–weed interactions (see later) developed by Kropff and Spitters (1992) was used to analyse the effect of shading by panicles in rice at different positions in the canopy (Kropff, Cassman and van Laar, 1993). The model demonstrated that the existing variation in panicle height caused differences in light interception resulting in more than 15% differences in daily photosynthetic rates during the grain filling period, assuming that the panicles do not photosynthesize.

AN ECOPHYSIOLOGICAL MODEL FOR INTERPLANT COMPETITION

In the past three decades, many mechanistic or ecophysiological simulation models for crop growth and production have been developed. These models simulate crop growth on basis of the response of physiological processes to the environment (cf. de Wit *et al.*, 1978; Penning de Vries and van Laar, 1982; Jones and Kiniry, 1986; Jones *et al.*, 1989). The approach taken to simulate the growth of a mixture of species is very similar to the approaches used in simulation of the growth of monoculture crops. The single difference is the calculation procedure for resource capture. Thus, competition for resource capture is also simulated on a per area basis in these models. They account for vertical heterogeneity in the canopy, but it is assumed that the leaves are horizontally homogeneously distributed. The general structure of the modelling approach to simulate competition for light and

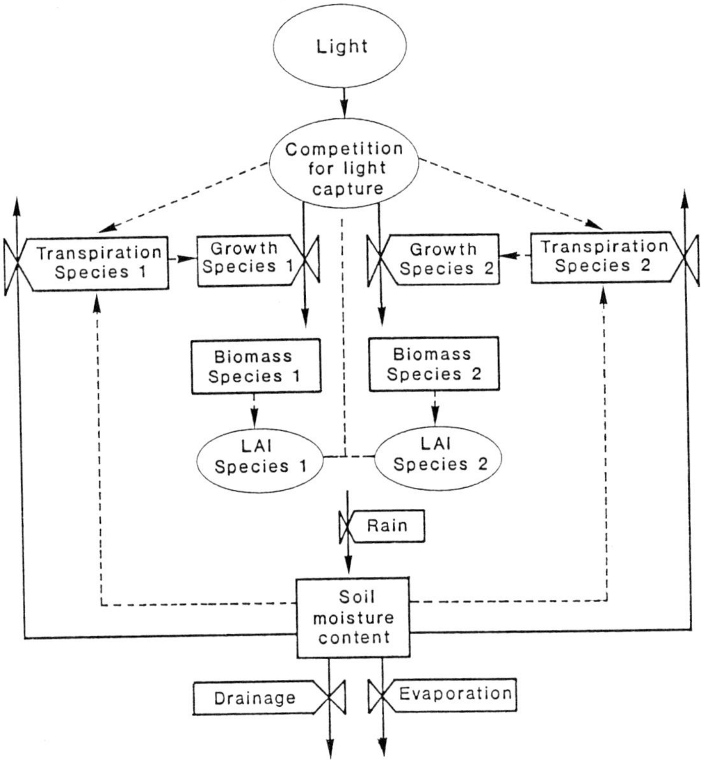

Figure 13.5 General structure of the ecophysiological model for interplant competition (INTERCOM)

water is given in *Figure 13.5*. Under favourable growth conditions, light is the main factor determining the growth rate of the crop and its associated weeds. From the leaf area indices (*LAI*) of the species and the vertical distribution of their leaf area the light profile within the canopy is calculated. In these functions, it is assumed that the leaves are distributed over the total height of the plants according to a parabolic function. In a canopy with a mixture of species, the net light flux at height h can be calculated from:

$$I_h = (1 - \rho)\, I_0 \exp(-\Sigma\, (k_j\, L_{h,j})) \tag{8}$$

where

I_h is the net flux (PAR) at height h (J m^{-2} ground s^{-1}),

I_0 is the flux at the top of the canopy (J m^{-2} ground s^{-1}),

$L_{h,j}$ the cumulative leaf are index of species j above height h (m^2 leaf m^{-2} ground),

ρ the reflection coefficient of the canopy, and

k_j the extinction coefficient of species j.

The leaf areas ($L_{h,j}$), weighted by the extinction coefficients (k_j), are summed over the $j = 1 \ldots n$ plant species in the mixed vegetation.

The light absorbed by species i at a height h in the canopy ($I_{a,h,i}$, J m^{-2} leaf s^{-1}) is obtained by taking the derivative of Equation 8 with respect to the cumulative leaf area index:

$$I_{a,h,i} = -\, dI_{h,i} \,/\, dL_i = k_i\,(1 - \rho)\, I_0 \exp(-\Sigma\,(k_j\, L_{h,j})) \qquad (9)$$

The same equation is used to calculate the radiation absorbed by stems and reproductive organs, i.e. these organs are regarded as different species. The diffuse and the direct flux have different extinction coefficients, resulting in different light profiles within the canopy for diffuse and direct radiation. Therefore, three different radiation fluxes are distinguished: the component of direct light, the total direct flux and the diffuse flux.

The CO_2 assimilation-light response of individual leaves follows a saturation type of function, characterized by the initial slope (the initial light use efficiency) and the asymptote (A_m) and is described by the negative exponential function (Goudriaan, 1982):

$$A_h = A_m\,(1 - \exp(-\epsilon\, I_a \,/\, A_m)) \qquad (10)$$

where

A_h is the gross assimilation rate (gCO_2 m^{-2} leaf s^{-1}),
A_m the gross assimilation rate at light saturation (gCO_2 m^{-2} leaf s^{-1}),
ϵ the initial light use efficiency (gCO_2 J^{-1}), and
I_a is the amount of absorbed radiation (J m^{-2} leaf s^{-1}).

From the absorbed light intensity at height h for one of the species, the assimilation rate of species i at that specific canopy height can be calculated with Equation 10. This procedure is followed for sunlit and shaded leaves separately. Based on photosynthesis characteristics of single leaves, the photosynthesis profile of each species in the mixed canopy is obtained. Integration over the height of the canopy and over the day gives the daily CO_2 assimilation rate for each species.

Profiles of absorbed radiation, CO_2 assimilation and the cumulative CO_2 assimilation of a mixture with two species with the same characteristics and the same leaf area index (LAI=2) but a different height are illustrated in *Figure 13.6*.

As in the monoculture model, the amount of assimilates, produced per day is converted to glucose production by multiplying by 30/44 (molecular weights of CH_2O/CO_2). After subtraction of respiration requirements for maintenance of the species, the net daily growth rate in kilograms dry matter per hectare per day is obtained using a conversion factor. The dry matter produced is partitioned among the various plant organs, using partitioning coefficients that are introduced as a function of the phenological development stage of the species. Phenological development rate is tracked in the model as a function of ambient daily average temperature. When the canopy is not yet closed, leaf area increment is calculated from daily average temperature (Kropff and Spitters, 1992). When the canopy closes, the increase in leaf area is obtained from the increase in leaf weight using the specific leaf area (SLA, m^2 leaf kg^{-1} leaf). Integration of daily growth rates of the organs and leaf area results in the time course of LAI and dry weight in the growing season. Plant height development follows an S-shaped pattern throughout the growing season. Height is obtained by integration of the height growth rate.

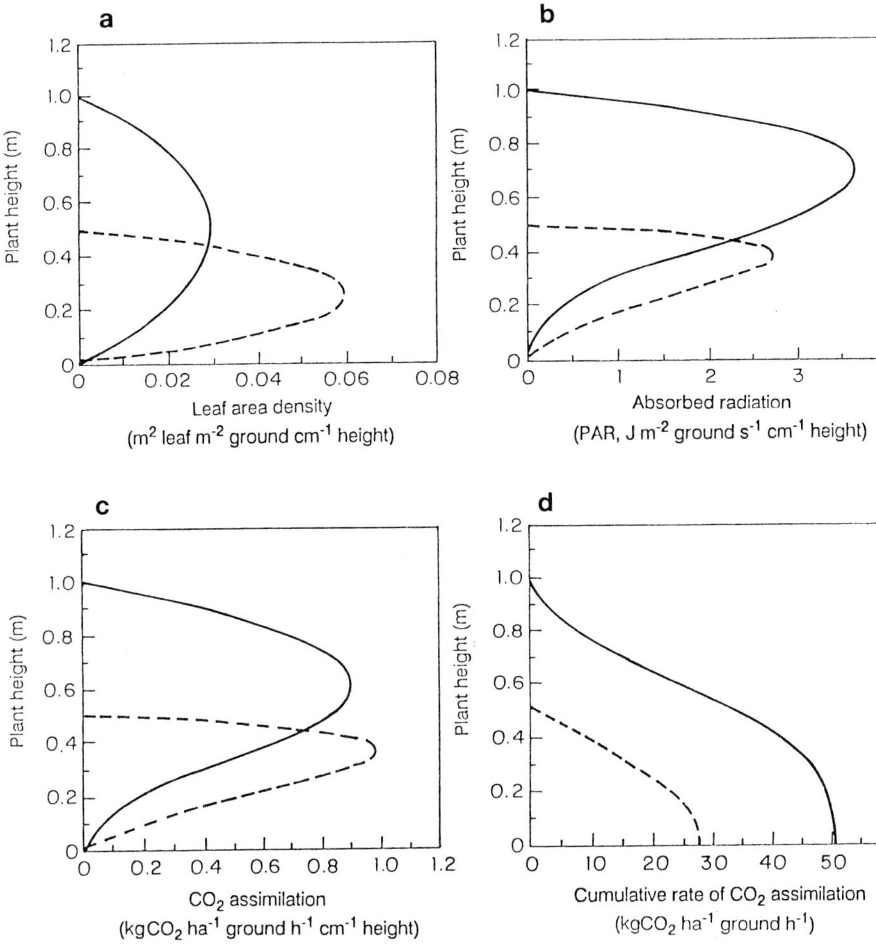

Figure 13.6 Simulation of competition for light using a parabolic leaf area distribution. (**a**) Leaf area distribution over height for two identical species with both LAI=2 and plant height of 1 m (solid line) and 0.5 m (broken line). (**b**) Profile of absorbed visible radiation (PAR) for both species in mixture with 300 J m^{-2} s^{-1} diffuse radiation as input. (**c**) CO_2 assimilation profile in the canopy for both species (kgCO_2 ha^{-1} h^{-1} cm^{-1} height). (**d**) The cumulative CO_2 assimilation rate counted from the top of the canopy (kgCO_2 ha^{-1} ground h^{-1})

In competitive situations however, height growth can be strongly reduced by shortage of assimilates as a result of shading. In crop–weed competition studies this is the case for late emerging weeds. In the model, an empirical function was introduced that limits the Specific Stem Length (SSL, m of height kg^{-1} stem) to a maximum value which depends on plant height. This function is based on the response observed in *Chenopodium album* L. growing in a wide range of competition situations (*Figure 13.7*), showing *C. album* to have high phenotypic plasticity with respect to height growth.

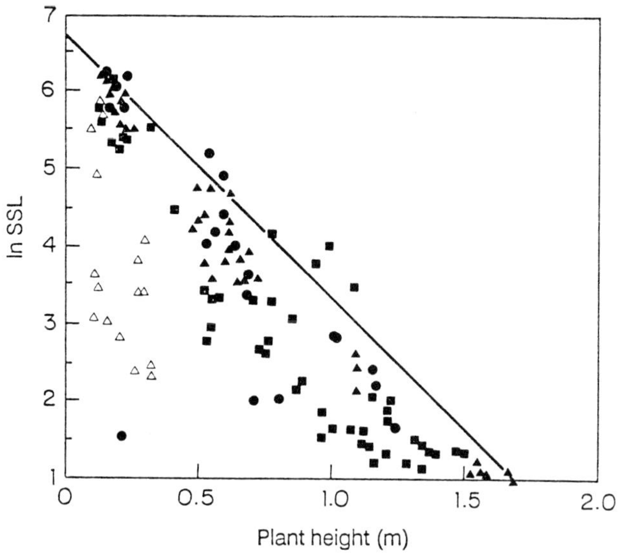

Figure 13.7 The relation between the Specific Stem Length (SSL) and plant height for *Chenopodium album* L., measured in different experiments in monoculture and competition situations. Symbols indicate different growth stages (Δ: 500 < temperature sum (*ts*) < 500; ■ : 600 < *ts* < 800; ● : 100 < *ts* < 1300; ▲ : 1300 < *ts* < 1500). The line indicates the maximum SSL possible in relationship to height

Plants growing in a mixture in a water-limited production situation compete for water and generally also for light during periods when water is not limiting growth. The importance of competition for water is determined by the length, severeness and timing of the drought period(s). Competition mechanisms for water differ principally from competition for light. Competition for light is a process of direct competition for resource capture, with instantaneous consequences for the growth rate of the plants. The amount of absorbed radiation generally limits the photosynthesis of closed leaf canopies in potential production situations. Light absorption by neighbouring plants reduces the rate of CO_2 assimilation and growth of the plant. When plants compete for water, the competition effect is not instantaneous, but delayed. Water requirements differ between the species because they are determined by the amount of absorbed radiation, temperature, vapour pressure deficit and species characteristics. The amount of available soil moisture will be reduced during the growing season if rainfall and other processes that increase soil moisture content, cannot meet the water losses by evapotranspiration and drainage. Therefore, plant transpiration in a period when water is not limiting growth, can affect the growing situation later in the season. Especially when the growth cycles of species in a mixture differ, it may happen that an early maturing species does not suffer from water stress itself, but will increase the water stress effects for the later maturing species if water stress occurs later in the growing season. For example, in one of the experiments with maize and *Echinochloa crus-galli* L., severe drought occurred during stem elongation of the maize. The competing *Echinochloa* plants had almost completed their life cycle by then, and were hardly affected by the drought. Water competition effects on maize, however, were very strong, as the mixtures transpired more water than the monoculture maize which had an open canopy for a long period, but in which a crust on the soil prevented evaporation (Kropff, Weaver and Smits, 1992).

Because both light and water competition are important in water-limited production situations, competition effects have to be studied in an integrated way. That can be done by incorporating modules that simulate processes related to the water balance in the canopy and the soil in the competition model for potential production. The processes that have to be simulated explicitly are evaporation and transpiration. The transpiration requirements of the species are calculated based on their share in absorbed radiation, and actual transpiration is simulated by using a reduction function that is a fraction of soil moisture content.

In many production situations, the supply of macronutrients (N, P, K) limits the growth of the crop or vegetation for at least part of the growing season. A mechanistic understanding of the soil and plant processes related to availability, uptake and use of these nutrients is difficult to obtain. Mechanistic simulation models have been developed for nitrogen effects, but their predictive capability is still limited (van Keulen and Seligman, 1987). Relatively simple approaches are, however, available (cf. Janssen *et al.*, 1990) and may be used to obtain more insight into processes related to competition for nutrients between plants.

In production situations where mobile soil elements like water and nitrates are not limiting growth, root length density has hardly any effect on the total uptake of these elements by the crop (van Keulen and Seligman, 1987). Therefore, the total uptake by mixed vegetation can be calculated in a similar way to the procedure described for the monoculture situation.

However, when the nutrient supply from the soil cannot meet demand, uptake by a species in mixed vegetation will be related to its share in the total effective root length. Below-ground competition for soil elements is modelled in analogy with competition for light. The fraction of nutrient ions that is taken up by a species is related to its share in the total root system. It is important to note that the relative, rather than the absolute, effective root length of a species determines its competitive ability. This approach could also be used for the water competition model. A species with an extensive root system, relative to its demand, is able to meet its demand at a lower soil nitrogen supply. At limited soil supplies, the soil nitrogen which is not used by such a species is distributed over the other species. For further details we refer to Kropff and Spitters (1992).

PERFORMANCE OF THE ECOPHYSIOLOGICAL MODEL FOR INTERPLANT COMPETITION

Most parts of the ecophysiological model have been evaluated and tested for monoculture crops (Penning de Vries and van Laar, 1982; Rabbinge, Ward and van Laar, 1989). The ecophysiological competition model was tested firstly with data from competition experiments with maize (*Zea mays* L.), yellow mustard (*Sinapis arvensis* L.) and barnyard grass (*Echinochloa crus-galli* L.) (Kropff *et al.*, 1984; Spitters and Aerts, 1983; Kropff, Weaver and Smits 1992). In these studies, the difference in the effect of barnyard grass on maize production between field experiments in two successive years was only partly explained by the model due to severe drought stress in one experiment that influenced competitive interactions. The analysis by Kropff, Weaver and Smits (1992) showed that the simulation of leaf area development and maize height deviated strongly from observed values in the dry year. When measured leaf area and height data were input in the simulation model, dry matter production and yield loss were simulated accurately (Kropff, Weaver and Smits, 1992). Irreversible effects of drought stress – such as strongly reduced stem elongation and death of leaves or the whole plant – were not included in the model. Quantitative information of extreme stress effects on physiological processes is lacking. In a subsequent study, the ecophysiological model was parameterized for sugar beet and the weed lambsquarters (*Chenopodium album* L.) and used to analyse data from five field experiments (Kropff, 1988). The experiments were conducted at the same site in Wageningen for three years. Weed densities ranged from 5.5 to 22 plants/m^2 and the period between crop and weed emergence ranged from 0 to 30 days. Yield loss in sugar beet ranged from –6 to 96%. Measured inputs in the model included daily weather variables measured at a nearby station (maximum and minimum temperature, total global radiation, rainfall, humidity and windspeed), weed density and dates of crop and weed emergence. Data on species characteristics were derived from the literature, unpublished results of experiments and three field experiments with sugar beet and lambsquarters (Kropff *et al.*, 1992). Soil parameters such as soil moisture content at permanent wilting point, field capacity and rootable depth were experimentally determined. To assess the explanatory power of the ecophysiological model, the same set of species-specific parameter values was used in the simulation of all five experiments. Since the simulation model explained 98% of the variation between experiments, the model was used to analyse the contribution of the different factors to differences in yield loss between experiments. Analysis (by changing the values of variables step by step) showed that variation between the experiments was mainly due to differences in the period between crop and weed emergence. Additional validation of the ecophysiological model was performed using data from critical period experiments with the same species (Weaver, Kropff and Groeneveld, 1992).

This model was evaluated subsequently (Kropff, Weaver and Smits 1992) using data with tomato (*Lycopersicon esculentum* L.) – redroot pigweed (*Amaranthus retroflexus* L.) and tomato – eastern black nightshade (*Solanum ptycanthum* L.) (Weaver, Smits and Tan, 1987) in Canada. Yield loss – weed density responses, effects of transplanting *versus* direct seeding and effects of weed-free or weed-infested periods (critical period experiments) were simulated accurately.

Sensitivity analysis showed that morphological species characteristics (height growth development and leaf area development) are the most significant plant factors determining competitive relationships in favorable growing conditions (Kropff *et al.*, 1992). Physiological characteristics such as photosynthetic rates or respiration characteristics are less important.

The ecophysiological model described, was developed on basis of general crop growth models, and competition was introduced by distributing the resources light, water and nutrients over the species in a mechanistic way. In principle, the physiological processes that determine plant growth are the same, irrespective of whether a species is grown in monoculture or in competition with other species. However, in competition situations, where one is interested in the biomass per individual or per species, and not in the total biomass per unit of area, the attributes determining resource interception become more important because they regulate the distribution of the limiting resources between the competing plants. For a monoculture crop in the linear growth phase, light capture hardly increases with LAI above 4, and plant height does not affect light capture at all. In a mixture, however, small differences in plant height or leaf area development may cause dramatic changes in the competitive relationships, although total biomass production may be more or less the same. The attributes that determine the capacity of a species to capture resources, such as starting position, plant height, leaf area dynamics and root morphology, therefore, need special attention in a competition model.

PRACTICAL APPLICATION

Although complex models for interplant competition in weedy crops will not be a practical tool in weed management because of the detailed input requirements, the models can be used to derive and pre-test simple approaches that may be promising for practical application. The ecophysiological model was used to analyse the possibility of predicting yield loss early in the season, based on observations of the relative leaf area index of the weeds (Kropff, 1988). A very close relationship was predicted between relative leaf area index of the weeds and yield loss for a wide range of weed densities and periods between crop and weed emergence. Based on these results, an alternative regression model for early prediction of crop losses by weed competition was introduced by Kropff and Spitters (1991). Their model was directly derived from the well-tested hyperbolic yield density model (Equation 6). This model relates yield loss to relative weed leaf area index (L_w expressed as weed leaf area index /crop+weed leaf area index) shortly after crop emergence, using the 'relative damage coefficient' q as the single model parameter:

$$Y_L = q\, L_w \,/\, (1 + (q - 1)\, L_w) \tag{11}$$

It was demonstrated that this model also accounts for the effect of the period between crop and weed emergence besides weed density (Kropff and Spitters, 1991). *Figure 13.8* shows the fit of the model to data from field experiments in which the effect of *C. album* on sugar beet was studied for different periods of weed emergence and densities (Kropff, 1988). A strong advantage of the approach is that separate flushes of the weeds do not have to be handled separately, since they are included in the total value of relative leaf area of weeds, if the weeds emerge before the time of observation.

An extended version of Equation 11, developed in analogy with Equation 7, also accounts for a limited maximum yield loss caused by weeds (m):

$$Y_L = q\, L_w \,/\, (1 + (q\,/m - 1\,)\, L_w) \tag{12}$$

Equation 12 gave a more accurate description of competition effects for several situations (Kropff *et al.*, unpublished results). In conclusion, the relative leaf cover–yield loss regression model accounts for the effect of weed densities, different flushes of weeds and the period between crop and weed emergence. However, it should be noted that the effect of other factors, such as transplanting shock or severe water stress, is not accounted for in such simple regression models (Kropff and Spitters, 1991)

To implement the approach in practical decision making, a methodology has to be developed that enables simple determination of the relative leaf area in the

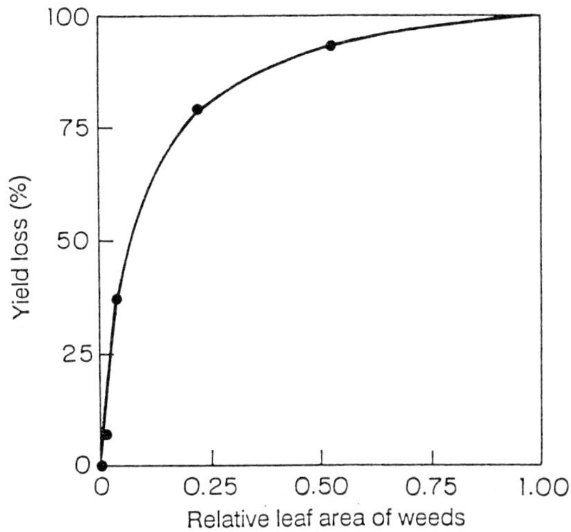

Figure 13.8 Yield loss in sugar beet related to relative leaf area of the weed *Chenopodium album* L. determined 30 days after sugar beet emergence. (After: Kropff and Spitters, 1991)

field, e.g. by estimating relative leaf cover. Preliminary studies showed that in early stages of the crop, the leaf area index is closely related to leaf cover. The potential use of the relative leaf cover model in practical weed management will be increased when weeds of different species with similar competition characteristics can be grouped. Then, the model could be parameterized per species group. Applicability will also be improved by identifying the time window for decision making with respect to specific problem weeds, because of the time dependence of the relative damage coefficient (Kropff and Spitters, 1991).

THE COMPLEMENTARITY OF ECOPHYSIOLOGICAL AND DESCRIPTIVE APPROACHES TO QUANTIFY INTERPLANT COMPETITION

The descriptive or phenomenological approaches and ecophysiological approaches to quantify interplant competition are complementary. The ecophysiological model is very comprehensive, directed towards understanding of the basic principles governing interplant competition. It is, therefore, intended as a research tool. Such a model facilitates the study of plant attributes determining the competitive ability of plants, the improvement of the efficiency of intercropping systems, the evaluation of newly developed weed control strategies and the development of improved descriptive approaches. The simple approaches, however, are more tailored for use as practical tools for on-farm decision making in weed management. They are easy to parameterize and require simple observations.

References

Ampong-Nyarko, K. and De Datta, S.K. (1989) Ecophysiological studies in relation to weed management strategies in rice. In *Proceedings of the Asian-Pacific Weed Science Society*, Twelfth Conference, Seoul, pp 31–45

Berendse, F. (1985) Een simulatiemodel voor de stikstofkringloop en de concurrentie tussen plantesoorten in vochtige heidevegetaties. *The Utrecht Plant Ecology News*, Report No. 1, Utrecht University, the Netherlands

Caldwell, R. and Hansen, J. (1990) Simulation of intercropping with IBSNAT models. *Agrotechnology Transfer*, **12**, 1–4

Connolly, J. (1986) On difficulties with replacement-series methodology in mixture experiments. *Journal of Applied Ecology*, **23**, 125–137

Cousens, R. (1985) An empirical model relating crop yield to weed and crop density and a statistical comparison with other models. *Journal of Agricultural Science*, **105**, 513–521

Cousens, R., Brain, P., O'Donovan, J.T. and O'Sullivan, A. (1987) The use of biologically realistic equations to describe the effects of weed density and relative time of emergence on crop yield. *Weed Science*, **35**, 720–725

Darwinkel, A. (1978) Patterns of tillering and grain production of winter wheat at a wide range of plant densities. *Netherlands Journal of Agricultural Science*, **26**, 383–398

Firbank, L.G. and Watkinson, A.R. (1987) On the analysis of competition at the level of the individual plant. *Oecologia (Berlin)*, **71**, 308–317

Firbank, L.G. and Watkinson, A.R. (1990) On the effects of competition: from monocultures to mixtures. In *Perspectives on Plant Competition* pp 165–192 Edited by J.B. Grace and D. Tilman. New York: Academic Press

Goudriaan, J. (1982) Some techniques in dynamic simulation. In *Simulation of Plant Growth and Crop Production*, pp 66–84 Edited by F.W.T. Penning de Vries and H.H. van Laar. Simulation Monographs, Wageningen: Pudoc

Grace, J.B. and Tilman, D. (Eds) (1990) *Perspectives on Plant Competition*. San Diego: Academic Press 484 pp

Graf, B., Gutierrez, A.P., Rakotobe, O., Zahner, P. and Delucchi, V. (1990) A simulation model for the dynamics of rice growth and development. Part II. The competition with weeds for nitrogen and light. *Agricultural Systems*, **32**, 367–392

Håkansson, S. (1983) Competition and production in short-lived crop–weed stands. *Sveriges Landbruks University, Uppsala, Sweden. Report* No.127, 85 pp

Harper, J.L. (1977) *The Population Biology of Plants*. London: Academic Press, 892 pp

Hodges, T. and Evans, D.W. (1990) Light interception model for estimating the effects of row spacing on plant competition in maize. *J. Prod. Agric.* **2**, 190–195

Janssen, B.H., Guiking, F.C.T., van der Eijk, D., Smaling, E.M.A., Wolf, J. and van Reuler, H. (1990) A system for quantifying evaluation of the fertility of tropical soils (QUEFTS). *Geoderma*, **46**, 299–318

Jones, C.A. and Kiniry, J.R. (1986) *CERES-Maize: a simulation model of maize growth and development*. College Station, Texas: Texas A and M University Press

Jones, J.W., Boote, K.J., Hoogenboom, G., Jagtap, S.S. and Wilkerson, C.G. (1989) SOYGRO V5.42, soybean crop growth simulation model: user's guide.

Florida Agricultural Experiment Station Journal No. 8304. Gainsville, Florida: Agricultural Engineering and Agronomy Department, University of Florida

Keulen, H. van and Seligman, N.G. (1987) *Simulation of Water Use, Nitrogen and Growth of a Spring Wheat Crop*. Simulation Monographs, Wageningen: Pudoc, 310 pp

Kira, T., Ogawa, H. and Sakazaki, N. (1953) Intraspecific competition among higher plants. I. Competition-yield-density interrelationships in regularly dispersed populations. *Journal of Institute for Polytechnics, Osaka City University*, Series **D 4**, 1–16

Kropff, M.J. (1988) Modelling the effects of weeds on crop production. *Weed Research*, **28**, 465–471

Kropff, M.J., Cassman, K.G. and van Laar, H.H. (1993) Quantitative understanding of the irrigated rice ecosystem for achieving high yield potential in (hybrid) rice. In *Proceedings of the International Rice Research Conference*, 21–25 April, 1992. International Rice Research Institute, Los Banos, Philippines (in press)

Kropff, M.J. and Spitters, C.J.T. (1991) A simple model of crop loss by weed competition from early observations on relative leaf area of the weeds. *Weed Research*, **31**, 97–105

Kropff, M.J. and Spitters, C.J.T. (1992) An eco-physiological model for interspecific competition, applied to the influence of *Chenopodium album* L. on sugar beet. I. Model description and parameterization. *Weed Research*, **32**, 437–450

Kropff, M.J., Spitters, C.J.T., Schnieders, B.J., Joenje, W. and de Groot, W. (1992) An eco-physiological model for interspecific competition, applied to the influence of *Chenopodium album* L. on sugar beet. II. Model evaluation. *Weed Research*, **32**, 451–463

Kropff, M.J., Vossen, F.J.H., Spitters, C.J.T. and de Groot, W. (1984) Competition between a maize crop and a natural population of *Echinochloa crus-galli* (L.). *Netherlands Journal of Agricultural Science*, **32**, 324–327

Kropff, M.J., Weaver, S.E. and Smits, M.A. (1992). On the use of eco-physiological models for crop–weed competition. 1. Relations between weed density, relative time of weed emergence, relative leaf area and yield loss. *Weed Science*, **40**, 296–301

Pacala, S.W. and Silander, Jr J.A. (1987) Neighborhood interference among velvet leaf, *Abutilon theophrasti*, and pigweed, *Amaranthus retroflexus*. *OIKOS*, **48**, 217–224

Pearcy, R.W., Tumosa, N. and Williams, K. (1981) Relationships between growth, photosynthesis and competitive interactions for a C_3 and C_4 plant. *Oecologia (Berlin)*, **48**, 371–376

Penning de Vries, F.W.T. and van Laar, H.H. (1982) Simulation of growth processes and the model BACROS. In *Simulation of plant growth and crop production* pp 114–135 Edited by F.W.T. Penning de Vries and H.H. van Laar Simulation Monographs, Wageningen: Pudoc

Rabbinge, R., Ward, S.A. and van Laar, H.H. (Eds) (1989) *Simulation and Systems Management in Crop Protection*. Simulation Monographs 32, Wageningen: Pudoc 420 pp

Radosevich, S.R. and Holt, J.S. (1984) *Weed Ecology: Implications for Vegetation Management*. New York: John Wiley, 265 pp

Rimmington, G.M. (1984) A model of the effect of interspecies competition for light on dry-matter production. *Australian Journal of Plant Physiology*, **11**, 277–286

Ryel, R.J., Barnes, P.W., Beyschlag, W., Caldwell, M.M. and Flint, S.D. (1990) Plant competition for light analyzed with a multispecies canopy model. I. Model development and influence of enhanced UV-B conditions on photosynthesis in mixed wheat and wild oat canopies. *Oecologia*, **82**, 304–310

Shinozaki, K. and Kira, T. (1956). Intraspecific competition among higher plants. VII. Logistic theory of the C-D effect. *Journal of Institute for Polytechnics, Osaka City University*, Series **D 7**, 35–72

Silander, Jr J.A. and Pacala, S.W. (1985) Neighborhood predictors of plant performance. *Oecologia (Berlin)*, **66**, 256–263

Spitters, C.J.T. (1983a) An alternative approach to the analysis of mixed cropping experiments. I. Estimation of competition effects. *Netherlands Journal of Agricultural Science*, **31**, 1–11

Spitters, C.J.T.(1983b) An alternative approach to the analysis of mixed cropping experiments. II Harvestable yield. *Netherlands Journal of Agricultural Science*, **31**, 143–155

Spitters, C.J.T. and Aerts, R. (1983) Simulation of competition for light and water in crop–weed associations. *Aspects of Applied Biology*, **4**, 467–484

Spitters, C.J.T., van Keulen, H. and van Kraalingen, D.W.G. (1989) A simple and universal crop growth simulator: SUCROS87. In *Simulation and Systems Management in Crop Protection*, pp 147–181 Edited by R. Rabbinge, S.A. Ward and H.H. van Laar. Simulation Monographs, Wageningen: Pudoc

Spitters, C.J.T., Kropff, M.J. and de Groot, W. (1989) Competition between maize and *Echinochloa crus-galli* analysed by a hyperbolic regression model. *Annals of Applied Biology*, **115**, 541–551

Suehiro, K. and Ogawa, H. (1980) Competition between two annual herbs *Atriplex gmelini* C.A. Mey and *Chenopodium album* L. in mixed cultures irrigated with seawater of various concentrations. *Oecologia*, **45**, 167–177

Taylor, D.R. and Aarssen, L.W. (1989) On the density dependence of replacement-series competition experiments. *Journal of Ecology*, **77**, 975–988

Tilman, D. (1988) Plant strategies and the dynamics and structure of plant communities. *Monographs in Population Biology* No.20. Princetown, New Jersey: Princeton University Press, 360 pp

Trenbath, B.R. (1976) Plant interactions in mixed crop communities. In *Multiple Cropping*. pp 129–169 Edited by M. Stelly. American Society of Agronomy Special Publication No. 27

Vandermeer, J. (1989) *The Ecology of Intercropping*. Cambridge: Cambridge University Press, 237 pp

Watkinson, A.R. (1980) Density-dependence in single-species populations of plants. *Journal of Theoretical Biology*, **83**, 345–357

Weaver, S.E., Smits, N. and Tan, C.S. (1987) Estimating yield losses of tomato (*Lycopersicon esculentum*) caused by nightshade (*Solanum* spp) interference. *Weed Science*, **35**, 163–168

Weaver, S.E., Kropff, M.J. and Groeneveld, R.M.W. (1992) Use of eco-physiological models for crop–weed interference: The critical period of weed interference. *Weed Science*, **40**, 302–307

Wilkerson, G.G., Jones, J.W., Coble, H.D. and Gunsolus, J.L. (1990) SOYWEED: A simulation model of soybean and common cocklebur growth and competition. *Agronomy Journal*, **82**, 1003–1010

Willey, R.W. (1979a) Intercropping – its importance and its research needs. Part I. Competition and crop yield advantages. *Field Crop Abstracts*, **32**, 1–10

Willey, R.W. (1979b) Intercropping – its importance and its research needs. Part I. Agronomic relationships. *Field Crop Abstracts*, **32**, 73–85

Wit, C.T. de, van Laar, H.H. and van Keulen, H. (1979) Physiological potential of crop production. In *Plant breeding perspectives*, pp 47–82 Edited by J. Sneep and A.J.T. Hendriksen. Wageningen: Pudoc

Wit C.T. de (1960). *On Competition*. Agricultural Research Reports 66.8, Wageningen: Pudoc 82 pp

Wit, C.T. de, Goudriaan, J., van Laar, H.H., Penning de Vries, F.W.T., Rabbinge, R., van Keulen, H., Sibma, L. and de Jonge, C. (1978) *Simulation of Assimilation, Respiration and Transpiration of Crops*. Simulation Monographs, Wageningen: Pudoc 141 pp

Wright, A.J. (1981) The analysis of yield-density relationships in binary mixtures using inverse polynomals. *Journal of Agricultural Science (Cambridge)*, **96**, 561–567

Yoda, K., Kira, T., Ogawa, H., Hozumi, K. (1963) Self-thinning in overcrowded pure stands under cultivated and natural conditions. *Journal of Biology, Osaka City University*, **14**, 107–129

Yoshida S. (1981) *Fundamentals of Rice Crop Science*. Los Banos, Philippines: International Rice Research Institute, 269 pp

14

COMPLEMENTARITY OF RESOURCE USE IN INTERCROPPING AND AGROFORESTRY SYSTEMS

C.K. ONG
International Centre for Agroforestry Research, PO Box 30677, Nairobi, Kenya

C.R. BLACK
**Department of Physiology and Environmental Science, University of Nottingham, Sutton Bonington Campus, Loughborough, Leics. LE12 5RD, UK*

Introduction

> '*Probably the most common cause of higher yields from intercropping over sole cropping is the improved use of environmental resources. Put very simply, if component crops in an intercropping system use resources differently than when grown together, the crops complement each other and make better overall use of resources than when grown as separate sole crops*'.
>
> (Willey *et al*., 1986).

The quest for a mechanistic understanding of resource utilization in intercropping systems began in 1978 when Robert Willey and his co-workers at the International Crops Research Institute for the Semi-Arid Tropics (ICRISAT) at Hyderabad, India joined forces with John Monteith's team at Sutton Bonington, UK to investigate a pearl millet–groundnut system (Gregory and Reddy, 1982; Marshall and Willey, 1983). The same analytical approach has also been used more recently to examine how far the basic principles established in intercropping research can be extended to agroforestry systems (Monteith, Ong and Corlett, 1991; Ong *et al*., 1991). It is therefore noteworthy that ecologists are now taking a somewhat similar interest in their attempts to determine the role of competition for resources in natural plant communities (Grime, this volume). The importance of resource capture is well recognized in ecology and plant competition has been defined as 'the tendency of neighbouring plants to utilize the same resources and that success is largely a reflection of the capacity of resource capture' (Grime, 1979). It is therefore appropriate that agronomists began intercropping research by borrowing many of the established ecological techniques (e.g. replacement series experiments) and methods of analyses (e.g. relative crowding coefficient (de Wit, 1960) and competition index (Donald, 1963)) to determine intra- and inter-specific competition between crops (reviewed by Willey, 1979).

The concept of complementarity of resource utilization is central to the assumption that mixtures of crops may lead to increased productivity in both intercropping and agroforestry systems. In terms of competition, complementarity may be said to occur when each species experiences less competition when grown with other species. This narrow definition of complementarity is not really useful as it simply reflects the *outcome* in terms of growth or survival for each species; it is more helpful to define complementarity in terms of the *efficiency* with which mixtures of species capture and utilise *limiting* growth resources. For example, if a slow-growing crop such as pigeonpea, which is very inefficient in using moisture and intercepting radiation during the early part of the growing season but is efficient in using stored moisture after the rains, is grown with a fast growing crop like maize, the overall use of water and interception of radiation is greatly increased (Sivakumar and Virmani, 1980). In terms of available resources, complementarity occurs when a mixture of species results in the capture of a *greater* fraction or quantity of the limiting growth resource(s) and/or the *more effective* utilization of the same resource(s) (i.e. more biomass produced *per* unit of resource captured) than when the same species are grown as sole crops. In agroforestry, increased resource capture may result from reduction of water and soil loss, utilization of off-season rainfall when cropping is not possible, recycling of nutrients from deeper horizons in the soil profile and other factors. The more effective use of captured resources may result from improvements in microclimate, greater light and water use efficiency and prevention of damage by strong winds or desiccation.

This paper describes how the principles of resource capture (Monteith, this volume) may be used to establish the origins of improved yields in intercropping systems and recent attempts to extend such approaches to agroforestry in order to design improved agroforestry systems.

Lessons from ecology

The vast majority of studies on plant competition (recently summarized in a volume edited by Grace and Tilman, 1990) concern the outcome of competition on the relative abundance of two or more species; much less emphasis has been placed on traits or situations that promote complementarity in resource use. Recent controversy concerning the mechanistic understanding of competitiveness has continued to concentrate on the behaviour of the 'winner' in the contest as either the species with 'the ability to exploit resources rapidly' (Grime, 1979) or 'the ability to tolerate a low level of limiting resource' (Tilman, 1988). There is a similar pre-occupation with the 'outcome' of competition in weeds research. To understand whether two or more species complement each other in terms of resource use, it is important to determine the major limiting factor and to know how effectively resources are used.

Nevertheless, there are several important lessons from ecology which could assist in the search for complementarity of resource utilization in mixed communities. The first is the importance of phenological differences between species. For example, in temperate zones it is well-established that the common nettle, *Urtica dioica* and the moss, *Brachythium rutabulum*, grow together because they have different temperature optima for growth (Grime, 1979). Not surprisingly, phenological differences became a major area of interest in intercropping research, as is typified by the sorghum–pigeonpea system described by Natarajan and Willey (1980). In agroforestry, the best example of phenological differences is the well known *Acacia albida* (now *Faidherbia albida*) and pearl millet system of sub-saharan Africa, in which the *Acacia* sheds its leaves for a large part of the rainy season, providing the millet or other understorey crops with a long period free of competition to establish a full canopy and an extensive root system.

An important question is whether competitiveness and complementarity are species or site specific. The ecological literature clearly indicates that competitiveness between species is strongly influenced by the date of sowing of each species, nutrient status, density and watering regimes (Connolly, 1986). However, several reports suggest that the relative competitiveness of certain species is maintained at both high and low nutrient status, e.g. between *Lolium perenne* and *Phalaris tuberosa* (Donald, 1958). In the intercropping literature, there is evidence for competitiveness or yield advantages which may be either insensitive to watering regime (e.g. pearl millet–groundnut, Willey *et al.*, 1986) or sensitive (e.g. pigeonpea–upland rice, Ramakrishna, Ong and Reddy, 1992).

Another ecological observation that may have important implications for resource sharing by mixed agricultural communities is the importance of size-asymmetry in determining competitiveness (Goldberg and Werner, 1983). This is particularly important in agroforestry because the tree component becomes progressively larger with time, with an increasingly large root system and canopy which capture an increasing proportion of the major limiting resources.

When reviewing the ecological evidence concerning the relative importance of above- and below-ground interactions, Snaydon and Harris (1981) concluded that there was overwhelming evidence that below-ground interactions should have high priority in research on intercropping. Techniques for separating above and below-ground interactions between species are now well established under both field and laboratory conditions. For example, as early as 1933, Watt and Fraser investigated competition for below-ground resources between Scots pine and grasses by digging trenches to different depths. Understorey grasses and wood sorrel grew better with deep trenches and the addition of distilled water failed to remove the competitive effect of the trees. Watt and Fraser concluded that the most likely explanation was nutrient competition. Their trenching techniques have now been successfully employed in agroforestry research to demonstrate the effects of root competition (Singh, Ong and Saharan, 1989). More ingenious segregation experiments have been devised for glasshouse trials to separate above- and below-ground effects; these include the study by Groves and Williams (1975) of competition between subterranean clover and skeleton weed in Australia.

Application of resource capture principles to intercropping research

SPATIAL COMPLEMENTARITY

The main difficulty encountered in extending the principles of resource capture to intercropping lies in the need to quantify resource capture by the two (or more) species involved. Separate measurements of resource capture are particularly important when there is a large overlap in the period of active growth of the crops involved. For example, in a pearl millet–groundnut intercrop of 80/105 days duration, the onset of the linear growth phase occurs between 35 and 40 days in both crops and the overlap lasts for about 40 days. The development of leaf area index is also very similar in both crops, peaking at 50–55 days (*Figure* 14.1). This intercropping system has been the subject of a long series of intensive studies at ICRISAT and is therefore worthy of further consideration here.

Marshall and Willey (1983) measured the transmission of radiation using a single quantum sensor placed above the taller pearl millet crop and another placed on an

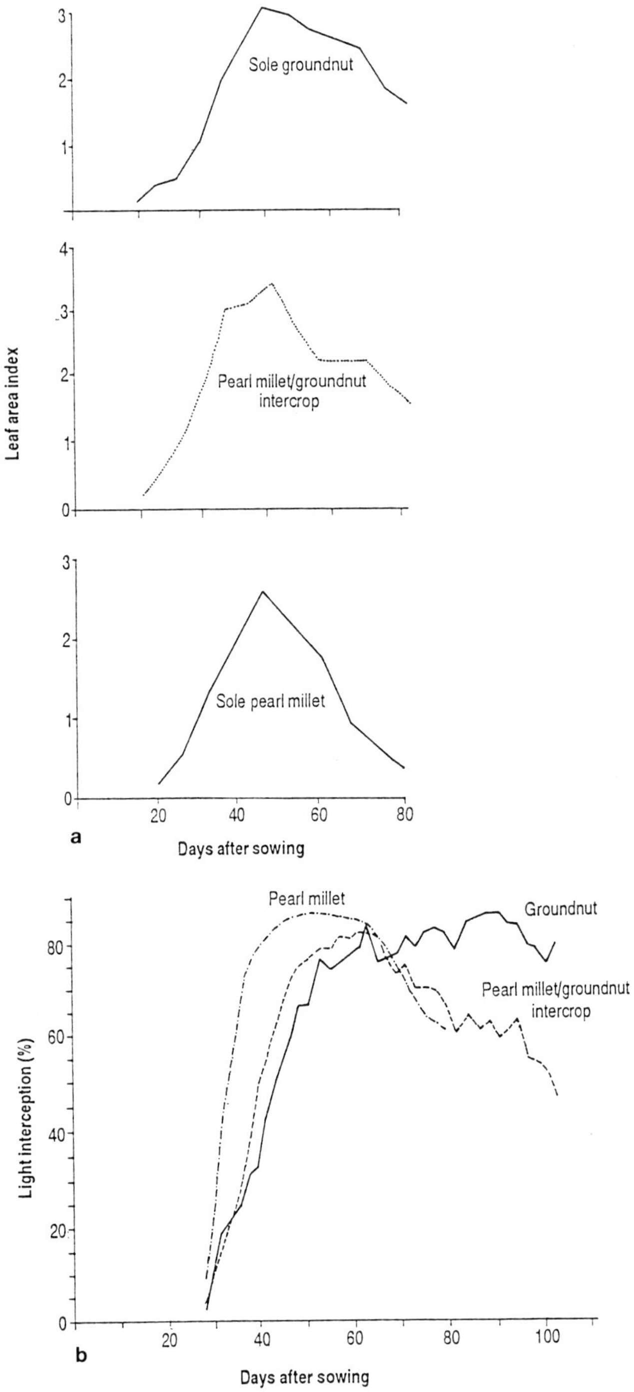

Figure 14.1 Leaf area index (**a**) and light interception (**b**) in sole pearl millet and groundnut, and in a pearl millet–groundnut intercrop grown on Alfisols at ICRISAT Centre, rainy season, 1978 (After Reddy and Willey, 1981)

aluminium track located either at the soil surface or above the shorter groundnut crop. The quantity of radiation intercepted by the canopies was measured using tube solarimeters located at the same positions. Marshall and Willey expressed radiation interception by the intercrop in terms of the quantity intercepted per row of each crop relative to that in the corresponding monocrop, a term they called the Resource Capture Ratio (RCR). They also calculated the conversion efficiency ratio (CER), which is the ratio of the biomass produced per megajoule of radiation intercepted by each species. The product of the proportion of the intercrop contributed by each species (P) and the ratios, RCR and CER, is the Land Equivalent Ratio (LER), or intercropping advantage of that species, i.e.:

$$\text{LER} = P \times \text{RCR} \times \text{CER}.$$

LER was defined by Willey (1979) as the relative land area under a sole crop required to produce the yields achieved in intercropping irrespective of plant populations.

Table 14.1 summarizes Marshall and Willey's (1983) results, which showed that the two major mechanisms leading to the yield advantage in the intercrop were an increase in interception by pearl millet (112% more) and the increased CER of groundnut (46%).

Table 14.1 RESOURCE CAPTURE RATIO AND CONVERSION EFFICIENCY RATIO OF A PEARL MILLETT-GROUNDNUT INTERCROP (AFTER MARSHALL AND WILLEY, 1983)

	Pearl millet		*Groundnut*
Crop proportion (*P*)	0.25		0.75
Resource capture ratio (RCR)	2.12		0.73
Conversion efficiency ratio (CER)	0.97		1.46
Component land equivalent ratio (LER)	0.51		0.80
Combined land equivalent ratio		1.31	

At that time it was not clear why the CER of groundnut was so greatly enhanced, but subsequent work by Stirling *et al.*, (1990) at ICRISAT showed that the groundnut canopy was light saturated in full light (15 MJ m^{-2} d^{-1} total radiation), but that at lower radiation levels (8 MJ m^{-2} d^{-1}) the efficiency of conversion was increased from about 1.00 to 2.36 g MJ^{-1}. Such severe shading of groundnut appeared to have no significant effect on the partitioning of dry matter to pods.

The analysis of resource capture in terms of ratios and as a function of the proportion of each crop in the mixture assumes a *linear* relationship between resource capture and plant population. This is seldom true for crops such as pearl millet, which has a tremendous capacity to compensate for low populations by increased tillering. For example, Azam-Ali, Gregory and Monteith (1984) found that total dry matter production and intercepted radiation in pearl millet stands in Niger were almost identical at populations of about 3 and 6 plants per square metre. Several workers have emphasised the need to ensure that intercrop comparisons should be made only when the respective sole crops are grown at their appropriate optimum population (Willey, 1979). Nevertheless, an accurate assessment of intercropping advantages in resource capture is not possible using the RCR calculation unless the 'parabolic' curve or asymptotic response of each crop is known.

Another method of interpretation is to compare the *total* quantity of resource used by the intercrop and the respective sole crops (grown at their optimum populations) instead of RCR. For example, comparison of the total intercepted radiation of pearl millet–groundnut intercrops over a 4-year period (including 1978) revealed that on average the intercrop intercepted 8% more radiation than sole groundnut, but 7% less than sole pearl millet (Ong, Subrahmanyam and Khan, 1991). These latest findings at ICRISAT support the earlier conclusion of Marshall and Willey (1983) that the light conversion efficiency, *e*, was generally increased in the intercrops, but that such increases were insufficient to explain the intercropping advantage when rainfall was below 600 mm. During the period 1985–1988, which included three consecutive dry years with less than 450 mm per season, the pattern of rainfall appears to have been a crucial factor in determining the intercropping advantage. Unpublished data at ICRISAT have confirmed that the intercropping advantage of sorghum–pearl millet intercrops varied from 1.0 to 1.8 when the timing of irrigation was changed but the total amount applied (470 mm) was similar.

Because of the short growing season in semi-arid areas, there is little difference in the timing of maturity in component crops such as pearl millet and groundnut, and so they make their major demands on resources at similar times during the season. As described above, the total quantity of resource captured, whether of light, nutrients or water (Reddy and Willey, 1981), is relatively similar in intercrops and sole crops, but the efficiency of utilization of the resources captured is increased in the former. This form of complementarity in resource use is known as spatial complementarity.

TEMPORAL COMPLEMENTARITY

Temporal complementarity in resource use occurs when each component of a mixed community captures resources at different times during the season. This is perhaps the most common cause of higher productivity in intercropping and is largely due to the capture of *more* resources rather than a change in the efficiency of utilization in dry matter production (Willey *et al.*, 1986). The best example of temporal complementarity is the groundnut–pigeonpea system, which combines differing durations to maturity of 105 days (groundnut) and 180 days (pigeonpea).

Figure 14.2 illustrates the seasonal timecourses of dry matter accumulation and fractional light interception in a groundnut–pigeonpea experiment at ICRISAT Centre. Sole groundnut reached its maximum light interception (*c.* 80%) by about 45–50 days, while the much slower-growing pigeonpea took 90–100 days. During the first 100 days there were negligible differences in the patterns of light interception between the sole and intercrop groundnut, the dominant crop in the mixture. However, dry matter production by groundnut was reduced by 24% at maturity. Fractional light interception fell to 50–60% when the groundnut was harvested but then remained relatively constant until the pigeonpea was harvested. Intercropped pigeonpea produced 72% of the dry matter obtained

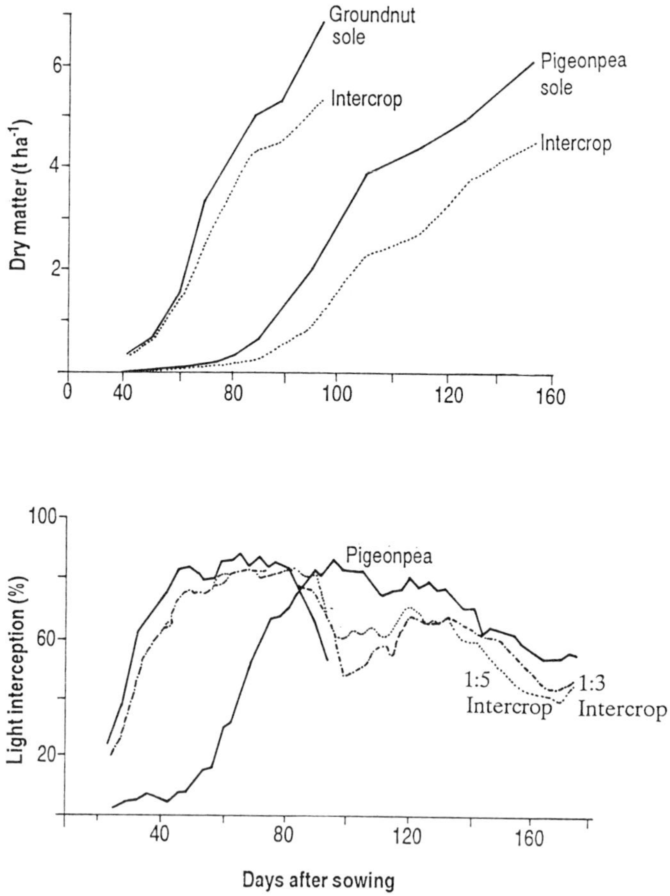

Figure 14.2 Dry matter accumulation and light interception in a groundnut–pigeonpea intercrop (After Willey *et al.*, 1986)

from sole pigeonpea. Total intercepted radiation for the intercrop was 15% greater than for the sole pigeonpea and twice that of the sole groundnut. However, the 15% increase in total intercepted radiation was insufficient to explain fully the intercropping advantage of 48% for total dry matter and 58% for grain yield. The conversion efficiency of intercepted radiation was similar in both the intercropping and sole cropping systems. Part of the explanation of the intercrop advantage lies in the higher partitioning of dry matter to the grain in intercropped pigeonpea (harvest index of 0.30) than in sole pigeonpea (0.26). The remainder of the apparent intercropping advantage is probably an artifact created by expressing the relative performance as the ratio of intercropped yield to sole crop yield. Pigeonpea is extremely plastic and shows an asymptotic relationship between yield and plant population (Rao and Willey, 1983).

Similar evidence was obtained for a sorghum–pigeonpea system at ICRISAT, in which total intercepted radiation was increased by only 8% relative to the sole pigeonpea, but the estimated intercropping advantage ranged from 40 to 78%

(Natarajan and Willey, 1980). Estimates of water use and nutrient uptake showed negligible differences between the intercropped and sole pigeonpea.

In view of the difficulty of reconciling these apparently large intercropping advantages with the much smaller increases in total resource capture, in this case of light, it is necessary to explore further the basis for comparing intercrop and sole crop performance. Intercropping advantage is commonly expressed in terms of land equivalent ratio (LER, Willey, 1979). However, the calculation of component LERs for each species in an intercrop depends on the choice of appropriate *standard sole crops* (Mead and Willey, 1980). The sorghum–pigeonpea system has been investigated in great detail at ICRISAT Centre, and the evidence obtained should facilitate the meaningful interpretation of resource capture. An important factor is the yield–population response curve of specific crops in both sole and intercrop situations (*Figure 14.3*).

The yield–population response for sorghum shows that the optimum population is about 8–10 plants per square metre and that intercropping with pigeonpea causes only a 6% reduction in sorghum yield. In other words, sorghum is highly dominant in the mixture and its yield is reduced significantly (by 20%) only at the lowest population (5 plants per square metre), when the leaf area is insufficient to utilize available radiation fully. Thus, provided the population of intercropped sorghum exceeds 10 plants per square metre, the 'expected' resource capture should be *c.* 90–95% of that in sole sorghum, as opposedd to the RCR of 0.67 calculated on a *per row* basis (*Table 14.2*), using the method described by Marshall and Willey

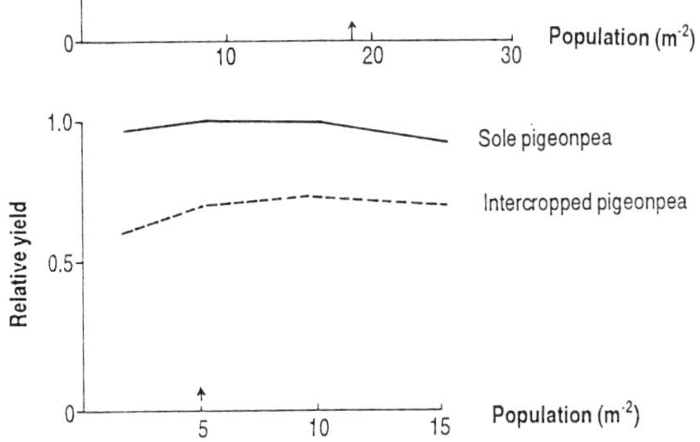

Figure 14.3 Yield–population response of sole and intercropped sorghum and pigeonpea (Adapted from Willey, 1979; Natarajan and Willey, 1980; Rao and Willey, 1983). Note: The curves for pigeonpea are the averages of three consecutive years, during which rainfall ranged from 908 mm in 1978 to 600 mm in 1980. The intercrop was grown with a constant row arrangement of 2 sorghum : 1 pigeonpea. Arrows indicate standard sole populations in resource capture studies at ICRISAT

Table 14.2 COMPARISON OF EXPECTED AND ACTUAL CAPTURE OF LIGHT AND YIELD OF A SORGHUM–PIGEONPEA INTERCROP (RECALCULATED FROM NATARAJAN AND WILLEY, 1985)

	Sorghum	*Pigeonpea*	*Total*
Expected ratio			
a) Per row (*P*)	0.67	0.33	1.00
b) Yield–population	0.96	0.68	1.64
Actual resource capture			
a) Per row (RCR)	1.43	2.21	1.69*
b) Yield–population	0.94	0.73	1.67
Land equivalent ratio			
a) Biomass	0.73	0.63	1.38
b) Grain	0.76	0.83	1.59

*Calculated at *P* × RCR

(1983). In fact, the values for total radiation intercepted by intercropped and sole sorghum were virtually identical (Natarajan and Willey, 1985), and close to those predicted from the population response (*Figure 14.3*), whereas the true RCR was 1.43 (*Table 14.2*).

The yield of sole pigeonpea is even more insensitive to variation in plant population over the range 4–13 plants per square metre, but at the lowest population the intercropped pigeonpea yield was reduced by 40% compared with sole pigeonpea. Therefore, at 5 plants per square metre or beyond, the 'expected' resource capture of intercropped pigeonpea should be *c*. 68% of sole pigeonpea (*Table 14.2*). Measurements of total intercepted radiation *after* sorghum harvest confirmed that interception by intercropped pigeonpea was 73% of that in sole pigeonpea, although interception by the intercrop over the season as a whole was only 8% greater than in the sole pigeonpea (Natarajan and Willey, 1980). Values calculated on a per row basis suggest an 'expected' resource capture of 0.33 and an actual RCR of 2.21 (*Table 14.2*). In summary, calculations of RCR should be based on yield–population response curves and not on a per row basis, regardless of whether replacement or additive experiments are involved.

Separating above and below-ground interactions

In an international workshop on intercropping held at ICRISAT in 1979, Snaydon and Harris (1981) argued that below-ground interactions between plants are more important than those above-ground in producing yield advantages in intercropping, 'though they may cause lower yield in some circumstances'. They found ten published studies which compared the two forms of interaction and concluded that below-ground interactions were more 'intense' than those above-ground in eight of the studies. In addition, five unpublished studies also indicated that below-ground interactions were greater than above-ground interactions. Various approaches were used, based mostly on modifications of the root segregation technique pioneered by Donald (1958) or, more indirectly, by varying the supply of water or nutrients which may be limiting. Strangely, there are surprisingly few published studies of intercropping that have examined the relative importance of above and below-ground interactions.

Willey and Reddy (1981) modified the root segregation technique for field use by installing vertical polythene barriers to a depth of 1 m at ICRISAT Centre to separate the roots of pearl millet and groundnut. The intercrop was grown as a replacement experiment consisting of one row of pearl millet to three rows of groundnut, similar to that described by Marshall and Willey (1983). The digging of trenches *per se* and root segregation had negligible effects on the growth of either crop or on the intercropping advantage, suggesting that below-ground competition was relatively unimportant. Willey and Reddy concluded that the yield advantage obtained in this system was due to beneficial above-ground interactions. This is perhaps not surprising since earlier studies by Gregory and Reddy (1982) indicated that rooting density in the same intercropping system was similar to that observed in the sole stands (LER=1.10). Furthermore, there is little prospect of nitrogen transfer from the root system or decaying foliage of groundnut to pearl millet since groundnut is harvested 25–35 days later than pearl millet. If the leguminous component of an intercrop were to mature earlier than the cereal, there would be a better prospect for nitrogen transfer to the cereal component.

The potential benefit of N-transfer from short-season legumes to cereals has been explored systematically at the International Rice Research Institute (IRRI) in the Philippines (Morris, 1986). At IRRI, the main objective of intercropping research is to maximize the contribution of biologically-fixed N by short-season cowpea and mungbean (60–65 days crop duration) for use by upland rice (115–120 days crop duration), because farmers seldom apply nitrogen fertilisers in this cropping system. When intercropped with cowpea or mungbean, the N-content of rice consistently increased as the legume replaced up to 50% of the rice, but beyond 50% the increased competitiveness of cowpea offset the benefit of N-transfer (Siri-Udompas and Morris, 1990).

Vertical root barriers installed to a depth of 0.5 m were used to separate above and below-ground interactions in a mungbean–upland rice system (Aggarwal *et al.*, 1992). A separate treatment (SV) was included to quantify the advantage to rice of the increase in soil volume available per plant following mungbean harvest by removing every third row of rice in a sole rice treatment at the time of mungbean harvest in the intercrop treatment. The results showed that the intercropping advantage ranged from 20 to 70%, which is consistent with results commonly reported for this intercropping system. The N-uptake and grain yield of rice were substantially reduced by the root barriers, but the yield of intercropped rice was similar to that of the SV treatment, indicating that intercropped rice benefited greatly from improved resource capture resulting from root exploration of the soil volume vacated by the intercropped mungbean. Aggarwal *et al.* concluded that the presence of mungbean *did not* increase the supply of N to the rice above that expected in the absence of mungbean. Nevertheless, this study shows that intercropping *did* result in more effective utilization of below-ground resources.

Because of the recent origin of formal agroforestry research, even less information is available concerning the relative importance of above and below-ground interactions. The vast majority of agroforestry literature is confined to studies of alley or hedgerow cropping, in which food crops are grown in alleys formed by hedgerows of trees or shrubs (Kang, Grimme and Lawson, 1985). Although alley cropping was developed for the humid tropics, primarily as an alternative to shifting cultivation and to improve soil fertility, researchers have extended it to the semi-arid tropics

(Singh *et al.*, 1989). Alley cropping experiments in the semi-arid tropics of India have consistently shown a considerable reduction (30–90%) in crop yield when the alley width was less than 5 m (Singh *et al.*, 1989; Rao, Sharma and Ong, 1990). It has been postulated that competition for moisture between the roots of the trees and crops or shading by the trees was responsible for restricting crop growth in the alleys.

Vertical polythene barriers installed to a depth of 0.5 m were again employed to examine below-ground interactions between *Leucaena leucocephala* hedgerows and three annual crops (sorghum, cowpea, castor) at the Central Research Institute for Dryland Agriculture (CRIDA), Hyderabad, India (Singh, Ong and Saharan, 1989). The alleys were initially 5 m wide but after two disastrous seasons when yield reductions reached 60%, even in the best crop combination, every alternate hedgerow was removed to create alleys 10 m wide. Microclimate, rooting profiles and soil moisture content were monitored in treatments with or without root barriers and in control sole crops. Areas of crop closely adjacent to the hedgerow experienced intense shading (30–85%) for much of the growing period, but the presence of root barriers was sufficient to avoid reductions in cowpea and sorghum yield (*Figure 14.4*). For the longer duration castor (180 days), the barrier was unable to prevent a substantial yield reduction close to the hedge, probably because a large part of the reproductive period occurred during the dry season when water was being drawn from below 0.5 m, where there would have been competition between tree and crop roots.

The importance of below-ground interactions in this system is in sharp contrast to a previous study of intercropped millet and groundnut (Willey and Reddy, 1981)

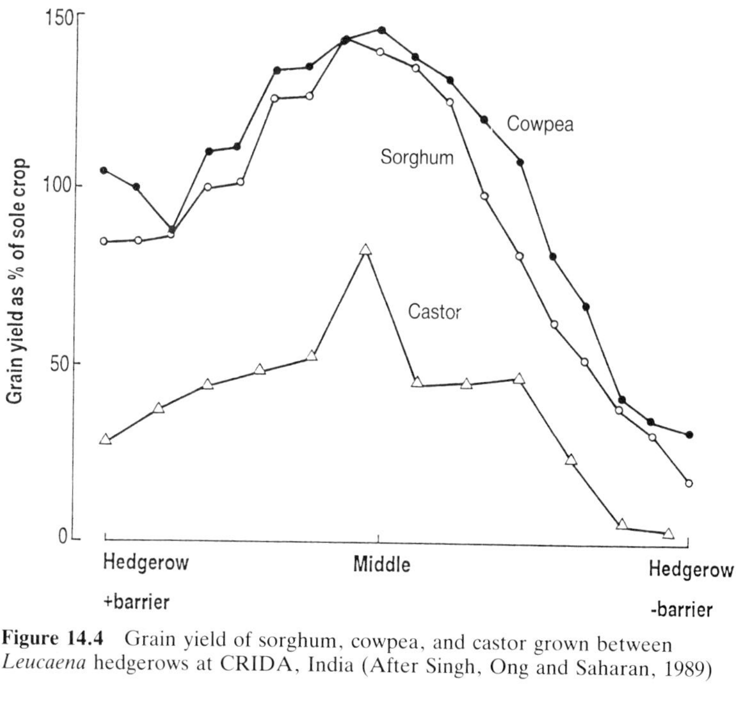

Figure 14.4 Grain yield of sorghum, cowpea, and castor grown between *Leucaena* hedgerows at CRIDA, India (After Singh, Ong and Saharan, 1989)

in the same environment. A major difference between intercropping and alley cropping systems is that in intercropping the establishmentof the root systems commences at the same time in each component (Gregory and Reddy, 1982), but in agroforestry the root system of the tree is already well established when the crop is sown. Consequently, in dry environments the crops are less competitive below-ground than the trees. However, it was not possible to conclude from the study of Singh, Ong and Saharan (1989) that competition was for water alone, although soil moisture content was slightly higher in the root segregation treatment.

An experiment was recently initiated at ICRAF's field station at Machakos, Kenya to distinguish between above and below-ground competition using vertical root barriers installed to 1 m or supplementary irrigation during dry periods to alleviate competition for moisture. Competitive effects on maize extended to a distance of 5 m from the *Leucaena* trees in both the wet and dry growing seasons (*Figure 14.5*). The presence of vertical partitions had a negligible effect on competition with maize, suggesting either that the barriers were ineffective, or that below-ground competition was unimportant. Modification of rainfall interception or shading by *Leucaena* grown either as hedgerows or as upperstorey trees was restricted to an area extending 1–2 m from the trees, suggesting that below-ground competition was the cause of yield reductions beyond 2 m. Supplementary irrigation was effective in preventing any reduction in crop yield, even immediately adjacent to the trees where the crop was shaded to 60% of full sunlight. This is the most convincing evidence demonstrating the importance of competition for

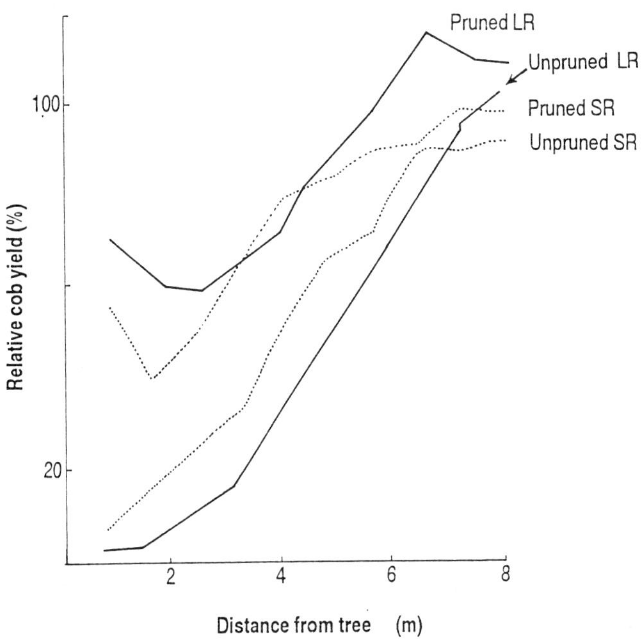

Figure 14.5 Effects of distance and shoot pruning on competition between *Leucaena leucocephala* and maize at Machakos, Kenya during the short rains (SR) and long rains (LR), 1991

moisture between trees and crops. These results, showing that root competition may extend to at least 5 m from the trees, cast doubt on the validity of many agroforestry experiments in which the sole crop control is not sufficiently far from the nearest trees to ensure there is no significant interference. For example, the installation of a 0.5 m deep polythene root barrier in an alley cropping trial on an alfisol at ICRISAT Centre increased the yields of sole groundnut by 32% and alley-cropped groundnut by three- to four-fold (Rao *et al.*, 1992). Clearly, for agroforestry systems to be successful in the semi-arid tropics, tree species with less aggressive rooting habits must be found to minimize the negative effects of competition for below-ground resources.

Application of resource capture principles to agroforestry

ICRISAT's agroforestry programme evolved from the intercropping experience of Robert Willey and his colleagues, and the results of resource capture studies between 1985 and 1989 have been summarized by Monteith, Ong and Corlett (1991) and Ong *et al.* (1991). Both reports were based largely on the work of Janet Corlett (1989) at ICRISAT Centre, in which pearl millet was grown between hedges of *Leucaena* on a shallow alfisol. Briefly, *Leucaena* hedgerows were spaced 3.3 m apart and pruned to a height of 0.7 m before sowing millet, 29 days after sowing (DAS) and at final harvest at 85 DAS, when the millet and *Leucaena* were both about 1.5 m tall.

Figure 14.6a shows the interception of light in the sole stands over an 11–month period. Maximum fractional light interception in the sole *Leucaena* was 0.6, as compared with 0.7 for sole pearl millet, but because the *Leucaena* retained some of its leaves during the dry season, it intercepted twice as much radiation as the sole millet over the entire measurement period (*Table 14.3*). *Figure 14.6b* compares the total interception of light by the agroforestry system as measured, and as predicted on the basis of interception by the sole crops. These calculations show that interception by the alley-cropped millet was only 40% of that in the sole millet stand, as opposed to the 'expected' value of 70% based on its contribution of five rows out of seven (5 rows millet: 2 rows *Leucaena*). Nevertheless, the alley system greatly increased total interception (810 MJ per square metre) during the rainy season and produced 2–4 t ha^{-1} more than either of the sole stands.

The efficiency of energy conversion, *e*, was generally low in all stands, presumably because of the severe drought experienced (273 mm during the rainy season, which is 36% below normal). In the alley crop, *e* for millet was 27% greater than in the sole crop, possibly because the reduced light saturating photon flux density for photosynthesis associated with drought was reached less frequently under conditions of partial shade. However, the overall *e* value for the alley system was only 7% greater than in sole millet. During the rainy season the value for *Leucaena* was less than expected for a C_3 species in the absence of stress, but was surprisingly close to that of sole millet. During the dry season the *e* value for *Leucaena* was substantially reduced to 0.12–0.15 g MJ^{-1} at a time when evaporative demand was greatest (saturation vapour pressure deficit (SD)$\simeq$4 kPa) and leaf senescence was extensive.

In summary, the agroforestry system was clearly more productive than either of the sole stands primarily because of its greater interception of radiation, since the higher *e* value of alley-cropped millet was insufficient to offset its greatly

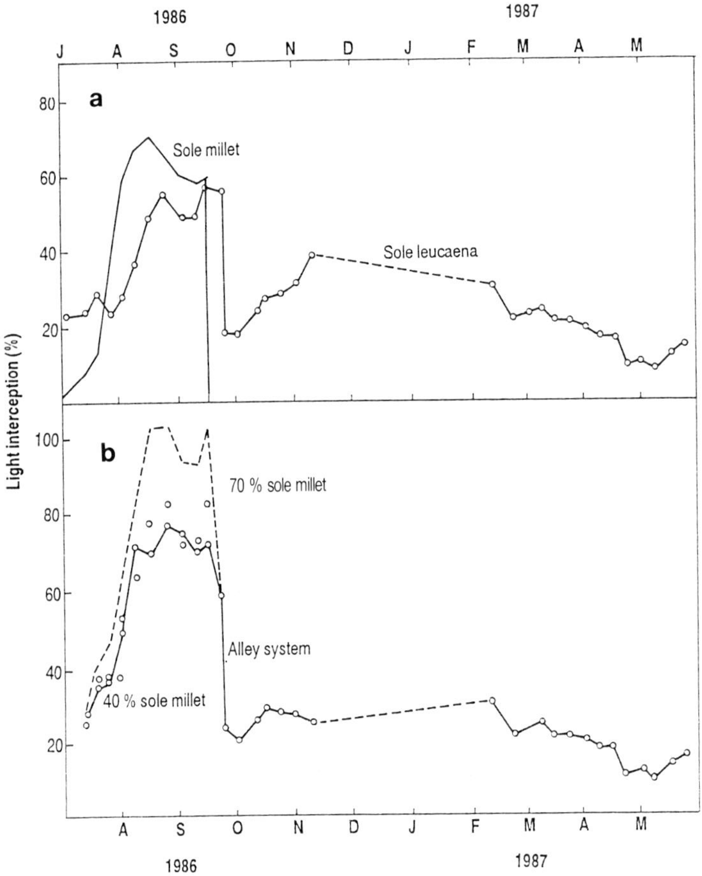

Figure 14.6 Fractional light interception by (**a**) sole stands of *Leucaena* and pearl millet and (**b**) the alley system (solid line) and estimates based on interception by its components, assuming that millet intercepted either 70% or 405 of the sole millet value (Monteith, Ong and Corlett, 1991)

reduced interception of radiation. No accurate measurements of transpiration were made in this study, but estimates of soil evaporation and crop transpiration based on fractional light transmission and potential evaporation suggest that soil evaporation was 20–24% lower in the alley system than in either of the sole stands, but that transpiration was more rapid in the sole millet stand because of its larger leaf area (Corlett, 1989).

The focus of research at ICRISAT Centre subsequently turned to examining the water use of agroforestry systems that are less competitive to crops, for example, dispersed or scattered plantings of trees and other perennial species which are slower growing than *Leucaena* but are compatible with dryland crops and efficient in using the major limiting resource for crop productivity in the semi-arid tropics, namely water (Ong *et al.*, 1992). The first objective was to test the hypothesis that annual crops cannot fully utilize available water and that combinations of crops with trees can greatly enhance water use. This would be particularly important

Table 14.3 INTERCEPTED (TOTAL) SOLAR RADIATION, ABOVE-GROUND BIOMASS AND LIGHT CONVERSION EFFICIENCY (*e*) IN AN ALLEY CROPPING SYSTEM, HYDERABAD, INDIA (AFTER MONTEITH, ONG AND CORLETT, 1991)

	Intercepted radiation (MJ m^{-2})	*Biomass* (t ha^{-1})	*e* (g MJ^{-1})
Rainy season (Jul–Aug 1986)			
Sole millet	581	4.7	0.81
Alley millet	300	3.1	1.03
Sole *Leucaena leucocephala*	520	4.0	0.77
Alley *Leucaena leucocephala*	510	4.0	0.77
Total alley system	810	7.1	0.88
Dry season (Sep 1986–Jun 1987)			
Sole *Leucaena leucocephala*	1270	1.5	0.12
Alley *Leucaena leucocephala*	1160	1.7	0.15
Annual value (Jul 1986–Jun 1987)			
Total alley system	1970	8.8	0.45

when the main objective of agroforestry is to extend the growing period into the dry season to produce fresh fodder when demand is most acute. *Figure 14.7* illustrates the soil moisture profiles under various annual cropping systems and sole *Leucaena* during the dry period (details of planting arrangements and productivity are given by Rao, Sharma and Ong, 1990). Sorghum (100 days duration) pigeonpea (180 days duration) and a sorghum–pigeonpea intercrop all left more residual water in the profile than the sole *Leucaena*. Measurements of moisture profiles in 4.4 m wide alleys between *Leucaena* hedgerows, which were either planted with annual crops or left fallow in both the rainy and dry seasons (*Figure 14.8*), showed that the alley crop utilized considerably more moisture than the fallow alleys, indicating that the roots of fallow *Leucaena* failed to utilize available water effectively within the alleys. Competition for moisture may be accentuated by interception losses (estimated to range from 10 to 30% of rainfall), since even in the rainy season the profile was much drier under alley crops than in the fallow alley treatment.

Further research at ICRISAT Centre has demonstrated the great potential for increasing the productivity of agroforestry systems by utilizing off-season rainfall, when there is no scope for cropping annual species. For example, on the alfisols of the Deccan plateau of India, a traditional intercrop of sorghum–pigeonpea utilises only about 41% of the annual rainfall, while the rest is lost as soil evaporation and runoff (26%) or deep percolation (33%). Measurements of transpiration by 1–year-old perennial pigeonpea grown as the tree component of an agroforestry system with groundnut were made using the heat balance approach (Ishida, Campbell and Calissendorff, 1991). Total annual transpiration of the pigeonpea amounted to 887 mm, or 84% of the annual rainfall, twice the value reported for the most productive intercrop system. A substantial proportion of the total transpiration (47%) occurred during the January–June period, when only 211 mm of rain was received, indicating that 205 mm was extracted from soil reserves. In contrast, in the wet season during July–November there was an excess of 420 mm, which was lost as runoff and percolation. Measurements of runoff in a separate trial, again involving a perennial pigeonpea–groundnut system, indicated that sole

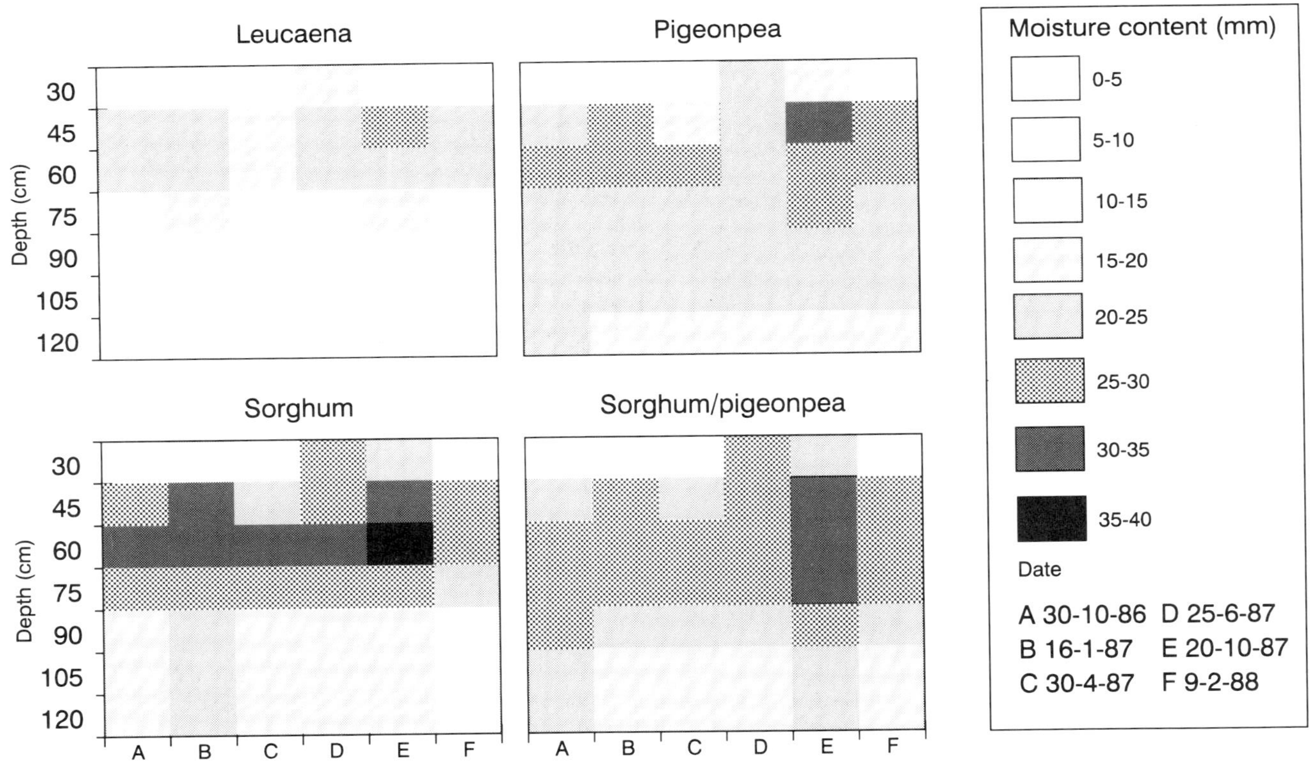

Figure 14.7 Soil moisture profiles (mm) during the post-rainy season on a shallow Vertisol under three annual cropping systems and sole *Leucaena* during two dry seasons (1986–1988) at ICRISAT Centre, India. Sorghum and pigeonpea were harvested in October 1986 and January 1987 respectively, after which the soil in all treatments was left fallow. Dark areas indicate high moisture content (After Ong *et al.*, 1992)

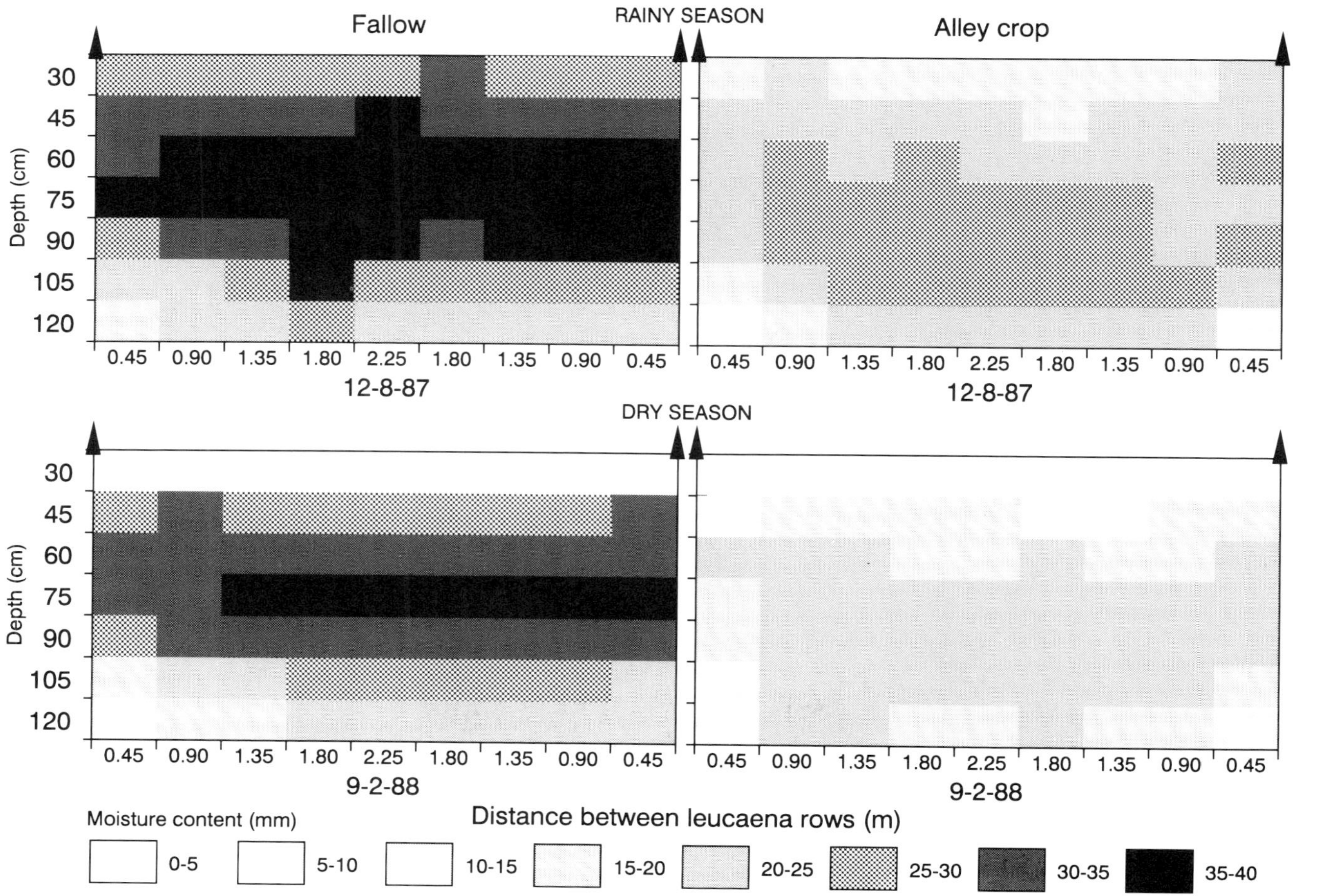

Figure 14.8 Soil moisture profiles (mm) in alleys between *Leucaena* hedgerows under annual crops of left fallow during the rainy and dry seasons at ICRISAT Centre, India. Dark areas indicate high moisture content. Arrows indicate positions of hedgerows (After Ong *et al.*, 1992)

pigeonpea is ineffective in reducing runoff (up to 30% runoff during high intensity storms) relative to sole groundnut or the pigeonpea–groundnut mixture, in which a maximum runoff of 5% was recorded during the same period.

Thus, the potential of agroforestry for increasing the utilization of rainfall is substantial. In much of semi-arid India, tree prunings are too valuable as fodder to be used as mulch, and so the protective role of mulching has seldom been investigated. In sharp contrast, the beneficial effects of mulch using crop residues and woody materials, in terms of moisture conservation, soil temperature amelioration and nutrient enhancement, have been well-documented elsewhere (see review by Huxley (1986) on its relevance to agroforestry). Huxley emphasized the need to determine the benefits of relatively small additions of plant residues to physically poor and/or nutrient depleted soils, especially when only limited quantities of prunings are available in dry areas. In such situations, the maximum benefit will be obtained by using plant residues which provide the greatest amelioration per unit of biomass of organic matter applied, i.e. increased efficiency in resource conservation.

Opportunities for designing better intercropping and agroforestry systems

Attempts at ICRISAT and elsewhere to use resource capture principles to understand the complex mechanisms of plant interactions have a more practical goal, which is to develop a scientific basis for designing more productive and sustainable agricultural systems. An important outcome of these studies is to provide general principles or guidelines to agronomists and plant breeders in their efforts to develop improved technologies for specific environments. The fact that such information is available does not necessarily mean that the recipients will automatically adopt it in their attempts to develop improved production systems. More often, such principles are used to explain when and why certain efforts failed to improve upon existing systems! It is therefore worth while examining some of these lost opportunities, as well as describing new opportunities.

A major concern in the improvement of crop genotypes for intercropping is the question of whether performance in sole cropping is well correlated with intercrop yields (Francis, Flor and Temple, 1976). The resource use studies of pigeonpea–sorghum intercrops described earlier have clearly indicated that pigeonpea performance in intercrops depends largely on its ability to recover from intense competition from the sorghum and intercept radiation *after* the sorghum is harvested. Therefore, intercrop performance is unlikely to be closely related to sole crop yield. Even before this evidence was obtained, there were already convincing grounds for believing that there is no relationship between intercropping advantage and sole crop yields of pigeonpea (Rao *et al.*, 1981). Nevertheless, efforts continued to develop more productive pigeonpea genotypes when grown as sole crops in the hope that these would perform better than local varieties. A similar danger exists in the selection of fast growing multipurpose tree species for potential use in agroforestry systems in the absence of trials to test their competitiveness with understorey crops. These dangers are well appreciated at ICRAF.

In intercropping and agroforestry research, a common objective is to produce a full yield of the staple or economically important crop. For example, farmers in the Deccan plateau of India or in the semi-arid district of Machakos, Kenya, tend to sow only a small proportion of pigeonpea with their sorghum or maize crops to ensure a reliable cereal yield, although a higher pigeonpea population would provide a greater overall intercropping advantage. When the price of pigeonpea grain recently exceeded that of sorghum due to a fall in pigeonpea production in the previous year, farmers in India responded rapidly by increasing the population of pigeonpea by two- to three-fold. In some intercropping and agroforestry systems the benefits of intercropping are achieved with a substantial reduction in the yield of the major crop, usually a cereal, as in the case of intercropped upland rice and pigeonpea (Jena and Misra, 1988) and alley cropped *Leucaena* and dryland crops (Rao, Sharma and Ong, 1990). Here is clearly an important opportunity to design improved systems which minimize the negative effects of competition and maximize the positive effects of intercropping.

Research on upland rice–pigeonpea systems at ICRISAT Centre has indicated that substantial yield advantages can be achieved without a major sacrifice in rice yield (Ramakrishna, Ong and Reddy, 1992). Because modern upland rice varieties have erect leaves, they intercept a maximum of 60% of incident radiation within 2 months of sowing, but are sensitive to low light and shading during the reproductive phase. It was found that the most compatible pigeonpea genotypes have a compact canopy with an extinction coefficient (K) of around 0.64, and are of a similar height to the rice. Differences in crop duration appeared to be relatively unimportant. As substantial variation in K values (0.45–0.78) exists within available pigeonpea genotypes, there is therefore considerable opportunity to select for compatible genotypes. Similarly, research at IRRI has confirmed that a compact canopy of short season cowpea and mungbean is more important for compatibility with upland rice than the ability of the legume to accumulate N (Siri-Udompas and Morris, 1990).

There is increasing evidence from agroforestry research that the beneficial effects of mulch for associated crops is seldom directly related to the *quantity* of biomass produced by the trees or to the nutrient quality of the mulch (unpublished ICRAF reports). These results suggest that the beneficial effects of mulch may be offset by negative below-ground interactions which may operate simultaneously or at different times during the season. Future research should place greater emphasis on minimizing negative below-ground interactions which may be less immediately obvious than above-ground interactions, but may be more extensive than previously realized. The limited information on the rooting profiles of useful agroforestry species suggests that considerable opportunities exist for obtaining compatibility with crops (Hariah and van Noordwijk, 1986). Equally important is the recent availability of relatively inexpensive and reliable techniques for observing and confining roots, the absence of which has previously discouraged investigations of the role of roots in resource capture.

Agroforestry research currently places far too much emphasis on alley cropping as a means of improving soil fertility, irrespective of rainfall and soil type. Evidence is accumulating in both India and Africa to show that the advantages of alley cropping become marginal compared with a cereal–legume rotation when annual rainfall is below 1000 mm. However, alley cropping has a clear advantage over annual cropping on sloping land, where soil loss and runoff are an important consideration. Even in the semi-arid tropics where rainfall is usually inadequate for crop growth, erosive rainfall is a major problem. For example, at the onset of the long rains of 1990, a 50 mm h^{-1} storm fell at Machakos, Kenya on a saturated soil on a 14% slope. Soil loss on an annual crop control was 34 t ha^{-1} or higher, while erosion on a *Cassia siamea* alley system ranged from 0.2 to 0.5 t ha^{-1} (Kieppe, unpublished). Other agroforestry systems which combine grasses and trees on contours are now being actively explored at various ICRAF sites as a means of controlling soil erosion on hillsides. The potential of agroforestry for erosion control on hillsides becomes apparent within 2–3 years, when sloping land is transformed into a series of terraces. However, the successful adoption of such agroforestry systems depends largely on the prospects for minimizing or eliminating the competition between trees or tree–grass mixtures and the annual crops growing on the terraces.

There are other opportunities for applying resource capture principles to the improvement of intercropping and agroforestry systems, including planting arrangement and microclimatic improvement, but those reviewed above represent some of the most challenging possibilities.

Conclusions

The above account of complementarity in resource utilization in intercropping and agroforestry systems has been drawn almost exclusively from ICRISAT's experience, which represents the most systematic and extensive information available for mixed agricultural stands. Not all of the general principles that govern the substantial yield advantage of intercropping over sole cropping can be transferred to agroforestry (see Ong (1991) for a fuller discussion). The greatest difference is the importance of root interactions in agroforestry, as was discussed earlier. There also appears to be less prospect of improving the efficiency of energy conversion (i.e. spatial complementarity) because the understorey crops in most agroforestry combinations are generally C_4 cereals, which can compete more effectively with trees than C_3 species, but are very sensitive to shading. The evidence of Corlett, Ong and Black (1989) suggests that the improvement in *e* of alley-cropped pearl millet was insufficient to compensate for the substantial reduction in total intercepted radiation. However, it is possible that an improvement in *e* can be achieved by a better dispersion of light within the canopy, a postulated advantage of multi-storey systems in the humid tropics (Nelliat, Bavappa and Nair, 1974).

Microclimatic improvement has been cited as a cause of the high partitioning of dry matter to grains in pearl millet–groundnut intercrops (Willey *et al.*, 1986; Harris, Natarajan and Willey, 1987) on the basis of results obtained during the post-rainy season, when temperature and radiation levels are much higher than during the rainy season. However, there was no evidence of a consistent increase in the harvest index of groundnut during the rainy season, suggesting that caution is necessary when extrapolating the results of post-rainy season trials (Ong, Subrahmanyam and Khan, 1991). An analysis of microclimatic improvements in agroforestry systems also suggests that the effects are small when temperature and humidity are already favourable for plant growth (Corlett, Ong and Black, 1989). However, major benefits of microclimatic modification may be expected from windbreaks in areas where temperature exceeds the optimum for growth ($\simeq$ 36°C) and saturation deficit exceeds 2.5 kPa, as in the Sahel (Brenner, 1991). In this study, the benefits of shelter for the growth of pearl millet were apparently closely related to reductions in saturation deficit and avoidance of supra-optimal temperatures, since water was not conserved behind the windbreaks.

The volume of agroforestry research has grown exponentially since 1985 as various international and national research institutes include agroforestry research in their programmes. With many field experiments in progress in a range of climatic and geographic regions, the scientific understanding of agroforestry practices and their potential benefits will soon begin to match the wisdom of farmers and pastoralists, whose extensive knowledge of agroforestry is sadly poorly documented.

Acknowledgements

The authors are grateful to the staff of the Cropping Systems Section at ICRISAT for their considerable help in the field measurements and to the postgraduate and post-doctoral visitors, C.M. Stirling, J.E. Corlett, F.M. Marshall, J.N. Daniels and J.C.W. Odongo, for their valuable contributions.

References

Aggarwal, P.K., Garrity, D.P., Liboon, S.P. and Morris, R.A. (1992) Resource use and plant interactions in a rice-mungbean intercrop. *Agronomy Journal*, **84**, 71–78

Azam-Ali, S.N., Gregory, P.J. and Monteith, J. L. (1984) Effects of planting density on water use and productivity of pearl millet (*Pennisetum typhoides*) grown on stored water. II Water use, light interception and dry matter production. *Experimental Agriculture*, **20**, 215–224

Brenner, A. (1991) Millet growth behind shelter in the Sahel. *Agroforestry in the UK*, **2**, 30–38

Connolly, J. (1986) On difficulties with replacement series methodology in mixture experiments. *Journal of Applied Ecology*, **23**, 125–137

Corlett, J.E. (1989) *Leucaena/millet alley cropping in India: Microclimate and productivity*. Ph.D. Thesis. University of Nottingham, UK

Corlett, J.E., Ong, C.K. and Black, C.R. (1989) Microclimatic modification in intercropping and alley cropping systems. In *The Applications of Meteorology to Agroforestry Systems Planning and Management*, pp 419–430 Edited by W.S. Reifsnyder and T.O. Darnhofer. Nairobi: International Council for Agroforestry Research

Donald, C.M. (1958) The interaction of competition for light and nutrients. *Australian Journal of Agricultural Research*, **9**, 421–435

Donald, C.M. (1963) Competition among crop and pasture plants. *Advances in Agronomy*, **15**, 1–118

Francis, C.A., Flor, C.A. and Temple, S.R. (1976) Adapting varieties for intercropping systems in the tropics. In *Multiple Cropping*, pp 235–253 Edited by R.I. Papendick, P.A. Sanchez and G.B. Triplett. ASA Special Publication No. 27. Madison, Wisc: ASA

Goldberg, D.E. and Werner P.A. (1983) Equivalence of competitors in plant communities: a null hypothesis and a field experimental approach. *American Journal of Botany*, **70**, 1098–1104

Grace, J.B. and Tilman, D. (1990) *Perspectives on Plant Competition*. San Diego: Academic Press

Gregory, P.J. and Reddy M.S. (1982) Root growth in an intercrop of pearl millet–groundnut. *Field Crops Research*, **5**, 241–252

Grime, J.P. (1979) *Plant Strategies and Vegetation Processes*. London: Wiley

Groves, R.H. and Williams J.D. (1975) Growth of skeleton weed (*Chondrilla juncea* L) as affected by growth of subterranean clover (*Trifolium subterraneum* L) and infection by *Puccinia chondrilla* Bubak and Syd. *Australian Journal of Agricultural Research*, **26**, 975–983

Hariah, K. and Van Noordwijk, M. (1986) Root studies of a tropical ultisol in relation to nitrogen management. *Report No. 7–86*, 4. Soil Fertility Institute, Haren, Netherlands, pp. 116

Harris, D., Natarajan, M. and Willey, R.W. (1987) The physiological basis for yield advantage in a sorghum–groundnut intercrop exposed to drought. I Dry matter production, yield and light interception. *Field Crops Research*, **17**, 259–272

Huxley, P.A. (1986) Rationalising research on hedgerow intercropping: an overview. *ICRAF Working Paper No. 40*, pp 148. Nairobi: ICRAF

Ishida, T., Campbell G.S. and Calissendorff C. (1991) Improved heat balance method for determining sap flow rate. *Agricultural and Forest Meteorology*, **56**, 35–48

Jena, D. and Misra, C. (1988) Effect of crop geometry (row proportions) on the water balance of the root zone of a pigeonpea and rice intercropping system. *Experimental Agriculture*, **24**, 385–391

Kang, B.T. Grimme, H. and Lawson, T.L. (1985) Alley cropping sequentially cropped maize and cowpea with leucaena on a sandy soil in southern Nigeria. *Plant and Soil*, **85**, 267–277

Marshall, B. and Willey R.W. (1983) Radiation interception and growth in an intercrop of pearl millet–groundnut. *Field Crops Research*, **7**, 141–160

Mead, R. and Willey R.W. (1980) The concept of a 'Land Equivalent Ratio' and advantages in yields from intercropping. *Experimental Agriculture*, **16**, 217–228

Monteith, J.L., Ong, C.K. and Corlett, J.E. (1991) Microclimatic interactions in agroforestry systems. *Forest Ecology and Management*, **45**, 31–44

Morris, R.A. (1986) Upland rice farming systems. In *Progress in Upland Rice Research*, pp 119–131. Los Banos, The Philippines: International Rice Research Institute

Natarajan, M. and Willey R. W. (1980) Sorghum-pigeonpea intercropping and the effects of plant population density. II Resource use. *Journal of Agricultural Science, Cambridge*, **95**, 59–65

Natarajan, M. and Willey, R.W. (1985) Effect of row arrangement on light interception and yield in sorghum–pigeonpea intercropping. *Journal of Agricultural Science, Cambridge*, **104**, 263–270

Nelliat, E.V., Bavappa, K.V. and Nair, P.K.R. (1974) Multi-storeyed cropping: A new dimension in multiple cropping for coconut plantations. *World Crops*, **26**, 262–266

Ong, C.K. (1991) Interactions of light, water and nutrients in agroforestry systems. In *Biophysical Research for Asian Agroforestry*, pp 107–124 Edited by M.E. Avery, M.G.R. Cannell and C.K. Ong. New Delhi: Winrock International

Ong, C.K., Corlett, J.E., Singh, R.P. and Black, C.R. (1991) Above and below-ground interactions in agroforestry systems. *Forest Ecology and Management*, **45**, 45–57

Ong, C.K., Odongo, J.C.W., Marshall, F. and Black, C.R. (1992) Water use of agroforestry systems in semi-arid India. In *Growth and Water Use of Forest Plantations*, pp 347–358 Edited by I.R. Calder, R.L. Hall and P.G. Adlard. Proceedings of the International Symposium on the Growth and Water Use of Forest Plantations, Bangalore, 7–11 February, 1991. Chichester: Wiley

Ong, C.K. Subrahmanyam, P. and Khan, A.A.H. (1991) The microclimate and productivity of a groundnut/millet intercrop during the rainy season. *Agricultural and Forest Meteorology*, **56**, 49–66

Ramakrishna, A., Ong, C.K. and Reddy, S.L.N. (1992) Canopy duration and structure of pigeonpea intercropped with upland rice. *Experimental Agriculture*, **28**, 295–307

Rao, M.R., Ong, C.K., Pathak, P. and Sharma, M.M. (1992) Productivity of annual cropping and agroforestry systems on a shallow alfisol in semi-arid India. *Agroforestry Systems*, **15**, 51–64

Rao, M.R., Sharma, M.M. and Ong, C.K. (1990) A study of the potential of hedgerow intercropping in semi-arid India using a two-way systematic design. *Agroforestry Systems*, **11**, 243–258

Rao, M.R. and Willey, R.W. (1983) Effects of pigeonpea plant population and row arrangement in sorghum/pigeonpea intercropping. *Field Crops Research*, **7**, 203–212

Rao, M.R., Willey, R.W., Sharma, D. and Green, J.M. (1981) Pigeonpea genotype evaluation for intercropping. In *International Workshop on Pigeonpea*, pp. 263–270. ICRISAT Centre, Hyderabad, 15–19 December, 1980. Hyderabad: ICRISAT

Reddy, M.S. and Willey, R.W. (1981) Growth and resource use studies in an intercrop of pearl millet–groundnut. *Field Crops Research*, **4**, 13–24

Singh, R.P., Van Den Beldt, R.J., Hocking, D. and Korwar, G.R. (1989) Alley cropping in the semi-arid regions of India. In *Proceedings of Alley Cropping Workshop*, pp 108–122 Edited by B.T. Kang and L. Reynolds. International Institute of Tropical Agriculture (IITA), Ibadan, Nigeria, March 10–14, 1986. Ibadan: IITA

Singh, R.P., Ong, C.K. and Saharan, N. (1989) Above and below-ground interactions in alley-cropping in semi-arid India. *Agroforestry Systems*, **9**, 259–274

Siri-Udompas, C. and Morris, R.A. (1990) Effects of crop proportion on intercropped upland rice and cowpea. 2. Nitrogen yields. *Field Crops Research*, **25**, 233–246

Sivakumar, M.V.K. and Virmani S.M. (1980) Growth and resource use of maize, pigeonpea and maize/pigeonpea intercrop in an operational research watershed. *Experimental Agriculture*, **16**, 377–386

Snaydon, R.W. and Harris, P.M. (1981) Interactions below-ground – the use of nutrients and water. In *Proceedings of International Workshop on Intercropping*, pp 188–201. ICRISAT Centre, Hyderabad, 10–13 January, 1979. Hyderabad: ICRISAT

Stirling, C.M., Williams, J.H., Black, C.R. and Ong, C.K. (1990) The effect of time of shading on development, dry matter production and light-use efficiency in groundnut (*Arachis hypogaea L.*) under field conditions. *Australian Journal of Agricultural Research*, **41**, 633–644

Tilman, D. (1988) *Plant Strategies and the Dynamics and Structure of Plant Communities*. Monograph on Population Biology, Princeton University Press

Watt, A.S. and Fraser, G.K. (1933) Tree roots and the field layer. *Journal of Ecology*, **21**, 404–414

Willey, R.W. (1979) Intercropping – its importance and research needs. II Agronomy and research approaches. *Field Crop Abstracts*, **32**, 78–85

Willey, R.W., Natarajan, M., Reddy, M.S. and Rao, M.R. (1986) Cropping systems with groundnut: Resource use and productivity. In *Agrometeorology of Groundnut*, pp 193–205. Proceedings of an International Symposium, ICRISAT Sahelien Centre, Niamey, Niger, 21–26 August, 1985. Patancheru, India: ICRISAT

Willey, R.W. and Reddy, M.S. (1981) A field technique for separating above and below ground interactions in intercropping: an experiment with pearl millet–groundnut. *Experimental Agriculture*, **17**, 257–264

de Wit, C.T. (1960) On competition. *Verslag Landbouwkundig Onderzoek*, **66**, 1–82

15

RESOURCE CAPTURE BY ARABLE CROPS

R.K. SCOTT
Department of Agriculture and Horticulture, University of Nottingham, Sutton Bonington Campus, Loughborough LE12 5RD, UK

K.W. JAGGARD
AFRC Institute of Arable Crops Research, Broom's Barn Experimental Station, Higham, Bury St. Edmunds, Suffolk IP28 6NP, UK

R. SYLVESTER-BRADLEY
ADAS Soil and Water Research Centre, Anstey Hall, Trumpington, Cambridge CB2 2LF, UK

Introduction

This chapter is prepared at a time when the producers of arable crops in north-west Europe face a number of new challenges. There will be lower prices and area based 'compensation' in terms of the payment for putting land out of production; so called 'set aside' will comprise some 15% of cereal land. This will require new management practices and create an opportunity for crops for non food uses. The prices of agrochemicals are likely to continue to increase and restrictions on the use of nitrogen fertilizer and pesticides may well be extended. Water is likely to be more scarce and therefore more expensive; there will be increasing concern about pollution, particularly of nitrogen compounds, both in drainage water and to the atmosphere.

Our objectives in preparing the chapter are to exploit understanding of the principles of resource capture to the benefit of the arable sector (scientists, policy makers, advisers and growers) in today's rapidly changing circumstances. The consideration of resource capture has been restricted to north-west Europe and to light, nitrogen and water and does not consider the capture of other nutrients or carbon dioxide, nor the impact of pollutants (Chapter 11). In relation to the specific resources, there is no restriction on the degree to which light energy should be exploited: if agriculture fails there is no other potential beneficiary. For nitrogen, effective capture is imperative in order to minimize waste and pollution; for water the priority is to gain maximum productivity consistent with minimum usage.

The chapter is based on work with sugar beet and wheat, particularly at Broom's Barn. It begins by examining the relationship between light capture and productivity of sugar beet, the arable crop that continues to build up yield for as long as the growing season allows. It extends this analysis to examine the way fertilizer nitrogen brings benefit, first for a vegetative crop, sugar beet, then for a seed crop, wheat; and continues by considering how light, nitrogen and water usage are inextricably linked and the inevitable disadvantages from excessive use of both nitrogen and water by arable crops generally.

Determination of yield

LIGHT USE AND DRY MATTER PRODUCTION

In a paper to the Royal Society, Monteith (1977), synthesized ideas about light use and crop growth that had been discussed by crop physiologists (Warren Wilson, 1967; Watson *et al.*, 1972) for several years. He was able to show that in the temperate climate of Britain there was a strong relationship between the amount of light energy intercepted by foliage during the growing season and the dry matter yield of a wide range of crops. Monteith recognized that a single relationship between intercepted radiation, growth and yield was a considerable simplification. In the UK, it often does apply because cloud cover is usually such that canopies are seldom light saturated for prolonged periods, and water stress is rarely severe. The great merit in the analysis was that it allowed at least this one variable in the crop environment to be isolated so that the effects of other variables could be analysed quantitatively.

A solid framework for unravelling the effects of various weather features on crop growth was created when it became possible to measure the short-term changes in weather and the minute-by-minute responses of the crop. A comprehensive study of a field crop's response to its environment was made by Biscoe *et al.* (1975) at Nottingham with spring barley and then with wheat; a principle conclusion was that growth rate was directly proportional to radiation absorbed (Gallagher and Biscoe, 1978). A direct extension of those programmes but with sugar beet, was made by Glauert (1983) who used an enclosure method to monitor gas exchange.

Figure 15.1 shows one of Glauert's relationships between irradiance and the amount of carbon dioxide fixed by photosynthesis. It is striking how tightly coupled the two variables were; both increased progressively through the morning to decline later in the day. There was a distinct and immediate decrease in assimilation whenever the sun was obscured by cloud. When the data were plotted to show the response of CO_2 uptake to radiation, there was an increase over the whole range (0–800 W m^{-2} total radiation) but with a diminishing response. *Figure 15.1* also shows a displacement in values for CO_2 uptake at the same irradiance; values in the morning were greater than in the afternoon. This shift was most pronounced on bright days and was probably a response to water stress in the afternoon closing stomata on the large, outer leaves.

Glauert measured CO_2 uptake from mid June until early December. *Figure 15.2* shows that the photosynthesis–light response curve was maintained in an essentially similar position until September. Only then did the diminished responsiveness of an ageing canopy become evident. Sugar beet continues to produce leaves throughout its life, but despite this continued production, the overall age of the leaf surface eventually increases because new leaves expand little and contribute only a small part to the light intercepting area (Milford *et al.*, 1985). After remaining stable through June, July and early August, the net photosynthesis at 500 W m^{-2} then declined throughout autumn to about half its summer-time value. Although temperature also declined over this period, cooler conditions were not the cause. At intervals from mid June until mid October, Glauert made sets of observations with the day-time temperature in his enclosure set at 20°C. Photosynthesis was maintained at the same rates during these episodes as when temperature followed the external air. Photosynthesis–light response curves were

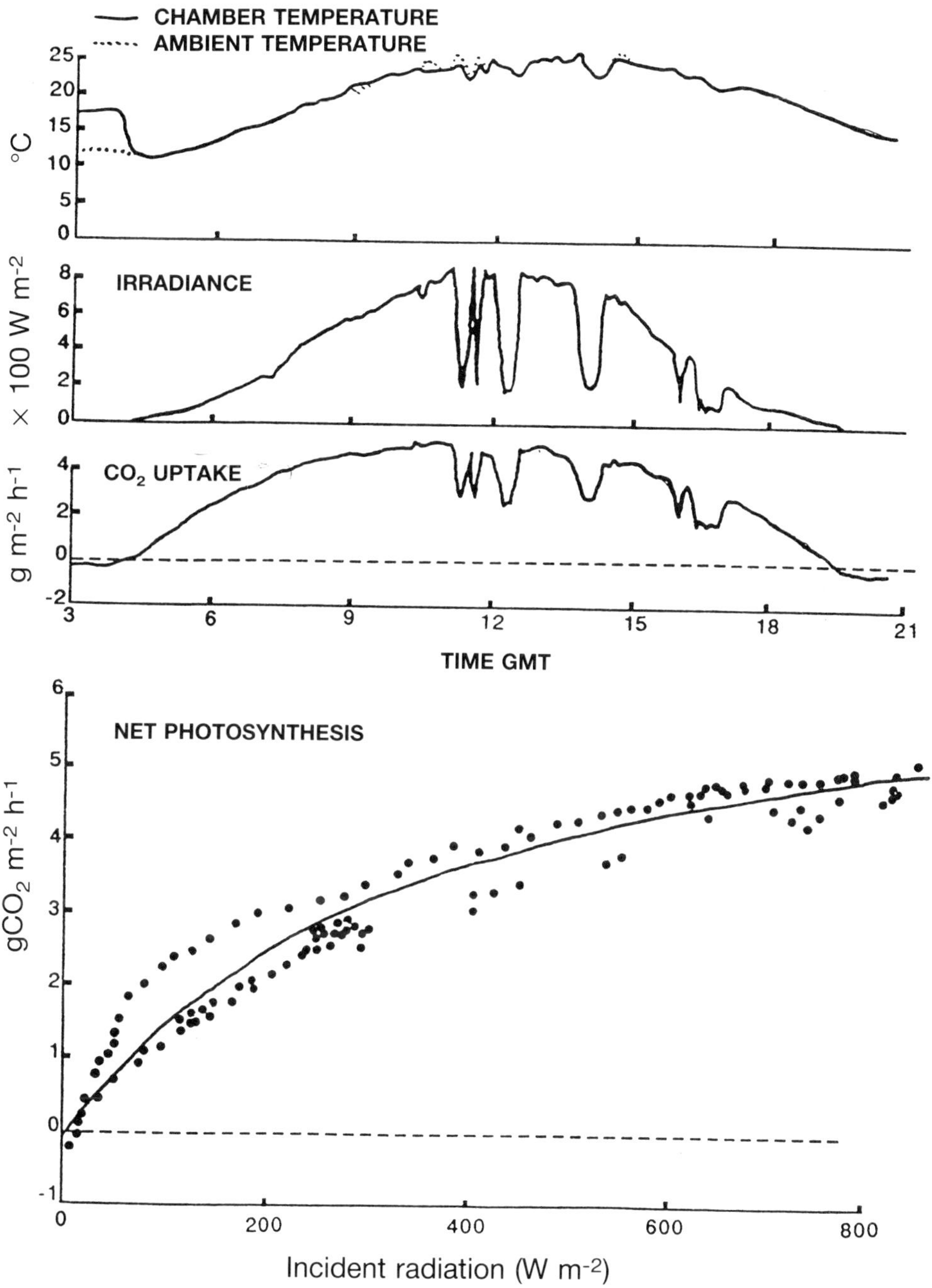

Figure 15.1 Changes in temperature, irradiance and carbon dioxide uptake by a sugar beet crop during 24 July 1980. (After Glauert, 1983)

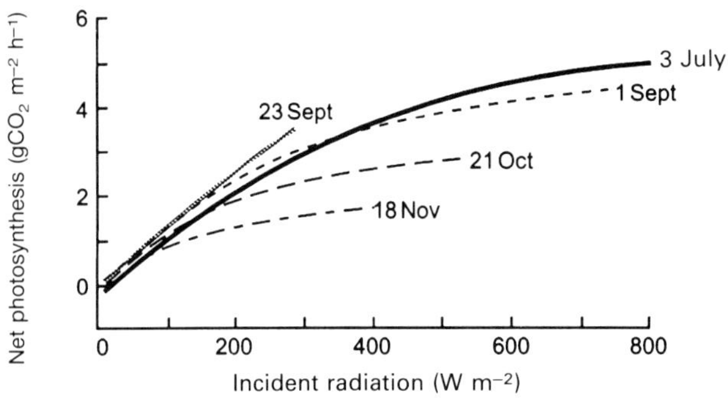

Figure 15.2 Relationships between net photosynthesis and incident radiation for a sugar beet crop on various dates during summer and autumn

similar on 15 November, a warm day when the temperature did not fall below 12°C, and on the following bright, cold day when temperature never exceeded 8°C (*Figure 15.3*). It was not until 29 November, when the temperature in the gas exchange system never exceeded 2°C, that the response to light was diminished. Clearly, the diminished responsiveness of the canopy late in the season was controlled internally by the plant rather than by external temperature. Ontogenetic drift in the efficiency of the canopy apparently occurred later in sugar beet than in lettuce (Wheeler *et al.*, 1993).

Glauert's (1983) system enabled him to estimate dark respiration from mid June until early December. The average loss of dry matter during the night was estimated at 2 g m^{-2} day^{-1}, and there was no clear evidence of systematic change as the season progressed, despite increases in the amount of respiring biomass and a longer dark period. To compensate, weather became cooler and the increasing fraction of the biomass stored in the tap root probably had a low rate of maintenance respiration. The rate of dark respiration was related to the amount of dry matter assimilated during the previous day (*Figure 15.4*).

The daily increment in dry matter can be estimated from the measured amount of carbon dioxide taken up by the crop during a day. Glauert plotted estimates

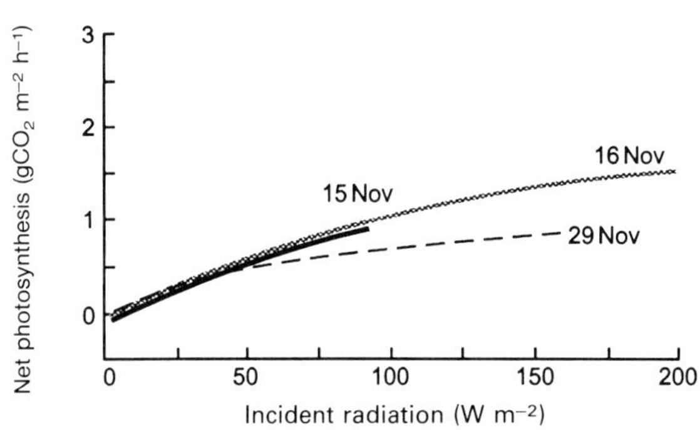

Figure 15.3 Relationships between net photosynthesis and incident radiation of a sugar beet crop on a warm, dull day (15 November), a clear, cool day (16 November) and a cold day (29 November)

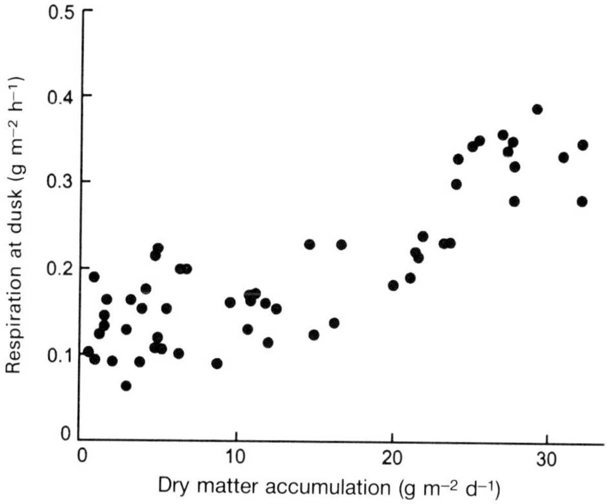

Figure 15.4 Correlation between amount of CO_2 fixed during a day and the rate of respiration of a sugar beet crop in the 2 hours after dark. (After Glauert, 1983)

of daily dry matter production on 60 days throughout the growing season against the amount of radiant energy intercepted by the foliage during those days (*Figure 15.5*). On some days the Biomass Radiation Coefficient was maintained close to 1.9 g MJ^{-1} of total radiation intercepted, a large figure surprisingly independent of temperature, plant size and the age of the crop. Smaller values of the coefficient on very sunny days can be attributed to light saturation (*Figure 15.1*). For the whole period the mean coefficient was 1.72 g MJ^{-1}. The clear implication from *Figure 15.5* is that biomass production during the season

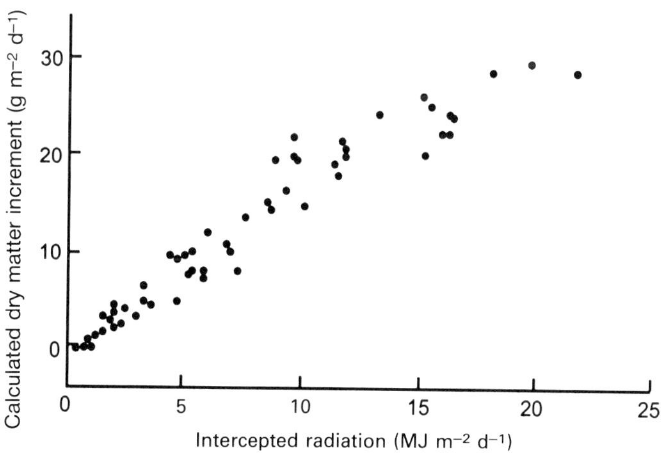

Figure 15.5 The correlation between the increment of dry matter (calculated from daily integrals of the CO_2 uptake) and daily solar radiation intercepted by the foliage of a sugar beet crop throughout summer and autumn, 1980. (After Glauert, 1983)

was directly related to the amount of radiation intercepted by the foliage between sowing and harvest.

There must always be concern about the validity of extrapolating to the outdoor environment from measurements made on enclosed plants. Light quality was slightly affected by the enclosure, and whereas there was satisfactory control of temperature, the air within the enclosure was often more humid than the air outside. Glauert checked whether these differences were critical. He estimated biomass production by integrating net carbon fixation as carbohydrate equivalents on a 24 hour basis, and compared the accumulated total with weight of crop from conventional growth analysis over the whole season. He found remarkable agreement; 18.0 t ha^{-1} estimated from gas exchange and 18.3 ± 0.48 t ha^{-1} from growth analysis. All other things being equal, the gas exchange estimate should somewhat exceed the growth analysis value because the latter excludes dry matter in the fibrous root system and in leaves shed before final harvest. When these have been measured they have added 4% (Brown *et al.*, 1987) and 9.6% to the biomass value.

To examine the effect of seasonal variation in temperature and radiation on growth and yield of sugar beet, a crop has been grown at Broom's Barn in a standard way each year since 1978. In order to minimize genetic influences, seed was drawn from only two batches; the change was made to maintain relevance to contemporary cultivars. The consistent objectives were to sow as soon as the soil was fit once the risk of excessive bolting had passed, then to harvest late and by hand so that seasonal influences were complete and not masked by differences in harvesting efficiency. All crops were irrigated. Given the standardization, the range of biomass (17–27 t ha^{-1}) and sugar (8–15 t ha^{-1}) yields is striking – greater than would be elicited by drastic modifications to husbandry within a season.

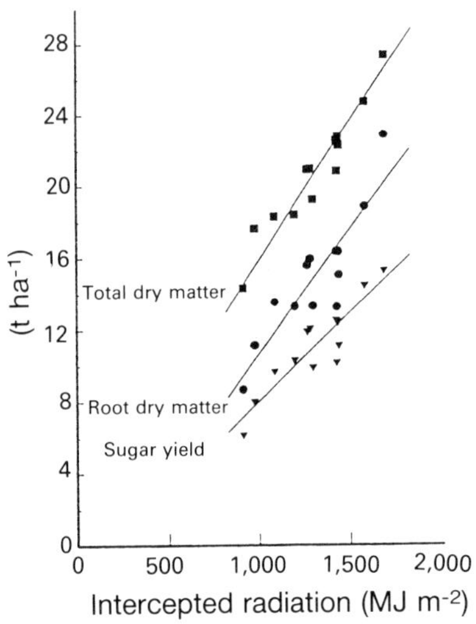

Figure 15.6 Relationships between amount of solar radiation intercepted by the foliage throughout the growing season and total dry matter (■), root dry matter (●) and sugar (▼) yields. Each data point represents the yield of a sugar beet crop grown with recommended husbandry practices, including irrigation, at Broom's Barn between 1978 and 1990

Figure 15.6 shows the relationships between sugar-beet yield and radiation intercepted by foliage between sowing and harvest. The relationship between root dry matter at harvest and light energy intercepted during growth is also shown. The fitted line is parallel to that for biomass, with a displacement representing the mass of foliage which averaged about 5.5 t ha^{-1}, irrespective of biomass or intercepted light energy. Thus, increases in season long radiation interception increased root yield proportionately. The lines for root and sugar yield diverge, reflecting the very stable proportion (70–72%) accounted for by sugar within the dry weight of the storage root. The coefficient representing the conversion of intercepted light to sugar is 0.97 g MJ^{-1}. From *Figure 15.6* it is evident that the ratio of the weight of root to weight of total biomass tends to increase as the seasonal radiation interception and yield increase. This also occurs in relation to sugar, but less markedly.

In England, the range in radiation interception season-long is 1000–1700 MJ m^{-2}. In looking for causes, our analysis supports Watson's (1956) contention at the third of this series of Easter Schools that: 'The key factor determining seasonal variation in yield was the degree to which crops intercepted June and July radiation'. In the present series (*Figure 15.6*) the crops that gave large yields were those that made exceptional use of May and June radiation. From July onwards, crops were always able to intercept 85% of incident radiation and to grow late into the autumn. However, until July, sowing date and spring temperatures control the rate of canopy closure (Milford *et al.*, 1985). Before Midsummer Day the percentage interception by the canopy ranged, over the 13–year period, from 31 to 72%. The origin of the largest yield, in 1982, can be traced directly to early canopy closure following a 4–week period in late May and early June when mean air temperatures were 4°C above average.

LIGHT CAPTURE AND NITROGEN FOR SUGAR BEET

It is commonplace for agronomists to think in terms of direct relationships between crop yield and nutrient requirements. For instance, the production of one extra tonne of wheat grain is often taken to indicate the need for an extra 30 kg of nitrogen (Remy and Viaux, 1982). Although such associations can sometimes be found, they are by no means universal. Certainly, it does not appear to hold for sugar beet: Last *et al.* (1983) showed that the sugar yield of well-grown crops differed by a factor of almost two at approximately the optimal nitrogen uptake of 200 kg ha^{-1}. At first sight this may seem surprising because yield relates to radiation interception, and nitrogen is known to stimulate leaf growth. Can relationships between yield, nitrogen uptake and light capture be unravelled? Growers' decisions would be greatly assisted if a prescription could be devised which translated the canopy area necessary for adequate light interception to a minimum quantity of nitrogen that a crop must contain.

Nitrogen deficiency slows the rate of growth, reduces leaf area and produces pale foliage (Milford *et al.*, 1985) but, surprisingly, in sugar beet it seems to have no effect on the light-to-dry matter conversion coefficient. This is shown in *Figure 15.7*, from an experiment at Broom's Barn where light interception throughout the season was reduced from 1400 to 1000 MJ m^{-2}, and where dry matter yield was reduced from 22.3 to 15.8 (± 0.49) t ha^{-1} by withholding nitrogen fertilizer. Linear regression was used to estimate separate conversion coefficients

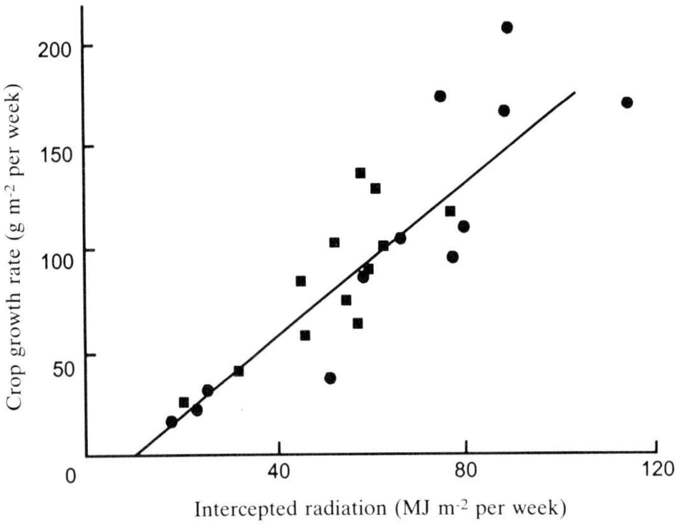

Figure 15.7 Relationship between weekly rates of dry matter production and the amounts of solar radiation intercepted by irrigated sugar beet crops grown with (●) and without (■) nitrogen fertilizer

for the two fertilizer treatments; they were not significantly different (1.67 and 1.61 ± 0.121 g MJ^{-1} with and without fertilizer respectively).

Measurements at Broom's Barn show that the relationship between the proportion of solar radiation intercepted (I_L/I_o) and the expansion of the intercepting surface i.e. leaf area index (L) of sugar beet crops grown at the recommended plant populations may be expressed as:

$$I_L/I_o = [0.32 + (1{-}0.32)\tau]^L$$

where τ is a mean transmission coefficient (Monteith, 1965). Using the mean biomass radiation coefficient quoted earlier (1.72 g MJ^{-1}), and a typical pattern of expansion of L with time in a summer, and average radiation for eastern England, net dry matter production for each unit of L can be calculated. *Figure 15.8* shows that if a fourth unit of L is added, then the additional light energy intercepted is insufficient to cover the original investment in creating the additional leaf area (Chapter 3). This simple analysis suggests that the optimum L value for sugar beet is about three, similar to the findings of Watson (1956) and Goodman (1968) both of whom used very different methods of analysis. In fact, *Figure 15.8* probably understates the detrimental effect of additional leaf because it makes no allowance for the large amount of carbohydrate which must be respired when leaf is created (S.P.Long, personal communication). In summary, *Figure 15.8* shows that, for root and sugar yield, it would be inappropriate to apply nitrogen fertilizer to beet crops to produce L values of more than three unless this was a precursor needed to ensure full interception of light late in the season.

Since yield is so strongly controlled by the rate of canopy closure, an essential part of the rationale for nitrogen nutrition is to ensure that the crop's ability to respond to spring and early summer temperatures is not limited by N shortage. Milford *et al.* (1985) demonstrated how the optimal application of nitrogen

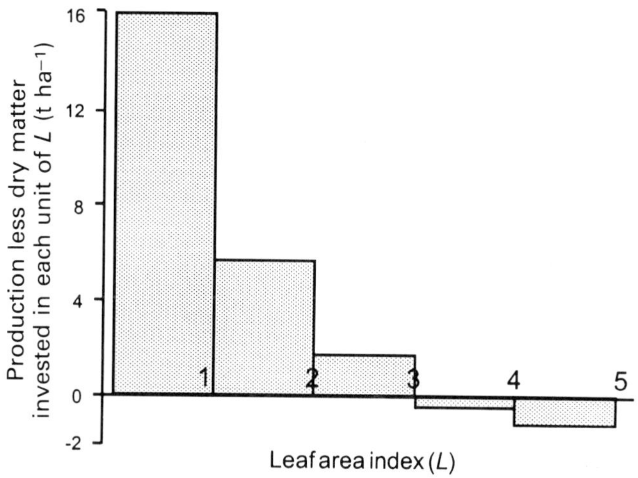

Figure 15.8 Calculated net changes in biomass productivity with each increment of leaf area index of sugar beet crops

maintained a prolonged period of leaf expansion at a rate dictated by ambient temperature (*Figure 15.9*). However, nitrogen availability appears to exert an absolute control over canopy expansion: data from crops grown at different sites, in several seasons and from plots with or without nitrogen fertilizer show L to be closely and linearly related to nitrogen uptake (*Figure 15.10*). Typically, beet crops contained 4 g of N per square metre of leaf lamina (40 kg ha^{-1}). Thus, a crop with $L = 3$ must contain 120 kgN ha^{-1}, and ideally this represents the state of the crop at the end of June or early in July.

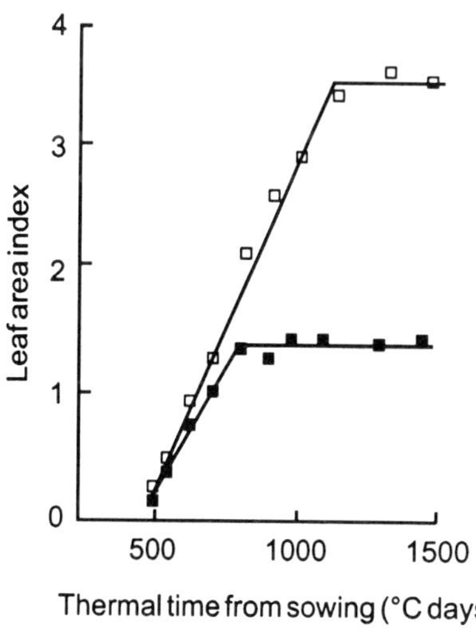

Figure 15.9 Relationships between thermal time and leaf area index in two sugar beet crops at Broom's Barn, one grown without nitrogen fertilizer (■) and the other with a recommended dressing of 125 kgN ha^{-1} (□). (After Milford *et al*., 1985)

After the full canopy has been produced the crop needs access to sufficient nitrogen for maintenance of the canopy and growth of the storage root. Usually these requirements are met from two sources. The first is from mineralization of soil organic matter. The beet crop is an efficient scavenger of soil N, commonly reducing the mineral N content of soil to only 20–30 kg ha^{-1} in the top 1 m by the end of the growing season. The second source is nitrogen remobilized from within the plant. Bürcky and Biscoe (1983) found that as much as 75 kgN ha^{-1} could be transferred from old to new leaves and to the storage root.

When large amounts of nitrogen are available during July and August, then the rapid uptake typical of early summer continues (Armstrong *et al.*, 1986) and in consequence the crop apportions much of its assimilate to the nitrogen-rich parts of the plants i.e. the foliage. Since the extra leaf leads to only trivial increases in dry matter production (*Figure 15.8*), yields of roots and sugar can decrease as a consequence of the shift in dry matter distribution (Draycott and Webb, 1971).

In summary, the beet crop must contain about 120 kgN ha^{-1} by the beginning of July. Subsequent growth can rely on continued uptake of soil N, uptake of freshly mineralized N or on N remobilized from within the plant. However, if N is too freely available at this stage, the crop will apportion a large amount of

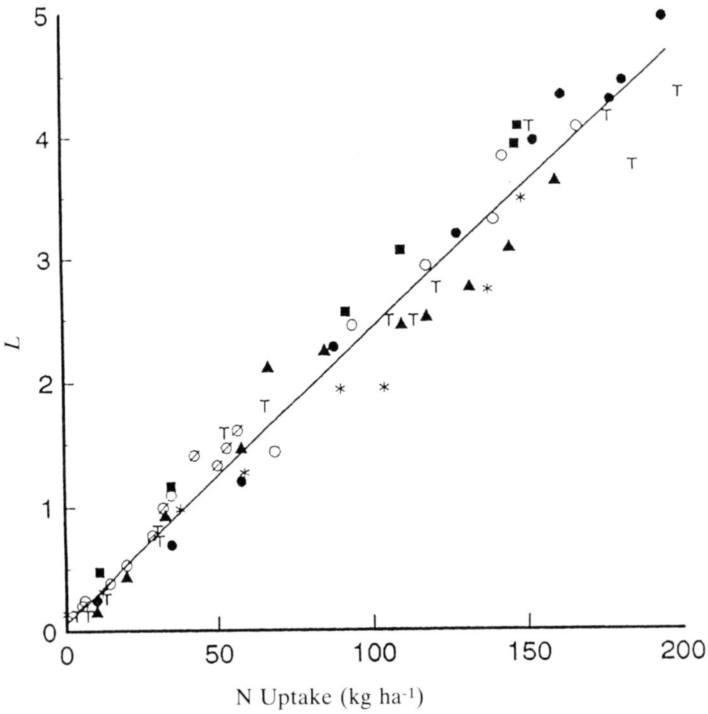

Figure 15.10 Relationship between leaf area index (L) of expanding canopies and the amounts of nitrogen which sugar beet crops contained each year from 1978 to 1982 (★, ○, ▲, ●, ■, respectively). One of the crops in 1979 (∅) was not given nitrogen fertilizer, and additional crops were grown in nitrogen-rich sites in 1979 and 1980 (T)

dry matter to production of surplus, N-rich foliage; sugar yield will be reduced as a consequence. So the essential features are: (1) to supply sufficient nitrogen early so as not to restrict the rate of canopy closure; but (2) to avoid, so far as possible, adding to the N available within the system from August onwards because it is usually adequate and frequently in excess of that required to maintain productivity in terms of sugar. Current fertilizer practice on mineral soils is normally to apply 100–120 kgN ha^{-1} by the four-leaf stage (mid to late May). However, precise fertilizer practice depends upon quantifying the extent to which the soil will meet the crop's requirement for N and the extent to which fertilizer N can supplement any shortfall.

Armstrong *et al.* (1986) reported experiments over a range of sites where standard practices were adopted. It was possible to calculate that the proportion of soil plus fertilizer nitrogen taken up by the crop ranged from 63 to 87% and averaged 70% by mid July, when the crops contained 110–130 kgN ha^{-1}. Thus the total supply of N necessary to provide for development of a canopy of optimal size appears to be about 170 kg ha^{-1}. Some of this is available in the soil before the crop's main uptake phase in May and June. Over winter and especially in the moist, warming and freshly tilled soils of sugar beet fields in spring, nitrogen is released by mineralization of organic residues from the last crop and from older organic matter in the soil. In NW Europe, cereals most commonly precede beet and measurements of available nitrate and ammonium in such conditions show 50–70 kgN ha^{-1}, amounts tending to be more on heavier and deeper soils (Vaidyanathan *et al.*, 1987).

We conclude that, through an analysis of the way interception of solar radiation determines crop growth, the way nitrogen uptake governs expansion of the intercepting surface, and how soil N supply relates to N uptake, the common practice of providing sugar beet with 100–120 kg ha^{-1} of nitrogen from fertilizer (ADAS, 1988) appears to be justified: it provides for a canopy of optimum size for the interception of solar radiation.

LIGHT CAPTURE AND NITROGEN FOR WHEAT

Is it possible to define the nitrogen requirement of wheat in a similar way? Such a definition would have considerable potential because wheat receives large dressings of nitrogen fertilizer and occupies a large proportion of the arable land; better understanding should lead to better means of accommodating the variable conditions in which the crop is grown, and possibly to new and unconventional practices with economic and environmental advantages over existing techniques of fertilizer application.

Wheat's growth pattern is very different from that of sugar beet. Wheat is sown in autumn and can generate a full canopy before the spring, but it dies before the summer's major input of solar radiation has passed. Beet's storage root grows throughout its life and constitutes a fairly stable portion at harvest (*Figure 15.6*), but in wheat the grain is formed in a short final phase, and harvest index varies more. Also with beet, the root may contain 40% of the crop's nitrogen at harvest, whereas the N in grain of wheat commonly constitutes 80% of total uptake by the shoot.

Despite these differences, the growth of wheat can be analysed using principles that apply for sugar beet. The growth of wheat relates closely to the amounts of solar radiation it intercepts (Gallagher and Biscoe, 1978; Monteith and Elston, 1983; Sylvester-Bradley *et al.*, 1990). The challenge is thus to trace the mechanisms whereby (1) energy conversion relates to canopy size, (2) canopy expansion and senescence are influenced by N, and (3) nitrogen uptake is influenced by fertilizer, in a way that quantitatively accounts for the 'linear plus exponential' pattern of yield response which has become so familiar to soil scientists through the multitude of experiments to determine empirically the N fertilizer requirements (e.g. George, 1984).

Two examples of such experiments are taken from Broom's Barn within the same rotation as the studies on sugar beet. From *Figure 15.11* it is clear that wheat requires more added nitrogen than sugar beet, that yield is more than doubled by fertilizer, but that production of the last tenth of the yield requires more than half of the N applied. The implication is of inefficiency; our analysis should indicate its source.

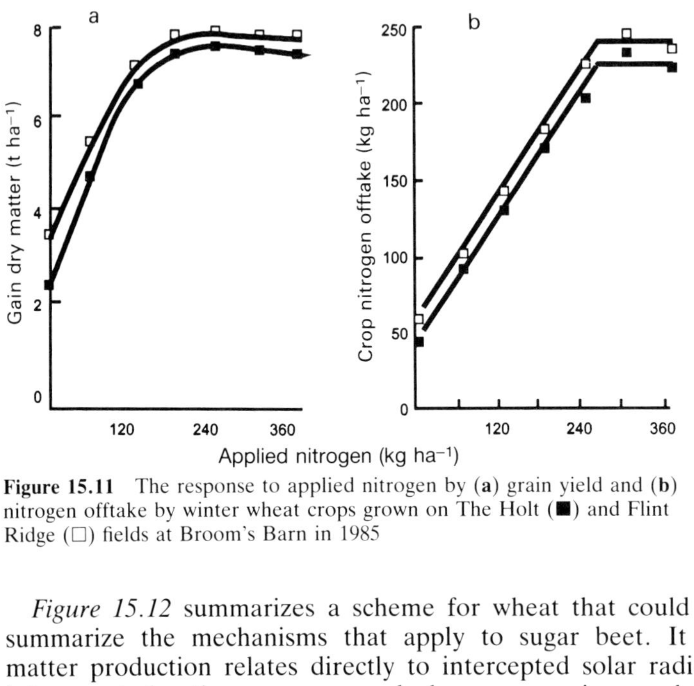

Figure 15.11 The response to applied nitrogen by (**a**) grain yield and (**b**) nitrogen offtake by winter wheat crops grown on The Holt (■) and Flint Ridge (□) fields at Broom's Barn in 1985

Figure 15.12 summarizes a scheme for wheat that could equally be taken to summarize the mechanisms that apply to sugar beet. It traces the way dry matter production relates directly to intercepted solar radiation, how radiation is intercepted as the canopy expands, how canopy size may be governed by uptake of N and how crop recovery of fertilizer N remains stable over a wide range of N amounts.

The relationships between crop growth, incident energy and the intercepting surface are commonly summarized in terms of a conversion coefficient and an extinction coefficient, even though there is good evidence that the 'constants' show some variation. For sugar beet, the conversion of intercepted radiation to biomass does not seem to change over the range of foliar N concentration likely to be experienced in arable fields. On the limited evidence available (Sylvester-Bradley *et al.*, 1990), wheat may convert intercepted radiation to biomass at a lower rate where nitrogen is deficient. However, this complicating effect needs further

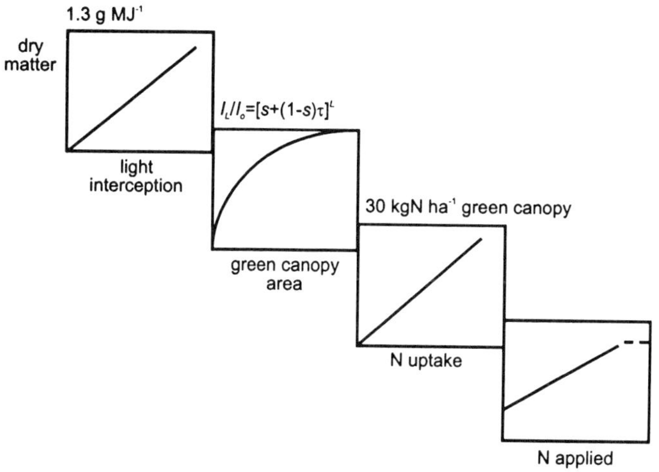

Figure 15.12 Four relationships accounting for the diminishing response in wheat production (as dry matter) to fertilizer nitrogen. Taking applied N as the initial independent variable, each dependent variable becomes the independent determinant of the adjacent relationship. Relationships are discussed in the text

examination to determine whether it will occur only at nutrient concentrations below those of interest to any farmer. Experiments at Broom's Barn show that, for wheat, a green area index of 5 must be achieved to intercept 85% of light, whereas sugar beet needs only an index of 3.

The novel step in *Figure 15.12* is the relationship between nitrogen uptake and green area index. For sugar beet there was a striking constancy, with 40 kgN uptake required per hectare of green area as shown in *Figure 15.10*. *Figure 15.13* shows that for wheat sown at Broom's Barn either in September or November, and with any of three contrasting nitrogen regimes, a similarly close proportionality was maintained throughout the phase of canopy expansion, but in this case with 30 kgN per hectare of green area (Sylvester-Bradley *et al.*, 1990). Surprisingly, the relationship was maintained as the wheat canopy senesced but the line was displaced towards the *x* axis, indicating retention of nitrogen in the dead foliage. Thus it would appear that the wheat canopy should contain 150 kgN ha^{-1} N if a green area index of 5 is to be maintained, rather more than for sugar beet.

The last step in the scheme of *Figure 15.12* is to relate a shoot content of 150 kgN ha^{-1} to amounts of N needed as fertilizer. Wheat appears able to recover aquantity of nitrogen from the soil equivalent to that measured as available in autumn (Vaidyanathan *et al.*, 1987). Thus, in circumstances where 70 kgN ha^{-1} is available, 80 kg ha^{-1} must be derived from fertilizer to balance the shortfall. Prescription of a quantity of fertilizer then depends upon the extent to which wheat will recover the N applied. In the two fertilizer response experiments (*Figure 15.11b*) the apparent recovery was 50 and 55%. More generally, Bloom *et al.*, (1988) found there was an unfortunately wide range of apparent recovery, from 43 to 88%, and were unable to account for the variability. If we adopt the rather uncertain average of 60%, 133 kg ha^{-1} of fertilizer nitrogen would be required to provide an uptake of 80 kg ha^{-1}. This is more than for sugar beet in similar circumstances but less than is now advised.

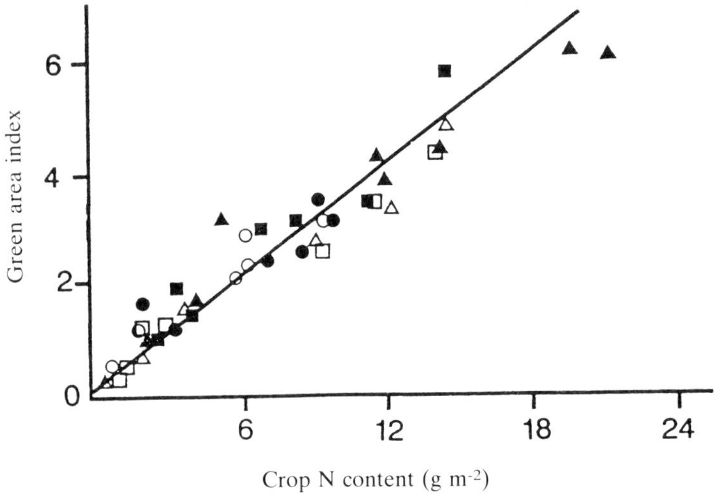

Figure 15.13 Relationship between the nitrogen content of wheat shoots (excluding ears) and the green area index (excluding ears) from crops grown at Broom's Barn with 90 (●,○,), 200 (■,□), or 350 (▲,△) kgN ha^{-1}, sown either in September (closed symbols) or November (open symbols). (After SylvesterBradley *et al.*, 1990)

Clearly, it is essential to say with more certainty whether a particular crop will show an apparent recovery of 43 or 88%. The proportion of fertilizer nitrogen recovered by a crop may sometimes be influenced by loss processes such as leaching and denitrification but primarily it appears to be governed by competition for nitrogen between the crop and the soil biomass (Recous *et al.*, 1988). Current investigations on soils of markedly different history and with a range of organic matter content aim to improve predictability. In the meantime, it may prove advantageous to apply fertilizer as a sequence of small doses so that the efficiency of recovery can be determined before later doses are applied.

However, there is a further factor that may explain why the scheme for sugar beet does not appear to form a realistic basis for judging fertilizer requirements of wheat: whereas the storage root of beet makes only a modest demand for nitrogen, 75–85% of the nitrogen taken up by wheat is in the grain. Major relocation of N from the foliage of wheat exerts a pronounced negative influence on the capacity of the crop to assimilate carbon (Sinclair and deWit, 1975), mainly by diminishing its green surface area. Accumulation of nitrogen by the ears occurred at an approximately constant rate and for the most part appeared to be balanced by decreases in N content of the leaves and sheaths. If energy conversion is to be maintained during the crucial phase when grains must fill with starch, either the canopy must contain some surplus nitrogen, or some must be supplied from elsewhere than the foliage. At present wheat in the UK is commonly fertilized to exceed a green area index of 5 by late May, despite the disadvantage that such large canopies are particularly prone to lodging and foliar disease.

Is it possible that fertilizer applications could be delayed? Although nitrogen uptake in most treatments of the experiments ceased around the time of anthesis and continued slowly only where large and late applications of N had been given, this may well be because it was too dry for efficient uptake from the surface stratum of soil where fertilizer N resides. In these circumstances foliar feeding may be appropriate, to provide the N for grain growth directly and prolong the photosynthetic activity of the flag and second leaf. Sprays of aqueous urea are used quite widely during grain filling to boost grain protein concentrations of breadmaking varieties, and crop recovery of this nitrogen can be as good as from fertilizing the soil with solid forms (Powlson *et al.*, 1987; Cross, 1992). So there is justification for exploring this as a new technique for fertilizing cereals.

To summarize, a scheme can be traced for wheat (*Figure 15.12*) and sugar beet that shows the steps from applying nitrogen to generating biomass. This appears to account for the familiar way that crop responses diminish with increasing nitrogen (George, 1984) in terms of diminishing increases in intercepted light and a limitation of N uptake with large N supplies. For the adviser and grower, this scheme offers a way to reconcile applications of nitrogen with effects on growth.

LIGHT CAPTURE, NITROGEN SUPPLY AND WATER USE

Recently there has been increasing awareness that water is a scarce resource over the important arable areas of England. Demand for public abstraction has increased by 2% per year since 1960, and this trend is projected to continue over the next 30 years (Anon., 1992). In addition, the use of water for irrigation has doubled since 1976 and in the dry, arable areas this demand is very significant; in the May to August period of a dry year, abstraction for irrigation is almost half the normal public demand (Grange, 1992). The National Rivers Authority's Anglian, Thames and Southern regions (broadly, land to the east of a line drawn between the Humber and the Solent) represent most of the arable land in England. Here, current abstraction for public water supply is approximately equal to the Hydrologically Effective Rain fall available during a 'drought', a

Table 15.1 THE EFFECT OF APPLIED NITROGEN ON GRAIN YIELD, WATER USAGE AND RESIDUAL NITROGEN IN THE SOIL AT HARVEST

				At harvest	
N rate ($kg\ ha^{-1}$)	*Grain yield* ($t\ ha^{-1}$)	*Biomass* ($t\ ha^{-1}$)	*Water used* (mm)	*Mineral N* ($kg\ ha^{-1}$)	*Soil water* ($mm\ m^{-1}$)
Sutton Bonington, Nottinghamshire					
0	5.1	10.8	189	–	–
100	8.3	15.6	229	–	–
200	8.2	18.0	242	–	–
S.E.	±0.16	±0.72			
Beeston, Norfolk					
0	3.2	–	–	53	152
40	4.0	–	–	45	122
80	5.4	–	–	43	113
120	6.6	–	–	33	121
180	7.4	–	–	44	106
240	7.5	–	–	53	89
300	7.4	–	–	76	89
S.E.	±0.20	–	–	±8.5	±10.1

condition that occurred in all the years 1988–1992. These values take no account of the demands for irrigation nor the environmental demand for water resources to keep streams and rivers flowing and to preserve wetland habitats. Clearly, the margin between demand and supply is on a knife-edge in drier than average years. As a consequence, a time will come when arable farmers will be required to justify their production practices in terms of the efficient use of water, just as they are having to do now for fertilizers and pesticides.

In Chapter 8 attention is drawn to the tight link between CO_2 assimilation, dry

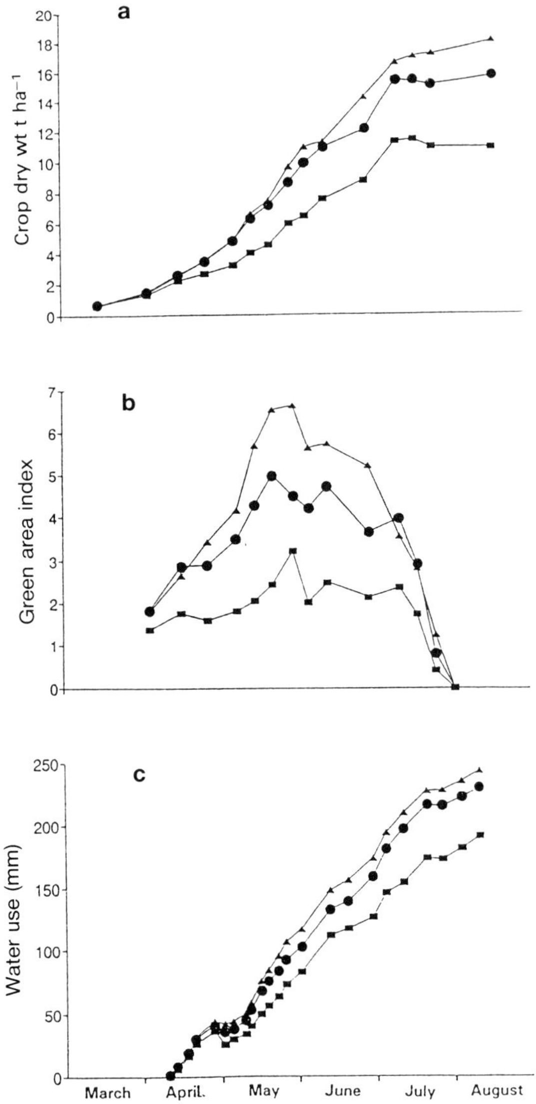

Figure 15.14 Changes during growth of crop dry weight, green area index and water use by winter wheat crops grown at Sutton Bonington and given nil (■), 100 (●) or 200 (▲) kgN ha^{-1}. Data kindly supplied by D.T. Stokes

matter production and transpiration. Reference was made earlier to the adverse effects of a plentiful supply of nitrogen on dry matter distribution in sugar beet; the shift from storage root to foliage actually decreased yield. Where such effects occur, water use efficiency (expressed in terms of sugar rather than biomass) is depressed. *Table 15.1* shows that this can also occur with winter wheat. At Sutton Bonington, maximum grain yields were achieved by applying 100 kgN ha^{-1}; supplying 200 kgN ha^{-1} increased green area during May (*Figure 15.14*), stimulated the production of extra biomass but not grain (harvest indices were depressed from 0.55 with the optimum fertilizer to 0.49 at 200 kgN ha^{-1}) and an extra 15 mm of water was used. In a similar experiment on poorer soil at Beeston in Norfolk, grain yield did not respond to the application of more than 180 kgN ha^{-1}, and measurements of the water present in the soil profile immediately after harvest showed that there also the crop used extra water where nitrogen was applied in excess of the optimum.

While the sugar-beet canopy is incomplete, water use per unit land area is a larger fraction of its potential value than is the interception of light (*Figure 15.15*). By the time the crop has 25% leaf cover, it is already transpiring at more than 50% of the potential rate. Thus crops whose leaf expansion is limited by availability of N when the canopy is incomplete will use more water per unit of energy intercepted than those that quickly form a closed canopy. This pattern of inefficient use of water was confirmed in an experiment where sugar-beet crops receiving either zero or 120 kgN ha^{-1} produced total dry matter at 4.0 and 5.6 kg per tonne of water respectively (unpublished data).

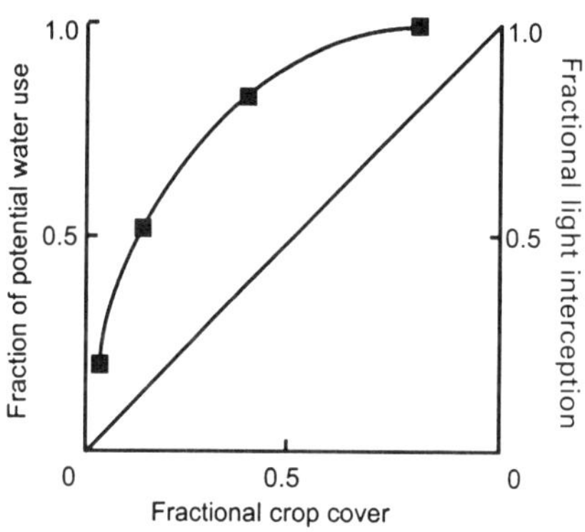

Figure 15.15 Relationships between fractional crop cover and, light interception and fractional cover and potential water use by sugar-beet crops. The relationship with crop cover is from Steven *et al.*, (1986) and the water use data is calculated from Brown *et al.* (1987)

There is evidence of a similar effect in the Sutton Bonington wheat experiment, where the water use efficiency on a dry matter basis increased from 5.7 to 6.8 kg t^{-1} with the first increment of nitrogen, and on a grain basis from 2.7 to 3.6 kg t^{-1}. This is in agreement with the findings of Green *et al.* (1983). On a biomass basis, the application of the next increment of N also increased the water use efficiency but this was not reflected in the grain yield. Day *et al.* (1978) showed, for unstressed barley, that evaporation was about 75% of the potential rate when fractional ground cover was 50%. Thus, for sugar beet and cereals, both shortage and oversupply of nitrogen decrease the efficiency with which water is used in the production of sugar and grain.

Success in the quest for heavier yields since the 1960s has inevitably placed a greater strain on water resources. The production of leafier crops through the use of fertilizers and other agrochemicals has taken its toll of water resources. Moreover, the trend to sow during autumn rather than spring (particularly the shift to winter barley and winter rape, both of which are established early in autumn) has increased the use of water still further. As measures to control production and avoid pollution are introduced to north-west Europe, it will be important to gauge and offset their potential impact on inefficient water use. Perhaps the introduction of set-aside offers a potential counterbalance, but the land needs to be managed to make minimal demand on water resources.

Avoiding pollution

NITROGEN FERTILIZER, WATER USE AND DRAINAGE

Amounts of nitrogen available in the soil profile immediately after harvest of crops that have received a wide range of nitrogen fertilizer are often small and relatively constant up to a fertilizer dose at or near the economic optimum. But soil nitrogen

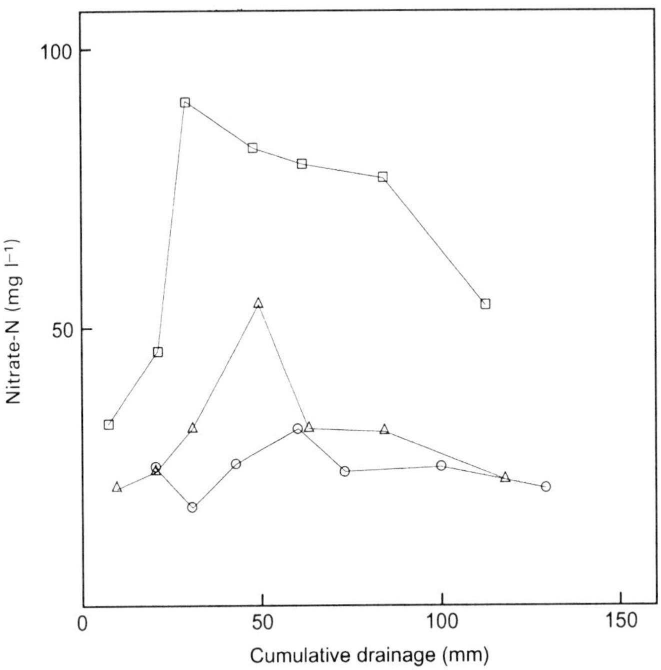

Figure 15.16 Relationships between the quantity of cumulative drainage and the nitrate N concentration of the drainage water of a sandy loam soil following a cereal crop grown with 50 (○), 150 (△) or 250 (□) kgN ha^{-1}. Data kindly supplied by M.A. Shepherd

is increased sharply where too much fertilizer was applied. During the growing season some of the applied nitrogen becomes incorporated into the soil biomass, but this is released as nitrate when the soil is re-wetted in autumn. Increases in fertilizer thus result in a so-called 'surplus nitrate curve' (Addiscott *et al.*, 1991).

Over the range of applied nitrogen in a fertilizer experiment, the increased leafiness of the crop increases transpiration and results in a drier soil at harvest; it appears that dry matter production and water use increase beyond the point where grain yield increases. The net effect of this on the pattern of drainage is shown in *Figure 15.16*; increasing use of nitrogen increased the concentration of N in the leachate and curtailed the amount of water available for recharging the aquifer. Of particular concern is that the curtailed through flow excluded that part of the drainage which contained least N. In other experiments (e.g. at Beeston, Norfolk) the effect of nitrogen application on cumulative drainage was more marked; 63 mm less drainage when 300 kg ha^{-1} was applied.

After harvest the combination of changes in inorganic nitrogen and water remaining in the soil profile produces a marked increase in the concentration of nitrate in the soil solution. For example, at Beeston (*Table 15.1*), the concentration changed from 133 mg l^{-1} at nil fertilizer to 160 mg l^{-1} at 180 kgN ha^{-1} and to 369 mg l^{-1} at 300 kgN ha^{-1}. The precise detail of how these concentrations translate to affect the drainage concentrations shown in *Figure 15.16* is not known. However, those patterns must result from the combined effects on the nitrogen residue at harvest, differences in N release from the soil organic matter as the soil re-wetted, and a concentrating of the soil solution as a result of increased water use.

If, on a national scale, the amounts of nitrogen applied to cereals during the 1970s and 1980s are compared with that removed by the crops, then the difference has been increasing and, by implication, the residue has been getting progressively larger (Sylvester-Bradley *et al.*, 1987). It would be instructive to examine the impact of changes in crop type and crop management on water usage and the concentration of nitrogen in drainage water, particularly in the last decade, because there has been a succession of dry winters.

CROPS TO CAPTURE NITRATE

As farming throughout the European Community has to coexist with the recently imposed limit of 50 mg l^{-1} of nitrate in potable water (*Drinking Water Directive, 80/778/EC*), there will be increasing requirements for bare land to be occupied by short-season crops grown specifically to capture nitrogen that would otherwise be prone to leaching. The principal niche for such crops is between cereal harvesting (July/August) and the sowing of spring crops, particularly spring barley, sugar beet, potatoes and some vegetables. Much current effort is devoted to finding the best species and the best management for each climate and crop rotation.

The previous section highlighted an issue which should be taken into account in selecting a species and its management: nitrogen must be effectively trapped with minimal use of water, so that aquifers and streams can be recharged as far as possible, and with water from the 'tail' of the cumulative drainage/nitrogen concentration curve that contains little nitrate (*Figure 15.16*). In addition, minimal water use by the catch crop will minimize the risk that the following crop is sown into a soil not restored to field capacity. Taking a typical soil at the harvest of a wheat crop as an example, the mineral nitrogen content is likely to be of the order of 50 kg ha^{-1} (Davies and Rochford, 1992) and the aim is to recover a large proportion of this. For most candidate species an uptake of 50 kg ha^{-1} would be achieved if L reached about 2 (D.J.C. Grindlay, personal communication).

During the growth of an effective catch crop, a stage is reached when most of the nitrogen in the system is already within the plants and further biomass production, that inevitably is at the expense of water, simply dilutes the nitrogen in the plant tissue and raises the C:N ratio, perhaps to 30 (Allison, Brown and Pilbrow, 1993). Since plant dry matter with a ratio of more than about 20 will tend to immobilize rather than release mineral nitrogen if incorporated into the soil (Jenkinson, 1985), there is a danger that too bulky a catch crop, left to grow for too long, would not only use water unnecessarily, but would deprive the following crop of mineral N.

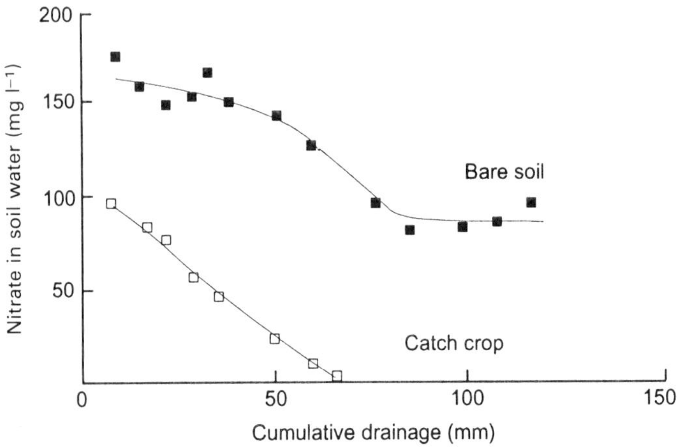

Figure 15.17 Relationships between cumulative drainage and the nitrate concentration of the soil solution at 60 cm depth under a bare fallow or a rye catch crop in Lincolnshire. (After Shepherd, Davies and Johnson (1993)

Figure 15.17 shows the effect of a catch crop of rye grown after winter wheat on the curtailment of drainage. The example in *Figure 15.17* implies that there was some scope to delay sowing or arrest the growth of the catch crop and hence increase drainage and benefit the overall water balance. There may be considerable disadvantages in destroying the crop early, particularly using methods that disturb the soil, because trapped nitrate may re-mineralize in time to leach. Mineralization of residues with a small C:N ratio can be surprisingly rapid, even at winter temperatures pertaining in the UK. The cover crop might be destroyed chemically or mechanically, but ideally its date of sowing would be judged to allow just enough time for N uptake before low temperature curtailed growth. In practice, it appears that unless a large N residue is known to be present, it may be best to delay sowing cover crops until the first half of September (Davies and Rochford, 1992), accepting the risks that leaching will occur before N uptake is complete, or that surface soil dryness will impair establishment.

Arable cropping in prospect

The principles governing resource capture outlined in this paper and shown to relate to conventional commercial production will be even more relevant to meeting the novel challenges that face farming in the future. There is little time to work out the best ways of curtailing over-production, controlling pollution by nutrients and pesticides, conserving water and fossil fuel resources.

With as much land now set aside as devoted to growing all sugar beet, potatoes and horticultural crops, the way this land is managed could have major implications for nitrogen capture and water use. Can a species composition and management regime be defined for set-aside land to maximize nitrogen capture and minimize water use?

If the concern was simply with water economy, the best management would be a bare fallow, and the worst, continuous green cover. But land acquires mineral nitrogen in rain and particulate deposits, amounting to about 45 kg ha^{-1} (Goulding, 1990). Bare fallow has the disadvantage that any mineral nitrogen in the soil, including that which has come from the atmosphere, is vulnerable to leaching. The right compromise may be to allow the establishment of minimal ground cover to absorb and retain the nitrogen deposited. Clearly it would be desirable to have species with large amount of nitrogen per unit leaf area (i.e. species akin to sugar beet rather than wheat), with a root system that ramified the surface layer of the soil to take up nitrogen, and with few deep roots to avoid 'mining' water. Ideally, water lost by transpiration should not greatly exceed the loss from a bare soil surface. To this end it would seem best to avoid having a cover of live vegetation during the period of maximum transpiration rate in midsummer.

Alternative uses for set-aside land are being sought, with the scope for 'energy' crops being of particular interest. Whether the favoured candidate is coppice or a perennial grass (Anon, 1991) it will be important to develop systems in which all the freshly deposited mineral nitrogen is taken up by a crop that retains the nitrogen it traps to minimize the risk of winter leaching. The presence of a canopy when the soil surface is regularly re-wetted in spring and autumn may ensure that water is transpired instead of just evaporating from the soil.

Perhaps the real environmental cost of nitrogen could be controlled if the energy consumed to fix atmospheric nitrogen chemically and to manufacture fertilizers were required to come from current primary production. If all the 500 000 or so hectares of land set aside in the UK produced only 13 t ha^{-1} dry matter per year, it would contain sufficient energy to manufacture all the 1.6 million tonnes of fertilizer nitrogen now used [assuming that 51 GJ are needed per tonne of nitrogen made into ammonium nitrate fertilizer (MacKenzie, personal communication), and that the gross energy of crop biomass at 80% dry matter is 13 GJ t^{-1}]. Because of the effort required to grow energy crops and make fertilizer, legumes and animal manures would acquire a much greater value as fertilizers, and crop requirements for N would be questioned more closely.

Certainly, there is a heightened need to understand farming systems in terms of their capture of energy, carbon, nitrogen and water. Interest in 'sustainability' has accented the requirement not only to maximize efficiencies but to ensure that they balance. In future, it is possible that more holistic approaches will integrate individual disciplines, and that there will be full recognition of the close interdependence of arable agriculture with the livestock industries for which many crops are grown.

Acknowledgements

Much of the sugar beet research described in this paper was financed by the Sugar Beet Research and Education Fund. The research on wheat was funded by the Home-Grown Cereals Authority and MAFF. R.S-B.'s preparation of the paper was funded by MAFF.

References

ADAS (1988) *Fertilizer recommendations for agricultural and horticultural crops.* MAFF Reference Book 209. London: HMSO, 193 pp

Addiscott, T.M., Whitmore, A.P. and Powlson, D.S. (1991) *Farming, Fertilizers and the Nitrate Problem.* Oxford: C.A.B. International

Allison, M., Brown, S. and Pilbrow, J. (1993) Nitrogen catch crops and sugar beet. *British Sugar Beet Review*, **61**(2), 16–19

Anon. (1991) Non-food uses of agricultural products. *House of Lords Paper 26, Select Committee on the European Comunities, 7th Report, Session 1990–91*, 104–108. London: HMSO, 217 pp

Anon. (1992) *Water Resources Development Strategy: a discussion document.* Peterborough: National Rivers Authority

Armstrong, M.J., Milford, G.F.M., Pocock, T.O., Last, P.J. and Day, W. (1986) The dynamics of nitrogen uptake and its remobilization during growth of sugar beet. *Journal of Agricultural Science, Cambridge*, **107**, 145–154

Biscoe, P.V., Gallagher, J.N., Littleton, E.J., Monteith, J.L. and Scott, R.K. (1975) Barley and its environment. IV. Sources of assimilate for the grain. *Journal of Applied Ecology*, **12**, 295–318

Bloom, T.M., Sylvester-Bradley, R., Vaidyanathan, L.V. and Murray, A.W.A. (1988) Apparent recovery of applied N by winter wheat. In *Efficiency of Nitrogen Use*, pp 27–37 Edited by D.S Jenkinson and K.A. Smith. London: Elsevier

Brown, K.F., Messem, A.B., Dunham, R.J. and Biscoe, P.V. (1987) Effect of drought on growth and water use of sugar beet. *Journal of Agricultural Science, Cambridge*, **109**, 421–435

Bürcky, K. and Biscoe, P.V. (1983) Stickstoff im Rubenblatt und N-Translokation aus alternden Blättern. Proceedings of the 46th Winter Congress of the *International Institute for Sugar Beet Research*, Brussels, pp 63–75

Cross, R.B. (1992) Crop recoveries and soil residues from nitrogen applied as prills or sprays to boost grain protein concentrations of winter wheat grown for breadmaking. *Aspects of Applied Biology 30, Nitrate and Farming Systems*, 431–434

Davies, D.B. and Rochford, A.D.H. (1992) Effect of husbandry on nitrate leaching from a chalk soil. *Aspects of Applied Biology 30, Nitrate and Farming Systems*, 271–274

Day, W., Legg, B.J., French, B.K., Johnston, A.E., Lawlor, D.W. and Jeffers, W.De.C. (1978) A drought experiment using mobile shelters: the effect of drought on barley yield, water use and nutrient uptake. *Journal of Agricultural Science, Cambridge*, **91**, 599–623

Draycott, A.P. and Webb, D.J. (1971) Effects of nitrogen fertilizer, plant population and irrigation on sugar beet. I Yields. *Journal of Agricultural Science, Cambridge*, **76**, 261–267

Gallagher, J.N. and Biscoe, P.V. (1978) Radiation absorption, growth and yield of cereals. *Journal of Agricultural Science, Cambridge*, **91**, 47–60

George, B.J. (1984) *Design and interpretation of nitrogen response experiments*. In *The Nitrogen Requirement of Cereals*, MAFF/ADAS Reference Book 385, pp 133–150. London: HMSO

Glauert, W. (1983) *Carbon exchange of a sugar beet crop through a season*. Ph.D. Thesis. University of Nottingham

Goodman, P.J. (1968) Physiological analysis of the effects of different soils on sugar beet crops in different years. *Journal of Applied Ecology*, **5**, 339–357

Goulding, K.W.T. (1990) Nitrogen deposition to land from the atmosphere. *Soil Use and Management*, **6,** 61–63

Grange, P. (1992) The water resources function of the NRA and the present drought situation. *British Sugar Beet Review*, **60**(1), 5–8

Green, C.F., Vaidyanathan, L.V. and Hough, M.N. (1983) An analysis of the relationship between potential evaporation and dry-matter accumulation for winter wheat. *Journal of Agricultural Science, Cambridge*, **100**, 351–358

Jenkinson, D.S. (1985) How straw incorporation affects the nitrogen cycle. In *Straw, Soils and Science*. London: AFRC Publications

Last, P.J., Draycott, A.P., Messem, A.B. and Webb, D.J. (1983) Effects of nitrogen fertilizer and irrigation on sugar beet at Broom's Barn, 1973–8. *Journal of Agricultural Science, Cambridge*, **101**, 185–205

Milford, G.F.M., Pocock, T.O., Jaggard, K.W., Biscoe, P.V., Armstrong, M.J. and Goodman, P.J. (1985) An analysis of leaf growth in sugar beet. IV. The expansion of the leaf canopy in relation to temperature and nitrogen. *Annals of Applied Biology*, **107**, 335–347

Monteith, J.L. (1965) Light distribution and photosynthesis in field crops. *Annals of Botany*, **29**, 17–37

Monteith, J.L. (1977) Climate and the efficiency of crop production in Britain. *Philosophical Transactions of the Royal Society of London*, **B281**, 277–294

Monteith, J.L. and Elston, J. (1983) Performance and productivity of foliage in the field. In *The Growth and Functioning of Leaves*. pp 499–518 Edited by J.E. Dale and F.L. Milthorpe. London: Cambridge University Press

Powlson, D.S., Poulton, P.R., Penny, A. and Hewitt, M.V. (1987) Recovery of ^{15}N-labelled urea applied to the foliage of winter wheat. *Journal of the Science of Food and Agriculture*, **41**, 195–203

Recous, S., Fresneau, C., Faurie, G. and Mary, B. (1988) The fate of labelled ^{15}N urea and ammonium nitrate applied to a winter wheat crop. *Plant and Soil*, **112**, 205–224

Remy, J.C. and Viaux, Ph. (1982) The use of nitrogen fertilisers in intensive wheat growing in France. In *Proceedings of the Fertiliser Society*, London **211**, pp 67–92

Shepherd, M.A., Davies, D.B. and Johnson, P.A. (1993) Minimizing nitrate losses from arable soils. *Soil Use and Management*, **9,** 94–99

Sinclair, T.R. and De Wit, C.T. (1975) Photosynthate and nitrogen requirements for seed production by various crops. *Science*, **189**, 565–567

Steven, M.D., Biscoe, P.V., Jaggard, K.W. and Paruntu, J. (1986) Foliage cover and radiation interception. *Field Crops Research*, **13**, 75–87

Sylvester-Bradley, R., Addiscott, T.M., Vaidyanathan, L.V., Murray, A.W.A., and Whitmore, A.P. (1987) Nitrogen advice for cereals: present realities and future possibilities. *Proceedings of the Fertiliser Society*, London **263**, 36 pp

Sylvester-Bradley, R., Stokes, D.T., Scott, R.K. and Willington, V.B.A. (1990) A physiological analysis of the diminishing responses of winter wheat to applied nitrogen. 2. Evidence. *Aspects of Applied Biology, 25, Cereal Quality*, **II**, 289–300

Vaidyanathan, L.V., Sylvester-Bradley, R., Bloom, T.M. and Murray, A.W.A. (1987) Effects of previous cropping and applied N on grain nitrogen content of winter wheat. *Aspects of Applied Biology 15, Cereal Quality*, 227–237

Warren-Wilson, J. (1967) Ecological data on dry matter production by plants and plant communities. In *The Collection and Processing of Field Data*, pp 77–123 Edited by E.F. Bradley and O.T. Denmead. New York: Interscience

Watson, D.J. (1956) Leaf growth in relation to crop yield. In *Growth of Leaves* pp 178–190 Edited by F.L. Milthorpe. Proceedings of the Third Easter School of Agricultural Science, University of Nottingham. London: Butterworths

Watson, D.J., Motomatsu, T., Loach, K. and Milford, G.F.J. (1972) Effects of shading and of seasonal differences in weather on the growth, sugar content and sugar yield of sugar-beet crops. *Annals of Applied Biology*, **71**, 159–185

Wheeler, T.R., Hadley, P., Ellis, R.H. and Morison, J.I.L. (1993) Changes in growth and radiation use by lettuce crops in relation to temperature and ontogeny. *Agricultural and Forest Meteorology*, **66**, 173–186

16

RESOURCE CAPTURE BY GREENHOUSE CROPS

W. DAY
Silsoe Research Institute, Wrest Park, Silsoe, Bedfordshire, UK

Introduction

Production of any crop is dependent upon effective capture of a range of resources, the most important of which are light, CO_2 and nutrients. Profitable production depends upon this capture being efficient. The major processes that are involved in resource capture are the same for all crops, though the relative importance of the processes will differ between crops. The main differences between greenhouse crops and field crops are associated with the impact of the greenhouse structure, and the strategy for environmental control.

The relationship between efficient capture, profitable production and the decisions about structure and environmental control will be the theme of this chapter. This relationship will be examined in the light of our ability to define quantitative relationships between the various components, and then to optimize the resulting equations in order to maximize profitable production.

This approach requires the development of mathematical models that accurately describe the various processes involved. Considerable progress has been made in such modelling over recent years. However there are many assumptions and shortcomings in the models that are currently available. Physiologists may well criticize the simplicity of many of the relationships, and horticulturalists question their accuracy when dealing with a number of the practical issues that arise in protected crop production. However it is important to recognize the significance of this approach. First it ensures that we are looking for the underlying basis for crop responses to resource availability. This is vital if we are to ensure that the results of scientific research have a broad impact across crops and climates. Second, for greenhouse production in particular, the opportunities for using new engineering technologies in the design or optimization of the crop's environment will specifically require quantitative models to be available – and to be reliable.

This paper will consider three aspects of the problem for greenhouse crops, and illustrate the principles by referring to three examples of resource capture. The aspects involved are the effect of the greenhouse structure on the environment within the house, the form of the crop response to that environment, and the ways in which the grower can seek to optimize resource capture given knowledge about crop and environment responses. These aspects are important in capture of light, CO_2 and nutrients, and ultimately the decisions on the optimal approach to profitable production will need to be a compromise between the different requirements for each resource.

Light capture

Much work has been done on the detailed photosynthetic response of greenhouse plants to irradiance, but this chapter will concentrate on the more general relationship between growth or yield of the crop and solar radiation receipts. As for many other crops, work on tomatoes and on cucumbers has shown that fruit production follows solar radiation receipts very closely (Cockshull, 1988; Cockshull, Graves and Cave, 1992). The difference between receipts and interception for long season greenhouse crops such as tomatoes is likely to be very stable because of the layering technique used which ensures that the light intercepting canopy looks very similar throughout the growing period.

So the first line model of crop response to light availability is simply a straight line, with slope largely independent of time. The grower is first able to make use of this information when deciding about the structure he wishes to use, and its cladding. If we consider the greenhouse as an engineering device to protect the crop from wind and cold, then the progressive reduction that has been seen in recent years in the number and dimensions of the structural elements in the greenhouse roof is entirely logical. For each 1% more radiation that enters the greenhouse, the grower can expect 1% more yield.

This qualitative approach to defining light capture by the greenhouse structure has been made quantitative over recent years, particularly through the work of Critten (1986) on modelling light transmission by structures. The three examples that follow show different aspects to the question of how to optimize light capture.

Critten (1988) has shown that many of the details of light penetration into the greenhouse can be understood by considering the component parts of the transmission problem. Primary reflection at the outside surface of the roof, and secondary reflections from inside the house can each influence the proportion of the incident beam that reaches the crop and can therefore be captured. As the reflection coefficient varies with angle of incidence, the optimum angle of the roof slope, which determines how much radiation is reflected back, away from the house, is dependent on solar elevation – and thus influenced by both latitude and time of year.

Bailey and Richardson (1990) have considered the competing demands that are likely to determine the optimum roof angle for a gothic-arch shaped film-plastics clad house. For UK cropping, the optimal slope for light penetration was shown to be around 30°, but with a very flat optimum. In this study the influence of roof angle on wind loading, and hence on the requirement for structural strength, and on heating cost were also considered. This reinforced the estimate of optimal slope as around 30°, and showed much clearer disadvantages of higher and lower slopes. These considerations are important because the greenhouse is not just a structure to let in maximal radiation. It has to be strong enough to resist wind and other structural loading, and the cost of this strength is a direct constraint on optimization of structural design.

The second example takes further the question of the detailed angular distribution of the light received. At high latitudes, much of the winter and early spring radiation is received from the sun at low angles (*Figure 16.1*). This

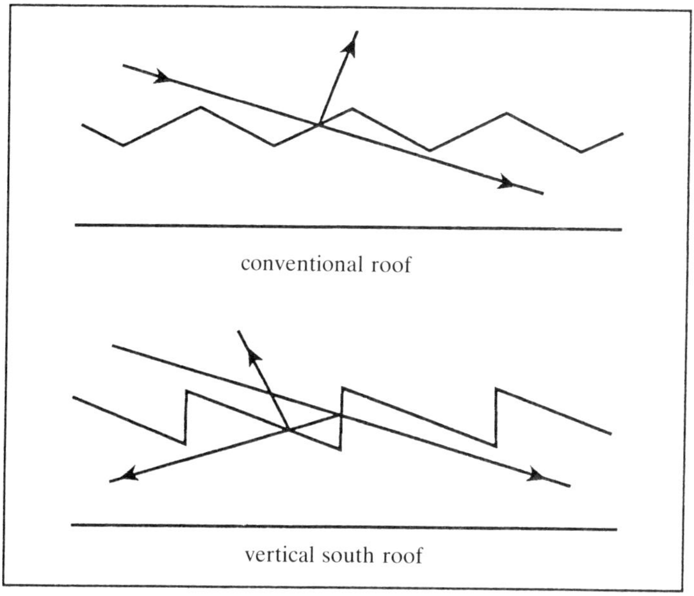

Figure 16.1 How changes in the slope of the south-facing greenhouse roof can influence the fate of reflected solar radiation. (Bailey and Critten, 1985, by kind permission of the publisher)

presents the opportunity for radically different greenhouse structures that attempt to capture a larger proportion of the radiation. The vertical south roof (VSR) greenhouse is an example (Bailey and Critten, 1985). The principle behind this design is that, as the primary reflection is responsible for a potential loss of around 10% of the solar energy, it would be advantageous to ensure that this reflection directs the radiation towards the crop if at all possible. Through much of the early growing season, the low sun is reflected downwards from the vertical face of the south side of the roof – and the low angle sun beam that penetrates this face will then be reflected on to the crop by the inside of the north face of the roof.

At this point it is worth considering whether all light is equal or whether some is more equal than others. The advantage of the VSR is in the early season – could this be outweighed by any (minor) disadvantage in light transmission in high summer (Critten, 1987)? Two points would indicated that this is unlikely. First there is some evidence that the proportionality of yield benefit and radiation received does not hold strictly. Thus in Cockshull's shading experiment (Cockshull, Graves and Cave, 1992), the heavy shading treatment had a higher constant of proportionality than the unshaded, during the summer, indicating that high light intensities were used less efficiently. There may also be some indication from the relationship between fruit yield and the radiation incident in the fruit growing period (the preceding 10 weeks) that yield was less dependent on light availability under high radiation receipts (*Figure 16.2*). But potentially as important as these physical and biological considerations is the economic one – the market price is greater in the spring, so that any increase in light capture in the winter and early spring is likely to have an extra benefit to total profitability.

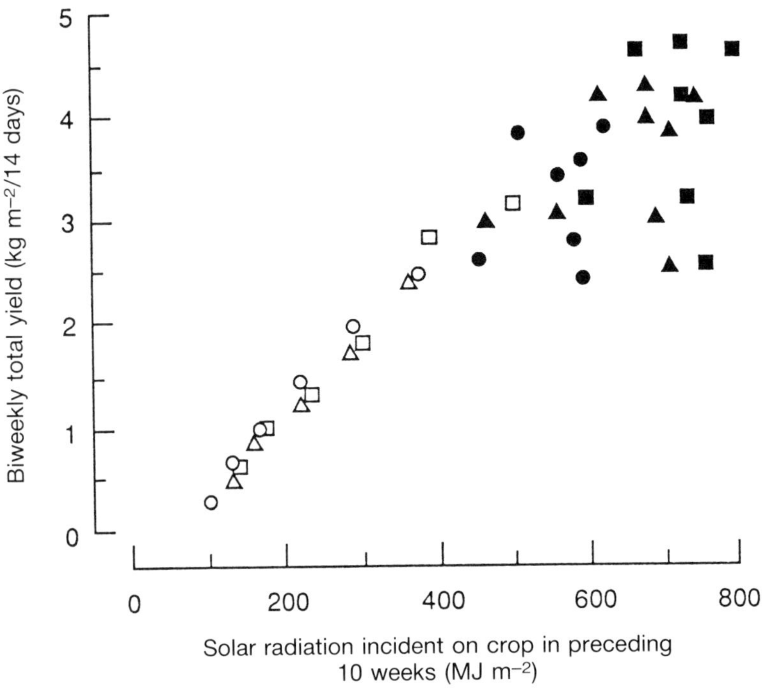

Figure 16.2 The relationship between total fruit yield (in 2–week periods) and solar radiation incident in the 10 weeks preceding picking, for unshaded, lightly shaded and heavily shaded treatment (□, △, ○ respectively) in the first 12 weeks of harvest, or the last 16 weeks (■, ▲, ●) (Cockshull, Graves and Cave, 1992, by kind permission of the publisher)

The third example relates to covering materials, and brings in the important issue of the cost required to maintain optimal growing temperatures in the greenhouse. After the energy crises in the 1970s, when the cost of heating fuel increased dramatically, alternative greenhouse covering materials to glass (or single layer plastic) were tried. However, it soon became apparent that the saving in energy cost from using various double glazing materials was more than offset by the reduction in light transmission of these materials. The reduction is typically between 6 and 11% of the available light. Bailey and Hunter (1988) have used models of greenhouse environment and the '1% for 1%' yield loss model to show that high values for energy costs must exist if these energy saving measures are to be cost effective.

These examples show that key engineering concerns about protected crop structures can be responsive to information from the physiologist on the process of light capture. The relationship used here is a very simple one, and more complex equations for crop response to incident light may be important for future developments.

The light saturation of the photosynthetic rate at high irradiance and its interaction with CO_2 concentration are important when considering CO_2 capture and this will be considered in the next section. At a further level of complexity, there is likely to be an interaction between temperature and light responses. Temperature influences a whole range of developmental processes such as leaf expansion, flower initiation, and flower and fruit development. The integrated effect of these differences has been shown to be that productivity, of cucumber for example, is linearly related to 24–h mean temperature (Cockshull, 1988). As productivity is also linearly related to radiation receipts, this suggests that one can be traded against the other. If this is so, energy-saving measures that reduce light transmission should be assessed against the heating cost for a suitably raised temperature that gives an equal boost to productivity. However other changes, in total season productivity and quality, may follow such a temperature increase. A more detailed understanding would be needed before long-term changes in temperature regime would be considered by the grower to be practically proven.

The practical demonstration in these studies that crop productivity is effectively stable under short-term fluctuations of temperature, provided that the long-term (up to 6 days) mean is constant and the absolute deviations are not more than about 3°C, does have an important message for optimal resource capture, where heat is also treated as a resource in the overall cost equation. For example, the rate of heat loss from the greenhouse is directly dependent on the heat transfer resistance between the house and the outside air. This resistance varies significantly with wind speed, so that for the same internal–external temperature difference, heat input required will be much greater at higher wind speeds (Bailey, 1985). It is therefore more cost effective to adjust the setpoint temperature to match the expected wind conditions, provided the biological capacity for temperature integration is not exceeded.

CO_2 capture

The capture of carbon dioxide by the greenhouse crop has probably been modelled more intensively than any other crop process. A wide range of equations have been developed to describe the photosynthetic response to CO_2 and to light (e.g. Acock *et al.*, 1978; Nederhoff *et al.*, 1989). Particular issues for CO_2 capture in protected crop production depend upon the mass transfer coefficient and additional CO_2 supply.

The major source of CO_2 for photosynthesis is the atmosphere outside the greenhouse. The CO_2 can reach the inside of the house by leakage and by ventilation. Both of these routes will also provide conduits for heat loss. So, as was seen with light capture, concern to minimize heat loss – in this case by sealing the greenhouse more tightly against leakage – will have a potentially deleterious effect on CO_2 supply to the crop. An approximate expression for the concentration difference if CO_2 uptake were independent of CO_2 concentration is:

$$\Delta C = \frac{A.P}{V_g.v} \tag{1}$$

where P is the rate of CO_2 uptake (g m^{-2} h^{-1}) per unit surface area A (m^2) of the greenhouse floor, V_g is the greenhouse volume (m^3) and v the ventilation rate expressed in air changes per hour. Typical values for v range from 1 per hour for leakage from a well-sealed unventilated house to 60 per hour for a fully ventilated house. Therefore for a house with ratio of floor area to volume of 0.4 m^{-1}, a CO_2 uptake rate of 5 g m^{-2} h^{-1} could give a concentration deficit of 2 g m^{-3} (1000 p.p.m.) between outside and the inside of a sealed house and 15 p.p.m. for a well ventilated house. As the external CO_2 concentration is only around 340 p.p.m., this calculation suggests that the well sealed house can maintain less than one-third of the maximum potential crop photosynthetic rate.

Of course the real position is more complicated, as high irradiance will be associated with increased temperature so that ventilation will generally be required in conditions that would permit such rapid CO_2 uptake rates. But there are many situations in which CO_2 depletion will occur to levels at which photosynthetic rates are reduced.

Before coming on to the possibility of additional CO_2 supply, it is also worth noting that conditions inside the greenhouse are also not conducive to maximal CO_2 capture rates by leaves. Air movement rates in greenhouses are low – typically around 0.1–0.2 m s^{-1}, though influenced by external wind speed if the ventilators are open, and by natural convective movement when heat input is high such that the heating pipe temperature is much greater than the air temperature. With wind speeds this low, the boundary layer resistance to mass transfer is significant when compared with the internal resistance to CO_2 assimilation inside the leaf.

Again the grower of protected crops has the option of manipulating crop capture of CO_2 by introducing forced air movements to decrease this boundary layer resistance. The challenge to the engineer is to do this efficiently – remembering that the benefit of improved CO_2 supply to the leaf surface has to be set against the cost of providing and running fans, and that excessive air movement close to the fan can damage the crop (Fernandez and Bailey, 1993).

The other opportunity to improve CO_2 capture by greenhouse crops is by provision of supplementary CO_2, a practice used widely by growers. Pure CO_2 is used extensively for this purpose and this poses a relatively straightforward issue of judging how to maximize the return. This can be considered formally by looking at a cost–benefit equation – for example for a single harvest crop

$$R = S_2W - S_1 \int (v(C_i - C_o) + P_c)\, dt \qquad (2)$$

where R is the net return, S_2 is the unit value of the crop and W its yield, v the ventilation rate, S_1 the unit cost of CO_2, C_i and C_o the CO_2 concentration inside and outside the greenhouse, and P_c the CO_2 uptake rate by the crop.

The control levels that growers use have been established largely on the basis of crop experiments at fixed set-point values. The use of high concentrations (1000 p.p.m.) is generally confined to periods when the ventilators are closed, and the set point is progressively reduced as the ventilators are opened, often with the ultimate objective of maintaining ambient CO_2 levels at full ventilation. But clearly Equation 2 provides the opportunity for a more precise calculation of the best concentration to use under any specific environmental conditions. In principle CO_2 capture could be optimized on a short time scale throughout each day to ensure that the best return to the grower is maintained. The various terms in Equation 2 indicate what is involved in such an optimization.

The second term on the right-hand side is the cost of supplying the supplementary CO_2. It clearly requires both the ventilation rate and the photosynthetic rate to be known – and, more exactingly, to be predicted ahead for the rest of the growing season. This issue of forecasting is an underpinning requirement in all approaches to optimization of resource capture. It is likely to be separable into a long-term forecast, which will generally be based on average conditions experienced in previous seasons, and a short-term forecast based on an extrapolation of current weather conditions, taking on board any trends discernible.

Ventilation rate can be estimated in a number of ways – for example, on the basis of the energy balance of the greenhouse, or from knowledge of the relationship between ventilation rate, vent aperture and external wind (Fernandez and Bailey, 1992). Predictions of ventilation rate can be made on the basis of the same type of energy balance calculation, by estimating the rate of air exchange required to achieve the greenhouse set point temperature, using forecast environmental conditions.

Photosynthetic rate appears in the cost term in Equation 2, but is more prominent in the first term on the right-hand side which represents the benefit to the grower. It can be quite well estimated using available models. Though photosynthetic response to added CO_2 is the primary part, the partitioning of that photosynthate to harvestable yield is also important, as is the expected unit value of that yield when the crop comes to market. The equation as written refers to a single harvest crop such as lettuce, which Critten (1991) has analysed to show the potential benefit from dynamic CO_2 optimization. His analysis showed that around a 10% improvement in grower's net return might be achieved by this method. It also made clear that there is a significant misconception in the view that maintaining the greenhouse CO_2 concentration at the external ambient, even under high ventilation rates, is always justified. The argument is based on the fact that there is no net loss of supplementary CO_2, and so no real cost. But of course the free CO_2 that would have come through the ventilators, is not received. So all growth has to be paid for. In Critten's modelling, the value of this external CO_2 is emphasized by the (unrealistic) suggestion that extraction and sale of CO_2 from the greenhouse air was sometimes the best option!

At present such mathematical optimization studies have largely been desk exercises, based on models alone. Current work for tomatoes, both here and in the Netherlands, is looking to take the question of dynamic optimal control of CO_2 concentration in the greenhouse a step further. Multiple harvest crops such as tomatoes add a further dimension to the problem of predicting the value of the product – yield is spread over a prolonged period within which the market price changes. Thus we need to know not only the increase in photosynthetic uptake of CO_2, and its partitioning to harvestable yield, but also the time when any increase in uptake appears as yield, and the likely market price at that time.

As well as these major issues of the estimation of resource capture and its responsiveness to changing resource concentration, it is also vitally important to bear in mind the real environment in the greenhouse and the degree to which it is controllable. It is entirely feasible to calculate the instantaneous optimum concentration at very short time intervals, say one minute, though errors involved in uncertainty of forecast variables can impose a significant range on this optimum. However it is not possible to change the physical environment on that time scale (Chalabi and Fernandez, 1992). The time constant for decreasing CO_2 concentration is typically half to one hour, and dependent on the rate of ventilation. It is under conditions of low ventilation rate, when the time constant is longest, that the benefits from adjusting the concentration are likely to be most important.

It is also important to recognize that the models of the greenhouse and crop system that are generally being used are lumped models. They assume that the internal environment in the house is effectively uniform so that CO_2 concentration, for example, can be represented by a single value. In most greenhouses, the rates of air movement are very low, and considerable gradients can develop – for example Fernandez and Bailey (1993) quote differences of 155 p.p.m. (about a mean of 630 p.p.m.) between sampling points in a greenhouse without any imposed air recirculation. These differences were dramatically reduced by using fans, to around 20 p.p.m.. But there is a cost penalty, and so the importance of variability in the concentration must be assessed in the development of optimal control concepts.

These questions of how to achieve dynamic optimum concentration are currently being studied from a variety of standpoints. Scientifically, research is still needed to show how, even whether, true optimal values can be estimated effectively and in real time, given the current form of our knowledge and the resulting models of the various processes involved. Practically, growers are already being offered the option of one form of optimal control. This is particularly the case in the Netherlands, where the supplementary CO_2 supply is often from the flue gases of the boiler, rather than pure CO_2 that is often used in the UK, and that is the basis of the research approach at Silsoe. Using flue gases changes the whole basis of the optimization, as the CO_2 is effectively free provided the heat is required in the greenhouse, or can be stored for later use. Optimization therefore involves a range of other issues, such as the sizing and cost of heat storage facilities.

Nutrient capture

The final example of resource capture concerns nutrients, particularly calcium, whose uptake and movement is greatly affected by the rate of transpiration. The physiological basis of any relationship here is much less well defined, and so the concept of optimizing resource capture is based on much weaker models of the underlying process. However the example does indicate how control of transpiration, rather than of the absolute level of greenhouse humidity, is one of the approaches likely to be developed over the coming years (Stanghellini and van Meurs, 1992).

Calcium is an essential element that is needed in comparatively large quantities to maintain the structural integrity of cells and to regulate various metabolic processes. It is generally immobile in the phloem, and thus movement is almost exclusively in the xylem. The amount of calcium entering an organ is therefore dependent on the volume flow of water. In tomatoes, for example, calcium deficiency can lead to leaf scorch, reductions in effective and actual leaf areas, and is also associated with fruit disorders such as blossom end rot. Again the need to look closely at calcium capture by protected crops has been made more urgent by the development in modern greenhouses of energy saving measures, such as thermal screens, fixed screens and double glazing, as well as the general improvement in the leak-tightness of such houses (Cockshull, 1988). The consequence of these changes is an increase in the humidities generated in the houses, and associated decreases in transpiration.

The grower is aware of these problems, and attempts to control humidity in the house. The predominant mechanism for such control is ventilation, which in many situations means that heat is lost from the house in order to ensure humidities are not excessive. In fact it has been suggested that two-thirds of the ventilation for some crops is demanded by the need to control humidity rather than to limit temperature rise (J.C. Bakker, personal communication).

If the process can be defined more precisely, then it may well be possible to achieve the plants requirements for calcium transport more accurately and at less expense. For example, if the requirement can be specified in terms of a transpiration flux, particularly one that need only be achieved over a reasonable length of time (days) rather than instant by instant, then control of transpiration could be part of the environmental control strategy.

Analysis of experiments on tomato crops grown at a range of humidities (specified as vapour pressure deficits) has indicated one possible way in which optimal control may be approached (Jolliet *et al.*, 1993). The experiments involved various combinations of day and night vapour pressure deficits, and the consequent yield differences have been analysed in terms of the effects of the environmental treatments on transpiration. The results suggest that the differences in yield can be explained in terms of reductions in transpiration, and indicate that the largest benefit is achieved by increasing very low rates of transpiration, rather than making the same absolute increase at a higher rate.

If this kind of analysis is correct, it opens up a number of options for optimal control of transpiration. It would be possible to choose the time of day for control measures to increase transpiration so that the net cost was minimized. This will

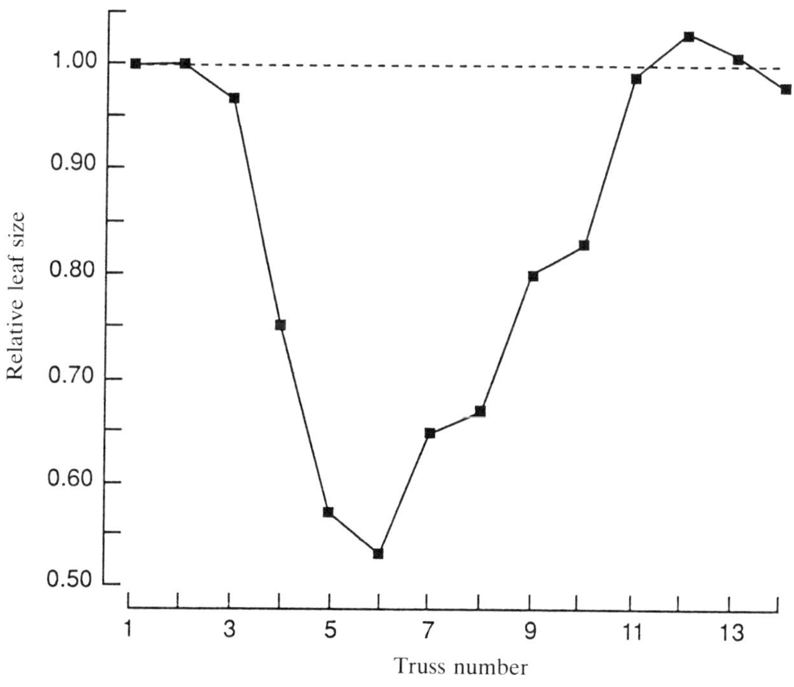

Figure 16.3 The effects of humidity (vapour pressure deficit, 0.1 kPa) on the size (length × breadth) of leaves below successive fruit trusses, relative to that in the low humidity treatment (vpd, 0.8 kPa). (After Holder and Cockshull, 1990, by kind permission of the publisher)

certainly require a clear understanding of the integration time of the crop for such adjustments in transpiration. It may also be possible to introduce new approaches to humidity control that are particularly effective at improving transpiration at the most sensitive times. Thus refrigerative dehumidification, powered by night-time cheap-rate electricity, could be used to enhance transpiration at a time when fluxes would otherwise be low. The heat generated by the process could be used to heat the house.

Again the effects of imposed air movement in decreasing the boundary layer resistance at the leaf surface, and hence enhancing transpiration, may also be a useful measure in optimizing resource capture.

Before leaving this final example, it is important to review the current status of the modelling of this system. The results used by Jolliet *et al.* (1993) are essentially empirical rather than mechanistic. Yield losses are observed, and can be related to the treatment, albeit expressed in terms of effect on the transpiration of the crop. Other studies (Holder and Cockshull, 1990) have indicated some of the mechanisms that may be involved, such as decreases in leaf expansion and hence light interception (*Figure 16.3*). But there is no quantitative model, and certainly not one that links to the real requirements for calcium uptake to maintain leaf growth, the impact of changes in transpiration on uptake, and the influence that other factors, such as changes in the calcium concentration in the nutrient solution, would have on the process. Some suggestions have been made by Aikman and Houter (1990) but further research is needed.

Conclusions

The examples used in this chapter show that there is a range of ways in which the protected crops grower can modify resource capture by his crops. If optimal resource capture is to be achieved, it is important to have quantitative models of the various processes involved.

It has been indicated that optimization is related directly to the return to the grower, and that is as it should be, if we are to expect improvements to be utilized. There may however be constraints on these optimizations that reflect other interests. A general constraint will be that the proposed changes to current practice will need to avoid unacceptable risks. This is not a very quantitative statement, nor even necessarily a quantitative concept. In transpiration control, for example, the grower might accept that the overall strategy could be to achieve a certain transpiration integral, but he would still wish to avoid very high humidity conditions at any time because of the risks of disease that might result.

In due course, there may also be specific constraints imposed from outside too. In the Netherlands, targets for energy use by the horticultural industry are being set. If these are implemented in any way that imposes direct constraints on the energy supply to the individual grower, then these constraints can be transferred directly into the optimization.

Finally, all the considerations here have started from an assessment of how capture of one resource can be optimized independent of the others. This is a necessary starting point, but it has been indicated that, in a number of areas, there is a direct interaction between optimal control for one variable and that for another. This is particularly the case if temperature integration and transpiration integration are acceptable assumptions for crop performance. If ventilation can be minimized during the period when there is the greatest benefit from CO_2 enhancement, for example, and transpiration can be enhanced in other periods or by methods that do not entail heat loss, the overall benefit may be considerable. These concepts of multivariable optimization and control based on models of the whole crop–greenhouse system are topics for the future, imposing significant demands on our scientific understanding of the physiological–engineering system we are trying to control and on control technology itself.

References

Acock, B., Charles-Edwards, D.A., Fitter, D.J., Hand, D.W., Ludwig, L.J., Warren Wilson, J. and Withers, A.C. (1978) The contribution of leaves from different levels within a tomato crop to canopy photosynthesis: An experimental examination of two canopy models. *Journal of Experimental Botany*, **29**, 815–827

Aikman, D.P. and Houter, G. (1990) Influence of radiation and humidity on transpiration: implications for calcium levels in tomato leaves. *Journal of Horticultural Science*, **65**, 245–253

Bailey, B.J. (1985) Wind dependent control of greenhouse temperature. *Acta Horticulturae*, **174**, 381–386

Bailey, B.J. and Critten, D.L. (1985) The vertical south roof multispan greenhouse. *Acta Horticulturae*, **170**, 193–199

Bailey, B.J. and Hunter, A. (1988) Plant response and energy use in five high thermal resistance greenhouses. *Acta Horticulturae*, **229**, 165–171

Bailey, B.J. and Richardson, G.M. (1990). A rational approach to greenhouse design, *Acta Horticulturae*, **281**, 111–117

Chalabi, Z.S. and Fernandez, E. (1992) Spatio-temporal responses of a glasshouse to gaseous enrichment. *Journal of Agricultural Engineering Research*, **51**, 139–151

Cockshull, K.E. (1988) The integration of plant physiology with physical changes in the greenhouse climate. *Acta Horticulturae*, **229**, 113–123

Cockshull, K.E., Graves, C.J. and Cave, C.R.J. (1992) The influence of shading on yield of glasshouse tomatoes. *Journal of Horticultural Science*, **67**, 11–24

Critten, D.L. (1986) General analysis of light transmission in greenhouses. *Journal of Agricultural Engineering Research*, **33**, 289–302

Critten, D.L. (1987) The transmissivity of E-W aligned models of a conventional and clerestory greenhouse under natural winter irradiance. *Journal of Agricultural Engineering Research*, **37**, 129–139

Critten, D.L. (1988) Direct sunlight losses in north-south aligned multispan greenhouses with symmetric roofs at UK latitudes. *Journal of Agricultural Engineering Research*, **40**, 71–79

Critten, D.L. (1991) Optimization of CO_2 concentration in greenhouses: a modelling analysis for the lettuce crop. *Journal of Agricultural Engineering Research*, **48**, 261–271

Fernandez, E. and Bailey, B.J. (1992) Measurement and prediction of greenhouse ventilation rates. *Agricultural and Forest Meteorology*, **58**, 229–245

Fernandez, E. and Bailey B.J. (1993) The use of fans in greenhouses. *Acta Horticulturae* (in press)

Holder, R. and Cockshull, K.E. (1990) Effects of humidity on the growth and yield of glasshouse tomatoes. *Journal of Horticultural Science*, **65**, 31–39

Nederhoff, E.M., Gijzen, H., Vegter, J.G. and Rijsdijk, A.A. (1989) Dynamic model for greenhouse crop photosynthesis: validation by measurements and application for CO_2 optimization. *Acta Horticulturae*, **260**, 137–147

Jolliet, O., Bailey, B.J., Hand, D.J. and Cockshull, K.E. (1993) Tomato yield in greenhouses related to humidity and transpiration. *Acta Horticulturae* **328**, 115–124

Stanghellini, C. and van Meurs, W.Th.M. (1992) Environmental control of greenhouse crop transpiration. *Journal of Agricultural Engineering Research*, **51**, 297–311

17

EXPLOITING RESOURCE CAPTURE – GRASSLAND

A. J. PARSONS
AFRC Institute for Grassland and Environmental Research, North Wyke Research Station, Okehampton, Devon, UK

Introduction – the importance of grasslands

Resource capture by grasslands plays a major role in both agricultural and natural ecosystems. It has been estimated that grasslands occupy some 23% of the land surface of the globe (Cragg, 1979), grasslands often replacing forest as land is cleared for agriculture. The most productive grasslands occur in cool-temperate, maritime zones such as the UK, New Zealand and the Netherlands where they make a substantial contribution to highly developed agricultural systems. In the UK, grasslands occupy some 60% of enclosed agricultural land and their economic output continues to exceed that of all other UK crops added together (see MAFF, 1990).

Over-familiarity with grasslands and their low standing biomass are two factors that might lessen their importance to many minds. But their low standing biomass disguises a considerable capacity for resource capture. Temperate grasslands in particular enjoy a continuity of leaf area in both space and time that can lead to rates of canopy photosynthesis that compare favourably with those of tropical forest (Woledge and Parsons, 1986). The low standing biomass is explained simply in that few grassland species accumulate large quantities of carbon (C) long-term in above ground structures (comparable to the stems or trunks of woody plants). In grasslands, the majority of resources captured by the canopy are soon returned to the soil surface as a litter of high carbon to nitrogen (C:N) ratio. Although carbon does not accumulate greatly in above ground structures, this flow of matter leads to a considerable accumulation of C below ground in soil organic matter (OM). Recent concern over the observed increase in atmospheric CO_2 concentration has highlighted the possible significance of the amount of C sequestered in soil OM beneath grasslands (e.g. 100–200 t ha^{-1}). This quantity is greater per unit land area than that in the leaves, trunks, roots and soil OM, combined, of tropical forest (Goudriaan, 1992). Some attempts to produce a balance of the components of the global C cycle have identified a 'missing sink' for carbon in the mid latitudes of the Northern Hemisphere (Watson *et al.*, 1990; Tans, Fung and Takahashi, 1990). It has been estimated that temperate grasslands may have played a significant role as a terrestrial net C sink, stimulated over the last 150 years by eutrophication, both by the intentional input of N as fertilizer, and by aerial deposition of nitrogenous pollutants (Thornley, Fowler and Cannell, 1991).

Over the past 30 years, a great deal has been learnt about the principles of resource capture by crops, as is apparent from the preceding chapters in this book. Research on the physiological basis of production in the grass crop has contributed substantially to this overall increase in knowledge. But, it is not the aim of this chapter to document the development of an understanding of resource capture in the context of grassland. This has been achieved elsewhere (see reviews by Robson, Ryle and Woledge, 1988; Parsons, 1988). The aim, here, is to describe how this understanding has been, or could be, exploited. At no time has the need for a detailed understanding of how this complex biological system functions been greater than now, when the aim is to develop a *sustainable* environmentally sensitive grassland agriculture.

Emphasis of research in resource capture in the grass crop

There are several features of resource capture 'peculiar' to the grass crop. First, many crops (e.g. cereals) are harvested once, at the end of a single extended period of growth. The act of harvest can have little affect on the amount grown. Some dryland and annual grass species may be allowed to grow in this way. But this is not the case for the many agriculturally significant cool-temperate perennial species. In these, the grass crop is harvested repeatedly, while it is still growing, and harvesting has a profound affect on the amount of grass produced. Second, in grasses the major harvestable component is the leaves, the photosynthetically active tissue itself. This puts the emphasis of attempts to optimize resource capture firmly on to the management of leaf area. Finally, the leaves of our major agricultural grassland species are highly ephemeral (having typical lifespans of some 30–35 days, see later). With such a rapid and continuous turnover of what is both the photosynthetic base for resource capture, and the harvestable component, exploiting the crop means maximizing the *net* capture of resources. That is, the crop must be managed as much to control losses by senescence (and abscission) as to control gross photosynthetic uptake and the gross production of new tissues (net carbon exchange, NCE). This chapter concentrates on the cool temperate perennial grasses, as these characterize the complex interaction between resource capture and harvest.

It is salutary to start by considering how effectively our increased knowledge of resource capture has been applied. Over the last 50 years, grass yields measured under standard management conditions have increased less than 10%. This can be seen, for example, by comparing the yield of the 'old' cultivar of *L. perenne*, 'S23' (introduced in 1936) with that of the variety 'Talbot' in the year when S23 was withdrawn from the recomended breeders lists (see NIAB, 1982/83). In 1991/92, the same small order of difference is apparent between 'Talbot' and the latest varieties (NIAB, 1991/92). Such a small increase in grass yields under a standard management may, arguably, be seen as evidence that little success has been enjoyed in applying our increased knowledge to enhance the photosynthesis and *potential* production of this crop (see also coments by Alberda, 1971). However in practice in the UK, grassland output has increased substantially over a similar period, stocking densities for example having approximately doubled (MAFF, 1926–1979; Lazenby, 1981). A proportion of this can be attributed simply to relieving nutrient limitations by the increased use of chemical fertilizers. But, from the mid 1950s to the present day, an understanding of the physiological principles of resource capture in the grass crop has also made a major contribution to increasing grass production by making it possible to identify the defoliation patterns necessary to control leaf area so as to optimize the balance of photosynthesis, growth, senescence and yield, and ensure that the potential for production is more reliably realized. Analyses of the sensitivity of agricultural grass yields to various management inputs highlight the significance of optimizing the control of sward state (e.g. leaf area) to this end.

Identifying limitations to resource capture in grassland

One of the major ways in which our knowledge of resource capture has been applied is in developing mathematical models of the growth of the grass crop. Over the last 30 years, physiological studies have lead to an extensive series of mechanistic models. These have provided the basis to explore quantitatively the role of a number of physiological features of grass in its response to environment and management (see Thornley and Verberne, 1989 and reviews by Thornley 1976; Thornley and Johnson, 1990). For consistency, this series of models is used to generate examples throughout this chapter.

LEAF AND CANOPY PHOTOSYNTHESIS

Clearly one of the major factors affecting the capture of resources by the grass crop, is the photosynthetic potential of its individual leaves. A detailed and exhaustive series of studies of the factors affecting the light saturated rates of net photosynthesis, A_{sat}, of grass leaves has highlighted a major limitation to photosynthesis in grasses. During *vegetative* growth, grass leaves expand from a meristem near the base of the sward canopy. The net photosynthetic capacity of the leaf area extruded is largely determined by the light environment into which this tissue emerges. Expansion in low irradiance reduces A_{sat} and the dark respiration rate of grass leaves (Treharne and Eagles, 1969; Wilson and Cooper, 1969a, c; Prioul, 1971; Woledge, 1971, 1973), and it decreases both stomatal and residual conductance (Prioul, 1971; Prioul, Reyss and Chartier 1975; Woledge, 1977; Woledge and Dennis, 1982). Leaves expanded in low light are thinner (Woledge, 1971); may have fewer stomata (Wilson and Cooper, 1969c) and fewer mesophyll cells per unit leaf area; fewer and smaller chloroplasts and a reduced activity of ribulose 1,5 diphosphate (RuBP) carboxylase/oxygenase (Prioul, Reyss and Chartier, 1975; Prioul, Brangeon and Reyss, 1980a). In swards regrowing from a low to a high leaf area, successive new leaves expand increasingly in shade and they therefore develop a low capacity for photosynthesis in the high light into which at least a portion of each leaf ultimately emerges (Woledge, 1973; Woledge and Leafe, 1976; and *Figure 17.1a*). It has been proposed that there is some capacity for 're-adaptation' of photosynthesis in portions of the leaf (Prioul, Brangeon and Reyss, 1980b), but studies on whole leaves show limited capacity for re-adaptation (Woledge, 1971).

During *reproductive* growth, the adverse effects of development in shade are avoided. The elongation of stems on reproductive grass tillers elevates growing leaves high into the canopy where they continue to develop a high A_{sat} (Woledge and Leafe, 1976; and *Figure 17.1b*). This has been shown to be due almost

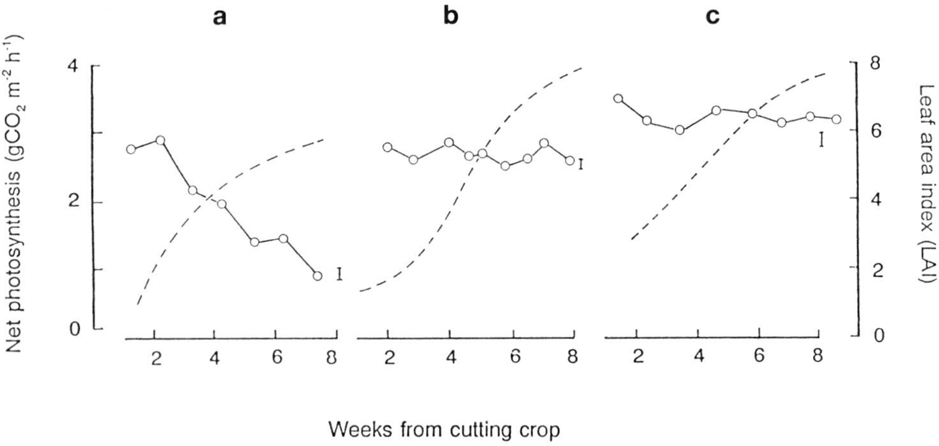

Figure 17.1 Photosynthetic capacity (measured at 250 J m^{-2} s^{-1} PAR) of successive newly expanded leaves (———) in a canopy, and canopy LAI (– – – –) in swards of (**a**) vegetative ryegrass, (**b**) reproductive ryegrass, (**c**) white clover. (After Woledge and Leafe, 1976; Dennis and Woledge, 1982)

entirely to the avoidance of development in shade (Woledge, 1977, 1978, 1979). Artificially shaded leaves on reproductive tillers develop a low A_{sat}: unshaded leaves on vegetative tillers develop a high A_{sat}, as high as that on reproductive tillers. This suggests that the increased 'sinks' for assimilates in reproductive tillers (or the lack of 'sinks' in vegetative tillers) play only a small role in determining the photosynthetic capacity, A_{sat}, of these leaves (Woledge, 1977). In some legumes, for example white clover, leaves also appear close to the base of the canopy, but the development of the photosynthetic capacity of the lamina does not take place until the leaflets unfold in high light near the top of the sward where they develop a high capacity for photosynthesis regardless of leaf area index (LAI) (*Figure 17.1c*). Clover leaves that are artificially shaded as they unfold do suffer a decrease in A_{sat} (Woledge, 1986) so a similar developmental mechanism is inferred, but in practice in the clover sward this phenomenon is avoided.

The decrease in leaf net photosynthesis due to the effects of development in shade may lead to a shortfall of some 30% in C fixation in the vegetative grass canopy (compared with that which would be achieved if A_{sat} remained constant) and this detracts significantly from dry matter production (Robson, 1973b). Offsetting the deleterious effects of shade on leaf photosynthesis has long been identified as a major criterion for breeders to increase canopy photosynthesis in grasses. Studies have shown a wide genetic variation in the response of leaf photosynthetic characteristics to development in shade (Wilson and Cooper; 1969c, Wilson 1981; and *Figure 17.2*). However, a major emphasis in breeding and selection programes has been to select for a high photosynthetic capacity *per se*. The studies by Wilson and Cooper (1969c) suggest that ryegrass genotypes that show less sensitivity to development in shade may have a lower capacity for photosynthesis when developed in high light. It has been argued that this may in part explain the difficulty in increasing grass yields by this route (Robson, 1980, 1981b).

The photosynthetic capacity of grass (and clover) leaves also declines steeply with leaf age (Woledge, 1971; 1972; 1986; Boller and Nosberger, 1989; and

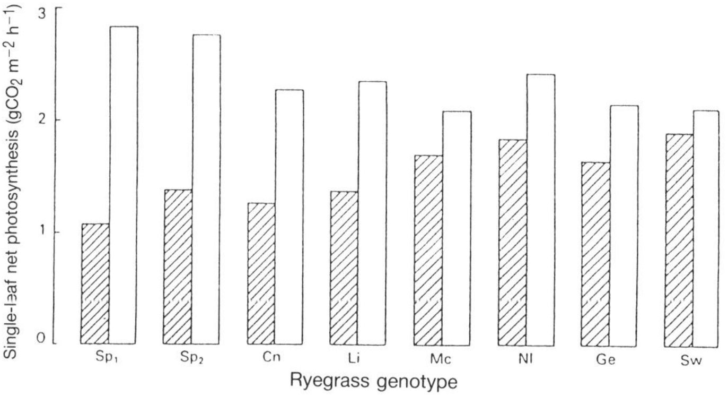

Figure 17.2 Genotypes of *Lolium perenne* differing in the sensitivity of their photosynthetic apparatus to development under shade. Net photosynthesis at 104 J m^{-2} s^{-1} PAR of leaves developed at 22 J m^{-2} s^{-1} (■) or 50 J m^{-2} s^{-1} (□) (Courtesy of Wilson 1981; data of Wilson and Cooper, 1969c)

Figure 17.3), though as is the case with the effects of shade, the decline in capacity to respond to low irradiances is less marked than the decline in A_{sat}. The rate of decline with age is affected by temperature: as temperature is increased, A_{sat} declines faster with age. However, the rate of appearance and longevity of leaves is also reduced (Woledge and Jewiss, 1969) and the number of live leaves per tiller may be affected little. In general, leaves continue to make a positive contribution to net photosynthesis up to the time of their programed senescence (seemingly regardless of their absolute longevity). But, the decline in A_{sat} with age can be accelerated by sustained exposure to very low light intensities e.g. $<$ 20 J m^{-2} s^{-1} PAR (Woledge, 1972; and *Figure 17.3*) and such heavily shaded leaves senesce prematurely.

The effects on canopy photosynthesis of the decrease in the photosynthetic capacity of leaves with age clearly depends on the proportion of the total leaf area in each age class, and where in the canopy (so at what operating irradiance) the age classes are to be found. Using parameters from the literature, the effect of three contrasting age structures, and of the position of the youngest leaves in the canopy, can be considered by comparing lines 1, 2, 3 and 6 in *Figure 17.4*. In *Figure 17.4*, the effects on canopy photosynthesis of the percentage proportion of leaf area in leaves L1 (growing leaf category), L2 (youngest fully expanded leaf category) and L3 (older leaves) are shown. The leaf age categories are assumed in the model to be either homogeneously mixed (H) through the depth of the canopy or layered (L), with the youngest category uppermost. The effects of age structure *per se* are relatively small, largely because only small differences in the proportion of leaves in different age classes are achieved in practice (Parsons, Johnson and Williams, 1988).

Finally, attempts have been made to modify resource capture in the grass canopy by selecting from genotypes that differ in canopy structure and so in the components of the coefficient of light attenuation, '*k*'. Grasses have been selected with different leaf angle, leaf length, rigidity, and tiller angle (see Rhodes, 1969, 1973, 1975; Rhodes and Mee, 1980). As described earlier in this book, the benefits of altering '*k*' depend on the LAI of the canopy. At low LAI, a 'prostrate' canopy (high values, e.g. k = 0.8) are advantageous to canopy

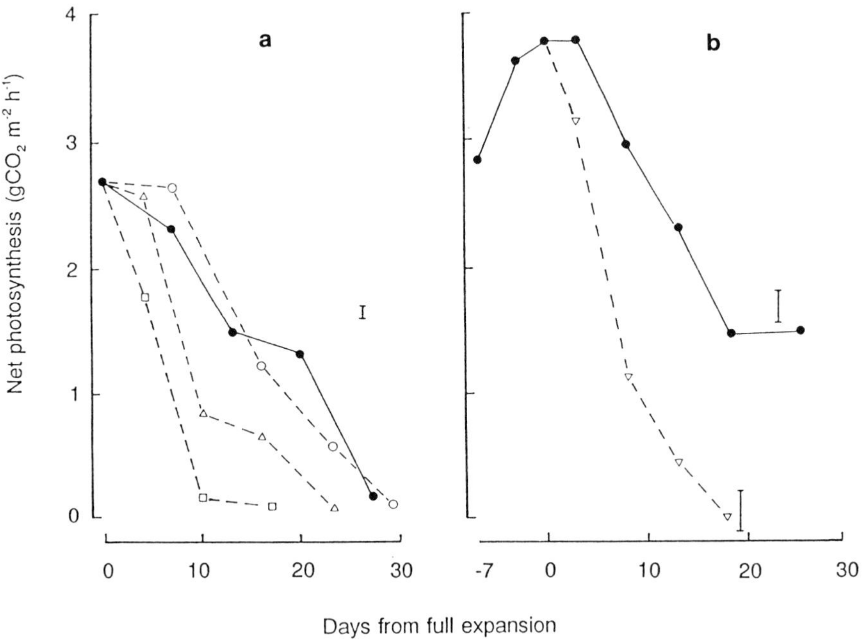

Figure 17.3 The effect of leaf age on the photosynthetic capacity of (**a**) perennial ryegrass and (**b**) white clover, measured at 175 J m^{-2} s^{-1} PAR, 20°C and a water saturation deficit of 3 g m^{-3}. All leaves were grown until fully expanded at 120 J m^{-2} s^{-1} PAR, then either kept at this irradiance (●) or artificially shaded to 57 (○), 20 (△) or 5 (□) J m^{-2} s^{-1}, or the natural shade in the regrowing sward (▽) (After Woledge, 1972; 1986)

photosynthesis as a large proportion of available light is intercepted. At high LAI, a more erect canopy (e.g. k = 0.3, 0.5) is more effective as light is distributed over a greater area of leaf albeit at lower mean intensity. But many species of grass show great natural plasticity and temporal variation in these characteristics. During regrowth in an intermittently defoliated vegetative crop, leaf angles have been observed to decrease from 65° to 70°, to less than 40°, (to the horizontal) as the canopy increases in LAI (de Wit, 1965; Alberda, 1966; Davies, 1977). Breeding for erect leaves would have greatest advantage in swards that experienced long periods at high LAI, particularly should the leaves otherwise 'fall over' and the canopy 'lodge'. Artificially compressing the upper layers of a sward of high LAI, while measuring canopy photosynthesis, shows that canopy photosynthesis can be reduced by as much as 20–30% (Leafe, Stiles and Dickinson, 1974; Sibma and Louwerse, 1977). Mathematical models also show how changes in sward structure can have as marked an effect in depressing canopy photosynthesis at high LAI as the effects of the development of leaves in shade (see line 5, *Figure 17.4*). But, later in this text an explanation is given of how the maintenance of high LAI does not give rise to maximum yield, so *sustained* periods of high LAI should in any case be avoided. Rather, intermittently defoliated swards fluctuate in LAI, and so in the requirements for leaf and canopy characteristics. Swards maintained at a relatively low LAI, for example under continuous grazing, themselves develop a relatively prostrate habit, low leaf angle being accentuated by the development

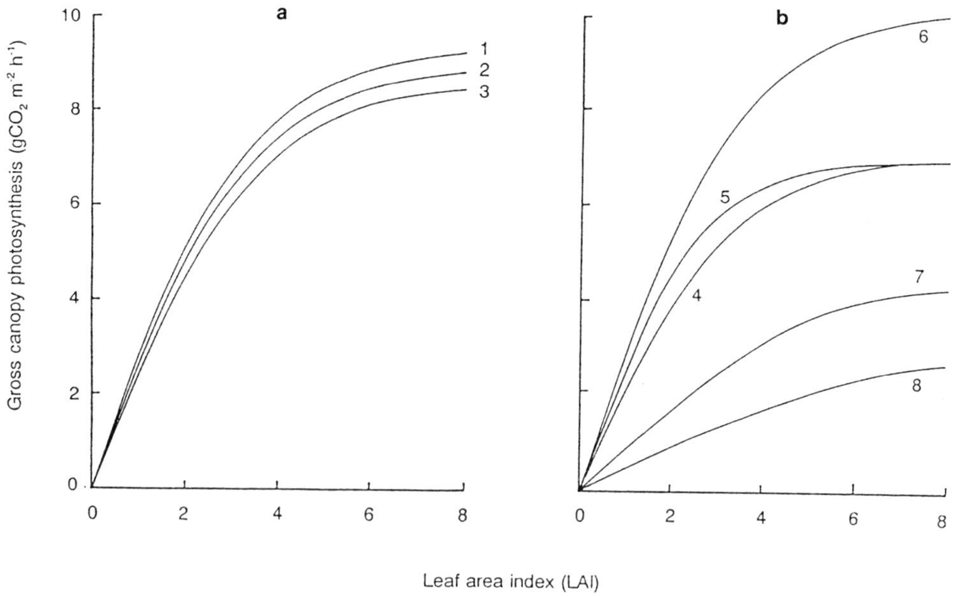

Figure 17.4 The use of a model to consider the role in canopy photosynthesis of: (**a**) leaf age structure (lines 1, 2, 3); (**b**) a reduction in leaf efficiency during development in shade (line 4); an increased prostration of the canopy (line 5); a distribution of young leaves near the top of the canopy (line 6); and a marked reduction in the light saturated gross photosynthetic capacity of leaves at 20°C, Pm (20), (lines 7 and 8) See *Table 17.1* below and text for explanation. (After Parsons, Johnson and Williams, 1988)

Table 17.1

Line in Figure 17.4	*Leaf age structure L1 : L2 : L3*			*Homogeneous/ layered*	*'k'*	*Effects of shade*	
1	40	40	20	H	0.5	–	
2	30	40	30	H	0.5	–	Effect of age structure
3	20	40	40	H	0.5	–	
4	20	40	40	H	0.5	Shade	Effect of shade
5	20	40	40	H	0.8	–	Effect of leaf angle
6	20	40	40	L	0.5	–	Effect of leaf age distribution
7	20	40	40	H	0.5	–	Pm (20) = 1.2 gCO_2 m^{-2} h^{-1}
8	20	40	40	H	0.5	–	Pm (20) = 0.6 gCO_2 m^{-2} h^{-1}

of prostrate tillers. The wide range of circumstances under which the grass crop may be managed greatly confounds attempts to improve growth by altering canopy characteristics. Though selected lines have shown yield differences of 10–20% (Rhodes and Mee, 1980), no new varieties have been released specifically based on these characteristics, though these desirable features have been carried forward into subsequent breeding programmes.

RESPIRATION

Of the 40–60 tonnes OM ha^{-1} per year estimated as fixed during 'gross' canopy photosynthesis (see Parsons and Penning, 1988), some 40–45% has been observed to be released during respiration of which approximately one half is associated with the synthesis of new tissues (*c.* 25% of 'gross' photosynthesis) and the remainder with the 'maintenance' of existing tissues. This is a carbon balance typical of many crop species. The realization that such a prodigious proportion of captured C was released from the crop, and so lost to production, in grasses has prompted attempts to select for low specific rates of respiration. Much of the pioneer work on respiration was carried out in grassland species (McCree and Troughton, 1966a; McCree, 1970; McCree, 1974; Penning de Vries, 1974). The component of respiration associated with the biosynthesis of new tissue is considered to proceed at close to the maximum efficiency that may be calculated given the biochemical pathways that are being followed (Penning de Vries, 1972, 1975). Reducing this component of respiration would mean reducing growth. However, there has been increasing evidence that components of the remaining respiration may be inefficient and therefore amenable to being reduced. Later evidence for the existence of an alternative oxidative pathway, producing only one ATP molecule *versus* three for the 'normal' cytochrome pathway, (Lambers, Noord and Posthumus, 1979; Lambers 1980; Lambers and Posthumus, 1980) recently enhanced optimism for breeding 'low' respiration varieties.

Wilson (1975) selected from within the variation existing in the cultivar *L. perenne* S23, to produce a ryegrass 'line' with a low rate of mature tissue (maintenance) respiration and, for comparison, a 'line' with a high rate of respiration. The low respiration line outyielded the high respiration line and, more importantly, the mean parent material by some 5–12% (Wilson and Robson, 1981; Robson, 1982a; Wilson and Jones, 1982). The benefits to yield were shown to be due *initially* to a greater retention of fixed C and so a greater crop weight and a larger population of tillers. In subsequent regrowths, the yield advantage was increased, as the low respiration line expanded leaf area (increased its light interception) more rapidly, from the increased tiller population (see Robson, 1982a, b; Wilson, 1982). Interestingly, the low and high respiration lines do not appear to differ in the degree of engagement of the alternative oxidative pathway (Pilbeam, Robson and Lambers, 1986). Preliminary field trials, using commercial testing procedures, initially showed yield differences between the respiration lines, but the benefits of low respiration (in one cultivar especially) diminished greatly in subsequent field tests. No new varieties have been released specifically on the basis of having reduced respiration.

Dynamics of the flow of matter in grassland

Little further progress can be made in identifying the limitations to resource capture in the grass crop, and what has or could therefore be done to overcome them, without understanding the *dynamics* of the uptake, flow and losses of matter in grassland species.

CONTINUOUS DEFOLIATION

The need to concentrate attention on the *net* capture of resources, and the principles for optimizing net capture in the grass crop, are most clearly seen in the case of continual defoliation (e.g continuous grazing). This is particularly so in situations where a near-constant LAI can be maintained. *Figure 17.5* shows diagramatically the rapid turnover of leaves (and associated sheaths) on tillers in a perennial ryegrass sward. Each tiller maintains close to three live leaves throughout the year (Davies, 1977). Each time a new leaf appears, which is approximately every 11 days during the main growing season (April to September in UK), the oldest leaf on that tiller dies. Thus the entire standing crop of leaf area effectively dies and is replaced each month. Different grassland species differ in the rate of turnover of leaves and in the number of live leaves maintained. This affects the balance of the mass flow described in *Figure 17.5*, but not the principles of it.

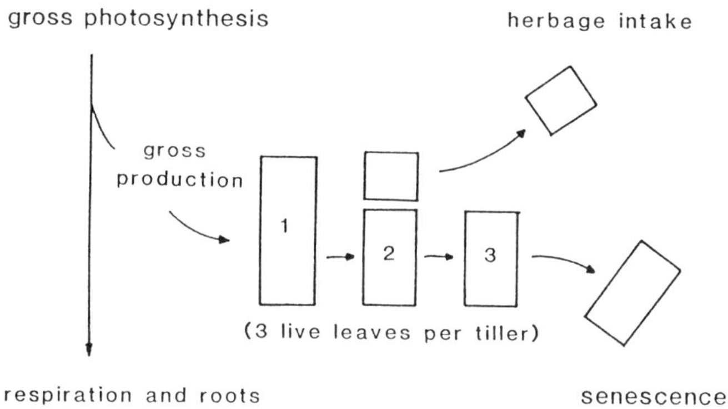

Figure 17.5 The flow of matter in a grazed grass sward. On each tiller the gross production of a new leaf is countered by the senescence (and abscission) of the oldest leaf, and the harvest (as intake) of a proportion of each new leaf that appears. Increasing the proportion harvested reduces the proportion lost to death but also reduces leaf area and the capacity to produce new leaves

Harvesting leaves, for example by continuous grazing, is essential to usefully intercept a proportion of this rapid turnover of mass. Increasing the *proportion* of each packet of tissue that is harvested, decreases the *proportion* of leaf tissue that passes to death and decay. However, as is apparent in *Figure 17.5*, increasing the intensity of utilization of leaves, in this way, inevitably reduces the *amount* of leaf area in the crop, and so may reduce photosynthesis and the rate at which new leaf tissue is produced (gross tissue production). Maximizing yield means seeking the optimum balance between retaining leaves in the crop to photosynthesize, and removing them to obtain a yield. That is, we require the optimal solutions to the mass flux equations represented by the physiological processes in *Figure 17.5*. Mechanistic mathematical models have been used for this purpose (Johnson and Parsons, 1985a). These make use of the fundamental principles of resource capture described throughout this book, and data on the values of parameters relevant to the grass crop, obtained from numerous physiological studies in controlled environments and in the field.

The effects of the intensity of defoliation, described here in terms of the leaf area index that the crop sustains, on the balance of these major physiological processes is described in *Figure 17.6* (Parsons *et al.*, 1983b). In a sward in which only a small proportion of the leaf area that is produced is harvested, the crop approaches a dynamic steady-state equilibrium at a high LAI in which gross photosynthesis is close to its maximum, some 50 tonnes OM ha^{-1} $year^{-1}$. The 'losses' of mass from the crop during the respiration of shoot tissues is also close to a maximum, though the proportion of gross uptake lost in respiration (*c.* 45%) is little altered by the intensity of harvest. The rate of gross production of new tissues (net carbon exchange, NCE) is also maximized, but, the *amount* harvested is necessarily very small: a large proportion of the tissue produced dies unharvested.

Increasing the intensity of defoliation, for example by increasing the rate of removal of leaf tissue per unit ground area by animals and so the proportion of leaf tissue removed, causes the crop to approach an equilibrium at a lower LAI (see *Figure 17.6*). The rate of photosynthesis and net carbon exchange is reduced, and therefore less new tissue is produced per unit ground area per unit

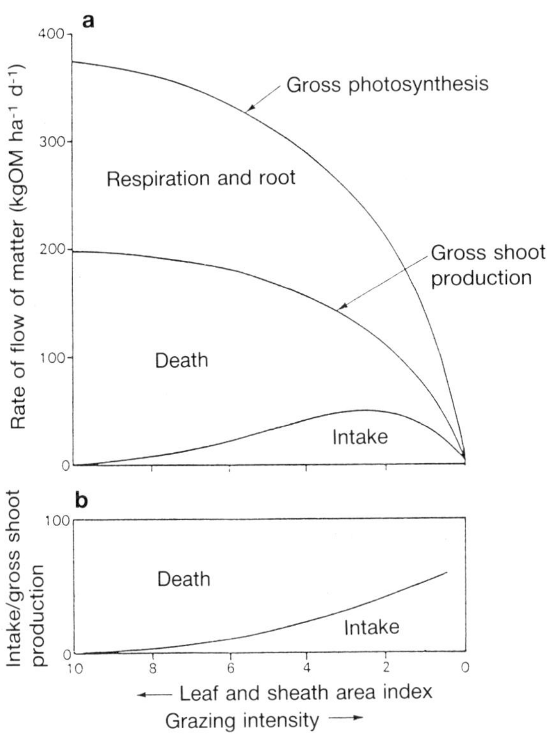

Figure 17.6 The effect of the intensity of grazing (the LAI at which the sward is maintained) on the balance of the physiological components of resource capture described in *Figure 17.5* (From Parsons *et al.*, 1983b)

time. However, the benefits of increasing the proportion of the flow of leaf tissue that is harvested outweigh the effects of reducing the amount of leaf grown, and the amount of leaf that may be harvested per unit ground area per unit time is actually increased. The maximum rate of yield that may be sustained is therefore achieved in a sward at a relatively low LAI, which is below the optimum for light interception and net carbon exchange, but which provides the optimum balance between the rates of photosynthesis, yield and leaf death.

At very high intensities of defoliation, the sward may come to equilibrium at a very low LAI. The efficiency of utilization will have improved in the sense that a greater *proportion* of the mass of leaves grown will have been removed, but, the *amount* removed per ha per unit time will have decreased. The general predictions of dynamic mechanistic models, such as that described by Johnson and Parsons (1985a) confirm the earlier accounts derived from field based observations of the net gas exchange, senescence, and the rate of removal (intake) of material by grazing animals (Parsons *et al.*, 1983a, b; King, Sim and Grant, 1984) and from studies of the appearance and turnover of leaves on individual tillers (Grant *et al.*, 1983).

The management of a relatively low leaf area to maintain maximum yield per hectare is also effective if sustaining a high average photosynthetic capacity per unit leaf area in the canopy. In swards maintained at low LAI, leaves continue to expand in high light and so develop a high capacity for photosynthesis (A_{sat}) (Parsons *et al.*, 1983a). Moreover, under continuous grazing, where each tiller may be defoliated approximately every 7–11 days (Hodgson, 1966; Hodgson and Ollerenshaw, 1969; but see Morris, 1969; Clark *et al.*, 1984) a high proportion of young leaves is observed (Parsons, Johnson and Williams, 1988). Repeated defoliation inevitably decreases the average age of leaves in the canopy, as once older leaves have been removed, leaf area can only be sustained by the production of new tissue. The management of leaf area therefore offers one of the most effective ways of optimizing the parameters of leaf and canopy photosynthesis in the grass crop within the scope of existing genetic material.

It has been argued that the high photosynthetic capacity of leaves, and the high proportion of young leaves, seen under continuous grazing 'compensates' for the lack of leaf area and that these 'compensatory' growth mechanisms increase grass growth under grazing. However, a high photosynthetic capacity and a large proportion of young leaves are not unique to this form of management, but are also seen during intermediate stages of regrowth following intermittent defoliation (see Parsons, Johnson and Williams, 1988). The question of 'compensatory growth' in response to herbivory is debated by Belsky (1986). However, the general relationship between the amount that may be harvested per unit ground area per unit time, and LAI, seen in *Figure 17.6*, arises even when (in a model) leaf photosynthetic capacity and leaf age structure are held constant. That is, in the absence of any directly compensatory mechanism. Certainly, canopy photosynthesis and growth would be reduced at low LAI if leaf photosynthesis and age structure did not respond as observed, but an increase in leaf area would still give rise to greater gross production. In the cool temperate perennial grasses (and in white clover), the explanation for a maximum harvested yield arising at an intermediate LAI remains that described in *Figure 17.6*, and Johnson and Parsons (1985a).

Although the above account is based on a limited range of species, it may be argued that the general principles for exploiting resource capture conveyed in *Figure 17.6* should apply for all crops in which the harvestable component is the major photosynthetic tissue, is continually harvested, and is short-lived. Moreover, because the relationships observed in *Figure 17.6* arise, basically, from the features of a simple dynamic system, they should apply for a wide range of edaphic conditions. The effect of nutrient input, for example, may be considered simply as a re-scaling of both axes.

INTERMITTENT DEFOLIATION

Under continuous grazing, the rates of the major components of resource capture may fluctuate only slowly: under intermittent defoliation these are likely to fluctuate markedly and, at the time of harvest, quite rapidly. The changes in the major components of carbon exchange during regrowth from a low to a high LAI, and the origin of changes in growth rate and the ceiling yield, have been described in detail in classic accounts by McCree and Troughton (1966b) and by Robson (1973a, b). These principles are displayed by a model in *Figure 17.7* for a grass crop growing in a constant daily light environment.

Starting from a low LAI, canopy 'gross' photosynthesis (P_{gross}) increases as carbon fixed on 'day one' is invested into new leaves which in turn intercept an increased proportion of the available light. This cycle of investment and benefit continues until 'gross' photosynthesis reaches a maximum rate. This occurs soon after all the available light is intercepted, or when the benefits of increasing light interception are offset by the decrease in the photosynthetic capacity of individual leaves. Thereafter, gross canopy photosynthesis may even decline, as successive new leaves develop a low A_{sat} in shade.

During regrowth, the rate of canopy respiration also increases as gross photosynthesis and the rate of biosynthesis increases, and there is an increase in the weight of live tissue to be maintained. However, in the cool temperate grassland species typified by perennial ryegrass that show a rapid turnover of leaves, a ceiling yield is *not* achieved because respiration rates come to equal the rates of uptake of carbon in photosynthesis as was once proposed (Donald, 1961). For a fascinating analysis of different previous philosophies of the origins of ceiling yield, see Robson (1973b), Robson, Ryle and Woledge (1988). Certainly, following a defoliation which removes virtually all effective leaf tissue, respiration rates may initially exceed the uptake of carbon in photosynthesis on a 24 hour basis, and the crop may make a net loss in weight (see Leafe, Stiles and Dickinson, 1974). New leaf tissue may nevertheless be produced, from 'reserves'. Initially these may be sugars, stored, in many cases, in the sheath bases (De Cugnac, 1931; Pollock and Jones, 1979; Volenec, 1986). 'Stuctural' material may also be used if the onset of a positive carbon balance is delayed (Davidson and Milthorpe, 1966a, b). The period over which biosynthesis can be supported from 'reserves' is very limited, and in any case the consumption of these resources is inevitably associated with a net loss of weight and so is not 'growth' in the comon sense. These processes lead to a 'lag' phase in the net accumulation of weight. But subsequently, as the crop accumulates leaf area, daily respiration returns to remain some 40–45% of 'gross' canopy photosynthesis (Robson, 1973b).

In these grassland species, a decline in the rate of accumulation of live tissue and a ceiling yield of live tissue are achieved because there is an increase in the rate of senescence of leaves, (per unit ground area) during the regrowth of the

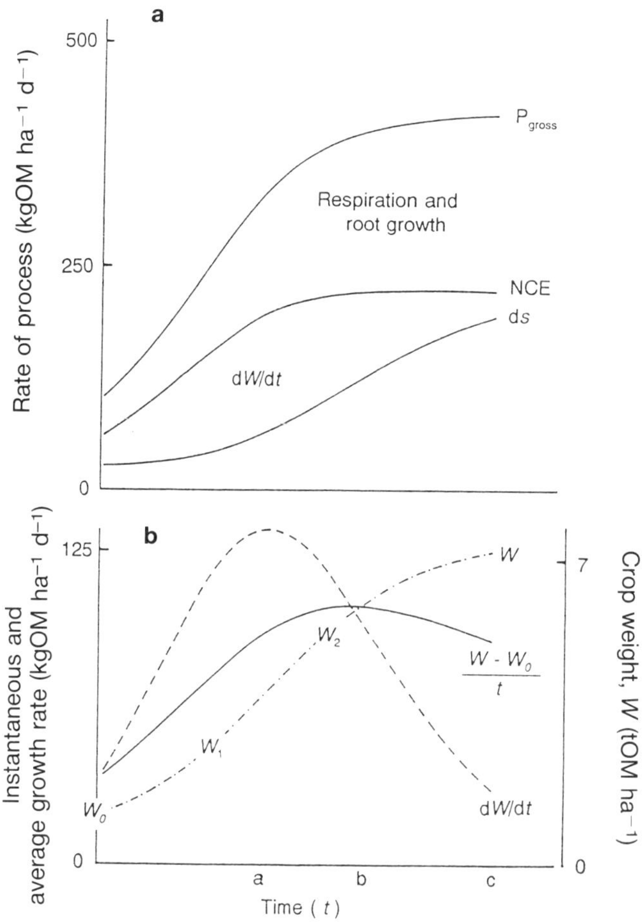

Figure 17.7 The use of a model to illustrate (**a**) the effect of the duration of regrowth on the major processes involved in the net accumulation of herbage: gross photosynthesis P_{gross}, gross (shoot) tissue production (net carbon exchange, NCE) and death (d*s*); and (**b**) the corresponding change in the instantaneous growth rate (d*W*/d*t*), the weight of the crop (*W*) and the average growth rate, $(W - W_0) / t$. See text for explanation of points 'a', 'b', 'c' and W_1, W_2. (After Parsons and Penning, 1988)

crop, that follows the increase in LAI, (d*s* in *Figure 17.7a*). It is tempting, in ecological terms, to regard the increase in senescence as evidence of density dependent mortality. But this is misleading. The number of live leaves, and the rate of loss of *leaves* per tiller is only partly affected by the rapid increase in LAI of the canopy. (A sustained shading of leaves to some < 20 J m^{-2} s^{-1} PAR can accelerate senescence, as mentioned earlier.) Rather, the increase in the rate of loss of matter in senescence largely reflects the increase in the size of leaves involved in the turnover of leaves on each tiller. During regrowth, changes in the rate of loss by senescence inevitably lag behind the changes in the rate of gross photosynthesis and gross tissue production (net carbon exchange, NCE) as, at any point in time, the size of leaves dying reflects the size of leaves produced previously (in this case some three leaf-appearance intervals previously). While current senescence remains smaller than current gross tissue production, the sward makes a net increase in live weight. But, once the rate of gross tissue production has reached a maximum, the rate of net accumulation of live tissue declines as the rate of senescence 'catches up' and comes to equal the rate of gross production of tissues (*Figure 17.7*). The crop therefore approaches a dynamic steady-state equilibrium at which the rates of photosynthesis and gross tissue production remain close to a maximum, but at which the rate of change in live weight is zero. The number and size of leaves dying is now as great as the number and size of leaves being produced. If it were not for the efficacy of soil flora and fauna, the rate of increase in total biomass (live plus dead material) would continue to accumulate at an only slightly abated rate (Robson, 1973a, b). Density dependent losses are seen in this crop, but they are of greater significance during periods of transition, for instance when a sward, which has developed a large population of small tillers under continuous grazing, is allowed to regrow to a high LAI (Parsons, Collett and Lewis, 1984). Under these circumstances, a large proportion of whole tillers (and so the leaves on them) are lost as 'self-thinning' occurs (see Ong, 1978), and the sward redevelops as a small population of much larger tillers.

The pattern of changes in the rate of gross photosynthesis during regrowth may be greatly modified by the pattern of changes in light energy receipt. If daily light energy receipt increases progressively during regrowth, as in spring, gross photosynthesis (per unit ground area) may continue to increase even after all (as a proportion) of the available light is intercepted. This in itself leads to a longer period of regrowth to a greater ceiling yield than in, say autumn, when the daily light energy receipt declines. This difference in growth rates and ceiling yield would be apparent even when the total light energy incident was the same in spring as in autumn (see Parsons, 1988) and can be shown to contribute substantially to the observed seasonal differences in yield of the grass crop under both cutting and continuous grazing (Johnson and Thornley, 1983; Johnson and Parsons, 1985b).

OPTIMUM TIMING OF HARVEST DURING REGROWTH

With a crop that has such a propensity for turning recently acquired resources into botanical litter, harvesting the sward effectively means identifying an optimal basis to interrupt the growth of the sward (and its development of a light intercepting canopy) in order to remove the optimum proportion of the live tissue before it dies. The optimal solutions to this problem under continuous defoliation have been described in *Figure 17.6*. Historically, much of the work in this area was carried out in the more complex context of intermittent defoliation.

From a knowledge of the interrelationship between the major components of the carbon balance of the grass crop during regrowth (*Figure 17.7a*), models may be used to identify a theoretical optimum timing of harvest (viz. duration of regrowth). This must be done for swards commencing regrowth from any state. Starting, first, with a sward regrowing following defoliation to a low LAI, harvesting at time 'c' (see *Figure 17.7b*) would clearly achieve close to the maximum yield from single period of regrowth (W–W_0 is greatest). However, in this case the instantaneous growth rate (dW/dt) would have been decreasing for some time, and would currently be close to zero. Harvesting the sward at time 'a'

(*Figure 17.7b*) would mean interrupting the growth of the sward at the time when the instantaneous growth rate was at a maximum, though the amount removed at harvest would be reduced. (In all cases here, it is assumed for now that the sward is returned at harvest to weight W_0.) But what is required to achieve the maximum rate of yield is to to consider the way the amount accumulated changes with the total time elapsed. That is, it is necessary to plot, and understand, changes in the *average growth rate*, $(W-W_0) / t$, during regrowth, and to interrupt each period of regrowth at the time when the *average growth rate* is at a maximum (Morley, 1968; Maeda and Yonetani, 1978; Parsons and Penning, 1988; Parsons, Johnson and Harvey, 1988). The changes in the average growth rate during regrowth following defoliation to a low LAI are also shown in *Figure 17.7b*. The average growth rate increases initially, but thereafter changes relatively gradually as the period of regrowth is extended. Nevertheless, a maximum average growth rate can be identified. This can be seen to occur (at time 'b', *Figure 17.7b*) after the time of the maximum instantaneous growth rate, but before the time of the ceiling yield.

To some it is counterintuitive that the overall rate of harvest of dry matter continues to increase even though a substantial period of declining instantaneous growth rates (from 'a' to 'b') has been endured. This however can be understood from *Figure 17.7b*. Clearly it is of benefit to allow the crop to continue to grow while the current instantaneous rate of growth (dW/dt) *exceeds* the current average growth rate, $(W-W_0) / t$, (even if dW/dt is declining), but not when the reverse is true. Hence, graphically, the optimum timing of harvest is apparent as the time at which the plots of dW/dt and $(W-W_0) / t$ intersect.

OPTIMIZING THE FREQUENCY AND SEVERITY OF DEFOLIATION

It is interesting to note that the agricultural literature makes much reference to studies of the 'frequency' and 'severity' of defoliation, as if emphasizing the act of harvest. Seeking an optimum basis to exploit the grass crop from a knowledge of the principles of resource capture places more emphasis on the leaf area index at the start of the growth period (rather than an almost indefinable concept of a 'severity' of defoliation), and the necessary duration of each subsequent individual growth period (rather than a set 'frequency' of harvest).

The effect of the LAI at the start of regrowth on the balance of the major processes involved in the net accumulation of dry matter is shown in *Figure 17.8 a–d* for swards regrowing from six contrasting initial LAIs of 0.5, 0.8, 1.1, 3.4, 5.3, and 6.8 (numbered 1 to 6 respectively). In swards regrowing from a greater LAI, the rates of gross photosynthesis and so gross tissue production (net carbon exchange) are initially higher (as more light is intercepted) and maximum rates of gross photosynthesis and gross tissue production (at full light interception) are achieved sooner (*Figure 17.8a*). The relationship between lines 1, 3 and 4 in *Figure 17.8a*, compares closely with the patterns of light interception observed by Brougham (1956, 1957) for swards regrowing after defoliation to '1, 3 and 5 inches' height. However, as was soon realized (Hunt, 1965; Smetham, 1975), rates of senescence are also greater at the start of regrowth from a high LAI, and the time when senescence rates come to equal the rate of gross tissue production is also sooner achieved (*Figure 17.8b*). The effect of the six initial conditions for regrowth on the instantaneous growth rates and the average growth rates are also plotted using the model, in *Figure 17.8 c, d*. When the sward starts regrowth

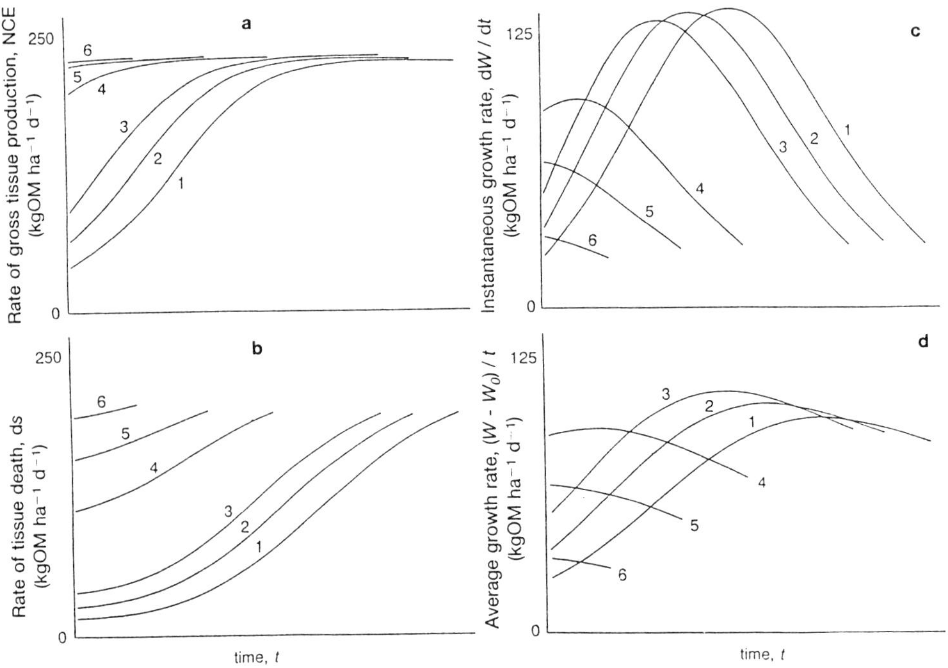

Figure 17.8 The effect of the 'severity' of defoliation (the initial conditions for regrowth) on (**a**) the rate of gross tissue production (net carbon exchange, NCE), (**b**) the rate of tissue death (d*s*), (**c**) the instantaneous growth rate (d*W*/d*t*) and (d) the average growth rate, $(W–W_0) / t$, as the duration of regrowth is extended over time (*t*). The initial leaf area indices for regrowth were 0.5, 0.8, 1.1, 3.4, 5.3 and 6.8 (numbered 1–6, respectively) (After Parsons, Johnson and Harvey, 1988)

from a high LAI, the optimum duration of regrowth, that is the time taken to achieve the maximum average growth rate, is reduced. Although in all cases in *Figure 17.8*, the instantaneous growth rate is positive (and so the swards increase in live weight), the average growth rate may actually decline. In extreme cases as in lines 5 and 6 (*Figure 17.8d*), it declines throughout the period of regrowth. To maximize the average rate of yield, these swards would have to be harvested again, immediately.

In *Figure 17.8* it can be seen that for any starting LAI, a wide range of average growth rates can be achieved, depending on how long the sward is allowed to regrow. In each case a maximum average growth rate can be identified, but a range of different maxima (depending on initial LAI) are evident. What is required is a means to characterize those combinations of initial LAI and duration of regrowth that achieve the overall maximum average rate of yield, and its basis in the principles of resource capture. This can be done, by replotting the 'data' from *Figure 17.8* in relation to the *average* LAI, $(L\text{-}L_0) / t$ sustained over the period of regrowth, as in *Figure 17.9*. The similarity between *Figure 17.9* (for intermittent defoliation) and *Figure 17.6* (for continuous defoliation) is self-evident.

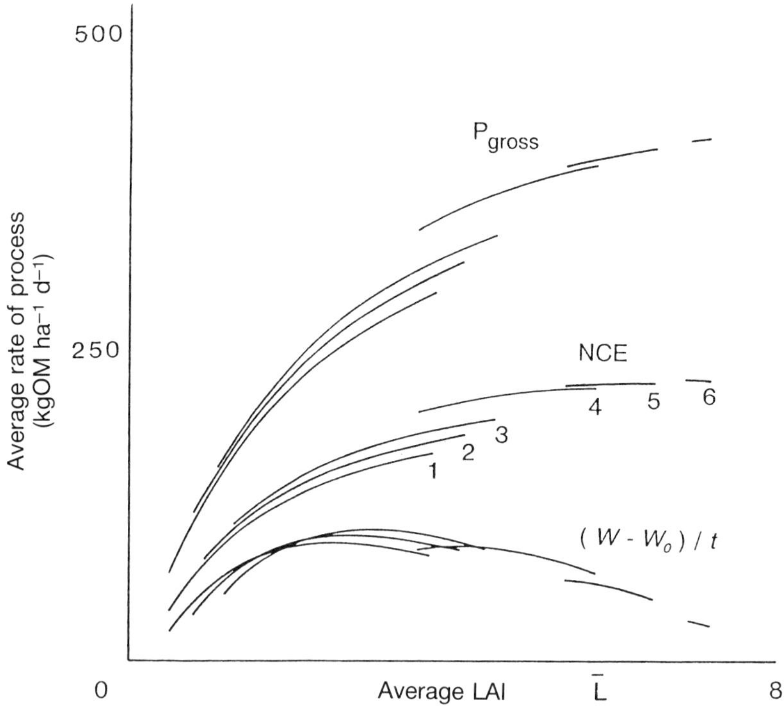

Figure 17.9 The balance between canopy gross photosynthesis (P_{gross}), gross (shoot) tissue production (net carbon exchange, NCE) and the average growth rate, $(W–W_0) / t$ for the six swards described in *Figure 17.8*, replotted in relation to the average LAI, $\bar{L}$, achieved from the start of regrowth to the time of harvest. (After Parsons, Johnson and Harvey, 1988)

PREVIOUS ANALYSES AND CONTROVERSY

The analysis above, based on identifying an optimum *average growth rate* in relation to an *average* LAI, should not be confused with previous analyses based on identifying an optimum *current* LAI for maximum *instantaneous* 'crop growth rate' (CGR) for a single period of regrowth (e.g. Donald and Black, 1958; Donald, 1961). In an early, but classic account, Brown and Blaser (1968) attempted to characterize CGR in terms of current LAI. However, they conveyed some controversy over the shape of the response of crop growth rate to LAI, plotting two alternative relationships. These appear remarkably similar to those plotted in *Figure 17.7b*. (The form of the relationships in *Figure 17.7b* changes very little in this case if growth rates are replotted on the basis of LAI rather than time.) It remains a possibility that one set of data reviewed in their paper represented the instantaneous growth rate, dW/dt: the other set the average growth rate, $(W–W_0) / t$. In either case, their data were not plotted against an *average* LAI.

Some early attempts to identify an optimum management for leaf area to maximize crop growth over a succession of periods of regrowth and defoliation were based on extrapolating from the relationship between instantaneous CGR and LAI observed in a single example: that of a crop regrowing from a low LAI ('severe' defoliation) to a ceiling yield ('infrequent' defoliation) for example as in

Figure 17.7b and line 2 in *Figures 17.8 and 9*. It was argued that, to maximize the average rate of growth, the crop should be harvested soon after the instantaneous growth rate had begun to decline (at W_2 in *Figure 17.7b*), but harvested 'leniently' (to W_1 in *Figure 17.7b*) see Booysen (1966); Harris (1978). This approach has featured widely in advisory literature, and the relationship between instantaneous growth rates and current LAI of an infrequently and severely defoliated sward is used to predict the rate of supply of resources from grasses to animals under a whole range of grazing pressures in accounts of herbivory in ecology (e.g. Noy-Meir, 1975). However, this denies there is a dynamic interaction between harvesting and regrowth, as to be generally valid it assumes that a sward regrowing from a high LAI (e.g. from W_1) would retrace the pattern of net accumulation seen in the more severely defoliated crop. This is unlikely to be the case, the crop growth rate at a given LAI depends on the current 'lag' between gross tissue production and senescence. Crop growth rate at a given LAI therefore differs depending on whether the crop is maintained at that LAI, or has regrown to pass up through that LAI (Johnson and Parsons, 1985a; Parsons, Johnson and Harvey 1988). This point is illustrated in *Figure 17.10a*. The instantaneous growth rates of a sward regrowing from a low to a high LAI are substantially greater, as the sward passes through a given LAI, than in a sward maintained at that LAI by continuous defoliation (dotted line). Leniently defoliated swards do not retrace the relationship between growth rate and LAI seen in the severely defoliated crop, (see *Figure 17.10a*). The difference in understanding seems subtle, but accounts of crop growth based on the dynamic balance of photosynthesis, growth, yield and senescence (rather than on a single empirical example) are eminently more flexible and can be used to consider all patterns of defoliation.

During 50 years of grassland research (1930–1980) there has been much controversy over the optimal method to harvest grass: continuously or intermittently; and at what 'frequency' or 'severity' (McMeekan, 1960; Campbell, 1961; Campling, 1975; and see Raymond, 1981). Numerous field trials were conducted, but very few

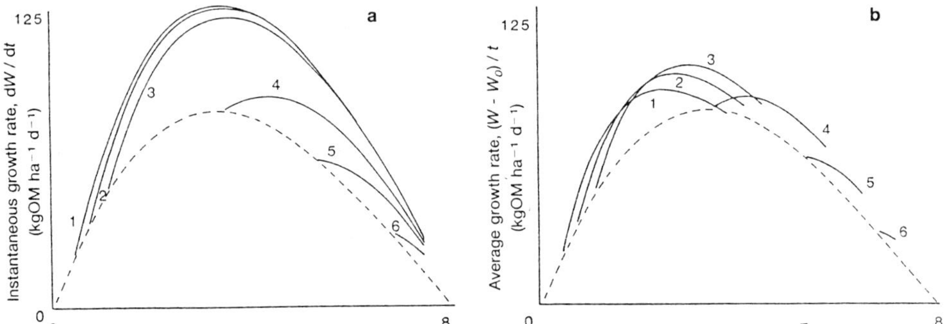

Figure 17.10 An analysis of the relationships between (**a**) the instantaneous growth rate (d*W*/d*t*) and leaf area index, (*L*) during regrowth following the six severities of defoliation described in *Figure 17.8*, and (**b**) the corresponding relationships between average growth rate, $(W–W_0) / t$) and average leaf area index ($\bar{L}$). In both (**a**) and (**b**) the dotted line describes the rate of intake achieved in swards maintained at each leaf area index. Note under steady state continuous grazing $L = \bar{L}$ and d*W*/d*t* = $(W–W_0) / t$. (After Parsons, Johnson and Harvey, 1988)

combinations of management can be observed at one time at a single site. Site to site and temporal variations have largely defeated an empirical approach to identifying optimal solutions. Analyses based on the principles of resource capture have arguably made far greater progress. As early as 1955, rational bases for defoliation managements to maximize resource capture were proposed, based on the principle of sustaining a high leaf area so as to maximize light interception (canopy photosynthesis could not be measured at that time), (see Brougham, 1958; Davidson and Donald, 1958). The significance of maximising *net*, rather than *gross* capture was soon appreciated as knowledge of the role of senescence and respiration developed. Despite this, controversy developed over the relative merits of 'frequent, lenient' and 'infrequent, severe' defoliation (see Brougham, 1956; Anslow, 1965, 1967; Brougham and Glenday, 1967; Anslow and Back, 1967). The analyses by a mechanistic model, benefitting greatly from the work of Brougham and Anslow, enable us to compare the productivity of a wide range of combinations of frequency and severity of defoliation. The models suggest that both frequent, lenient and infrequent, severe defoliation may theoretically achieve similarly high average rates of yield provided both achieve a similar average LAI over the cycle of regrowth and defoliation (see *Figure 17.10b*).

Intermittent defoliation may similarly be compared with continuous grazing. Yields under intermittent defoliation can be seen to be similar to those under continuous grazing when compared on the same fundamental basis, namely in terms of the average LAI sustained under each management (*Figure 17.10b*). Both Grant *et al.*, (1988) and Parsons and Penning, (1988) have recently retested the hypothesis that intermittent defoliation may give greater yields than continuous defoliation. It was argued that, as the increase in senescence lags behind the increase in gross tissue production during regrowth, senescence rates per unit net carbon exchange under intermittent defoliation should be less than under continuous defoliation, and therefore the rate of yield should be greater per unit LAI sustained. This effect is apparent using a model in *Figure 17.10b*, though small. Moreover, both authors argue how the theoretical benefits to intermittent defoliation are counteracted by progressive differences in morphology between the two managements, in particular, the decrease in the proportion of lamina to non-lamina tissue that arises imediately after a defoliation under intermittent cutting, or during the period of grazing 'down' in rotational grazing. When this is incorporated into a model, the benefits of intermittent defoliation diminish (Parsons, Johnson and Harvey, 1988). For many years, the detrimental effects on long-term yield of the deterioration in the morphology of a grass sward that arises when the sward is allowed to achieve a high LAI, and particularly of sustaining a sward at a high average LAI, have been recognized. This consideration raised the strongest argument against 'frequent', 'lenient' defoliation (Hunt, 1965; Tainton, 1974; Smetham, 1975).

Finally, on the generality of concepts of resource capture, it is fascinating to consider the similarity between the optimal pattern of defoliation proposed here to maximize net resource capture in agriculture, and the marginal value theorem (MVT) which is a basic tenet of optimal foraging in ecology (Charnov, 1976; Stephens and Krebs, 1986). Both are concepts which consider the optimal time for an 'animal' to move on from one 'patch' (cf. 'regrowth period') to the next as being when the rate of capture of resources on the current 'patch' (cf. dW/dt) has fallen below the maximum possible average rate of capture, $(W - W_0) / t$ in *Figure 17.7b*.

Some further bases for increasing resource capture

Accounts of the opportunities for increasing resource capture and manipulating yield in grasses, and the scientific efforts aimed at pursuing these objectives, have been presented by Robson (1980), Wilson (1981). Below, just two examples are mentioned of opportunities that have not been explored extensively.

ASSIMILATE PARTITIONING

In cereals, considerable increases in yield (50–60%) have been achieved by increasing the proportion of assimilated carbon partitioned to the harvestable component, the grain, and away from the non-harvestable components, particularly the reproductive stem (see Austin, 1978). But, there is far less scope for increasing yield in this way in grasses. In grasses, not only is the reproductive stem harvestable (albeit more readily by cutting than by animals, Rattray 1978), but it plays a major role in maintaining high A_{sat} in leaves in the infrequently harvested crop in spring (see Woledge, 1977, 1978, 1979). In grasses the only truly non-harvestable component is the roots, but these clearly also play a significant role in resource capture. Studies of the control of root:shoot partitioning suggest that the balance of root:shoot activity is unlikely to be far from optimal (see Brouwer, 1962; Davidson, 1969). Teleonomic models, based on this assumption, have increasingly become accepted (e.g. Johnson and Thornley, 1987; Thornley and Verberne, 1989) as a means to predict the response of root:shoot ratio to defoliation, nutrient and water availability in grasses. However, comparisons of the physiology of growth of different grassland species (e.g. ryegrass and white clover) combined with the use of mathematical models, suggest there are further means to increase the sustainable yield of grasses by modifying aspects of assimilate partitioning.

One possibility is to increase specific leaf area (SLA). Clover leaves, for example, have an intrinsically greater specific leaf area than do grass leaves, even when compared at similar N content (Parsons, Harvey and Woledge, 1991). An increased SLA can be shown to be of major benefit, *provided* that the rate of photosynthesis per unit leaf area is not correspondingly reduced. In clover, photosynthesis per unit leaf area is at least as great as that of grass of similar N content (Woledge and Dennis, 1982), the higher SLA in clover implying therefore a greater rate of photosynthesis per unit leaf *mass*. A species growing with these characteristics acquires a greater return in mass, per unit mass invested in leaf tissue, and this gives rise during uninterrupted growth to a greater rate of increase in leaf area and a greater ceiling LAI. The ceiling *mass* is also, but less markedly, increased. Even under continuous defoliation, a greater SLA is an advantage, though the advantage is less when a large proportion of new leaf is continually removed, and so a low LAI is sustained, than at high LAI. It has been estimated that a difference in SLA of the order of magnitude observed between grass and clover could lead to a greater increase in sustainable yield than a 25% increase in A_{sat} : one that is unlikely to be achieved (Parsons, Harvey and Johnson, 1991). Only recently have attempts been made to refocus attention on this aspect of the dynamics of resource capture (see Dijkstra, 1989). Success may depend on routinely reporting rates of leaf photosynthesis per unit leaf *mass* as well as per unit leaf *area*.

There is also variation between grassland species in the proportion of carbon invested into lamina *versus* the less photosynthetically active components of the shoot. This may be termed the '*incremental* leaf weight ratio' (ILWR) to distinguish the pattern of investment in new tissues from the ratio of lamina to non-lamina in the standing crop (the leaf weight ratio). Under continuous defoliation, the latter is more a reflection of the intensity of defoliation than of assimilate partitioning by the plant. In ryegrass this proportion (parameter 'ρ' in the models of Johnson and Thornley, 1983), is close to 0.7, regardless of the intensity of defoliation and so the size of the unit (lamina plus sheath plus internode) produced. In clover, the proportion of C invested into lamina is also close to 0.7 when under intensive defoliation a small unit (lamina plus petiole plus internode) is produced, but declines to <0.5 under lenient defoliation when larger units are produced (*Figure 17.11*; and see Parsons, Harvey and Woledge, 1991).

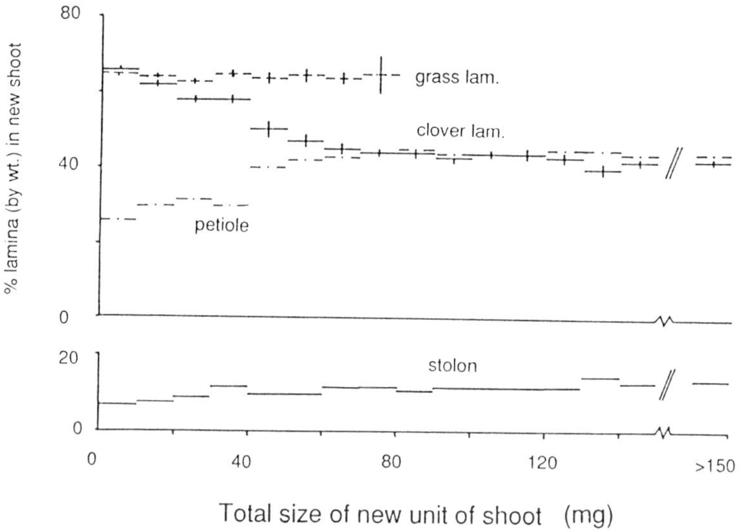

Figure 17.11 (**a**) The relationship between the proportion of lamina weight to total weight in the youngest expanded ungrazed units of grass and clover (incremental leaf weight ratio, ILWR) and the size (in steps of 10 mg) of the unit produced. For clover the proportion of petiole is also shown, and in (**b**) the proportion of stolon. (Courtesy of Parsons, Harvey and Woledge, 1991)

Both SLA and ILWR are parameters of biosynthesis that can readily be shown to affect the dynamics of crop growth and may be significant in understanding the balance of competition between species in mixed grass/legume swards (Parsons, Harvey and Johnson, 1991), but more effort is needed to explore the possibilities of exploiting these aspects of resource capture in grasslands.

OPPORTUNITIES TO MODIFY LOSSES BY SENESCENCE

Mathematical models may also be used to propose the benefits to resource capture of altering the rate of turnover, and so the longevity, of tissues in the shoot. This is more clearly understood in terms of the mass flux of shoot tissues, per unit ground area, rather than in terms of the way this is expressed in the number and rate of

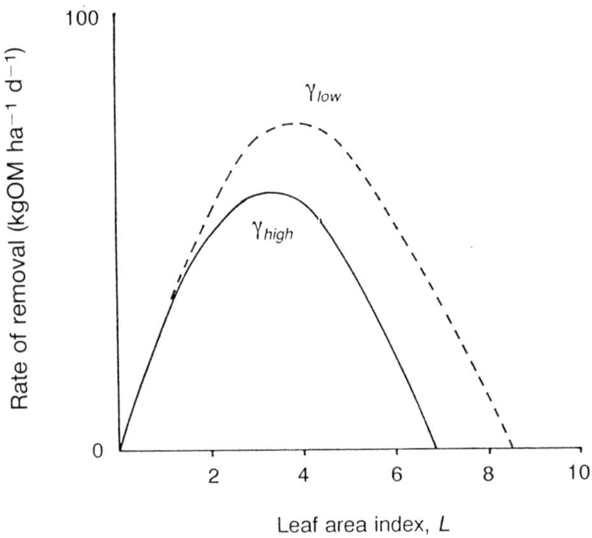

Figure 17.12 The effect on the rate of removal of material that may be sustained, in swards maintained at each of a range of LAI, of a change in the proportion 'γ' of the standing live weight lost per day in senescence. γ_{low} = 0.030 (longevity = 33.3 days/leaf) : γ_{high} = 0.0375 (longevity = 26.6 days/leaf)

turnover of leaves on tillers in the sward. Using the simplest possible model of tissue turnover, the shoot may be rationalized to a single category of live tissue of mass, W. The daily rate of senescence can then be expressed very simply as a function of the standing live weight: $ds/dt = \gamma W$; where γ is a temperature dependent rate constant and describes the proportion of the standing live tissue that senesces per unit time. Thus $1/\gamma$ is the mean longevity of a packet of shoot tissue. This very simple model, and the considerably more comprehensive models derived from it (Thornley and Johnson, 1990) suggest how grass crop growth is sensitive to changes in the parameter γ. In the absence of defoliation, then for small values of γ (greater longevity of tissues), W increases more rapidly to a higher ceiling yield. In the presence of continuous defoliation, a far greater rate of removal of tissue can be sustained when γ is small, while maintaining the sward at the same LAI, (*Figure 17.12*). Moreover, the maximum rate of harvest should be achieved at a greater LAI. This in itself may offer significant advantages, for when swards are maintained at the relatively low LAI that achieves optimum yield per hectare, large herbivores may be unable to acquire sufficiently large bites, and their intake rate may be depressed. Increasing the LAI optimum for yield per hectare may lead to a closer coincidence of optimal yield per hectare and yield per head.

It is confusing, but necessary, to consider what a change in mass flux, due to γ, implies in terms of the rate of turnover and number of live leaves per tiller. It is evident (*Figure 17.13*) that the sustainable benefits of modifying mass flux by reducing γ can be obtained both with and without altering the number of live leaves per tiller. Thus differences between species in the number of live leaves per tiller does not in itself imply differences in the specific rate of turnover of mass, γ. In *Figure 17.13* for example, the same mass flux is achieved where in (*a*) there

a

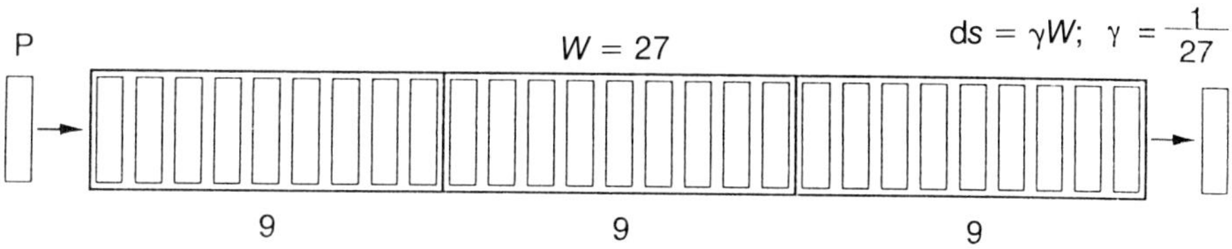

Number of live leaves = 3
Leaf app. = 1 leaf every 9 'days' (1/9 leaf day)
Longevity of a leaf (or a packet of tissue in a leaf) = 27 days
Senescence rate = 1 leaf every 9 'days' (1/9 leaf day)
Leaf size = integral of 9 days of rate P therefore large

b

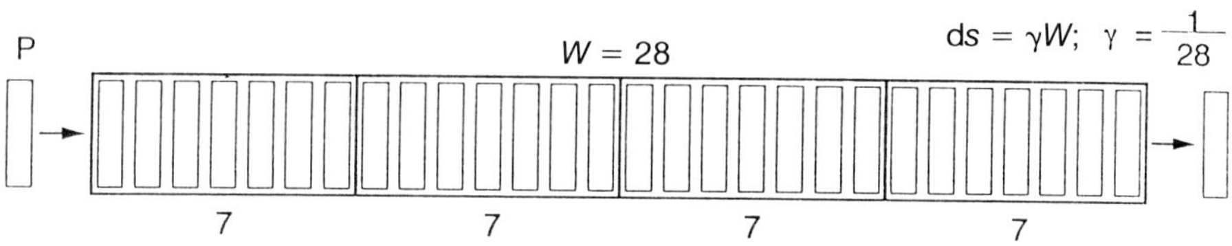

Number of live leaves = 4
Leaf app. = 1 leaf every 7 'days' (1/7 leaf day)
Longevity of a leaf (or a packet of tissue in a leaf) = 28 days
Senescence rate = 1 leaf every 7 'days' (1/7 leaf day)
Leaf size = integral of 7 days of rate P therefore small

Figure 17.13 Diagram to demonstrate how the rate of turnover and loss of *tissue*, γ, is not necessarily synonymous with the rate of turnover and loss of *leaves*. The small vertical boxes represent the daily inputs of mass; the contribution of these 'packets' of tissue to the standing live weight; and the daily loss of mass per tiller. The mass flows in (**a**) and (**b**) are virtually identical, as is the longevity of a leaf, despite the retention of four leaves per tiller in (**b**), and only three leaves per tiller in (**a**). In this case, leaf size compensates

are three live leaves per tiller, but in (*b*) there are four. The two examples differ simply in the size of the leaves involved. It is not therefore adequate to state the criteria for breeding as being to increase the number of live leaves per tiller. Nor is it adequate to increase the longevity of leaves *per se*, (this could be achieved for example simply by imposing a low temperature, and would not increase yield). The aim must be to reduce γ relative to the major parameters of crop growth.

There has been a great deal of research on the molecular and biochemical mechanism of senescence in grasses (see Thomas and Stoddart, 1980) but altering the programed pattern of production and senescence of leaves on grass tillers will be complex. It is salutory to consider that our many grassland species differ substantially in the number of live leaves and the rate of turnover of leaves per tiller, if they also can be shown to differ (more importantly) in the parameter γ, then it is notable that none of them appear to outyield *L. perenne*.

Stability of grassland systems

Because grasslands represent extensive natural ecosystems, and in many cases their use for agriculture is only marginally viable economically, the stability (cf.

sustainability) of performance of grasslands and the persistence of key species is a serious concern. It has been seen how grazing animals act to decrease leaf area and so the capacity for growth in the grass crop. However, the resulting reduction in leaf area, in turn, restricts the capacity of animals to remove leaf material from the sward. It decreases bite mass and there are limits to the capacity of animals to compensate for this by increasing bite rate, or by increasing the time spent grazing (Penning, 1986; Penning *et al.*, 1990; Penning, Rook and Orr, 1991). It may seem surprising that such a complex interaction could ever come to equilibrium. But, a wide range of models, based both in ecology and in physiology, suggest that this dynamic interaction may approach not one but a range of equilibria, but that not all these equilibria are stable.

Instability in plant–animal interaction is best explained by reference to Morley (1966), and Noy-Meir (1975), see *Figure 17.14*. Although their accounts do not consider the interaction dynamically, they illustrate graphically the principles involved. *Figure 17.14* describes an example of a relationship between the rate of supply of resources (grass growth) and LAI (see solid line). Superimposed on this is the rate of consumption, the intake response of animals (per ha) in relation to LAI. Intake responses of this general form have been indicated experimentally (Penning, 1986; Hodgson, 1986). Where these relationships intersect, the rate of removal of leaves exactly matches the rate of growth of leaves. At low stocking density only a single intersection is apparent. The stability of this equilibrium can be explored by considering what happens if the system is perturbed, in particular

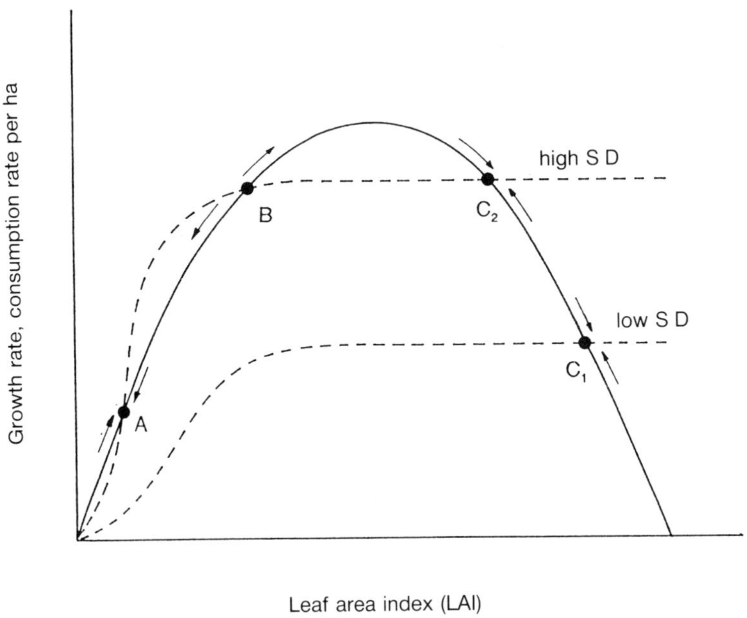

Figure 17.14 The relationship between consumption rate per hectare and LAI (dotted lines) at high and low stocking density (SD), superimposed on the relationship between herbage net growth rate and LAI (solid line). 'A' and 'C_1, C_2' indicate stable equilibria, 'B' is an unstable equilibrium. (After Morley (1966) and Noy-Meir (1975))

the direction of changes in resource supply relative to the rate of removal. If at 'C_1', LAI is decreased slightly, for whatever reason, the rate of supply of resources would increase relative to the rate of removal. The sward would consequently regrow towards the equilibrium at 'C_1'. If LAI was increased, the growth rate of herbage would decline relative to the rate of removal, and the system would now decline in LAI toward the equilibrium at 'C_1'. Hence, at low stocking densities, a single *stable* equilibrium at 'C_1' should arise.

At higher stocking densities, three points of intersection can arise (see *Figure 17.14*). The equilibria at 'A' and 'C_2', exactly like that at 'C_1', can be shown to be stable. However, there is an equilibrium at 'B' that is unstable. At 'B', a small increase in LAI would cause growth rate to increase more markedly than intake, and so the sward would continue to regrow until it came to equilibrium again at 'C_2'. A small decrease in LAI, from 'B', would cause growth to decline relative to the rate of removal, and the sward would continue to decline in LAI to the equilibrium at 'A'.

In *Figure 17.15*, a dynamic mechanistic model of resource capture and intake (Johnson and Parsons, 1985a) is used to illustrate the effect of a whole range of stock densities on yield per hectare, yield per animal, and the LAI that can be sustained. This analysis, unlike that in *Figure 17.14* does take account of the dynamic interaction between intake (harvest) and sward growth rate. The general form of these relationships is consistent with empirical accounts by Mott

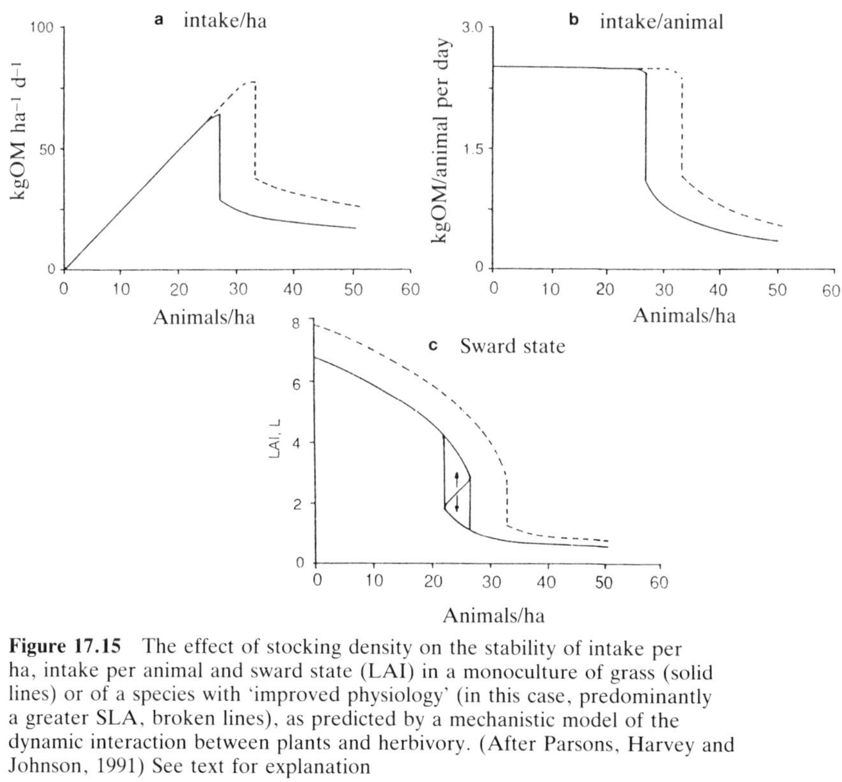

Figure 17.15 The effect of stocking density on the stability of intake per ha, intake per animal and sward state (LAI) in a monoculture of grass (solid lines) or of a species with 'improved physiology' (in this case, predominantly a greater SLA, broken lines), as predicted by a mechanistic model of the dynamic interaction between plants and herbivory. (After Parsons, Harvey and Johnson, 1991) See text for explanation

(1960) and Owen and Ridgman (1968), though by removing in the model the fluctuations in processes seen over time in the field, the features of stability are more apparent. The full features of instability are drawn only for the example of sward state (*Figure 17.15c*) to avoid over complicating the figure. In keeping with Morley and Noy-Meir, at low stock density, plant–animal interactions approach a single stable equilibrium at high LAI (regardless of the initial conditions for grazing) at which yield per hectare is small, although each animal achieves close to maximum intake. As stock density is increased, the sward comes to a new stable equilibrium at lower LAI. Yield per hectare increases (as explained in *Figure 17.6*) and intake per animal remains high. However, as stock density is increased further, the solutions enter a bi-stable region, where (consistent with Noy-Meir) two alternative stable equilibria may be approached depending on the initial conditions for grazing. The alternative stable solutions are separated by a highly unstable region. Given even higher stock densities, than in *Figure 17.15*, using parameters for the intake response of sheep grazing perennial ryegrass swards, the sward is not grazed to extinction but comes to equilibrium at an unproductive, low LAI.

It is notable, and relevant to the past success of grassland agriculture, that in cool temperate plant species the sward is unlikely to be grazed to extinction. Changes in the morphology of these species (a reduction in leaf and tiller size, combined with the formation of an increased population of small tillers) mean that intake by animals becomes severely restricted while a finite LAI remains. (Hence the intake response of animals to LAI is theoretically sigmoidal, as in *Figure 17.14*. An hyperbolic intake response would imply the sward may be grazed to extinction, see Noy-Meir, 1975.) The same cannot be said for many tropical and dryland grassland species. Some e.g. *Themeda triandra* have a poor capacity to adapt their morphology to retain leaf area under repeated defoliation (Mott *et al.*, 1992). Others may have the capacity to adapt, given time, but during a short growing season, a high intensity of defoliation when growth first begins can greatly limit the amount of material produced. This may be one explanation why cool temperate grasslands support highly stable, productive agricultural systems, whereas many tropical and rangeland grasslands are sensitive to overgrazing and degradation.

Models have also been used to consider the effects of altering the physiological characteristics of plant species on their stability of performance (see Parsons, Harvey and Johnson, 1991). An increase in SLA, for example, would enable an improved species (see dotted line, *Figure 17.15*) to come to equilibrium at a greater LAI, and to remain stable up to a greater stock density. Note that the model suggests that yield per hectare at a given stock density is *not* increased. In a plant species mixture grazed by a common overall stock density, one species may thus achieve a higher LAI than the other species in the mixture, and under certain circumstances be in a more stable state. Analyses such as these, based on a knowledge of the principles of resource capture applied to mixtures, may contribute to understanding how to maintain valuable plant species associations, such as that between grasses and legumes.

In recent years there has been a resurgence of interest in using measures of sward state (e.g. sward height or LAI) rather than stocking density, as the basis for characterizing and controlling grazing intensity both in experimentation and in practice (Hodgson, 1985; Parsons and Johnson, 1985). There are sound reasons for this. It has been seen in this chapter that sward state (particularly average LAI) is fundamental to the capture of resources by the grass crop, and it has been shown to be equally fundamental to controlling the intake behaviour of grazing animals (Penning, 1986; Hodgson, 1986). The value of manipulating plant–animal interactions from a basis of sward state, rather than stocking density, is also evident in the context of the stability of grazing. If guidelines for management were based on stock density, then, in the 'improved' plant species shown in *Figure 17.15*, a different guideline (for a greater stock density) would have to be given. Simple differences in fertilizer input or environmental conditions (such as the seasonal reduction in light energy receipt, see Johnson and Parsons, 1985a) would equally decrease the number of animals that may be sustained by the reduced growth rate of the sward and again changes in the guidelines for stock density would be required. By contrast, guidelines based on sward state (LAI) need change much less. In both cases in *Figure 17.15* (and in the case of a reduced light energy receipt, see Johnson and Parsons, 1985a), maintaining the sward at an LAI close to 3.0 would ensure a highly productive and stable output. Besides, if the fundamental source of resource capture, the sward LAI, is being monitored and controlled, the situation need never arise where an inappropriate stock density is present.

Conclusions

Our knowledge of resource capture in the grass crop has progressed substantially in the last 30 years. This has led to relatively small increases in the *potential* production of grassland. Despite its familiarity, the grass crop is complex. Increasing output is certainly not a matter of maximizing light interception and photosynthesis. But, our knowledge of the physiological principles of resource capture has arguably played a major role in providing the basis for controlling leaf area to optimize the balance of photosynthesis, growth, senescence and yield. Given the diversity of the response of the grass crop to its environment, and the wide range of environments in which a perennial crop must be managed as the growing season progresses, an understanding of these principles is invaluable in ensuring that the *potential* for production is routinely and reliably realized in *practice*. Practical guidelines for management based on controlling leaf area have been advocated extensively for systems of grazing and cutting, and particularly to integrate grazing and conservation on farms. Having the knowledge to control the level of agricultural output and to minimize instability and risk will be as important in future in a time of reducing demand for agricultural products as it was during the past drive for increased production. With the probability of global climatic change, there should be an increasing, not diminishing, interest in maintaining a sound understanding in whole crop physiology, to form the basis of a sustainable grassland agriculture.

References

Alberda, Th. (1966) Responses of grasses to temperature and light. In *The Growth of Cereals and Grasses*, pp 200–212 Edited by F.L. Milthorpe and J.D. Ivins. Proceedings of the 12th Easter School in Agricultural Science, Nottingham. London: Butterworths

Alberda, Th. (1971) Potential production of grassland. In *Potential Crop Production*, pp 159–171 Edited by P.F. Wareing and J.P. Cooper. London: Heinemann

Anslow, R.C. (1965) Grass growth in midsummer. *Journal of the British Grassland Society*, **20**, 19–26

Anslow, R.C. (1967) Frequency of cutting and sward production. *Journal of Agricultural Science, Cambridge*, **68**, 377–384

Anslow, R.C. and Back, H.L. (1967) Grass growth in midsummer and light interception and growth rate of a perennial ryegrass sward. *Journal of the British Grassland Society*, **22**, 108–111

Austin, R.B. (1978) Maximum yields of cereals as determined by plant type and environment. In *Maximizing Yields of Crops*, pp 18–24 Edited by P.F. Wareing and J.P. Cooper. London: HMSO

Belsky, A.J. (1986) Does herbivory benefit plants? A review of the evidence. *American Naturalist*, **127**, 870–892

Boller, B.C. and Nosberger, J. (1989) Photosynthesis of white clover leaves as influenced by canopy position, leaf age and temperature. *Annals of Botany*, **56**, 19–27

Booysen, P. de V. (1966) A physiological approach to research in pasture utilization. *Proceedings of the Grassland Society of South Africa*, **2**, 77–85

Brougham R.W. (1956) Effect of intensity of defoliation on regrowth of pasture. *Australian Journal of Agricultural Research*, **7**, 377–387

Brougham, R.W. (1957) Pasture growth studies in relation to grazing management. *Proceedings of the New Zealand Society for Animal Production*, **17**, 46–55

Brougham, R.W. (1958) Interception of light by the foliage of pure and mixed stands of pasture plants. *Australian Journal of Agricultural Research*, **9**, 39–52

Brougham, R.W. and Glenday, A.C. (1967) Grass growth in midsummer: a re-interpretation of published data. *Journal of the British Grassland Society*, **22**, 100–107

Brouwer, R. (1962) Distribution of dry matter in the plant. *Netherlands Journal of Agricultural Science*, **10**, 361–376

Brown, R.H. and Blaser, R.E. (1968) Leaf area index in pasture growth. *Herbage Abstracts*, **38**, 1–9

Campbell, A.G. (1961) A theoretical basis for grazing management. *Proceedings of the New Zealand Society for Animal Production*, **21**, 18–32

Campling, R.C. (1975) Systems of grazing management for dairy cattle. In *Pasture Utilization by the Grazing Animal*, pp 113–117 Edited by J. Hodgson and D.K. Jackson. Occasional Symposium of the British Grassland Society, No. 8. Hurley: British Grassland Society

Charnov, E.L. (1976) Optimal foraging, the marginal value theorem. *Theoretical Population Biology*, **9**, 129–136

Clark, D.A., Chapman, D.F., Land, C.A. and Dymock, N. (1984) Defoliation of *Lolium perenne* and *Agrostis* spp tillers, and *Trifolium repens* stolons in set stocked and rotationally grazed hill pastures. *New Zealand Journal of Agricultural Research*, **27**, 289–301

Cragg, J.B. (1979) Foreword. In *Grassland Ecosystems of the World*, pp xxi-xxv Edited by R.T. Coupland. International Biological Programme, Vol. 18, Cambridge: Cambridge University Press

Cugnac, A. De (1931) Recherches sur le glucides de graminées. *Annals Science Naturelles (Bot.)*, **13**, 1–129

Davidson, R.L. (1969) Effects of root/leaf temperature differentials on root:shoot ratios in some pasture grasses and clover. *Annals of Botany*, **33**, 561–569

Davidson, J.L. and Donald, C.M. (1958) The growth of swards of subterranean clover with particular reference to leaf area. *Australian Journal of Agricultural Research*, **9**, 53–72

Davidson, J.L. and Milthorpe, F.L. (1966a) Leaf growth in *Dactylis glomerata* following defoliation. *Annals of Botany*, **30**, 173–184

Davidson, J.L. and Milthorpe, F.L. (1966b) The effect of defoliation on the carbon balance in *Dactylis glomerata. Annals of Botany*, **30**, 185–198

Davies, A. (1977) Structure of the grass sward. *Proceedings of the International Meeting on Animal Production from Temperate Grassland*, pp 36–44. Dublin: An Foras Taluntais

Dennis, W.D. and Woledge, J. (1982) Photosynthesis by white clover leaves in mixed clover/ryegrass swards. *Annals of Botany*, **49**, 627–635

Dijkstra, P. (1989) Cause and effect of differences in specific leaf area. In *Causes and Consequences of Variation in Growth Rate and Productivity of Higher Plants*, pp 125–140 Edited by H. Lambers. The Hague: SPB Academic Publishers

Donald, C.M. (1961) Competition for light in crops and pastures. In *Mechanisms in Biological Competition*, pp 283–313 Edited by F.L. Milthorpe. Cambridge: Cambridge University Press

Donald, C.M. and Black, J.N. (1958) The significance of leaf area in pasture growth. *Herbage Abstracts*, **28**, 1–6

Goudriaan, J. (1992) Biosphere structure, carbon sequestring potential and the ^{14}C carbon atmospheric record. *Journal of Experimental Botany*, **43**, 1111–1120

Grant, S.A., Barthram, G.T., Torvill, L., King, J. and Smith, H.K. (1983) Sward management, lamina turnover and tiller population density in continuously stocked *Lolium perenne*-dominated swards. *Grass and Forage Science*, **38**, 333–344

Grant, S.A., Bartham, G., Torvell, L., King, J. and Elston, D. (1988) Comparison of herbage production under continuous stocking and intermittent grazing. *Grass and Forage Science*, **43**, 29–39

Harris, W. (1978) Defoliation as a determinant of the growth, persistance and composition of pasture. In *Plant Relations in Pastures*, pp 67–85 Edited by J.R. Wilson. Australia: CSIRO

Hodgson, J. (1966) The frequency of defoliation of individual tillers in a set stocked sward. *Journal of the British Grassland Society*, **21**, 258–263

Hodgson, J. (1985) The significance of sward characteristics in the management of temperate sown pastures. *Proceedings of the 15th International Grassland Congress, Kyoto*, pp 31–34

Hodgson, J. (1986) Grazing behaviour and herbage intake. In *Grazing*, pp 51–64 Edited by J. Frame. Occasional Symposium of the British Grassland Society, No. 19. Hurley: British Grassland Society

Hodgson, J. and Ollerenshaw, J.H. (1969) The frequency and severity of defoliation of individual tillers in set stocked swards. *Journal of the British Grassland Society*, **24**, 226–234

Hunt, L.A. (1965) Some implications of death and decay in pasture production. *Journal of the British Grassland Society*, **20**, 27–31

Johnson, I.R. and Parsons, A.J. (1985a) A theoretical analysis of grass growth under grazing. *Journal of Theoretical Biology*, **112**, 345–36

Johnson, I.R. and Parsons, A.J. (1985b) Use of a model to analyse the effects of continuous grazing managements on seasonal patterns of grass production. *Grass and Forage Science*, **40**, 449–458

Johnson, I.R. and Thornley, J.H.M. (1983) Vegetative crop growth model incorporating leaf area expansion and senescence, and applied to grass. *Plant, Cell and Environment*, **6**, 721–729

Johnson, I.R. and Thornley, J.H.M. (1987) A model of shoot:root partition with optimal growth. *Annals of Botany*, **60**, 133–142

King, J., Sim, E.M. and Grant, S.A. (1984) Photosynthetic rate and carbon balance of grazed ryegrass pastures. *Grass and Forage Science*, **39**, 81–92

Lambers, H. (1980) The physiological significance of cyanide-resistant respiration. *Plant, Cell and Environment*, **3**, 293–302

Lambers, H., Noord, R. and Posthumus, F. (1979) Respiration of *Senecio* shoots; inhibition during photosynthesis, resistance to cyanide and relation to growth and maintenance. *Physiologia Plantarum*, **45**, 351–356

Lambers, H. and Posthumus, F. (1980) The effect of light intensity and relative humidity on growth rate and root respiration of *Plantago lanceolata* and *Zea mays*. *Journal of Experimental Botany*, **31**, 1621–1630

Lazenby, A. (1981) British grasslands; past, present and future. *Grass and Forage Science*, **36**, 243–266

Leafe, E.L., Stiles, W. and Dickinson, S. (1974) Physiological processes influencing the pattern of productivity of the intensively managed grass sward. *Proceedings of the 12th International Grassland Congress, Moscow*, **1**, 442–457

McCree, K.J. (1970) An equation for the rate of respiration of white clover plants grown under controlled conditions. In *Prediction and Measurement of Photosynthetic Productivity*, pp 221–229 Edited by I. Malek. Wageningen: Pudoc

McCree, K.J. (1974) Equations for the rate of respiration of white clover and grain sorghum, as functions of dry weight, photosynthetic rate and temperature. *Crop Science*, **14**, 509–514

McCree, K.J. and Troughton, J.H. (1966a) Prediction of growth rate at different light levels from measured photosynthesis and respiration rates. *Plant Physiology*, **41**, 559–566

McCree, K.J. and Troughton, J.H. (1966b) Non-existence of an optimum leaf area index for the production rate of white clover grown under constant conditions. *Plant Physiology*, **41**, 1615–1622

McMeekan, C.P. (1960) Grazing management. *Proceedings of the 8th International Grassland Congress*, Reading, pp 21–27

Maeda, S. and Yonetani, T. (1978) Optimum cutting stage of forage plants. II. Seasonal changes in CGR and average productivity in Italian ryegrass population. *Journal of Japanese Society of Grassland Science*, **24**, 10–16

MAFF (1926–1979) *June Agricultural Censuses for England and Wales*. Ministry of Agriculture, Fisheries and Food. London: HMSO

MAFF (1990) *Agriculture in the United Kingdom: 1990*. Ministry of Agriculture, Fisheries and Food. London: HMSO

Morris, R.M. (1969) The pattern of grazing in continuously grazed swards. *Journal of the British Grassland Society*, **24**, 65–71

Mott, G.O. (1960) Grazing pressure and the measurement of pasture production. *Proceedings of the 8th International Grassland Congress*, Reading, pp 606–611

Mott, J.J., Ludlow, M.M., Richards, J.H. and Parsons, A.J. (1991) Effects of moisture supply in the dry season and subsequent defoliation on persistance of the savanna grasses *Themeda triandra, Heteropogon contortus* and *Panicum maximum*. *Australian Journal of Agricultural Research*, **43**, 241–260

Morley, F.H.W. (1966) Stability and productivity of pastures. *Proceedings of the New Zealand Society of Animal Production*, **26**, 8–21

Morley, F.H.W. (1968) Pasture growth curves and grazing management. *Australian Journal of Experimental Agriculture and Animal Husbandry*, **30**, 40–45

NIAB (1982/3;1991/2) *Recommended varieties of grasses and herbage legumes*. Farmers leaflet. Cambridge: National Institute of Agricultural Botany

Noy-Meir I. (1975) Stability of grazing systems: an application of predatory-prey graphs. *Journal of Ecology*, **63**, 459–481

Ong, C.K. (1978) The physiology of tiller death in grasses. I. The influence of tiller age, size and position. *Journal of the British Grassland Society*, **33**, 197–203

Owen, J.B. and Ridgman, W.J. (1968) The design and interpretation of an experiment to study animal production from grazed pasture. *Journal of Agricultural Science, Cambridge*, **71**, 327–335

Parsons, A.J. (1988) The effects of season and management on the growth of grass swards. In *The Grass Crop – the physiological basis of production*, pp 129–177 Edited by M.B. Jones and A. Lazenby. London: Chapman and Hall

Parsons, A.J., Collett, B. and Lewis, J. (1984) Changes in the structure and physiology of a perennial ryegrass sward when released from a continuous stocking management – implications for the use of exclusion cages in continuously stocked swards. *Grass and Forage Science*, **39**, 1–9

Parsons, A.J., Harvey, A. and Woledge, J. (1991) Plant/animal interactions in continuously grazed mixtures: 1. Differences in the physiology of leaf expansion and the fate of leaves of grass and clover. *Journal of Applied Ecology*, **28**, 619–634

Parsons, A.J., Harvey, A. and Johnson, I.R. (1991) Plant/animal interactions in continuously grazed mixtures: 2. The role of differences in the physiology of leaf expansion and of selective grazing on the performance and stability of species in a mixture. *Journal of Applied Ecology*, **28**, 635–658

Parsons, A.J., Leafe, E.L., Collett, B. and Stiles, W. (1983a) The physiology of grass production under grazing. 1. Characteristics of leaf and canopy photosynthesis of continuously grazed swards. *Journal of Applied Ecology*, **20**, 117–126

Parsons, A.J., Leafe, E.L., Collett, B., Penning, P.D. and Lewis, J. (1983b) The physiology of grass production under grazing. 2. Photosynthesis, crop growth and animal intake of continuously grazed swards. *Journal of Applied Ecology*, **20**, 127–139

Parsons, A.J. and Johnson, I.R. (1985) The physiology of grass growth under grazing. In *Grazing*, pp 3–13 Edited by J Frame. Occasional Symposium of the British Grassland Society, No. 19. Hurley: British Grassland Society

Parsons, A.J., Johnson, I.R. and Harvey, A. (1988) Use of a model to optimize the interaction between the frequency and severity of intermittent defoliation and to provide a fundamental comparison of the continuous and intermittent defoliation of grass. *Grass and Forage Science*, **43**, 49–59

Parsons, A.J., Johnson, I.R. and Williams, J.H.H. (1988) Leaf age structure and canopy photosynthesis in rotationally and continuously grazed swards. *Grass and Forage Science*, **43**, 1–14

Parsons, A.J. and Penning, P.D. (1988) The effect of the duration of regrowth on photosynthesis, leaf death and the average rate of growth in a rotationally grazed sward. *Grass and Forage Science*, **43**, 15–27

Penning de Vries, F.W.T. (1972) Respiration and growth. In *Crop Processes in Controlled Environments*, pp 327–347 Edited by A.R. Rees, K.E. Cockshull, D.W. Hand and R.G. Hurd. London: Academic Press

Penning de Vries, F.W.T. (1974) Substrate utilization and respiration in relation to growth and maintenance in higher plants. *Netherlands Journal of Agricultural Science*, **22**, 40–44

Penning de Vries, F.W.T. (1975) The cost of maintenance processes in plant cells. *Annals of Botany*, **39**, 77–92

Penning, P.D. (1986) Some effects of sward conditions on grazing behaviour and intake by sheep. In *Grazing Research at Northern Latitudes*, pp 219–226 Edited by O. Gudmundsson. New York: Plenum Publishing Corporation

Penning, P.D., Parsons, A.J., Orr, R.J. and Treacher, T.T. (1990) Intake and behaviour responses by sheep to changes in sward characteristics under continuous stocking. *Grass and Forage Science*, **46**, 15–28

Penning, P.D., Rook, A.J. and Orr, R.J. (1991) Patterns of ingestive behaviour of sheep continuously stocked on monocultures of ryegrass or white clover. *Applied Animal Behaviour Science*, **31**, 237–250

Pilbeam, C.J., Robson, M.J. and Lambers, H. (1986) Respiration in mature leaves of *Lolium perenne* as affected by nutrient supply and cutting. *Physiologia Plantarum*, **66**, 53–57

Pollock, C.J. and Jones, T. (1979) Seasonal patterns of fructosan metabolism in forage grasses. *New Phytologist*, **83**, 9–15

Prioul, J.L. (1971) Réaction des feuilles de *Lolium multiflorum* à l'éclairement pendant la croissance et variation des résistances aux éxchange gazeux photosynthétiques. *Photosynthetica*, **5**, 364–375

Prioul, J.L., Brangeon, J. and Reyss, A. (1980a) Interaction between external and internal conditions in the development of photosynthetic features in a grass leaf. I. Regional responses along a leaf during and after low-light or high-light acclimation. *Plant Physiology*, **66**, 762–769

Prioul, J.L., Brangeon, J. and Reyss, A. (1980b) Interaction between external and internal conditions in the development of features in a grass leaf. II. Reversibility of light-induced responses as a function of development. *Plant Physiology*, **66**, 770–774

Prioul, J.L., Reyss, A. and Chartier, P. (1975) Relationship between carbon dioxide transfer resistances and some physiological and anatomical features. In *Environmental and Biological Control of Photosynthesis*, pp 17–28 Edited by R. Marcelle. The Hague: Junk

Rattray, P.V. (1978) Pasture constraints to sheep production. *Proceedings of the Agronomy Society of New Zealand*, **8**, 103–108

Raymond, W.F. (1981) Grassland Research. In *Agricultural Research 1931–1981*, pp 311–323 Edited by G.W. Cooke. London: Agricultural Research Council

Rhodes, I. (1969) The yield, canopy structure and light interception of two ryegrass varieties in mixed culture and mono-culture. *Journal of the British Grassland Society*, **24**, 123–127

Rhodes, I. (1973) The relationship between productivity and some components of canopy structure in ryegrass (*Lolium* spp.). III Spaced plant characters, their heritabilities and relationship to sward yield. *Journal of Agricultural Science, Cambridge*, **80**, 171–176

Rhodes, I. (1975) The relationship between productivity and some components of canopy structure in ryegrass (*Lolium* spp.). IV Canopy characters and their relationship with sward yield in some intra population selections. *Journal of Agricultural Science, Cambridge*, **84**, 345–351

Rhodes, I. and Mee, S.S. (1980) Changes in dry matter yield associated with selection for canopy characters in ryegrass. *Grass and Forage Science*, **35**, 35–39

Robson, M.J. (1973a) The growth and development of simulated swards of perennial ryegrass. I. Leaf growth and dry weight changes as related to the ceiling yield of a seedling sward. *Annals of Botany*, **37**, 487–500

Robson, M.J. (1973b) The growth and development of simulated swards of perennial ryegrass. II. Carbon assimilation and respiration in a seedling sward. *Annals of Botany*, **37**, 501–518

Robson, M.J. (1980) A physiologists' approach to raising the potential yield of the grass crop through breeding. In *Opportunities for Increasing Crop Yields*, pp 33–49 Edited by R.G. Hurd, P.V. Biscoe and C. Dennis. Boston: Pitman Advanced Publishing Programme

Robson, M.J. (1981b) Potential production – what is it and can we increase it? In *Plant Physiology and Herbage Production* pp 5–18 Edited by C.E. Wright. Hurley: British Grassland Society

Robson, M.J. (1982a) The growth and carbon economy of selection lines of *Lolium perenne* cv. S23 with 'fast' and 'slow' rates of dark respiration. I. Grown as simulated swards during a regrowth period. *Annals of Botany*, **49**, 321–329

Robson, M.J. (1982b) The growth and carbon economy of selection lines of *Lolium perenne* cv. S23 with 'fast' and 'slow' rates of dark respiration. II. Grown as young plants from seed. *Annals of Botany*, **49**, 331–339

Robson, M.J., Ryle, G.J.A. and Woledge, J. (1988) The grass plant – its form and function. In *The Grass Crop – the physiological basis of production*, pp 25–83 Edited by M.B. Jones and A. Lazenby. London: Chapman and Hall

Sibma, L. and Louwerse, W. (1977) The effects of changes in crop structure on photosynthesis, dry matter production and chemical composition of *Lolium perenne. Proceedings of the 13th International Grassland Congress, Leipzig*, section 1, pp 2–12

Smetham, M.L. (1975) The influence of herbage utilization on pasture production and animal performance. *Proceedings of the New Zealand Grassland Association*, **37**, 91–103

Stephens, D.W. and Krebs, J.R. (1986) *Foraging Theory*. Princetown, New Jersey: Princetown University Press

Tans, P.P., Fung, I.Y. and Takahashi, T. (1990) Observational constraints on the global atmospheric CO_2 budget. *Science*, **247**, 1431–1438

Tainton, N.M. (1974) Effect of different grazing rotations on pasture production. *Journal of the British Grassland Society*, **29**, 191–202

Thomas, H. and Stoddart, J.L. (1980) Leaf senescence. *Annual Review of Plant Physiology*, **31**, 83–111

Thornley, J.H.M. (1976) *Mathematical Models in Plant Physiology*. London: Academic Press

Thornley, J.H.M., Fowler, D. and Cannell, M.G.R. (1991) Terrestrial carbon storage resulting from CO_2 and nitrogen fertilization in temperate grasslands. *Plant, Cell and Environment*, **14**, 1007–1011

Thornley, J.H.M. and Johnson, I.R. (1990) *Plant and Crop Modelling*. Oxford: Clarendon Press

Thornley, J.H.M. and Verberne, E.L.J. (1989) A model of nitrogen flows in grassland. *Plant, Cell and Environment*, **12**, 863–886

Treharne, K.J. and Eagles, C.F. (1969) Effect of growth at different light intensities on photosynthetic activity of two contrasting populations of *Dactylis glomerata* L. In *Progress in Photosynthesis Research*, pp 377–382 Edited by H. Metzner. Tubingen: International Union of Biological Science

Volenec, J.J. (1986) Non-structural carbohydrates in stem base components of tall fescue during regrowth. *Crop Science*, **26**, 122–127

Watson, R., Rodhe, H., Oeschger, H. and Siegenthaler, U. (1990) Greenhouse gases and aerosols. In *Climate Change*, pp 1–41 Edited by J.T. Houghton, G.J. Jenkins and J.J. Ephraums. IPCC Section 1. Cambridge: UNEP/WMO, Cambridge University Press

Wilson, D. (1975) Variation in leaf respiration in relation to growth and photosynthesis of *Lolium*. *Annals of Applied Biology*, **80**, 323–338

Wilson, D. (1981) The role of physiology in breeding herbage cultivars adapted to their environment. In *Plant Physiology and Herbage Production*, pp 95–108 Edited by C.E. Wright. Hurley: British Grassland Society

Wilson, D. (1982) Response to selection for dark respiration rate of mature leaves of *Lolium perenne* L. and its effect on growth of young plants. *Annals of Botany*, **49**, 303–312

Wilson, D. and Cooper, J.P. (1969a) Effect of light intensity and CO_2 on apparent photosynthesis and its relationship with leaf anatomy in genotypes of *Lolium perenne* L. *New Phytologist*, **68**, 627–644

Wilson, D. and Cooper, J.P. (1969b) Effect of temperature during growth on leaf anatomy and subsequent light-saturated photosynthesis among contrasting *Lolium* genotypes. *New Phytologist*, **68**, 1115–1123

Wilson, D. and Cooper, J.P. (1969c) Effect of light intensity during growth on leaf anatomy and subsequent light-saturated photosynthesis among contrasting *Lolium* genotypes. *New Phytologist*, **68**, 1125–1135

Wilson, D. and Jones, J.G. (1982) Effects of selection for dark respiration of mature leaves on crop yields of *Lolium perenne* cv. S23. *Annals of Botany*, **49**, 313–320

Wilson, D. and Robson, M.J. (1981) Varietal improvement by selection for dark respiration rate in perennial ryegrass. In *Plant Physiology and Herbage Production*, pp 209–211 Edited by C.E. Wright. Hurley: British Grassland Society

Wit, C.T. de (1965) Photosynthesis of leaf canopies. *Agricultural Research Report, No. 663*, pp 1–57. Wageningen: Pudoc

Woledge, J. (1971) The effect of light intensity during growth on the subsequent rate of photosynthesis in leaves of tall fescue (*Fescue arundinacea* Schreb.) *Annals of Botany*, **35**, 311–322

Woledge, J. (1972) The effect of shading on the photosynthetic rate and longevity of grass leaves. *Annals of Botany*, **36**, 551–561

Woledge, J. (1973) The photosynthesis of ryegrass leaves grown in a simulated sward. *Annals of Applied Biology*, **73**, 229–237

Woledge, J. (1977) Effects of shading and cutting treatments on the photosynthetic rate of ryegrass leaves. *Annals of Botany*, **41**, 1279–1286

Woledge, J. (1978) The effect of shading during vegetative and reproductive growth on the photosynthetic capacity of leaves in a grass sward. *Annals of Botany*, **42**, 1085–1089

Woledge, J. (1979) Effect of flowering on the photosynthetic capacity of ryegrass leaves grown with and without natural shading. *Annals of Botany*, **44**, 197–207

Woledge, J. (1986) The effect of age and shade on the photosynthesis of white clover leaves. *Annals of Botany*, **57**, 257–262

Woledge, J. and Dennis, W.D. (1982) The effect of temperature on photosynthesis of ryegrass and white clover leaves. *Annals of Botany*, **50**, 25–35

Woledge, J. and Jewiss, O.R. (1969) The effect of temperature during growth on the subsequent rate of photosynthesis in leaves of tall fescue (*Festuca arundinacea* Schreb.) *Annals of Botany*, **33**, 897–913

Woledge, J. and Leafe, E.L. (1976) Single leaf and canopy photosynthesis in a ryegrass sward. *Annals of Botany*, **40**, 773–783

Woledge, J. and Parsons, A.J. (1986) Temperate grasslands. In *Photosynthesis in Contrasting Environments*, pp 173–197 Edited by N.R. Baker and S.P. Long. Amsterdam: Elsevier

18

CAPTURE OF CARBON DIOXIDE BY A CONIFEROUS FOREST

P.G. JARVIS
Institute of Ecology and Resource Management, University of Edinburgh, Darwin Building, Mayfield Road, Edinburgh EH9 3JU, UK

Introduction

With the information presently available, it is difficult, if not impossible, to prepare a global carbon balance sheet with any degree of certainty (Jarvis and Dewar, 1993). In 1992, we are reasonably certain from both production and utilization data that fossil fuel consumption put *c.* 5.7 Gt of carbon into the atmosphere. To this source term must be added CO_2 generated as a result of tropical deforestation and other land use changes, such as enhanced burning of grasslands and drainage of mires, variously estimated as 1.5 to 3.5 Gt a^{-1}. Recent evaluation suggests a figure about 2.8 Gt a^{-1}, giving a total source term of 8.5 Gt a^{-1}. The rise in atmospheric CO_2 concentration accounts for only 3.2 Gt a^{-1} of this, leaving a substantial 'missing sink' of maybe 5 Gt a^{-1}. There is some dispute about how much of this 5 Gt may be retained in the oceans and how much by terrestrial ecosystems, current estimates for the oceanic sink varying between 0.5 and 3 Gt a^{-1}. The recent estimate by Tans, Fung and Takahashi (1990) puts the oceanic sink at the bottom of this range and thus implicates the terrestrial biota in a major way. Only if the contribution of tropical deforestation and other land use changes was negligible *and* the oceanic sink was at the upper end of this range would it be reasonable to ignore the rôle of terrestrial biota. Thus there is a *prima facie* case for investigating the net carbon assimilating capacity of major terrestrial ecosystems.

Forward modelling of the annual carbon dynamics of terrestrial ecosystems (Fung, Tucker and Prentice, 1987; D'Arrigo, Jacoby and Fung, 1987; Tans *et al.*, 1990), suggests a major rôle for the temperate and boreal forests of the northern hemisphere in sequestering carbon. Of greater significance, because of the greater sensitivity of the method, *inverse modelling* of the fortnightly global flask sampling data of atmospheric CO_2 composition, coupled with a two-dimensional atmospheric transport model, points decisively at a major sink for CO_2 in the middle latitudes of the northern hemisphere (Enting and Mansbridge, 1991). Both the small area of ocean relative to the area of land masses in these latitudes and the available data on CO_2 partial pressures in oceans and atmosphere indicate that the oceans are unlikely to be a sink of the magnitude suggested by inverse modelling in these latitudes. On the other hand, the most extensive terrestrial ecosystems in the northern hemisphere are the temperate (5.03×10^6 km^2) and boreal (7.16×10^6 km^2) forests and possible sinks are their vegetation and soils. Thus the evidence available points strongly to the hypothesis that the temperate and boreal forests in the mid-latitudes of the northern hemisphere comprise a major part of the 'missing sink'.

The area of productive woodland in the UK has been increasing steadily since 1950, generally at about 20 000 to 30 000 hectares per year. It has now reached 2.1 Mha (approaching 10% of the total land area). The larger part of this woodland is Sitka spruce (*Picea sitchensis* (Bong.) Carr.) much of which is now in an exponential phase of growth. The annual production of stem wood was about 4.6 Mm^3 a^{-1} in 1980 and is expected to double by the year 2005. The 1990 figure of 6.0 Mm^3 a^{-1} is equivalent to an annual growth in the total carbon stored of about 2.0 Mt. This figure is likely to reach 2.8 Mt a^{-1} in 2000, as shown in *Table 18.1*.

Table 18.1 ANNUAL SEQUESTRATION OF CARBON BY UK FORESTS BASED ON JARVIS (1981) AND FORESTRY COMMISSION (1990)

	Mt a^{-1}
1980	1.5
1990	2.0
2000	2.8

This projected increase in sequestration is a result of both a continuing increase in area of the forest estate and of increases in the rate of incorporation of carbon. Increases in the latter will result particularly from the use of better, ex-agricultural soils for a proportion of future planting, more intensive fertilization and better silvicultural management. However, the rate of increase in the incorporation of carbon may slow as a result of a swing towards the planting of a higher proportion of slower growing hardwoods in the next century.

The current rapid increase in sequestration of carbon in trees in the UK can largely be attributed to the expansion of the national forest estate over the last forty years and the relatively young age of most stands. It is questionable how long this rate of sequestration of carbon will continue as a steady state, sustainable forest is approached.

The *net* carbon balance of a forest depends on both the amount of carbon dioxide being assimilated by the stands of trees, and on the carbon dioxide emissions associated with decomposition and respiratory processes. The carbon dioxide assimilation depends upon the area and phase of growth of the stands of different species, the latter depending upon the age and physiological activity of the trees, the climate and the weather, and on the silvicultural system practised. The biological carbon dioxide emissions depend upon the application of fertilizers, the respiratory costs of both growth and maintenance and the allocation of carbon to components of the trees (roots and branches, *vis-à-vis* stems), production of litter and the turnover of organic matter in soil, brashings, lop and top etc. The annual net gain of carbon by a stand is the algebraic sum of uptake by the foliage and emissions by these various sources over day and night through the seasons of the year (*Figure 18.1*). While we have a basic understanding of some of the processes involved, our knowledge of others is somewhat sketchy.

There are few published measurements of net CO_2 influxes to coniferous forest (Jarvis, 1985), although there have been a number of studies of the energy budget

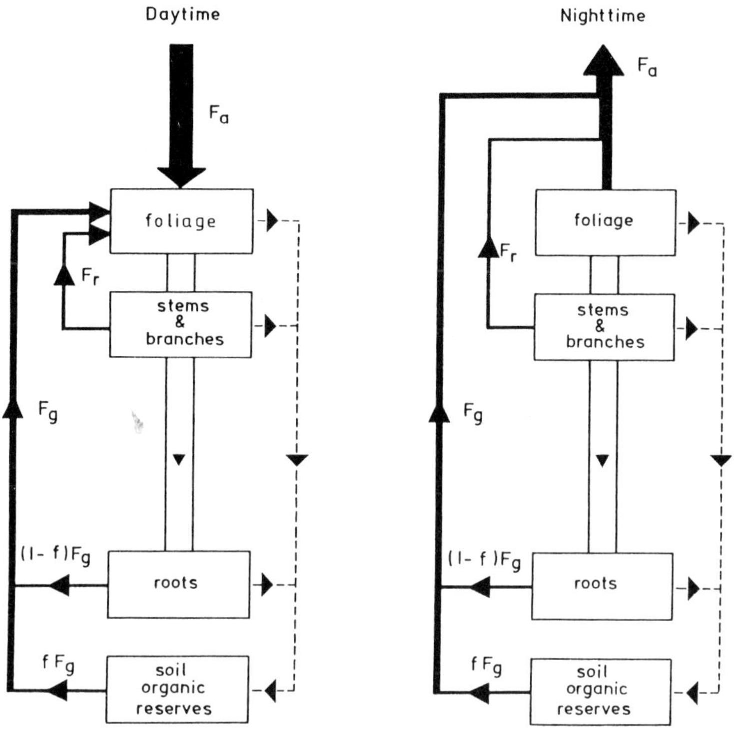

Figure 18.1 Diagram to show daytime and nighttime fluxes of CO_2 in the forest stand. F_a is the flux to or from the atmosphere above; F_g is the efflux from the ground made up of a component from the soil organic matter (fF_g) and a component from the roots ($1 - fF_g$); F_r is the efflux arising from respiration of the stems and branches.
The open pathways (═) indicate liquid phase transport in the phloem, the solid lines (—) show gas phase transport and the dashed lines (– – – –) litter transfers

and evaporation (see Jarvis *et al.*, 1976). Baumgartner (1956) was perhaps the first to publish measurements of CO_2 flux over a young Norway spruce forest in 1956, and although he continued with such measurements, little more was published. Denmead (1969) presented three days of CO_2 fluxes measured over *Pinus radiata* D. Don and some data for *Pinus taeda* L. have emerged from the US Deciduous Forest Biome in the International Biological Programme (IBP) project, and for *Pinus sylvestris* L. from the UK Institute of Hydrology/ Edinburgh University project at Thetford Forest (Jarvis *et al.*, 1976). A search of the literature appearing since that review has revealed only one significant study and that was carried out in the early 1980s, although published more recently (Price and Black, 1990, 1991).

The data presented here for a stand of pole-stage Sitka spruce at Fetteresso in the Forest of Mearns in Scotland, were collected during IBP over a wide variety of conditions in summer months and have not previously been published in detail, although some extracts have appeared. The initial objective was to determine rates of net CO_2 assimilation of the stand and to relate them to environmental variables and season, without interfering appreciably with the foliage, through the use of micrometeorological methods. The current objective is to prepare a carbon balance for stands of Sitka spruce in today's environment, with a view to modelling the likely consequences of the rise in atmospheric CO_2 concentration.

Stand carbon balance

The principal CO_2 fluxes that can readily be distinguished are shown in *Figure 18.1*. F_a is the CO_2 exchange between stand and atmosphere and represents a net gain or loss of CO_2 to the stand. F_r is the respiratory CO_2 efflux from the stems, branches and twigs of the trees. In the daytime this flux of CO_2 is usually recycled through the canopy but at night it represents a net loss from the stand.

F_g is the CO_2 efflux from the ground and consists of a component derived from respiration of the soil micro-organisms and a component derived from the roots. These two components are not readily distinguishable one from the other. In the short term, respiration of the micro-organisms represents a net transfer of CO_2 from soil to trees in daytime, whereas root respiration results in recycling of CO_2 during daytime: both processes result in loss of CO_2 from the stand at night. In the long term, much of the CO_2 resulting from the respiratory activity of the soil micro-organisms is derived from leaf and root litter and thus also represents CO_2 which is recycled in daytime but is lost to the stand at night. However, on many newly afforested sites, there may be *c.* 0.5 m of organic soil derived from marshy vegetation that occupied the site before afforestation. A considerable amount of the CO_2 evolved in soil respiration may be derived from this organic material with the result that there is a net transfer of carbon to the trees from the decaying residues of the vegetation that previously occupied the site, as well as a net loss of soil organic carbon from the site at night.

The fluxes of CO_2 recycled between these sources within the stand and the canopy diminish the influx of CO_2 from the atmosphere during the daytime and represent a loss at night. Their evaluation is, therefore, important to an appraisal of the canopy as an atmospheric CO_2 capture system.

In addition (not shown in *Figure 18.1*), there may be short-term gains or losses in the amount of CO_2 contained within the air column in the stand as a result of the photosynthetic and respiratory activities of the trees and soil, F_s.

The instantaneous CO_2 balance of the tree canopy can be expressed as:

$$F_a + F_g + F_r \pm F_s = F_t \qquad (1)$$

where F_t is the net assimilation rate of the canopy (all symbols for quantities are listed in Appendix 1).

The net gain of CO_2 by the *stand* is F_a. The instantaneous net gain of CO_2 by the trees (F_c) is $F_a + fF_g \pm F_s$, where f is that fraction of F_g which is derived from soil organic matter rather than from the roots. If the mass of soil organic matter is changing over time, fF_g represents a net transfer of C from soil organic matter to trees. When F_g is small relative to F_a (see below) and the reservoir of soil organic matter large as a result of previous accumulation, f may be approximated as unity. Over an entire day F_s is usually zero and while fluctuations in F_s from hour to hour may cause a certain amount of noise in F_a, F_s is generally small in oceanic coniferous forest and can be ignored. The net capture of CO_2 by the trees is then:

$$F_a + F_g = F_c. \qquad (2)$$

If, however, the mass of soil organic matter is constant with time or small, $F_a = F_c$.

The Site

The research site was at a height of 280 m, about 30 km SW of Aberdeen, Scotland (56° 58' N, 02° 24' W). The ground at the site slopes downwards to the NE with a slope of 7° and the surrounding terrain is rolling. A forest road passed through the site *c.* 60 m from the measurement masts. The stand consisted solely of Sitka spruce, planted without ploughing in 1948 at 1.5 m spacings. There were two 0.1 ha Forestry Commission sample plots in the vicinity of the masts, on which the trees had been pruned clean up to a height of 3 m. In 1971–73 there were about 3800 trees ha^{-1} and a totally closed canopy with no ground vegetation whatsoever. In early 1974 the plots and surrounding areas were thinned, half the trees and about a third of the basal area and leaf area being removed (*Table 18.2*). The litter layer was 5–10 cm deep under which there was a peaty soil over clay. The prevailing wind direction was SSW and the fetch in that direction was about 600 m.

Table 18.2 PROPERTIES OF THE STAND BEFORE AND AFTER THINNING AND THE AMOUNTS REMOVED. THE FIGURES IN PARENTHESIS ARE THE AMOUNTS REMOVED AS PERCENTAGES OF THE STAND BEFORE THINNING. DATA PROVIDED BY PROF. H.G. MILLER AND THE FORESTRY COMMISSION

	Before thinning autumn 1973	*After thinning spring 1974*	*Removed in thinnings*
No. of stems	3725	2080	1645 (44)
Dry mass of tree (t ha^{-1})	166.2	113.1	53.1 (32)
Volume (m^3 ha^{-1})	331	226	105 (32)
Basal area (m^2 ha^{-1})	50.6	35.8	17.2 (34)
Mass of needles (t ha^{-1})	14.9	10.4	4.4 (30)
Leaf area index	7.8	5.5	2.3 (30)
Mean tree height (m)	12.0	12.9*	

* Autumn 1975

Methods of CO_2 flux measurements

ABOVE THE CANOPY

The mean hourly influxes of CO_2 from above the Sitka spruce canopy were calculated from measurements of the CO_2 concentration gradient, dC/dz, above the forest, using the transfer coefficient K_E, which was obtained from the Bowen ratio energy-budget technique described in Appendix 2, as follows:

$$F_a = -K_E \rho_a \frac{dC}{dz}. \qquad (3)$$

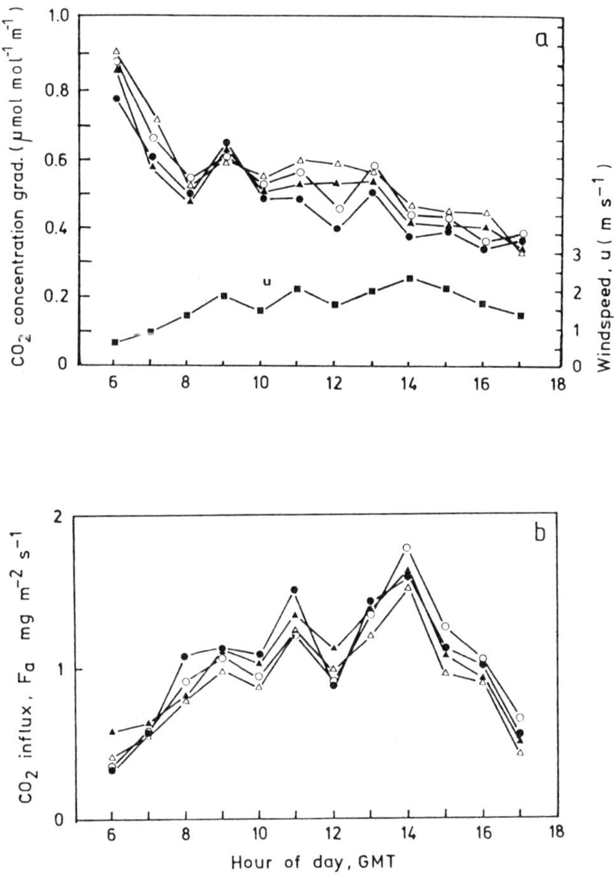

Figure 18.2 (**a**) Gradients of CO_2 concentration given by four gradient measuring systems on 17 August 1973 for each hour. Mast 1 carried system 1 (●) and 2 (○); Mast 2 carried system 3 (▲) and 4 (△). Windspeed (*u*) was measured at 6.5 m above the canopy. (**b**) Corresponding fluxes of CO_2 given by each gradient measuring system

The hourly gradients of CO_2 concentration given by the four replicated gradient measuring systems individually (see Appendix 3) are compared in *Figure 18.2a* for a typical day in 1973 and the corresponding fluxes are shown in *Figure 18.2b*. A similar level of agreement was obtained on about half the measurement days. While measurement of the CO_2 concentration gradient is quite accurate (± 0.025 μmol mol^{-1} per metre, see Appendix 3), the accuracy of the calculated fluxes is much less because estimation of K_E also depends on both the temperature and humidity gradients (see Appendix 2). For this reason agreement between the CO_2 fluxes was generally less good than between the water vapour fluxes.

The level of agreement between the four replicated measurements of F_a shown in *Figure 18.2b* is satisfactory, considering the various possible sources of error in deriving F_a (see Appendix 3). The two masts were 42 m apart and these data show that horizontal variation in the flux measurements was not serious. The overall error from horizontal variation and instrumental errors shown by the data in *Figure 18.2b* amounts to about 20%.

BELOW THE CANOPY

The efflux of CO_2 from the litter and humus layer was measured using a 25 dm^3 steel drum, 70 cm in diameter, open at one end, that was pushed into the ground (Ritchie, 1971). The CO_2 evolved was absorbed into $Ba(OH)_2$, which was later titrated against HCl. The temperature coefficient was determined using the same system in the laboratory, with litter and humus removed from the forest floor, to give the following expression for F_g:

$$F_g = 0.0112 + 0.00224\, T_g \text{ (mg m}^{-2}\text{ s}^{-1}) \quad (4)$$

where T_g is the hourly average ground temperature (°C) at a depth of 50 mm. F_g was only about 3% of typical midday values of F_a (~1 mg m^{-2} s^{-1}).

The CO_2 evolved as respiration by the trees, F_r, was not measured but was estimated by fitting the observations of F_c to a model, as described later.

Results

QUANTUM FLUX DENSITY

While solar irradiance is generally measured in energy budget studies and net radiation is required for the Bowen ratio method, CO_2 assimilation is a quantum dependent process. Both quantum flux density and solar irradiance were measured above the canopy over many months. For about 300 h between May and September, the relation between solar quantum flux density (Q_s, μmol m^{-2} s^{-1}) and solar irradiance (R_s, W m^{-2}) obtained was

$$Q_s = 2.06\, R_s \; (r^2 = 0.99)$$

or

$$1 \text{ μmol m}^{-2}\text{ s}^{-1} \equiv 0.485 \text{ W m}^{-2} \quad (5)$$

The mean fraction of irradiance in the waveband 400 to 700 nm, relative to the total solar irradiance, for summer conditions at Fetteresso was found by Landsberg, Jarvis and Slater (1973) to be 0.53 using two solarimeters (CM3, Kipp & Zonen, Delft), one of which was equipped with a hemispherical RG8 filter (Schott & Gen, Mainz, Germany).

THE RATE OF CO_2 ASSIMILATION BY THE CANOPY

The canopy assimilation rate was highest when R_s was in the range 300–600 W m^{-2} and D was less than 0.6 kPa: in these conditions F_c was about 1.3 mg m^{-2} s^{-1}. Some individual hourly values exceeded 2.0 mg m^{-2} s^{-1}, but it is not clear to what extent this represents error or genuine variability arising from a combination of

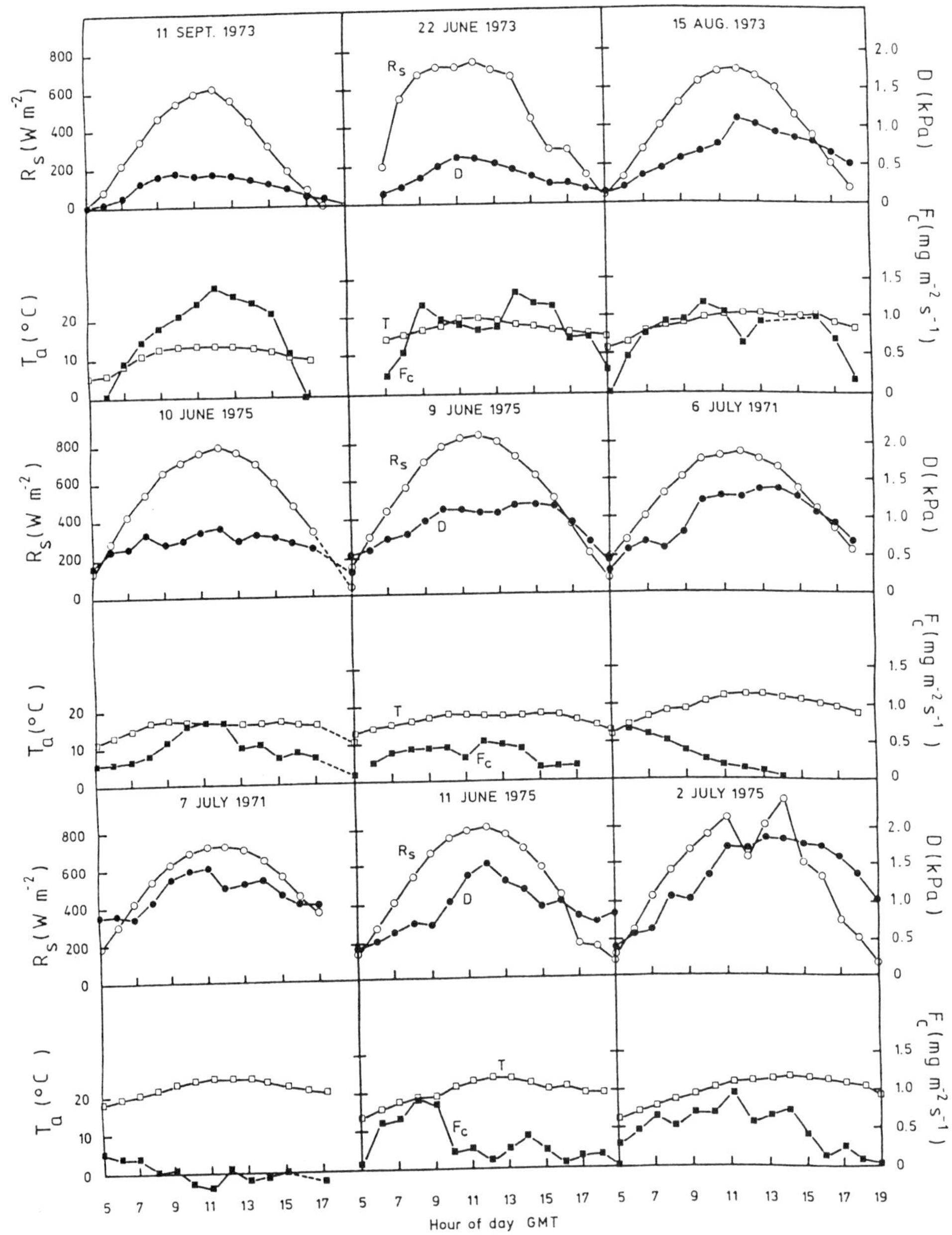

Figure 18.3 Examples of the diurnal course of CO_2 assimilation rate ($F_c = F_a + F_g$) for a number of fine days. R_s is solar irradiance, D is vapour pressure deficit and T_a is canopy air temperature

favourable conditions. Considerably smaller CO_2 fluxes occurred in the hours with the highest solar irradiances (>700 W m^{-2}): these are hours of both high temperature and large vapour pressure deficit.

DIURNAL COURSE OF CO_2 ASSIMILATION

Examples of the diurnal variation in CO_2 assimilation (F_c) on fine days are shown in *Figure 18.3*, together with solar irradiance, temperature, and vapour pressure deficit. The form of the diurnal curves varied widely. Very occasionally, F_c became extremely small, or negative, on particularly hot, dry days. Striking examples occurred on 6th and 7th July 1971 when F_c declined to approximately zero (*Figure 18.3*). Hence, the variations in F_c were investigated in relation to solar irradiance, ratio of diffuse : direct-beam radiation, air temperature, windspeed, wind direction, vapour pressure deficit and time of year. The two most influential non-stomatal, environmental variables were found to be solar irradiance (R_s) and vapour pressure deficit (D).

RANGE OF CANOPY STOMATAL RESISTANCE

Minimal midday values of canopy stomatal resistance, r_s, on fine sunny days were about 50 s m^{-1} at $D < 0.5$ kPa and $T_a \approx 10°C$. At higher temperatures and larger saturation deficits, r_s increased to 150 s m^{-1} and exceeded this value at the maximum temperatures and vapour pressure deficits attained (23°C and 1.8 kPa, respectively) (*Figure 18.3*).

INFLUENCE OF SOLAR IRRADIANCE AND VAPOUR PRESSURE DEFICIT

For 145 hours in 1975, class intervals of D and R_s were used to group the values of F_c, and medians were obtained for each group. Class intervals were unequal in order to obtain a sufficient number of values in each group (between 5 and 20). F_c was then plotted in relation to R_s for different classes of D (*Figure 18.4*). Each curve can be approximated by a rectangular hyperbola. A similar number of hours in 1973 were treated similarly, and showed a similar response to R_s but a less evident response to D because there were very few hours in which D exceeded 0.8 kPa.

Figure 18.4 shows hyperbolic relationships between F_c and solar irradiance or quantum flux density and a negative correlation with vapour pressure deficit. In general, CO_2 capture by the canopy was not saturated by solar radiation. Only at high temperatures and large vapour pressure deficits was saturation approached. The decrease in F_c with increasing vapour pressure deficit was sometimes approximately linear and sometimes closer to exponential.

Because E increased with increase in vapour pressure deficit and F_c decreased with increase in both vapour pressure deficit and temperature, the flux ratio E/F_c (i.e. the hourly water use efficiency) increased greatly from a midday period median value of *c*. 70 to 200 mole of transpired water per mole of net CO_2 assimilated at low vapour pressure deficits ($D < 0.7$ kPa, $T < 16°C$) to over 1000 at the occasional high vapour pressure deficits and temperatures

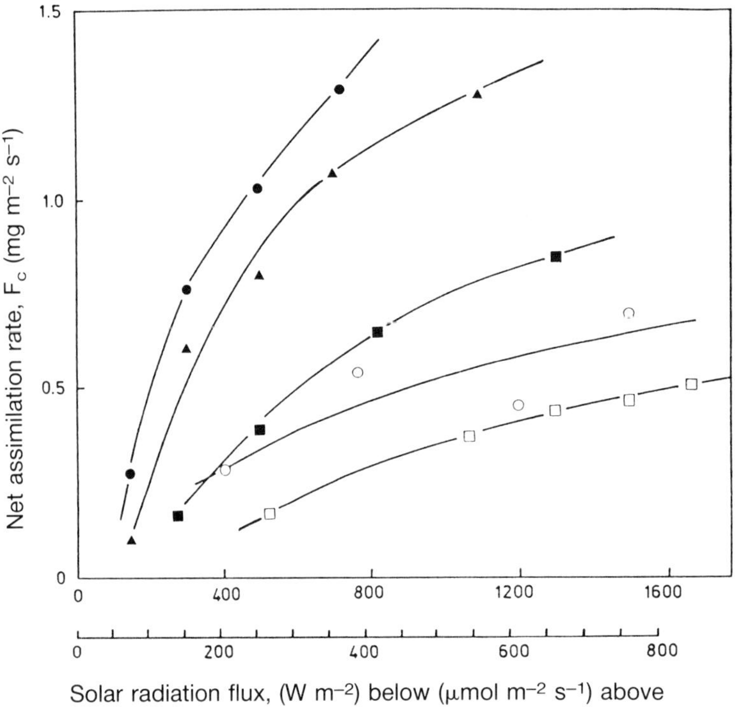

Figure 18.4 The relation between CO_2 assimilation rate, F_c, and solar irradiance, R_s, for different classes of vapour pressure deficit D: 0–0.2 kPa (●), 0.2–0.6 kPa (▲), 0.6–0.8 kPa (■), 0.8–1.0 kPa (○), 1.0–2.0 kPa (□). Points are medians of 6–20 points

($D > 1.5$ kPa, $T > 20$°C) that occur on exceptional days in Scotland (*Figure 18.5*). A representative value of 200 mole H_2O per mole CO_2 for Sitka spruce in this oceanic environment may be compared with values of 500 for deciduous forest and 1000 for irrigated C_3 crops in more continental climates but comparable data are scarce (Jarvis, 1986).

Decrease in F_c with increasing D was to be expected as a consequence of decrease in stomatal conductance at high values of D. However, the fall of F_c to zero on fine sunny days, and even, on some fine days, the occurrence of CO_2 efflux from the canopy (*Figure 18.3*), were larger effects than could be attributed to the increase in canopy resistance alone. In general, sunny days are days of both large vapour pressure deficit and high temperature. It seemed possible, therefore, that on sunny days, increase in respiration by the stems and branches of the trees, i.e. in F_r, and recycling of this respiratory CO_2 was sufficient to reduce F_a still further. To estimate F_r as a function of temperature, measured values of F_c were fitted to a simple model to test this hypothesis.

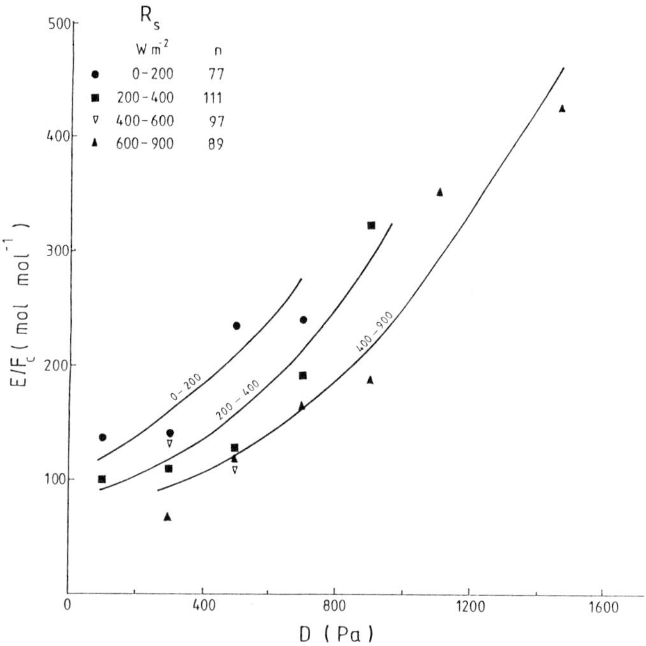

Figure 18.5 The relation between median hourly values of water use efficiency, E/F_c, expressed as a molar ratio, and vapour pressure deficit in three classes of solar irradiance. The curves are drawn by eye (redrawn from Jarvis, 1986)

MODEL OF RESPONSE OF CO_2 EXCHANGE TO SOLAR IRRADIANCE, TEMPERATURE AND STOMATAL RESISTANCE

An empirical model was required which would take into account the effects of irradiance and vapour pressure deficit on F_c and the effect of temperature on F_r. It was assumed that vapour pressure deficit would affect F_c through increase in stomatal resistance at large vapour pressure deficits.

From Equations 1 and 2 (ignoring F_s):

$$F_c = F_t - F_r. \tag{6}$$

The following 'big leaf' model of canopy CO_2 assimilation was assumed (Thornley, 1976; Jarvis, Miranda and Muetzelfeldt, 1985):

$$F_c = \frac{\alpha Q_s C_i \rho_c / r_m}{\alpha Q_s + C_i \rho_c / r_m} - F_r \tag{7}$$

where

$$C_i = C_a - F_c\, r'_s / \rho_c \tag{8}$$

and

$$F_r = F_{r0} \exp(\eta T_a). \tag{9}$$

(see Appendix 1 for definition of symbols).

This model assumes the usual rectangular hyperbolic relationship between canopy assimilation and quantum flux density, a linear dependence on CO_2 concentration and a mesophyll resistance in series with the stomatal resistance, and ignores a possible CO_2 compensation concentration. Any deviation from linearity of the response between F_c and C_a, or the presence of a CO_2 compensation concentration of *c.* 100 μmol mol^{-1}, is likely to be masked by variability in the data obtained using the Bowen ratio method.

Thus the model is based on the following four measured variables: incident quantum flux density (Q_s), air temperature at the height of maximum foliage density (T_a), stomatal resistance of the canopy for CO_2 (r'_s) and the influxes of carbon dioxide from the ground (F_g) and the atmosphere (F_a). Canopy stomatal resistance was obtained for each hour from the measured Bowen ratios and rearrangement of the Penman–Monteith equation (see Appendix 2). The model also contains the following three parameters: the mesophyll resistance (r_m), the rate of respiration by non-photosynthetic tissues at 0°C (F_{r0}) and the initial slope of the curve relating F_c to solar irradiance, i.e. the apparent photochemical efficiency of the canopy (α). The ambient CO_2 concentration (C_a) was taken as a constant value of 514 mg kg^{-1} (≡ 330 μmol mol^{-1} at 15°C, appropriate to 1972) and the temperature coefficient, η, was taken as 0.0693, i.e. F_r was assumed to double for every 10°C rise in temperature (see Jarvis and Leverenz, 1983).

This model was fitted to 280 sets of hourly data, after they had been grouped into classes and the medians found as shown in *Figure 18.4*, using a non-linear least squares regression procedure. The model accounted for no more than 50% of the variation in the data. Part of the reason for this is almost certainly the naturally occurring correlation between air temperature and vapour pressure deficit (T_a, D). As a result, the data for r'_s and T_a have a degree of interdependence and are not distributed uniformly over the variable space (r'_s, T_a). It is also likely that r_m was not constant over the range of temperature and vapour pressure deficit covered by the data and may also have changed seasonally as well. None the less, this kind of model has been found to fit data obtained in assimilation chambers extremely well (e.g. Charles-Edwards and Ludwig, 1974; Reed *et al.*, 1976; Watson, Landsberg and Thorpe, 1978), so the variation unaccounted for here may also derive in part from the accumulation of errors in the technique.

The parameters found are listed in *Table 18.3*. The estimate of F_{r0} is of particular interest since it is clear from Equation 7 that if F_r is sufficiently large, F_c may be reduced to zero.

Table 18.3 PARAMETERS OF THE MODEL RELATING CO_2 EXCHANGE BY THE CANOPY TO IRRADIANCE, AIR TEMPERATURE AND CANOPY STOMATAL RESISTANCE. FOR THE EXPRESSION OF α THE ENERGY FLUX DENSITY HAS BEEN CONVERTED TO THE FLUX DENSITY OF MOLE QUANTA USING EQUATION 5

Canopy parameters			
α	apparent photochemical efficiency	0.051	mol CO_2 per mol quanta
r_m	bulk mesophyll resistance	198	s m^{-1}
F_{r0}	rate of respiration at 0°C	0.41	mg m^{-2} s^{-1}

Discussion

Comparison of these values of F_c with results from other coniferous forest stands is limited by the shortage of reports of similar work. Unpublished data for *Pinus taeda*, *Picea abies* and *Pinus sylvestris* (Jarvis *et al.*, 1976) are generally also in the range 0.6–1.3 mg m^{-2} s^{-1}. Somewhat higher rates for *Pinus radiata* (1.0–3.5 mg m^2 s^{-1}) were obtained by Denmead (1969) in springtime (near Canberra in Australia). More recently, remarkedly similar time courses were obtained by Price and Black (1990, 1991) for a similar-aged stand of Douglas-fir (*Pseudotsuga menziesii*) on Vancouver Island, British Columbia. Their site was poor, the stand of low productivity, the leaf area index much less (~3.5) and maximum rate of CO_2 influx was only *c.* 0.5 mg m^{-2} s^{-1}. At their site F_c declined to zero in a very similar manner on days of high vapour pressure deficit (> 1.8 kPa) and high temperature (> 20°C).

Reduction in F_c in warm sunny conditions is the result of two separate processes: increase in stomatal resistance at high values of D and increase in rate of respiration by the stems, branches and twigs of the trees at high temperatures. Since vapour pressure deficit and temperature in the forest are strongly correlated, it is difficult to distinguish the effects of each individually on F_c. However, by inverting the model and putting in the derived parameters, the model can be used to evaluate the effects of quantum flux density, air temperature and canopy stomatal resistance on predicted values of $\hat{F}_c$. *Figure 18.6a* shows the idealized response of $\hat{F}_c$ to quantum flux density at different canopy stomatal resistances. The curves are generally similar to those in *Figure 18.4*. The fall in $\hat{F}_c$ over the lower quantum flux densities with increasing temperature is shown in *Figure 18.6b*. It is clear from the value of F_{r0} (Equation 9 and *Table 18.3*) that a substantial increase in respiration rates by the stems and branches at the higher temperatures accounts to a significant degree for the observations. Finally, *Figure 18.6c* shows the dependence of $\hat{F}_c$ on canopy stomatal conductance ($1/r_s$) at different temperatures. The shape of the curves is hyperbolic because of the presence of r_m in series with r'_s. The curves show clearly that for this reason F_c is not strongly dependent on canopy stomatal conductance until g_s falls below about 0.8 cm s^{-1} (i.e. r_s exceeds 125 s m^{-1}), but that the effect of respiration at high temperatures on $\hat{F}_c$ is considerable.

The model represents the data well. In saturating irradiance, the model predicts a canopy CO_2 assimilation flux density of *c.* 1.4 mg m^{-2} s^{-1} at a temperature of 10°C and a canopy stomatal resistance of 50 s m^{-1} (i.e. a vapour pressure deficit of <0.5 kPa). At 24°C and r_s = 200 s m^{-1}, $\hat{F}_c$ = 0.2 mg m^{-2} s^{-1}. Allocation of the reduction in F_c to vapour pressure deficit or temperature depends on the 'route' taken (increase in r_s followed by increase in respiration, or increase in respiration followed by increase in r_s). *Figure 18.6b* shows that in both cases the effect of temperature exceeds the effect of vapour pressure deficit and that approximately half to three quarters of the reduction in F_c is the result of the rise in temperature and quarter to half is the result of increase in vapour pressure deficit and consequent increase in r'_s.

Price and Black (1990, 1991) also concluded that the reduction in CO_2 influx to their stands of Douglas-fir on warm sunny days was caused both by stomatal closure, resulting from the high vapour pressure deficits, and by high respiration rates, resulting from the high temperatures. Although they made no quantitative evaluation of the relative contributions of these two processes to the fall in F_c, they

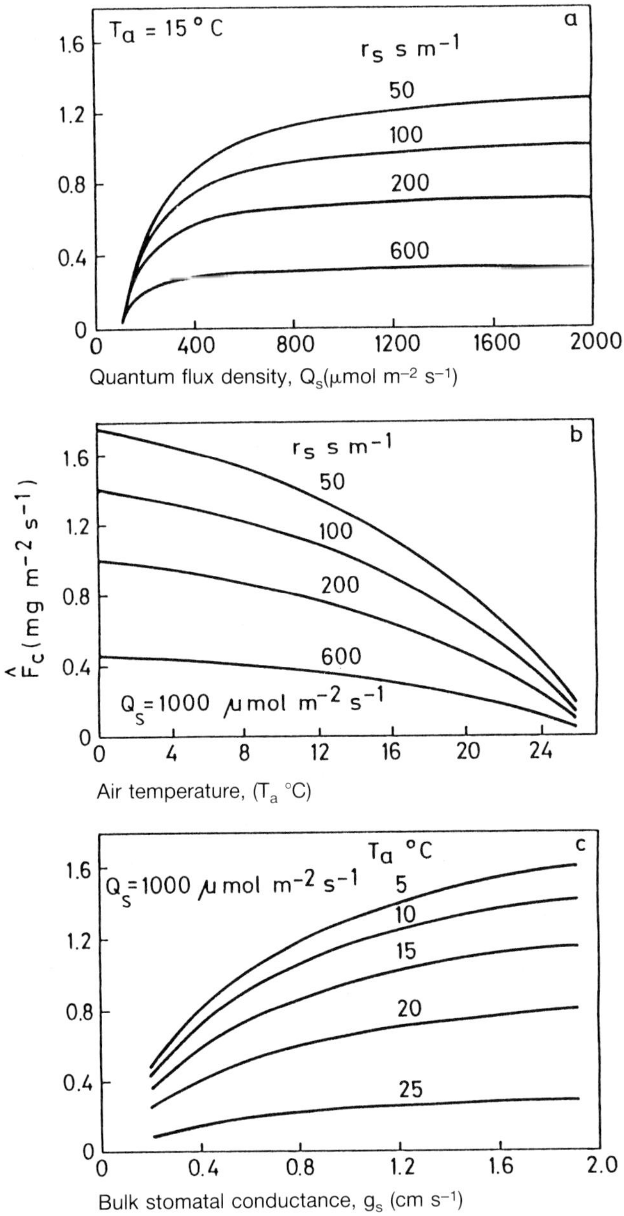

Figure 18.6 The relation between predicted CO_2 assimilation rate, $\hat{F}_c$, and (**a**) quantum flux density, Q_s, at four different canopy stomatal resistances, r_s, (**b**) air temperature, T_a, at four different canopy stomatal resistances, r_s, and (**c**) canopy stomatal conductance, g_s, at five different temperatures. The curves have been calculated using the model given in the text and parameters derived from the observations using the model

too concluded that the major part of the reduction was attributable to increase in stand respiration.

These predictions from the model are the expected results of scaling up leaf and tree assimilation and respiration properties. In general, mesophyll resistance of spruce *needles* is two to four times larger than stomatal resistance for CO_2, similar to the factor found here for canopy properties by optimization, and consequently increases in stomatal resistance would not be expected to reduce CO_2 assimilation proportionately, until the increases were large, e.g. a three-fold increase in r'_s from 96 to 288 s m^{-1} would be expected to decrease F_t from 1.75 to 1.06 mg m^2 s^{-1}, i.e. by only 40%.

Combining parameters for respiration of stems and large branches (but not twigs) from the literature (see Jarvis and Leverenz, 1983) with a *total* wood surface area index of 4.9 for the stand (Norman and Jarvis, 1975), leads to estimates of F_r at 23°C of between 0.5 and 1.1 mg m^{-2} s^{-1}, depending on the parameters taken; and if respiration of twigs were taken into account explicitly, these values would be up to 30% higher. Clearly, as $(F_r + F_g)$ approaches the assimilatory capacity of the needles, $F_a \rightarrow 0$, as on warm, dry days in eastern Scotland.

Conclusions

Capture of CO_2 by pole-stage stands of Sitka spruce was maximum on cool, humid, sunny days ($T_a \sim 10°C$, $D < 0.5$ kPa, $R_s > 350$ W m^{-2} or $Q_s > 700$ μmol m^{-2} s^{-1}). The median CO_2 influx in these conditions is *c.* 1.3 mg m^{-2} s^{-1}, but higher rates (> 2 mg m^{-2} s^{-1}) can occur at higher irradiance, provided that air temperature and vapour pressure deficit do not also increase. These rates of CO_2 capture are similar to rates achieved by C_3 agricultural crops in similar environments (Jarvis, 1981).

On warm sunny days, rates of CO_2 influx are much less and may go to zero or become negative at air temperatures > 24°C and vapour pressure deficits > 1.5 kPa, as a result of both increase in respiration by stems, branches and twigs, and stomatal closure. A model was used to show that recycling of CO_2 within the stand at high temperatures appears to have a larger effect in reducing CO_2 capture than stomatal closure. Thus, more effort should be put into measuring respiratory CO_2 fluxes *within* forest stands in relation to temperature in future studies. By contrast with annual agricultural crops, it is likely that as forest stands age net CO_2 capture falls to zero and eventually a net carbon loss may result (Jarvis and Leverenz, 1983). However, there are at present no studies of CO_2 flux over old-growth stands with which to make comparisons.

Acknowledgements

This work was carried out as part of a larger project when the author was at the Department of Botany, University of Aberdeen. The following participated in the project at various times: Dr G.B. James, Dr J.J. Landsberg, Dr S.J. Tajchman, Prof. H.G. Miller, Prof. J. M. Norman, Mr M.B. Slater, Mr B. Davidson, Mr M. Durrant and Mrs H. Talbot. Financial support was provided by the Natural Environment Research Council and the site was made available by the Conservator, Forestry Commission, East Scotland Conservancy. I thank all the above and many others for their help.

References

Baumgartner, A. (1956) Untersuchungen über den Wärmeund Wasserhaushalt eines jungen Waldes. *Berichte Deutschen Wetterdienstes*, **5**, 28, 4–53

Bradley, E.F. (1969) A small, sensitive anemometer system for agricultural meteorology. *Agricultural Meteorology*, **6**, 185–193

Cellier, P. and Brunet, Y. (1992) Flux-gradient relationships above tall plant canopies. *Agricultural and Forest Meteorology*, **58**, 93–117

Charles-Edwards, D.A. and Ludwig, J. (1974) A model for leaf photosynthesis by C_3 plant species. *Annals of Botany (London)*, **38**, 921–930

D'Arrigo, R., Jacoby, G.C. and Fung, I.Y. (1987) Boreal forests and atmosphere-biosphere exchange of carbon dioxide. *Nature, London*, **329**, 321–323

Denmead, O.T. (1969) Comparative micrometeorology of a wheat field and a forest of *Pinus radiata*. *Agricultural Meteorology*, **6**, 357–371

Denmead, O.T. and Bradley, E.F. (1985) Flux-gradient relationships in a forest canopy. In *The Forest-Atmosphere Interaction*, pp 421–422 Edited by B.A. Hutchison and B.B. Hicks, Dordrecht: Reidel

Enting, I.G. and Mansbridge, J.V. (1991) Latitudinal distribution of sources and sinks of CO_2: results of an inversion study. *Tellus* **43B**, 156–170

Forestry Commission (1990). *Forestry: Facts and Figures, 1989–1990*. Edinburgh: Forestry Commission

Fung, I.Y., Tucker, C.J. and Prentice, K.C. (1987) Application of advanced very high resolution radiometer vegetation index to study atmosphere-biosphere exchange of CO_2. *Journal of Geophysical Research*, **92**, 2999–3015

James, G.B. (1977). *Exchanges of Mass and Energy in Sitka Spruce*. Ph.D. Thesis. University of Aberdeen

Jarvis, P.G. (1981) Production efficiency of coniferous forest in the UK. In *Physiological Processes Limiting Plant Productivity*, pp 81–107 Edited by C.B. Johnson. London: Butterworth Scientific Publications

Jarvis, P.G. (1985) Transpiration and assimilation of tree and agricultural crops: the 'omega factor'. In *Trees as Crop Plants* pp 460–480 Edited by M.G.R. Cannell and J.E. Jackson. Abbots Ripton: Institute of Terrestrial Ecology

Jarvis, P.G. (1986). Coupling of carbon and water interactions in forest stands. *Tree Physiology*, **2**, 347–368

Jarvis, P.G. and Dewar, R.C. (1993) Forests in the global carbon balance from stand to region. In *Scaling Physiological Processes: Leaf to Globe* pp 191–221 Edited by J.R. Ehleringer and C.B. Field. New York: Academic Press

Jarvis, P.G., James, G.B. and Landsberg, J.J. (1976) Coniferous Forest. In *Vegetation and the Atmosphere* Vol. 2, pp 171–240 Edited by J.L. Monteith. London: Academic Press

Jarvis, P.G. and Leverenz, J. (1983) Productivity of temperate, deciduous and evergreen forests. In *Encyclopedia of Plant Physiology, New Series, Physiological Plant Ecology IV* Vol. 12D, pp 233–280 Edited by O.L.Lange, P.S. Nobel, C.B.Osmond and H. Zeigler. Berlin: Springer-Verlag

Jarvis, P.G., Miranda, H.S. and Muetzelfeldt, R.I. (1985) Modelling canopy exchanges of water vapour and carbon dioxide in coniferous forest plantations. In *The Forest-Atmosphere Interactions*, pp 521– 542 Edited by B.A. Hutchison and B.B. Hicks. Dordrecht: D. Reidel

Jones, J.I.P. (1964). A portable sensitive anemometer with proportional d.c. output and a matching wind velocity-component resolver. *Journal of Scientific Instruments* **42**, 414–417

Landsberg, J.J. and James, G.B. (1971) Wind profiles in plant canopies: studies on an analytical model. *Journal of Applied Ecology*, **8**, 729–741

Landsberg, J.J. and Jarvis, P.G. (1973) A numerical investigation of the momentum balance of a spruce forest. *Journal of Applied Ecology*, **10**, 645–655

Landsberg, J.J., Jarvis, P.G. and Slater, M.B. (1973). The radiation regime of a spruce forest. In *Plant Response to Climatic Factors*, Uppsala Symposium 1970, 411–418 Paris: UNESCO

Landsberg, J.J. and Thom, A.S. (1971) Aerodynamic properties of a plant complex structure. *Quarterly Journal of the Royal Meteorological Society*, **97**, 565–570

Monteith, J.L. (1965). Evaporation and environment. In *The State and Movement of Water in Living Organisms*, pp 205–234 Edited by G.E. Fogg. Symposia of the Society for Experimental Biology No 19

Monteith, J.L. (1975) *Principles of Environmental Physics*. London: Edward Arnold

Norman, J.M. and Jarvis, P.G. (1974) Photosynthesis in Sitka spruce. III. Measurements of canopy structure and interception of radiation. *Journal of Applied Ecology*, **11**, 375–398

Norman, J.M. and Jarvis, P.G. (1975) Photosynthesis in Sitka spruce. V. Radiation penetration theory and a test case. *Journal of Applied Ecology*, **12**, 839–879

Price, D.T. and Black, T.A. (1990) Effects of short-term variation in weather on diurnal canopy CO_2 flux and evapotranspiration of a juvenile Douglas-fir stand. *Agricultural and Forest Meteorology*, **50**, 139–158

Price, D.T. and Black, T.A. (1991) Effects of summertime changes in weather and root-zone soil water storage on canopy CO_2 flux and evapotranspiration of two juvenile Douglas-fir stands. *Agricultural and Forest Meteorology*, **53**, 303–323

Reed, K.L., Hamerley, E.R., Dinger, B.E. and Jarvis, P.G. (1976) An analytical model for field measurements of photosynthesis. *Journal of Applied Ecology*, **13**, 925–942

Ritchie, J. (1971) *An Investigation into the Carbon Dioxide Production of Spruce Litter*. Honours thesis, Aberdeen University Botany Department

Stewart, J.B. and Thom, A.S. (1973) Energy budgets in pine forest. *Quarterly Journal of the Royal Meteorological Society*, **99**, 154–170

Tanner, C.B. (1968) Evaporation of water from plants and soil. In *Water Deficits and Plant Growth. Vol. I. Development, Control and Measurement* pp 71–106 Edited by T.T. Kozlowski. Academic Press: New York

Tans, P.P., Fung, I.Y. and Takahashi, T. (1990) Observational constraints on the global atmospheric CO_2 budget. *Science*, **247**, 1431–1438

Thom, A.S. (1975) Momentum, mass and heat exchange of plant communities. In *Vegetation and the Atmosphere* Vol. 1, pp 57–109 Edited by J.L. Monteith. London: Academic Press

Thom, A.S., Stewart, J.B., Oliver, H.R. and Gash, J.H.C. (1975) Comparison of aerodynamic and energy budget estimates of fluxes over a pine forest. *Quarterly Journal of the Royal Meteorological Society*, **101**, 93–105

Thornley, J.H.M. (1976) *Mathematical Models in Plant Physiology*. London: Academic Press, pp 318

Watson, R.L., Landsberg, J.J. and Thorpe, M.R. (1978) Photosynthetic characteristics of the leaves of 'Golden Delicious' apple trees. *Plant, Cell and Environment*, **1**, 51–58
Webb, E.K., Pearman, G.I. and Leuning, R. (1980) Correction of flux measurements for density effects due to heat and water vapour transfer. *Quarterly Journal of the Royal Meterological Society*, **106**, 85–100

Appendix 1 List of Symbols

A	Available energy for sensible and latent heat transfer (W m^{-2})
C	Specific concentration of CO_2 (kg CO_2 per kg air). Subscript a at the height of maximum foliage density; i mean intercellular space
D	Water vapour pressure saturation deficit at the height of maximum foliage density (kPa)
d	Zero plane displacement for the wind profile (m)
E	Water vapour flux density (kg m^{-2} m^{-1})
e	Vapour pressure of water in the air (kPa)
F	Carbon dioxide flux (mg m^{-2} s^{-1}). Subscripts a from atmosphere, g from ground, r from stems and branches, s from storage in the stand air column, c from ground and atmosphere, t from all sources, 0 at zero °C
f	Fraction of CO_2 efflux from the ground that derives from microbial respiration in the soil (i.e. excluding roots)
G	Ground heat flux density (W m^{-2})
g_s	Bulk stomatal conductance of canopy referred to unit ground area ($= 1/r_s$) (m s^{-1})
H	Sensible heat flux density (W m^{-2})
K_E	Transfer coefficient for CO_2 obtained from energy budget method (m^2 s^{-1})
K_M	Transfer coefficient for momentum (m^2 s^{-1})
K_V, K_H, K_C	Transfer coefficients for water vapour, heat, and CO_2 (m^2 s^{-1})
k	von Karman's constant, 0.41
m	Mass of wood per unit volume of canopy (kg m^{-3})
Q_s	Quantum flux density above the canopy (μmol m^{-2} s^{-1})
q	Specific humidity (kg H_2O per kg air)
R_n	Net radiation flux density above the canopy (W m^{-2})
R_S	Solar irradiance above the canopy (W m^{-2})
r_a	Boundary layer resistance of canopy (s m^{-1}). Subscripts H sensible heat; V water vapour
r_i	Climatological resistance, $\rho_a c_p D/(\gamma A)$, (s m^{-1})
r_m	Canopy mesophyll resistance for CO_2 referred to unit ground area (s m^{-1})
r_s	Canopy stomatal resistance for water vapour referred to unit ground area (s m^{-1})
r'_s	Canopy stomatal resistance for CO_2 referred to unit ground area (s m^{-1})
S	Storage rate for sensible and latent heat in air and biomass (W m^{-2})
s	The rate of change of saturated vapour pressure of air with respect to temperature of air (kPa °C^{-1})

T	Temperature (°C). Subscripts a for air at the height of maximum foliage density, g for ground at a depth of 50 mm
T_w	Wood temperature (°C)
t	Time
u	Windspeed (m s^{-1}). Subscript h at canopy top
u_*	Friction velocity (m s^{-1})
z	Height above ground (m)
z_0	Roughness length for momentum (m)
z_H	Roughness length for heat (m)
α	Apparent photochemical efficiency of CO_2 fixation (dF_c/dQ_s at $Q_s \rightarrow O$)
β	Bowen ratio, $H/\lambda E$
Γ	Dry adiabatic lapse rate (0.0098 °C m^{-1})
γ	Psychrometric constant (0.066 kPa $°C^{-1}$)
θ	Potential temperature ($T_a + \Gamma_z$)
λ	Latent heat of vaporization of water (2.46×10^6 J kg^{-1})
μ	Energy of fixation of CO_2 (10^4 J kg^{-1})
ρ	Density. Subscript a of air (1.22 kg m^{-3} at 15°C); c of CO_2 (1.87 kg m^{-3} at 15°C)
τ	Momentum flux (N m^{-2})
η	Temperature coefficient of wood respiration (0.0693)

Appendix 2 THEORY

TRANSFER EQUATIONS

Both the Bowen ratio/energy balance method and the momentum balance method depend upon the application of turbulent transfer equations within the boundary layer above an extensive uniform area of canopy (Thom, 1975). It is assumed that temperature, humidity, and CO_2 concentration are functions of height only, so that horizontal advective fluxes can be ignored. The basic one-dimensional transfer equations are:

for water vapour,
$$E = -\rho_a K_V \frac{dq}{dz}; \tag{1A}$$

for sensible heat,
$$H = -\rho_a c_p K_H \frac{d\theta}{dz}; \tag{2A}$$

for carbon dioxide,
$$F_a = -\rho_a K_C \frac{dC}{dz}; \tag{3A}$$

and for momentum,
$$\tau = \rho_a K_M \frac{du}{dz}; \tag{4A}$$

where q is specific humidity and C is specific CO_2 concentration. All symbols are defined in Appendix I.

The analogous nature of the above equations is apparent since $\rho_a q$, $\rho_a c_p \theta$, $\rho_a C$ and $\rho_a u$ are effectively concentrations of water vapour, sensible heat, carbon dioxide, and momentum, respectively. The exchange coefficients K_V, K_H, K_C and K_M depend on the degree of turbulence and are constantly changing. Conventionally K_V, K_C and K_H are assumed equal (Thom, 1975), whereas K_M may be considerably different (Thom *et al.* 1975). The equations should only be used for average fluxes and average profiles over an appropriate period of time, such as the hour used here.

THE ENERGY BUDGET

Considering unit volume of canopy at an arbitrary height z, conservation of energy requires that:

$$\text{div}\, R_n + \text{div}\, H + \lambda \text{div}\, E - \mu\, \text{div}\, F + \rho_a c_p \frac{d\theta}{dt} + m\frac{dT_w}{dt} + \lambda \rho_a \frac{dq}{dt} = 0, \quad (5A)$$

where R_n, H, E and F are vector fluxes of net radiation, sensible heat, water vapour and CO_2 (Tanner, 1968).

Expanding Equation 5A in three dimensions, neglecting both horizontal components of the divergence terms (on the assumption of an extensive, horizontal, uniform canopy), integrating with respect to height from ground level to an arbitrary reference level (z) above the canopy and applying conservation of energy at the ground surface leads to (James, 1977)

$$R_n(z) + \text{H}(z) + \lambda\text{E}(z) - \mu F_a(z) + G + S(z) = 0,$$

where $S(z)$ is the summed storage of sensible and latent heat in trees and air below the height z.

Following convention and reversing the sign for R_n to make it positive for net radiation *input* to the canopy (Monteith, 1975) gives:

$$R_n = H + \lambda E - \mu F_a + S + G. \quad (6A)$$

Combining Equations 1A, 2A and 3A with Equation 6A and taking $K_H = K_V = K_C$, leads to:

$$K_E = \frac{R_n - G - S}{-\rho_a c_p \dfrac{d\theta}{dz} - \rho_a \lambda \dfrac{dq}{dz} + \rho_a \mu \dfrac{dC}{dz}}, \quad (7A)$$

where K_E is the 'energy budget' K that in practice replaces K_H, K_V and K_C in calculation of the fluxes.

The stomatal resistance of the canopy (r_s) was obtained by substitution into the Penman–Monteith combination equation (Thom, 1975; Jarvis *et al.*, 1976) to give:

$$r_s = [(s/\gamma)\beta - 1]\, r_{aV} + (\beta + 1) r_i, \quad (8A)$$

where s is the rate of change of the saturated vapour pressure of water with respect to temperature and r_i is the isothermal or climatological resistance (Monteith, 1965) defined by:

$$r_i = (\rho_a c_p/\gamma)(D/A). \qquad (9A)$$

where D is saturation deficit and following Stewart and Thom (1973), thc 'available energy' flux (A) is defined as:

$$A = R_n - G - S + \mu F_a = H + \lambda E. \qquad (10A)$$

The Bowen ratio (β) was found from simultaneous measurements of the vertical difference in air temperature ($\triangle T_a$) and water vapour pressure ($\triangle e$) over the same, 4 m, height intervals above the canopy, assuming $K_H = K_V$ as:

$$\beta = \gamma \triangle T_a / \triangle e \qquad (11A)$$

where γ is the psychrometric constant (Monteith, 1975).

The canopy boundary layer resistance for water vapour transfer r_{aV} was estimated in two ways: from wind profile parameters and from wind tunnel measurements. In the first method (r_{aV}) from the surfaces of the trees to the reference height, z, above was estimated as:

$$r_{aV} = r_{aM} + r_b, \qquad (12A)$$

where r_{aM} (= $u(z)/u^2_*$) is the aerodynamic resistance for momentum and is *c.* $10u_h^{-1}$ (Jarvis *et al.*, 1976). r_b is the excess resistance for heat or water vapour resulting from form drag at the surface (Thom, 1975), and is given by:

$$r_b = \ln(z_0/z_H)/(ku_*), \qquad (13A)$$

where z_0 and z_H are the roughness lengths for momentum and heat, respectively, and k is von Karman's constant (0.41). The ratio $\ln(z_0/z_H)$ was taken as 1.0 for the spruce plantation (Thom, 1975). The eddy velocity, u_*, was found from conventional analysis of wind profiles using:

$$u_* = ku(z)/[\ln(z - d)/z_0] \qquad (14A)$$

with values of $u(z)$, z_0 and the zero-plane displacement (d) estimated from wind profiles in near neutral conditions (Landsberg and Jarvis, 1973). The mean value of d derived from 66 mean hourly adiabatic profiles was 0.84 h where h is height of stand, and the mean value of z_0 was 0.03h (Jarvis e*t al.*, 1976). In the second method, r_{aV} was obtained by integrating the boundary layer resistance of shoots measured in a wind tunnel (Landsberg and Thom, 1971) with respect to shoot density and leaf area over the canopy to obtain (Landsberg and James, 1971):

$$r_{aV} = 2.3\ u_h^{-0.58}, \qquad (15A)$$

where u_h is windspeed at canopy top.

Both methods gave estimates of r_{aV} in the range 5–10 s m^{-1}. Because the boundary layer resistance of forest canopies is low (Jarvis *et al.*, 1976), Equation 8A for r_s is not very sensitive to the precise value of r_{aV}.

Appendix 3 Instrumentation

Gradients of windspeed, temperature, humidity and CO_2 are very small and extreme accuracy of the sensors, an order of magnitude higher than over field crops, is necessary (Jarvis *et al.*, 1976).

The micrometeorological data collection system comprised a 100-channel digital datalogger, environmental measuring instruments, two masts on which the sensors and sampling units were mounted, instrument huts at the foot of the masts and an instrument caravan.

The data were collected every 10 min by a datalogger and averaged over an hour for calculation of the fluxes of radiation, CO_2, sensible and latent heat. Other variables were also expressed as mean hourly values. For the period May to September 1973 and 1975, *c.* 100 and 150 sets of mean hourly values have been analysed for the dry canopy, respectively.

GRADIENT MEASUREMENTS

The sensors and sampling units to measure temperature, humidity and CO_2 gradients were mounted above the canopy on two 21 m tall masts (BICC Co Ltd. Thornton Heath, UK), 42 m apart. Each mast carried a pair of 10 m long horizontal booms at right angles to the prevailing wind, 0.5 m and 4.5 m above treetop height. Each pair of booms carried two gradient measuring systems (GMS), one to each side of a mast. Each GMS consisted of four air sampling units at each height, each of which also contained a ventilated, shielded, platinum resistance thermometer. Each group of four sampling units was connected together to provide spatial averaging. Gradients were measured differentially to avoid time lag effects.

Air temperature and differences between air temperatures at the two heights were measured twice in each 10 min scan by matched platinum in-glass resistance thermometers (type Hartglas P5, Degussa, Janau, Germany), embedded in wax, enclosed in stainless steel sheaths and exposed in shielded, ventilated units, with absolute or differential, four lead, Kelvin bridges (types E400A or E400C, Rosemount Engineering Co. Ltd., Bognor Regis, UK). The sets of thermometers were checked for zero and sensitivity regularly by placing them in holes in an insulated aluminium block at known temperatures.

The four air intakes of each GMS at each level were connected to a common butyl rubber tube (5 mm i.d.) that carried a continuous flow of air down to the foot of each mast, through two 10 dm^3 glass mixing vessels to dampen fluctuations in composition, and then to the humidity and CO_2 analysis instruments. The sample tubes were heated throughout with electrical heating tapes (GP, Hotfoil Ltd., Wolverhampton, UK), thermally insulated with sponge rubber (13 HO 42 Armaflex, Armstrong Cork Co. Ltd., Uxbridge, UK) to prevent condensation in the tubes, and encased in plastic pipe for protection.

Air samples from the top and bottom levels of each GMS were switched together with solenoid valves through the CO_2 or humidity analysis instruments for one and two minutes, respectively, so that six humidity and 12 CO_2 differential measurements were made each hour.

The four humidity gradients were measured sequentially, together with an instrument zero, with a pair of dewpoint meters (880, EG & G Inc., Waltham, USA) connected differentially, once every 10 minute scan. The differential dew point meter was calibrated using a standard humidity generator comprising copper condensing coils in a precision water bath at a range of temperatures.

The four carbon dioxide gradients were measured sequentially, together with an instrument zero and a calibration gas, using a differential infra-red gas analyser (URAS II, Hartmann & Braun (UK) Ltd., Northampton, UK) with the addition of narrow bandpass interference filters with peak transmission at 4.4 μm to minimise cross-sensitivity to changes in humidity, twice per 10 minute scan. The calibration gas was calibrated itself with a cascade of three gas mixing pumps (SA 18/3, SA 27/3F, H. Wösthoff, Bochum, Germany). Because humidity and CO_2 concentration were measured at constant temperature, no corrections to the fluxes for changes in air density with height (Webb, Pearman and Leuning, 1980) were made.

RADIATION BALANCE

The all-wave radiation balance, R_n, was measured with two Funk-type net radiometers (type S-1 and type E-1, Swissteco Pty. Ltd., Hawthorn, Australia), placed horizontally at a height of 4 m above the canopy. At periods of maximum net radiation the two instruments agreed within ±2.5%.

Shortwave (or solar) radiation (R_s) was measured at a height of 4 m above the canopy with two Moll-Gorczynski-type pyranometers (type CM3, Kipp and Zonen Delft, The Netherlands), one facing upward and the other facing downward. Quantum flux density (Q_s) was measured with an adjacent sensor (LI 191, LiCor Inc., Lincoln, Nebraska).

Soil heat flux, G, was measured with two pairs of heat flux discs (type S.R.I.9, Solar Radiation Instruments, Altona, Australia), positioned horizontally below about 20 mm of litter and humus. At periods of maximum soil heat flux (20 W m^{-2}), the two sets agreed within ±8%. Trunk, bark and soil temperatures were measured with copper-constantan thermocouples and a reference oven. Air temperature and humidity was also measured within the canopy at the height of maximum foliage density with ventilated, shielded wet and dry PRT psychrometers.

MOMENTUM BALANCE

Windspeeds were measured over a height interval of 6.5 m above the canopy using miniature cup anemometers (Bradley, 1969) logarithmically spaced at six heights (Rimco type R/AMI, Rauchfuss Inst. Div., Analite Pty. Ltd., Mitcham, Australia). Windspeeds were also measured at three heights above the canopy using two sets of Porton cup anemometers (Jones, 1964) (Type A100, Vector Instruments, Rhyl, UK). Mean hourly wind profiles for analysis were obtained by averaging 12 sets of windspeeds integrated over periods of 5 min (Landsberg and Jarvis, 1973). Wind direction was measured *c.* 6.5 m above the canopy.

ERROR ANALYSIS

The maximum likely errors attached to the measured mass fluxes depend on the absolute values of $\triangle C$, $\triangle T$, $\triangle e$, and D. The height interval of 4 m close to the canopy maximized $\triangle C$, $\triangle T$ and $\triangle e$ (Jarvis *et al.*, 1976) but is none the less in the zone suitable for application of the Bowen ratio (Denmead and Bradley, 1985; Cellier and Brunet, 1992). For typical values of the variables during the middle hours of the day, likely errors are listed in *Table 18.A1*.

Table 18.A1 ESTIMATES OF TYPICAL VALUES OF THE ERRORS ASSOCIATED WITH THE MEASURED AND DERIVED QUANTITIES FOR DRY AND WET CANOPIES, BASED ON AN ERROR ANALYSIS BY JAMES (1977)

Quantity	*Estimated error* ±%	
	Dry	*Wet*
R_n	5	10
G	15	15
A	6	10
$\triangle T$	5	10
$\triangle e$	12	20
$\triangle C$	5	8
β	6	20
H	5	10
λE	10	20
F_a	20	30
u_*	10	10
r_{aM}	20	20
r_{aV}	20	20
r_s	20	–
r_i	15	30

Satisfactory agreement was obtained among the latent heat fluxes calculated from each GMS. The differences were not systematic and agreement was usually within ±8%. Agreement among the sensible heat fluxes calculated for each GMS was usually ±5%.

19

RESOURCE REJECTION BY HIGHER PLANTS

HOWARD THOMAS
Cell Biology Department, AFRC Institute of Grassland and Environmental Research, Plas Gogerddan, Aberystwyth, Dyfed SY23 3EB, UK

Sequential leaf senescence

The throw-away lifestyle of higher plants depends on differential organ senescence. How did programmed disposability arise in evolution, what are its advantages and what price has to be paid for its benefits? A commonly-accepted cause of leaf senescence is declining photosynthesis, associated with environmental influences such as daylength or disease, or internal signals such as hormones or nutrient transport patterns (Thomas and Stoddart, 1980; Stoddart and Thomas, 1982; Thomas, 1992). A factor thought to be particularly important in progressive or sequential senescence is the shading of lower leaves by foliage above them in the canopy.

It is reasonable to suppose that a triggering influence is the photosynthetic compensation point. A deep layer of foliage may be located permanently beneath the lowest limit of the compensation point as it undertakes its daily migration through the canopy. These leaves will be discarded, usually following the initiation of senescence and recovery of mobilized materials, notably reduced nitrogen (Nobel, 1983). A simple calculation enables the progress of the compensation point through the canopy to be estimated. The Monsi exponential equation for the radiation profile in the canopy is:

$$F = \ln(Q_0/Q_F)/k \qquad (1)$$

where Q_0 is the flux density at the canopy surface, Q_F the flux density at leaf area index (L) = F, and k is the foliar absorption coefficient. L_c, the value of L at the compensation point (Q_F = 10 μmol m^{-2} s^{-1}) is given by:

$$L_c = \ln(Q_0/10)/k \qquad (2)$$

Figure 19.1a presents a typical curve of Q_0 throughout a cloudless late June day in England (data for Rothamsted taken from Monteith, 1973). Superimposed is the L_c curve for a crop with a k value of 1. If maximal penetration of the compensation point is an absolute limit below which leaves senesce and are lost, L for this crop should never exceed about 5.5.

Of course, these calculations are extremely crude. How frequently (and for how much of the growing season) would the Q_0 curve get anywhere near 2000 μmol m^{-2} s^{-1}? Not very often in the English climate. A figure of 10 μmol m^{-2} s^{-1} is on the low side for the flux at the compensation point. The *k* value only takes transmittance into account and makes no allowance for reflectance and other light-attenuating interactions between foliage and incident radiation (e.g. Goudriaan, 1988). These and other corrections would, however, tend to move the notional penetration profile towards the top of the canopy. It is suggested, therefore, that the curve illustrated in *Figure 19.1a* represents the most optimistic trajectory for foliar survival.

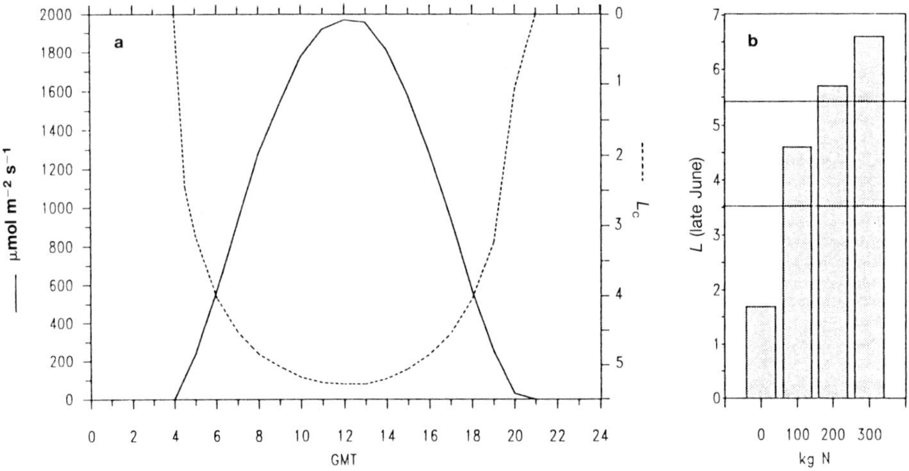

Figure 19.1 (**a**) Q_0, (——), the flux density at the canopy surface, during a cloudless late June day in England, and L_c, the estimated penetration of the photosynthetic compensation point into a canopy with a foliar absorption coefficient of 1. (**b**) Leaf area indices in June of potato crops grown with different levels of nitrogen fertilization (Firman and Allen, 1989). The horizontal lines represent the estimated maximal value of L_c (upper line) and the optimal leaf area index for tuber production (lower line)

How does the predicted maximal *L* match observed leaf area indices? The profile in *Figure 19.1a* represents the best case for a species with planofile leaves, such as potato. But *L* values well in excess of 5.5 have been observed in potatoes in late June, particularly when they have been grown under high nitrogen fertilization (*Figure 19.1b*). Although 5.5 is indicated as the cut-off point for such a crop, in terms of tuber yield optimal *L* is no greater than about 3.5 (Firman and Allen, 1989 – *Figure 19.1b*). If the true survival limit for foliage of this species is accepted to be somewhere between these two values, this would indicate that, in all but the lowest fertilizer regimes, the potato crop is carrying a substantial load of foliage in carbon overdraft.

Table 19.1 presents data for a different crop. Forage grasses exhibit a range of canopy geometries from prostrate to extreme erectness and this is reflected in the range of *k* values. Data are presented for two extreme *Dactylis glomerata* varieties and for *Phleum pratense* intermediate in *k* (Sheehy and Cooper, 1973). The highly erect S37 cocksfoot was an underachiever, but in S345 and S352 a

Table 19.1 CALCULATED MAXIMAL PENETRATION OF THE PHOTOSYNTHETIC COMPENSATION POINT EXPRESSED AS CUMULATIVE LEAF AREA INDEX COMPARED WITH OBSERVED LEAF AREA INDEX OF GRASS SPECIES WITH CONTRASTING CANOPY STRUCTURES (SHEEHY AND COOPER, 1973)

Species/variety	k	*maximal* L_c	*observed* L_{total}
Dactylis glomerata S37	0.23	23.0	11.2
Dactylis glomerata S345	0.91	5.8	14.9
Phleum pratense S352	0.55	9.6	14.5

Maximal L_c calculated as

$$\ln(Q_{max}/Q_{comp})/k$$

where Q_{max} and Q_{comp} are flux densities at full sunlight (2000 μmol m^{-2} s^{-1}) and the photosynthetic compensation point (10 μmol m^{-2} s^{-1}) respectively. *D. glomerata* cv. S345 is a selection with an extreme prostrate growth habit.

significant fraction of the canopy must have been in negative carbon balance. The gap between L_c and observed leaf area index is probably exaggerated because these grasses, unlike dicots such as potato, do not lose old leaves by abscission, so that senescent and dead tissue may contribute to the observed L.

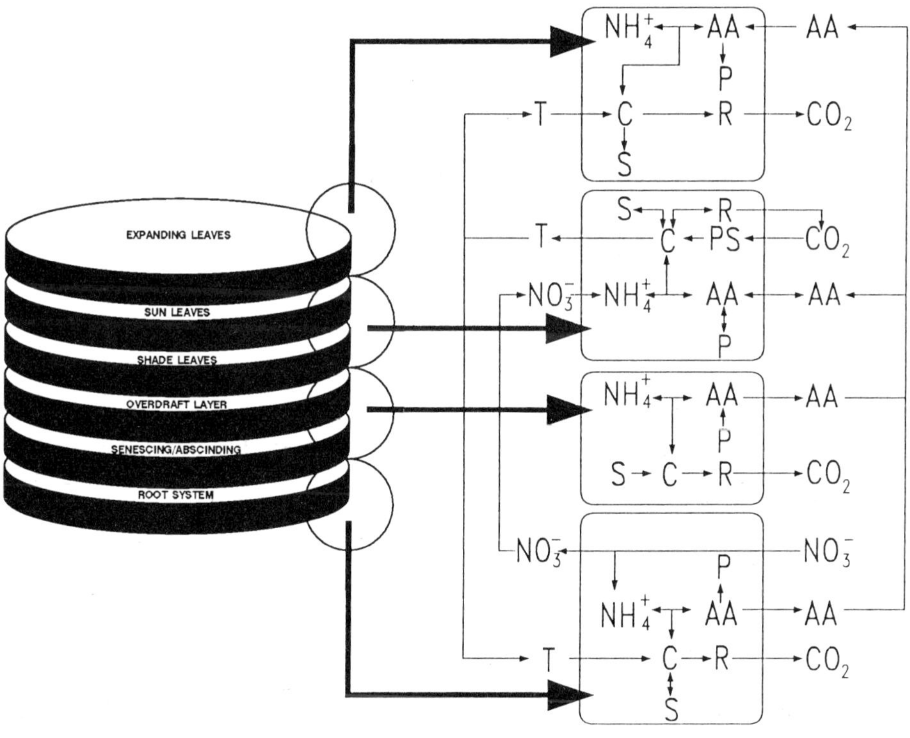

Figure 19.2 Stratification of canopy structure and function. AA, amino acids; C, carbon skeletons; P, proteins; PS, photosynthesis; R, respiration; S, structural and storage compounds; T, translocated carbon

Analyses of this sort generally fail to support the simplistic proposition that senescence is triggered by the altered light environment as leaves move deeper into the canopy. *Figure 19.2* is a picture of an idealized canopy, identifying a number of strata including a *carbon overdraft* layer. The model also indicates how the different layers are related metabolically and translocationally. Although this chapter does not intend to consider in detail the physiological state of foliage surviving at depth in the canopy, it is worth pointing out that deep shade, in contrast to darkness, has been observed to *prolong* the lifetime of leaves.

For example, *Figure 19.3* summarizes data for the fourth leaf of *Lolium temulentum* plants grown in a controlled environment under an 8 h daylength (Mae *et al.*, 1993). At full leaf expansion, 30 days after sowing, some plants were transferred from ambient 350 into 90 μmol m^{-2} s^{-1}. The greenness of shaded leaves was greatly extended (*Figure 19.3a*). Measurement of components of the photosynthetic apparatus revealed that light-harvesting (*Figure 19.3c, d*) and other constituents responded differentially to the treatment. Ribulose-1,5–bisphosphate

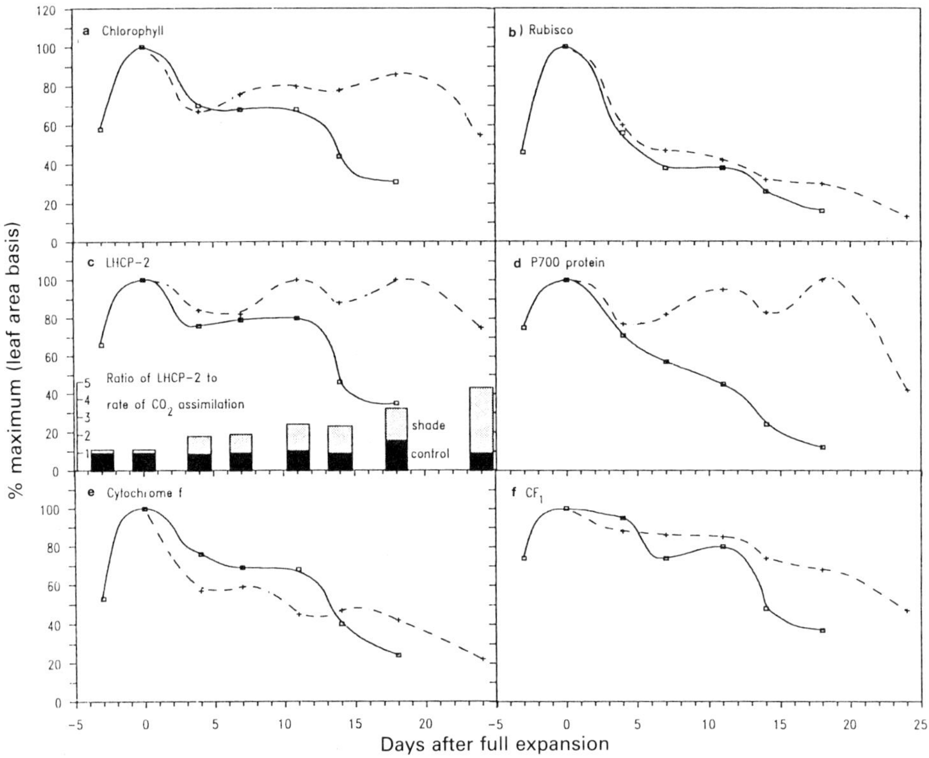

Figure 19.3 Components of the photosynthetic apparatus of *Lolium temulentum* fourth leaves from full expansion to senescence at 350 μmol m^{-2} s^{-1} (——) and after transfer at day 0 to 90 μmol m^{-2} s^{-1} (– – –). Proteins (**b-f**) were determined by quantitative immunoblotting using specific antibodies. Carbon exchange rate (**c**) was measured using an infra-red gas analyser (data of Mae *et al.*, 1993)

carboxylase (Rubisco) was not significantly stabilized by shading (*Figure 19.3b*). On the other hand the ratio of the light-harvesting complex of photosystem 2 to carboxylation activity greatly increased as shading was prolonged (*Figure 19.3c*). The authors interpret these data to indicate that leaves transferred to low light, even at maturity after growth has ceased and the photosynthetic apparatus is fully formed, are able to acclimate and to become, in effect, shade leaves *sensu* Boardman (1977). There is no reason to doubt that mature leaves moving down in the canopy can adjust their physiology to the changing environment. Moreover, far from hastening the onset of senescence, the shading they experience may even increase their life-expectancy.

Why should a leaf permanently beneath its compensation point be preserved by the plant? It must depend on whether there are survival benefits to be gained from expenditure of the energy necessary to retain it. It is not too difficult to think of ways in which such leaves could be useful. A variable burden of sub-compensation point leaves may act as insurance against defoliation and would be valuable for species growing in variable light environments. Leaves in carbon overdraft could also be sites of temporary storage. They may have value in contributing to the area of evaporative surface acting to pull water and nutrients out of the soil. Other roles which might also be considered are as decoys for predators and pathogens, as cooled surfaces for condensation and as counterweights reducing top-heaviness and susceptibility to physical damage. But the idea that plants can and do support organs in negative carbon balance may have more general significance. To develop this argument, it is necessary to introduce an evolutionary digression.

Evolutionary origin of the photosynthetic apparatus

The colonization of land by aquatic organisms is a familiar episode in the evolution of animals; anyone who has studied zoological form and function is equally familiar with the interpretation of many of the physiological characteristics of terrestrial animals in terms of adaptations to life out of water. But plant physiologists tend to overlook the fact that plants also made the same journey, from sea to land, some 500 million years ago, and that, like animals, the physiology of terrestrial plants has been greatly rejigged in the process.

When one considers the photosynthetic machinery of the higher plant cell, it is important to appreciate that this apparatus was evolved, and adapted to function optimally, under water, in relatively dim light and cool temperatures. Although there may have been very early evolutionary experiments with different ways of utilizing light energy biologically, it seems certain that the process in the vast majority, perhaps all, modern photosynthetic organisms evolved only once. There does not appear to have been a radical reworking of chloroplast organization and function during the great evolutionary crisis when plants colonized the land; in fact the basic design principles of photosynthetic membranes are remarkably conserved across the whole range of autotrophic organisms. For example, notwithstanding the vast taxonomic gulf, the photosynthetic reaction centre of the purple bacterium *Rhodopseudomonas* shows virtually complete polypeptide-for-polypeptide (and gene-for-gene) homology with photosystem 2 of higher plants (Marder and Barber, 1989).

As plants emerged from the sea they found themselves saddled with a primary metabolic process that was organized in a way quite inappropriate to meet the challenges of the new environment. In such circumstances, evolution does not usually take the course of fundamental redesign; instead, the fossil photosynthetic apparatus was subjected to a whole series of *post hoc* physiological fixes to allow it to work properly on land.

It seems that the earliest terrestrial plants encountered problems of superabundance. Having relocated to a high light, warm, CO_2-rich environment, they must have found themselves awash with carbon. This would explain the rapidity with which conquest of the land was followed by establishment of forests. Almost the moment plants made landfall, they became trees. Promiscuous photosynthesis which the early plants could not avoid must have presented severe physiological problems. It follows that carbon-dumping came to play a major part in the life of the higher plant at all levels.

To take an example from nutrition, higher plants generally have a voracious appetite for nitrogen. It is customary to consider nitrogen as a serious limiting factor for growth. Nitrogen assimilation processes are high consumers of energy and carbon skeletons. Therefore one could equally well regard nitrogen as a metabolic sink for excess carbon. It would make sense if nitrogen could take carbon out of the osmotic and metabolic picture by sequestering it in a high molecular weight form undergoing minimal turnover, perhaps as an enzyme with a rather low catalytic activity. If that enzyme could itself eject carbon from the cell as one of its properties, so much the better. All of which adds up to an accurate description of Rubisco. The photorespiratory function of this enzyme has the beneficial effect of efficiently re-emitting carbon at the point of fixation. According to this argument, trying to disable photorespiration is a risky, perhaps unattainable, goal for crop improvement. Apparent exceptions to the principle of carbon avoidance, such as the C_4 syndrome, may be regarded as instances of *post post hoc* add-ons. It has been proposed that the C_4 pathway is primarily an adaptation to water and high temperature limitation (Schuster and Monson, 1990). Resources that are discontinuously distributed in the environment such as phosphorus (spatially patchy) and water (availability fluctuating in time) really do seem to pose capture problems for plants, and physiological measures (mycorrhizal associations, for example) have clearly been adopted during evolution to cope with the limitations in supply. Carbon and nitrogen, which are everywhere, and just a few enzymic conversions away from the osmotic and metabolic milieu, are different.

Morphology also plays a role in dealing with carbon surfeit. What is a tree if not a colony of autotrophs living on the surface of a large deposit of excreted carbon in the form of cellulose? Plants are also great accumulators of reserve substances – starch, oils, sucrose. Generally this behaviour is interpreted as having arisen as a way of evening out supply–demand relations over time in a fluctuating environment. It could equally well be argued that reserves originated as carbon dumps and only subsequently became exploited as investments for the future. The list of carbon-consuming activities assumed by terrestrial plants is a long one. It includes the cyanide-resistant alternative pathway of respiration; futile cycles; the tendency to control the stoichiometry of components of complex cellular structures such as membranes via seemingly profligate post-translational turnover mechanisms rather than balanced gene transcription. Looked at this way, even roots invite reinterpretation. What better way of disposing of fixed carbon than by burying it where it cannot make more assimilatory biomass? The symbiotic root nodules of legumes can also be thought of as fulfilling a carbon-dumping requirement (40–50% of each day's photosynthate is processed by the nodules and about 50% of this carbon is respired away – Gordon *et al.*, 1987) while simultaneously enhancing the metabolic sink function of reduced nitrogen.

So this is the 'carbon-excess' model of higher plant evolution, and it has a number of important implications, not least in the area of agricultural genetics and biochemistry. One consequence is that notions of 'efficiency' and 'economy' applied to cellular energy or carbon relations, if they are meaningful at all, do not have the adaptive significance they may have for heterotrophs. In particular, to look at the physiology of fundamental plant processes from the standpoint of crop species and the demands and priorities of agricultural production is often to invert the imperatives which shaped the design in the first place. When this happens, attempts to 'improve' a particular attribute, such as breeding for increased carbon assimilation rate or lower photorespiration, will always end up swimming against the evolutionary tide – which may account for the conspicuous lack of success of such strategies.

Evolution of programmed senescence

Returning to discussion of the significance of photosynthetic compensation point in deciding the fate of leaves in the canopy, if the evolutionary argument is correct, and the origins of the photosynthetic apparatus really do imply a general requirement for gratuitous carbon disposal, then it would not be surprising to find a carbon overdraft horizon to be a common feature of many canopies. It may be significant that plant communities with a dense, closed top foliage layer – tropical forest, for example – often include a green understorey which supplements its meagre or negative carbon income by direct intake of organic material from the litter. A canopy is demographically structured like a community of individuals, so it is suggested that what is referred to here as the overdraft component is equivalent to the deep shade organotrophic substratum in tropical forests.

Let us consider further the disposable nature of plant parts. If green plants are, as suggested, dedicated to unloading surplus carbon by a whole range of means then such apparently profligate phenomena as autumnal leaf fall may be understandable. But when and how was this kind of life-style evolved? A set, or genetically-determined, lifespan was probably not a feature of the leaves of the earliest terrestrial plants, which were growing in virtually season-proof parts of a world into which evolution had yet to introduce fungal and invertebrate pathogens and predators. Nevertheless a deciduous *Glossopteris* flora was already well established in Gondwanaland during the Carboniferous era (Plumstead, 1958). Green tissues of modern cryptogams certainly have the capacity to run through a yellowing process indistinguishable in its major features from angiosperm foliar senescence (De Greef *et al.*, 1971; Behera and Biswal, 1990). It is supposed that the habit of leaf turnover evolved as seasonal factors including colder temperatures and decreased daylength episodes became part of the yearly cycle. As already discussed, the archetypal terrestrial plant form is the tree and smaller herbaceous species with shorter life-cycles developed from trees by colonization of more unstable habitats away from the even environment of the primaeval forest (Corner, 1964). That is, all herbs, including crop species, are juvenile trees. Grasses are neotenic bamboos. *Arabidopsis* is a chestnut suffering from arrested development.

Modern tropical forests may provide a glimpse of how life was in the earliest plant communities. Even here patterns of leaf turnover are well established (Longman and Jenik, 1987). Trees with continuously growing shoots shed leaves continuously. Typically leaf lifespans are in the range of 3 to 15 months. Flush growth is common (but often difficult to associate with any particular environmental cue). There are even some species, mostly monocots, which senesce in the monocarpic fashion. Although the bright autumnal colours of temperate deciduous woodlands are largely absent, ageing leaves of tropical forests usually lose chlorophyll and become yellowish and there is considerable recovery of mobile nutrients indicating that abscission is preceded by a recognizable senescence-like process (Richards, 1957; Chapin and van Cleve, 1989). The general conclusions to be drawn from considering the likely evolutionary origins of programmed senescence are that the phenomenon is an ancient and probably near-universal feature of the genetic repertoire of higher plants, but it is variable in its emphasis during the life-cycles of different species.

Senescence and carbon overcapacity

From the utilitarian agronomic perspective, the value of senescing leaves rests largely with their function in nutrient remobilization and the canopies of modern crop varieties generally combine a prolonged pre-senescence phase followed by a fast and efficient senescence that reallocates a high proportion of biomass in the harvested fraction. In general, non-domesticated species are less 'efficient' at recovering materials from leaves before or during senescence. In a sense, the foliage of grain crop species shares the desirable (from the point of view of agricultural production) physiological features of the seeds, sequestering raw mobilizable biomass to the point of hypertrophy. This connection may be a pleiotropic consequence of the operation of common genetic programmes (Thomas, 1984). Although there are exceptions, it is broadly true that plants yielding small seeds with meagre reserves do not display emphatic patterns of foliar senescence.

In the wider ecological context senescing leaves are significant in other ways. The striking colours of autumn foliage bear more than a passing resemblance to those of showy flowers. Similar phytochemical components occur in petals and senescing leaves – anthocyanins, carotenoid derivatives, flavonoids – and the organelles characteristic of the highly coloured state – chromoplasts and gerontoplasts – are closely related in the plastid family tree (Whatley, 1978). In many species, *Sorbus*, for example, the brilliant colours of senescing foliage appear to supplement the attractiveness of the brightly-pigmented fruit which ripens at that time, and may enhance the chances of fruit dispersal by animals. This has been referred to as 'fruit flagging' (Stiles, 1982). Even species which simply turn yellow without apparently accumulating new richly-coloured compounds may become more intensely photoactive in the non-visible regions of the spectrum. *Ginkgo*, for instance, accumulates a highly fluorescent compound in the epidermis, which acts like an optical brightener (Matile *et al.*, 1992 – *Figure 19.4*).

It is a short step, in developmental and evolutionary terms, from deploying highly visible leaves attractive to animal dispersers to organizing colourful leaf-like

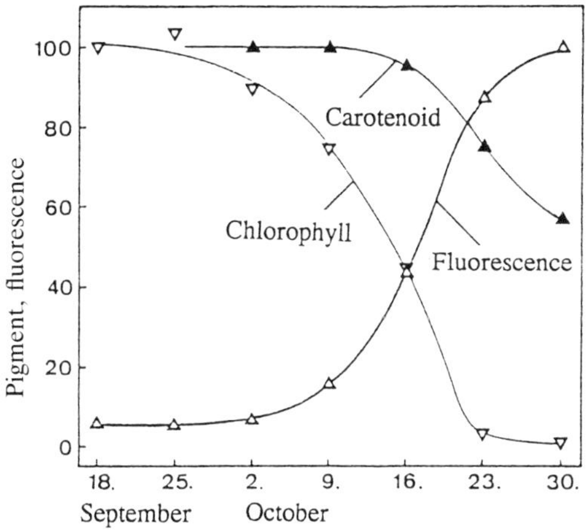

Figure 19.4 Chlorophyll and carotenoid degradation in leaves of *Ginkgo biloba* during autumn is accompanied by the appearance of intense fluorescence (λ_{ex} = 350 nm, λ_{em} = 530 nm). Data of Matile *et al.* (1992)

structures into insect-pollinated flowers. It has been seen that programmed foliar senescence was well established in cryptogams long before flowers appeared on the scene. Was the chloroplast-gerontoplast transition appropriated during the evolution of angiosperm flowers and fruits, which owe their colours to chromoplasts? It may be that the evolution of flower form and pigmentation originates with the characteristic metabolism of senescing leaves. In any case, the propensity for biochemical improvization, evident from the extent of secondary product elaboration, including foliar and floral pigmentation, throughout the plant kingdom, is perhaps another manifestation of carbon overcapacity, permitting speculative metabolic doodling which occasionally has a significant evolutionary payoff.

Conclusion

To summarize, it is argued that when aquatic plants took to the land they found themselves, like European farming under the Common Agricultural Policy, with a problem of surplus. The remedies are also comparable – storage, dumping, processing for new uses, freedom from subsistence, the pursuit of often frivolous objectives, great diversity of life-style, conspicuous consumption and complicity in the rise of opportunist parasites. Every time the sun comes up it presents the green plant with problems, but they are problems of excess and the legacy of half a billion years of evolution.

Acknowledgement

I am grateful to Alan Gay for helpful discussion.

References

Behera, Y.N. and Biswal, B. (1990) Leaf senescence in fern: effect of duration, intensity and quality of light. *Environmental and Experimental Botany*, **30**, 181–186

Boardman, N.K. (1977) Comparative photosynthesis of sun and shade plants. *Annual Review of Plant Physiology*, **28**, 355–377

Chapin, F.S. and van Cleve, K. (1989) Resorption of biomass and nutrients during autumn senescence. In *Plant Physiological Ecology. Field Methods and Instrumentation*, pp 186–207 Edited by R.W. Pearcy, J.R. Ehleringer, H.A. Mooney and P.W Rundel. London: Chapman and Hall

Corner, E.J.H. (1964) *The Life of Plants*. London: Weidenfeld and Nicolson

De Greef, J.A., Butler, W.L., Roth, T.F. and Fredericq, H. (1971) Control of senescence in *Marchantia* by phytochrome. *Plant Physiology*, **48**, 407–412

Firman, D.M. and Allen, E.J. (1989) Relationship between light interception, ground cover and leaf area index in potatoes. *Journal of Agricultural Science*, **113**, 355–359

Gordon, A.J., Mitchell, D.F., Ryle, G.J.A. and Powell, C.E. (1987) Diurnal production and utilization of photosynthate in nodulated white clover. *Journal of Experimental Botany*, **38**, 84–98

Goudriaan, J. (1988) The bare bones of leaf-angle distribution in radiation models for canopy photosynthesis and energy exchange. *Agricultural and Forest Meteorology*, **43**, 155–169

Longman, K.A. and Jenik, J. (1987) *Tropical Forest and its Environment*. 2nd Edn. London: Longman

Mae, T., Thomas, H., Gay, A.P. Makino, A. and Hidema, J. (1993) Leaf development in *Lolium temulentum*: photosynthesis and photosynethetic proteins in leaves senescing under different irradiances. *Plant and Cell Physiology*, **34**, 391–399

Marder, J.B. and Barber, J. (1989) The molecular anatomy and function of thylakoid proteins. *Plant Cell and Environment*, **12**, 595–614

Matile, P., Flach, B.M.P. and Eller, B.M. (1992) Autumn leaves of *Ginkgo biloba* L.: optical properties, pigments and optical brighteners. *Botanica Acta* **105**, 13–17

Monteith, J.L. (1973) *Principles of Environmental Physics*. London: Edward Arnold

Nobel, P.S. (1983) *Biophysical Plant Physiology and Ecology*. San Francisco, New York: W.H. Freeman

Plumstead, E.P. (1958) The habit of growth of Glossopteridae. *Transactions of the Geological Society of South Africa*, **61**, 81–94

Richards, P.W. (1957) *The Tropical Rain Forest. An Ecological Study*. Cambridge: Cambridge University Press

Schuster, W.S. and Monson, R.K. (1990) An examination of the advantages of C3–C4 intermediate photosynthesis in warm environments. *Plant Cell and Environment*, **13**, 903–912

Sheehy, J.E. and Cooper, J.P. (1973) Light interception, photosynthetic activity and crop growth rate in canopies of six temperate forage grasses. *Journal of Applied Ecology*, **10**, 239–250

Stiles, E.W. (1982) Fruit flags: two hypotheses. Fruit colors provide signals to dispersing birds. *American Naturalist*, **120**, 500–509

Stoddart, J.L. and Thomas, H. (1982) Leaf senescence. In *Encyclopedia of Plant Physiology. New Series Volume 14A. Nucleic Acids and Proteins in Plants I*, pp 592–636 Edited by D. Boulter and B. Parthier. Berlin: Springer

Thomas, H. (1984) Cell senescence and protein metabolism in the photosynthetic tissues of leaves. In *Cell Ageing and Cell Death*, pp 171–188 Edited by I. Davies and D.C. Sigee. Cambridge: Cambridge University Press

Thomas, H. (1992) Canopy survival. In *Crop Photosynthesis: Spatial and Temporal Determinants*, pp 11–41 Edited by N.R. Baker and H. Thomas. Amsterdam: Elsevier

Thomas, H. and Stoddart, J.L. (1980) Leaf senescence. *Annual Review of Plant Physiology*, **31**, 83–111

Whatley, J.M. (1978) A suggested cycle of plastid developmental interrelationships. *New Phytologist*, **80**, 489–502

ASSIMILATE PARTITIONING IN SPRING WHEAT UNDER MEDITERRANEAN CONDITIONS

J.P. ABREU, F.M. ABREU and G.S. CAMPBELL*
*Instituto Superior de Agronomia, Dept. Ciências do Ambiente, Tapada da Ajuda, 1399 Lisboa Codex, Portugal; *Washington State University, Department of Crop and Soil Sciences, Pullman, WA 99164–6420, USA*

Spring wheat (*Triticum aestivum* L., cv. Pesudo) was grown near Lisbon, for 2 years, at three water treatments. Dry weights (DM) were determined of green (GL) and yellow leaves (YL), 'stems' (ST) – where ear structures where included before ear emergence – and ears (ER). The fractional distribution of dry matter between above ground plant parts was plotted against a normalized developmental stage scale (de Vries and van Laar, 1982). The accumulated temperature at a given moment is divided by the thermal time required to reach an event. Emergence is event 0 and anthesis is 1; maturity (15% water content) corresponds to 2. Base temperatures of 3°C before anthesis and 8°C thereafter were used.

At maturity, DM ranged from 1.68 to 4.34 g per plant, and differences were noticeable after tillering. Nevertheless, fractional distribution of dry matter between the various plant parts was similar among the different soil water treatments, when plotted against developmental stage (*Figure 1*).

Before anthesis, the assimilates in excess of those required for crop maintenance are available for growth of roots and shoots, leaves and 'stems' (where ears are included). If the DM lost by senescent leaves is not taken into account, it can be shown that:

$$dW_c = \phi'_c \, d\tau W_t + \phi_c \, \epsilon \, dP_{an} \qquad (1)$$

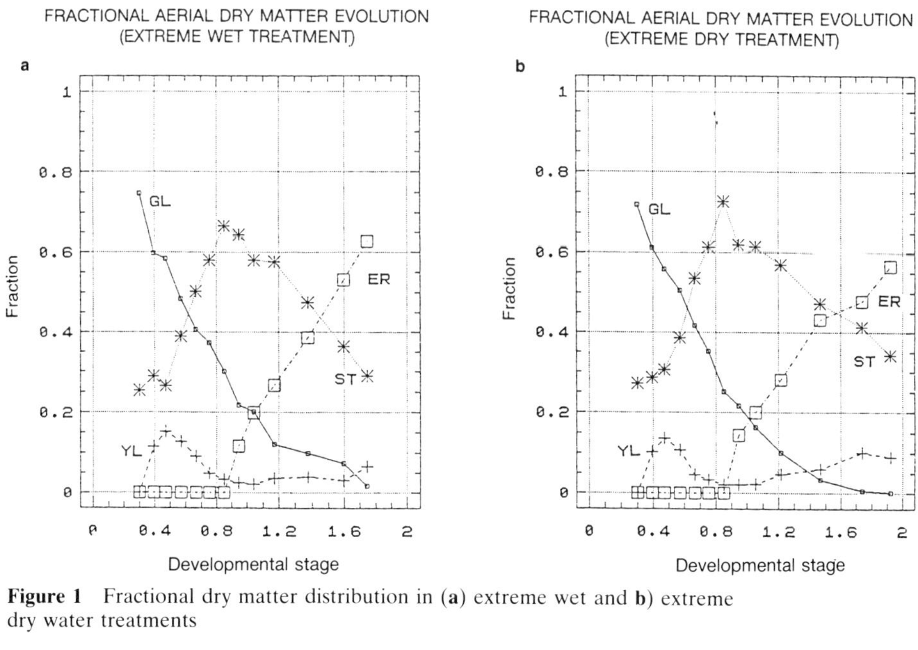

Figure 1 Fractional dry matter distribution in (**a**) extreme wet and **b**) extreme dry water treatments

where W_c is the component DM (i.e., leaves or 'stems'); ϕ_c, ϕ'_c is the fraction of the component and its derivative; τ is developmental stage; W_t is total aerial DM; ϵ is the conversion efficiency of dry matter; dP_{an} is the net photosynthate available for the aerial parts' growth.

A Richards growth equation was fitted to the ST and ER fractions against stage. The resulting equation is

$$\phi_{ST} = \frac{\phi_o \, \phi_f}{\left[\phi_o^n + (\phi_f^n - \phi_o^n)\, e^{-k\tau}\right]^{1/n}} \qquad (2)$$

where ϕ_{ST} is the fraction of 'stems'; ϕ_o, ϕ_f is initial (0.103) and final (0.847) fractions; and n, k are constants (4.125, 10.121).

After anthesis, current assimilates and reserves (mainly from stems) are almost exclusively diverted to grain growth. Ears grow almost linearly, depending on kernel growth and number.

Potential grain growth seems to be described by an unique curve, when plotted in this scale. Analysing the data of Sofield *et al.* (1977), the base temperature for grain filling duration was calculated from their Figure 7. Using this temperature, grain growth per degree day was almost constant, for two varieties at three temperatures below 25°C. Higher temperatures might be associated with a limiting assimilate flow.

Grain number is closely related to the DM of ear structures. At anthesis, these structures are about 20% of above ground biomass. The present data shows, for small and big ears alike, a kernel number per gram chaff of 79 at maturity. These observations agree with Rawson and Bagga (1979) and Fisher and Stockman (1980).

The above relations might be useful in modelling carbon allocation, under water limiting conditions.

References

Fischer, R.A. and Stockman, Y.M. (1980) Kernel number per spike in wheat (*Triticum aestivum* L.): responses to preanthesis shading. *Australian Journal of Plant Physiology*, **7**, 169–80

Rawson, H.M. and Bagga, A.K. (1979) Influence of temperature between floral initiation and flag leaf emergence on grain number in wheat. *Australian Journal of Plant Physiology*, **6**, 391–400

Sofield, I., Evans, L.T., Cook, M.G. and Wardlaw, I.F. (1977) Factors influencing the rate and duration of grain filling in wheat. *Australian Journal of Plant Physiology*, **4**, 785–797

de Vries, F.W.T.P. and van Laar, H.H. (1982) Simulation of growth processes and the model BACROS. In *Simulation of Plant Growth and Crop Production*, pp 114–135 Edited by F.W.T.P. de Vries and H.H. van Laar. Simulation Monographs, Wageningen: Pudoc

Acknowledgements

The authors thank JNICT (Junta Nacional de Investigaçao Científica e Tecnológica) and Sangari for the financial support, and their colleague Isabel Flores for the technical assistance.

A RADIATION CAPTURE MODEL OF CROP PLANT GROWTH

D.P. AIKMAN
Horticulture Research International, Littlehampton, UK

A mathematical model has been developed that gives good fits to experimental data in even-aged and mixed-age monocultures.

It is assumed that an isolated plant has a canopy spread area, which increases as the 2/3 power of plant weight, $A_c = k_1 w^{2/3}$, and that it has a total leaf area, which is proportional to its weight, $A_l = k_2 w$. Assume that, in the absence of any shading, the growth rate is proportional to leaf area and hence to weight, resulting in exponential growth with rate constant:

$$\alpha : \frac{dw}{dt} = \alpha w \qquad (1)$$

Within the zone area of the individual plant's canopy, the effective leaf area index (LAI), ξ, is given by $\xi = A_l/A_c = k_2 w (k_1 w^{2/3})^{-1} = (w/\beta)^{1/3}$ with $\beta = (k_1/k_2)^3$, being the weight at which the isolated plant has an effective LAI of value 1. The fraction of vertically incident light transmitted through a uniform random canopy is given by exp $(-\xi)$ (Monsi and Saeki, 1953). Hence a self-shading function, $f_s(\xi) = (1 - \exp(-\xi))/\xi$, is the fraction of incident light intercepted *per* unit leaf area. This function, giving an efficiency of use of leaf area, has the value 1 at $\xi = 0$, and decreases monotonically with increasing ξ. Metabolic maintenance may be included as negative term directly proportional to w.

Thus, allowing for self-shading and maintenance, the growth rate of the isolated plant is given by:

$$\frac{dw}{dt} = w\ (\alpha_1 f_s(w) - \alpha_2) = w\left(\alpha_1 \frac{1-\exp(-\xi)}{\xi} - \alpha_2\right) \qquad (2)$$

For low values of self-shading, this may be seen to approximate to a Richards (1959) equation. Taking the first terms of a series expansion of [2], one obtains:

$$\frac{dw}{dt} \approx (\alpha_1 - \alpha_2)\, w\left\{1 - \frac{\alpha_1 \xi}{2(\alpha_1 - \alpha_2)}\right\} = aw\left\{1 - \left(\frac{w}{b}\right)^{1/3}\right\} \qquad (3)$$

Assuming that mutual shading comes into operation with canopy closure, it will be phased in earlier at higher densities than at lower. After canopy closure, assume that the effective leaf area per plant is given by $(\phi_1 w)^\theta$. With an interplant spacing r in a square pattern, the crop LAI would be given by $x = (\phi_1 w)^\theta/r^2$. Thus, after canopy closure, mutual shading in an even-aged stand results in an efficiency of utilization of leaf area given by:

$$f_m(w) = \frac{1-\exp(-x)}{x} \qquad (4)$$

Modifying the basic plant growth equation, to allow for growth under mutual shading within the complete uniform canopy, gives:

$$\frac{dw}{dt} = w(\alpha_1 f_m(w) - \alpha_2) = w\left(\alpha_1 \frac{1-\exp(-x)}{x} - \alpha_2\right) \qquad (5)$$

A transition phase may be considered as occurring between the state in which the plants are growing independently, like self-shading isolated plants Equation 2, and one in which the plants mutually shade one another in a uniform closed canopy Equation (5). In a transition phase the plants will be competing but not yet experiencing the full effect of competition of a complete canopy. The transition will occur as A_c tends to r_2. A competition function that would provide such a composite behaviour is:

$$f_c(w) = \frac{1}{2}\left\{f_m + f_s + (f_m - f_s)\tanh\left[\phi_2\left(\frac{A_c}{r^2} - \phi_3\right)\right]\right\} \quad (6)$$

in which ϕ_2 controls the sharpness of the transition which occurs in the vicinity of $A_c = \phi_3 r^2$.

The resulting growth rate in even-sized monoculture is given by:

$$\frac{dw}{dt} = w(\alpha_1 f_c(w) - \alpha_2) \quad (7)$$

The behaviour after canopy closure will be similar to the expolinear model of Goudriaan and Monteith (1990).

With a large difference in the dates of emergence, a younger smaller plant will become over-shaded by its larger neighbours, causing one-sided competition: lesser difference would produce two-sided asymmetric competition. This may be modelled in a similar fashion.

This is an extension and improvement of an earlier model by Aikman (1991) that gave a good fit to data from an experiment on carrots (Benjamin, 1988).

References

Aikman, D.P. (1991) *Aspects of Applied Biology 26, The Art and Craft of Modelling in Applied Biology*, 122–34. Warwick: Association of Applied Biologists

Benjamin, L.R. (1988) *Annals of Botany*, **62**, 199–214

Goudriaan, J. and Monteith, J.L. (1990) *Annals of Botany*, **66**, 695–701

Monsi, M. and Saeki, T. (1953) *Japanese Journal of Botany*, **14**, 605–14

Richards, F. J. (1959) *Journal of Experimental Botany*, **10**, 290–300

GAP FREQUENCY AND CANOPY ARCHITECTURE OF SUGAR BEET AND WHEAT CROPS

F. BARET, M. STEVEN* and B. ANDRIEU†

*INRA Bioclimatologie, BP 91 84143 Montfavet Cédex, France; *Department of Geography, University of Nottingham, Nottingham NG7 2RD, UK; †INRA Bioclimatologie, 78850 Thiverval-Grignon, France*

Introduction

The gap frequency is the probability that a light beam impinging from above the canopy hits the soil. It has been recognized for a long time as the key variable directly governing the penetration of light in canopies. Nilson (1971) using a Markov chain model showed that the gap frequency P_0 is related to the leaf area index L through:

$$P_0(\theta,\phi) = \exp(-\lambda_0 . G(\theta,\phi) . L/\cos\theta) = \exp(-K(\theta,\phi) . L)$$

Where $K(\theta,\phi)$, θ and ϕ are respectively the extinction coefficient, the sun zenith and azimuth angles. $G(\theta,\phi)$, the projected area of the leaves in the direction (θ,ϕ) depends on the leaf angle distribution. The parameter λ_0 characterizes the mutual dependency of the positions of the leaves. When the leaves preferentially avoid themselves, the canopy is qualified as regular, and λ_0 is greater than one. Conversely, the leaves of clumped canopies preferentially overlap themselves, and λ_0 is lower than one. A random distribution of the leaves corresponds to the first term of a Poisson law, and λ_0 equals one. The parameter λ_0 may be evaluated by comparing the observed gap frequency to the theoretical gap frequency computed with the Poisson law using the observed leaf angle distribution. However, there is no theoretical background to explain variations of λ_0 with direction (θ,ϕ). Few values of λ_0 are reported in the literature and there are no experimental results to illustrate the variation of λ_0 with the azimuth. The purpose of this study is to investigate the directional variation of gap frequency over the whole hemisphere for sugar beet and wheat crops. A global and dynamical description is given of the directional variation of $P_0(\theta,\phi)$ rather than focussing on a given phenological stage and specific conditions, providing an operational tool to compute light penetration inside the canopy, throughout large portions of the growth cycle.

Methods and materials

The first experiment took place at Broom's Barn experimental farm in 1989 over sugar beet canopies. A large range of leaf area index was obtained by varying the sowing date, the plant density and the date of observation. The second experiment was conducted over wheat canopies at Montfavet during the same year. Three wheat crops were sown at the same time but with various row patterns: north–south rows; east–west rows; and randomly sown, without any row pattern. To evaluate $P_0(\theta,\phi)$ hemispherical photography was used (Baret *et al.*, 1992) which provides a convenient solution for all azimuths and zeniths. Once the hemispherical photographs were taken on each plot, the corresponding leaf area index was measured within one or two days.

Results and discussion

THE EXPONENTIAL LAW

For each direction (θ,ϕ) ($5<\theta<75°$, $5<\phi<355°$, $K(\theta,\phi)$ was adjusted using non-linear fitting techniques on the $P_0(\theta,\phi)$ values averaged for each plot. The 0.082 RMSE (root mean square error) computed over the whole data set leads to a value of 0.91 for the coefficient of determination (R^2). The spatial variability of the small sampling area used explains most of the scattering. Sample areas were 10° zenith by 10° azimuth sectors, with the camera positioned at a distance from the top of the canopy ranging from 0.3 m (wheat) to 0.75 m (sugar beet). It is concluded that the gap frequency follows the classical exponential law with the total leaf area index quite well throughout the growing cycle or when plant density or sowing dates are changed.

ROW EFFECT

When both wheat row treatments are pooled together, the RMSE does not increase significantly. Thus row orientation does not influence the structure of the canopy in these particular conditions where no severe wind effects were observed. The small increase of the RMSE when fitting P_0 independently of the azimuth also indicates that the row effect is tiny, although the gap frequency is slightly lower in the row direction, especially for small zenith angles. It is thus concluded that row direction does not significantly affect the azimuthal variation of gap frequency.

GAP FREQUENCY, LEAF ANGLE DISTRIBUTION AND DISPERSION COEFFICIENT

Following the former results, the extinction coefficients were evaluated for each zenith direction, independently of azimuth (*Figure 1*). The sugar beet canopy shows a more planophile behaviour with a quite flat extinction coefficient. Conversely, the sharp increase of gap frequency (or the associated extinction coefficient) observed over the wheat for large zenith angles are characteristic of erectophile leaves. These results are in good agreement with visual observations of the foliage structure of each crop. The leaf angle distribution was not accurately measured in this study. However, under the hypothesis of a constant value of the dispersion coefficient λ_0 with respect to the zenith angle θ, it is possible to use the Markov chain model and fit both the average leaf inclination (θ_l) and λ_0. An ellipsoidal distribution (Campbell, 1986) is assumed to represent the actual leaf angle distribution. The Poisson model performs well for sugar beet (λ_0 = 1.04, θ_l = 30.1°) but the wheat canopy has a regular behavior (λ_0 = 1.56, θ_l = 71.7°). For both canopies, the RMSE does not increase significantly.

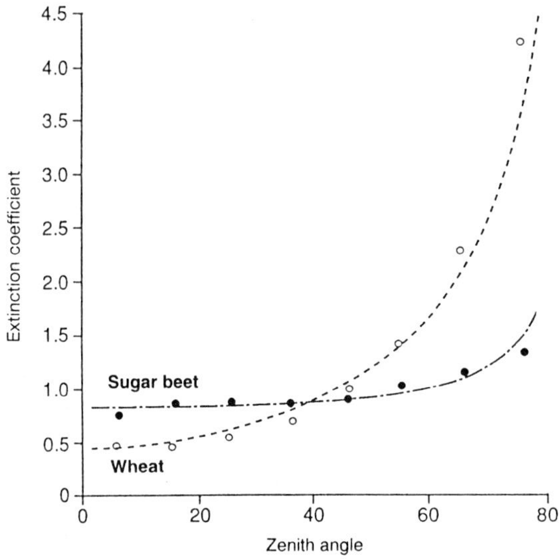

Figure 1. Variation of the extinction coefficient $K(q)$ as a function of the zenith angle q, when the canopies are assumed being azimuthally isotropic. The circles and crosses represent the adjusted values of the extinction coefficient. The solid lines correspond to the Markov chain model

Conclusion

This work provides an easy way to describe the gap frequency with as few input variables as the leaf area index and the type of canopy (in this case wheat or sugar beet). This is very useful when modelling the dynamics of the crop growth or the canopy spectral response. The present approach implicitly assumes some constant structural features of the canopies throughout their growth cycle or when the plant density varies, or at least a leaf orientation and spatial arrangement very strongly correlated with the leaf area index. However, in many cases, this simple approach will not apply, particularly for heliotropic crops, or when for example water stresses modify the potential scheme of the canopy development through structural changes such as occur with wilting.

References

Baret F., Andrieu B., Folmer J.C., Hanocq J.F. and Sarrouy C. (1992) Gap fraction measurement from hemispherical photography and its use to evaluate PAR interception efficiency. In *Proceedings of conference "Structure du couvert végétal et climat lumineux: Méthodes de caractérisation et applications"*, Edited by C. Varlet-Grancher and R. Bonhomme. Saumane, France, 23–27 September, 1991. (1992 in the press)

Campbell G.S. (1986). Extinction coefficients for radiation in plant canopies calculated using an ellipsoidal inclination angle distribution. *Agricultural and Forest Meteorology*, **36**, 317–321

Nilson T. (1971) A theoretical analysis of the frequency of gaps in plant stands. *Agricultural and Forest Meteorology*, **8**, 25–38

OZONE EFFECTS ON TWO BEAN CULTIVARS AT THE EBRO DELTA (SPAIN)

V. BERMEJO, R. ALONSO, S. ELVIRA, J.M. SALLERAS* and B.S. GIMENO
Ciemat, Instituto De Medio Ambiente, Av. Complutense 22, 28040 Madrid
**Migjorn S.A., Carretera del canal s/n, San Jaume d'Enveija, Tarragona, Spain*

Ozone levels at the Ebro Delta (north-east mediterranean coast of Spain) are high enough to produce effects on sensitive species. Since 1988, ozone effects on different crops have been assessed.

In May 1991, open-top chambers were installed at the Migjorn SA farm, on the Ebro Delta. Two bean cultivars were selected: cv. Lit (used in the European Open Top Chamber Programme) and cv. Win (commercial cultivar from the farm). Each chamber and plot was divided in two halves to study both cultivars simultaneously. Seeds were hand-sown in the soil of the chambers. Crop density was 50 plants m^{-2}. Standard farm practices were used, except for a trickle irrigation system.

Three treatments were used: charcoal filtered air (CF), non-filtered air (NF) and ambient air (AA). The experiment comprised three replicates for each treatment.

O_3, SO_2, NO and NO_2 concentrations, air temperature, relative humidity (RH), soil moisture and temperature, and photosynthetically active radiation (PAR) were measured continuosly.

The following parameters were measured: symptoms development; chlorophyll a, b and total content, ratio a/b and phaeophytinization quotient (PQa); carbon, nitrogen, hydrogen, phosphorus content and C:N ratio; number, fresh and dry weight of leaves, stems and pods (two green harvests); number and weight of seeds per pod and per plant (one dry harvest); and a daily cycle of photosynthetic activity, stomatal conductance, stomatal resistance and intercellular CO_2 concentration of cv. Lit (a portable photosynthetic system LICOR 6200 was used).

The seasonal 7 h mean ozone concentrations were 12, 40 and 48 nl l^{-1} in the CF, NF and AA treatments respectively (*Figure 1*). The efficiency of charcoal filters was around 70%.

Figure 2 shows the hourly mean of growing cycle of air temperature, RH, and PAR. Inside the chambers air and soil temperatures were 3–4°C higher than outside.

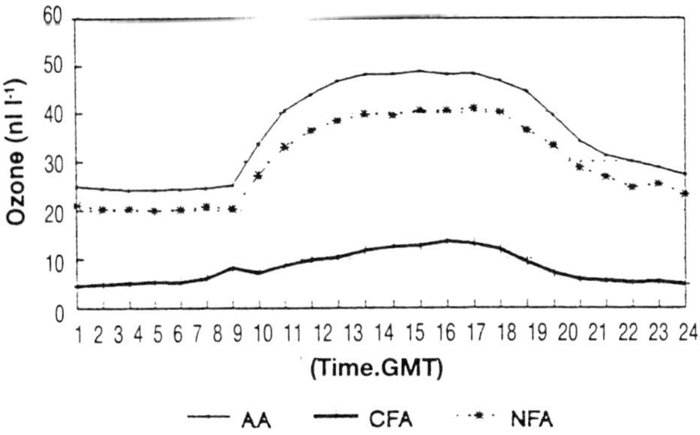

Figure 1 Ozone seasonal 7 h mean for ambient air, charcoal filtered air and non-filtered air treatments

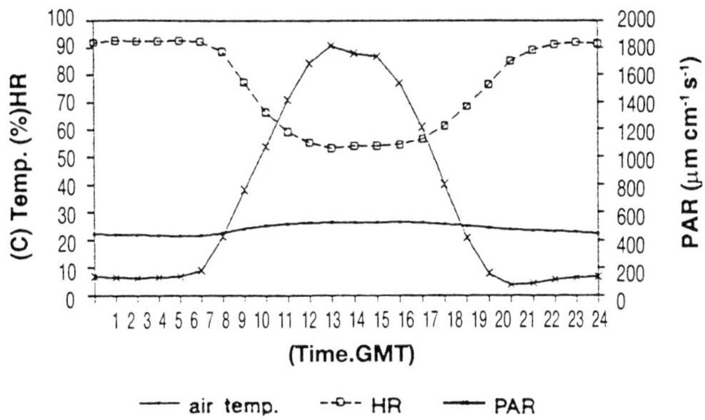

Figure 2 Hourly mean of air temperature, RH and PAR during the growing cycle

Results have shown a different ozone sensitivity for the two cultivars. Although neither cultivar has developed ozone visible symptoms inside the CF and NF chambers, their physiological behaviour was different, with Win more resistant than Lit.

Chlorophyll levels (a,b and total) in Lit were significantly decreased (26%)in NF and AA treatments compared with CF, at 14 days after emergence. In Win, on the same sample date, chlorophyll contents were (36%) significantly higher in CF chambers than in NF or AA plots ($P< 0.01$).

Carbon and nitrogen levels at the green harvest date in Lit were significantly higher, 4% and 17% respectively ($P < 0.01$) in CF chambers than in NF, and the C:N ratio was 14% lower. In Win, no significant ozone effects on the levels of the elements were observed.

At the green harvest, cv. Lit showed a significant reduction ($P < 0.01$) in the number of well developed pods in NF treatments compared with CF (12% reduction), but the greatest reduction was found in the AA plots (27% compared with CF). Win did not show ozone effects between treatments CF and NF, but AA plots had the lowest number of pods (9% to CF). In both cultivars, the percentage of non-developed pods was significantly lower ($P < 0.01$) in the CF treatment than in NF and AA, by 46% for Lit and 30% for Win.

A daily photosynthetic activity cycle was determined on Lit. Although photosynthetic rate was not significantly different between CF and NF chambers (except for 15:00 h GMT), a significant reduction of stomatal conductance was found in the plants exposed to ozone (*Figures 3* and *4*).

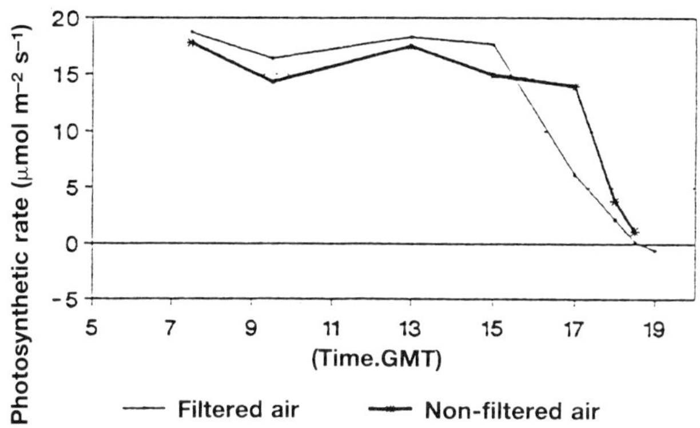

Figure 3 Daily cycle of photosynthetic activity for cv. Lit in filtered and non-filtered treatments

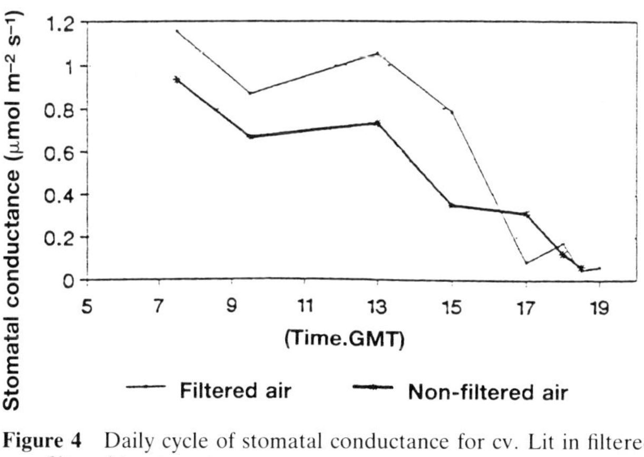

Figure 4 Daily cycle of stomatal conductance for cv. Lit in filtered and non-filtered treatments

INFLUENCE OF CULTIVAR AND NITROGEN FERTILIZATION ON INTERCEPTED RADIATION BY BARLEY IN DRYLAND CONDITIONS

C. CANTERO-MARTINEZ, J.M. VILLAR AND E. FERERES*
*U.P.C.-I.R.T.A. Rovira Roure, 177, 25006 Lleida, Spain; *Universidad de Córdoba, Dpto. Agronomía, Apdo.3048, 14008 Córdoba, Spain*

To determine cultivar adaptation of barley and the influence of nitrogen fertilization, an experiment in a dryland system in the north-east of Spain (Canos, Lleida) was carried out. The main climate pattern is Mediterranean with an average annual rainfall of 400 mm. The soil is classified as a xerocrept fluventic (Soil Survey, 1987). A factorial experiment with two different barley cultivars (Dobla and Tina) and two nitrogen levels of fertilization (40 and 100 kg ha^{-1}) was designed in randomized blocks with four replications. Growth of crop and solar radiation intercepted by the canopy was measured to evaluate the efficiency in radiation capture. Yield and yield components were determined at harvest. The two cultivars reached the same levels of intercepted radiation (*Table 1*). The higher values of extinction coefficient of Dobla cultivar demonstrated greater efficiency in capturing radiation (*Table 1*). At several stages of development (tillering, anthesis and maturity), the net assimilation rate (NAR) of Dobla was between 6 and 59% higher than that of Tina indicating greater photosyntetic efficiency of the canopy of Dobla cultivar. The yield results, as illustrated in *Table 2*, show that other factors were more important in its determination.

Table 1 INTERCEPTED RADIATION AND EXTINCTION COEFFICIENT OF TWO BARLEY CULTIVARS (DOBLA AND TINA) IN THREE STAGES

Year & Cultivar	*Tillering*		*Anthesis*		*Maturity*	
	I/Io(%)	K	I/Io(%)	K	I/Io(%)	K
1988						
Dobla	47 a	0.16 a	32 a	0.36 a	36 a	0.29 a
Tina	42 a	0.16 a	38 a	0.16 b	36 a	0.19 b
1989						
Dobla	66 a	0.30 a	49 a	0.40 a	42 a	0.33 a
Tina	63 a	0.36 a	41 b	0.32 b	46 a	0.26 b

Table 2 YIELD, YIELD COMPONENTS AND HARVEST INDEX OF TWO BARLEY (DOBLA AND TINA) CULTIVARS

Year & Cultivar	*Yield* (kg ha^{-1})	*HI*	*NEP*	*NGE*	*GW (mg)*
1988					
Dobla	3216 a	0.49 a	2.35 a	31.7 a	24.2 a
Tina	2602 b	0.33 b	2.20 a	18.2 b	27.7 a
1989					
Dobla	2186 a	0.44 a	1.17 a	26.7 a	26.0 a
Tina	1914 a	0.31 b	0.92 a	21.2 a	26.3 a

HI, Harvest index; NEP, number of ears per plant; NGE, number of grains per ear; GW, grainweight.

The data in *Table 3* show that the effect of nitrogen fertilization on radiation capture efficiency by the canopy was not simple. The analysis demonstrated the existence of a genotype–nitrogen interaction on radiation interception by the crop associated with the way the two genotypes responded to water availability during the year (Cantero-Martínez, 1989 and Villar, 1989). Thus, in this case it is difficult to explain the yields as a function of the radiation capture by canopy (*Table 4*).

Table 3 INTERCEPTED RADIATION AND EXTINCTION COEFFICIENT OF TWO BARLEY CULTIVARS (DOBLA AND TINA) AND TWO NITROGEN FERTILIZATION LEVELS (100 AND 40 kg ha^{-1}) AT THREE STAGES

Year & Treatment	*Tillering*		*Anthesis*		*Maturity*	
	I/Io (%)	K	I/Io (%)	K	I/Io(%)	K
1988						
N 100	47 a	0.16 a	34 a	0.26 a	34 a	0.29 a
N 40	44 a	0.18 a	34 a	0.20 b	37 a	0.19 b
Dobla 100	60 a	0.16 a	33 a	0.36 a	34 a	0.29 a
Dobla 40	44 b	0.18 a	32 a	0.22 b	37 a	0.20 b
Tina 100	40 b	0.16 a	36 a	0.16 b	34 a	0.19 b
Tina 40	44 b	0.19 a	36 a	0.19 b	37 a	0.19 b
1989						
N 100	62 a	0.32 b	42 b	0.36 a	20 b	0.30 a
N 40	66 a	0.43 a	49 a	0.34 a	29 a	0.31 a
Dobla 100	67 a	0.30 b	48 a	0.40 a	20 b	0.36 a
Dobla 40	64 a	0.47 a	50 a	0.53 a	22 b	0.26 b
Tina 100	48 b	0.36 b	36 b	0.32 b	19 b	0.26 b
Tina 40	69 a	0.40 ab	47 a	0.16 b	37 a	0.37 a

Table 4 YIELD, YIELD COMPONENTS AND HARVEST INDEX OF TWO BARLEY CULTIVARS (DOBLA AND TINA) AND TWO NITROGEN FERTILIZATION LEVELS (100 AND 40 kg ha^{-1})

Year & Treatment	*Yield* (kg/ha)	*HI*	*NEP*	*NGE*	*GW* (mg)
1988					
N 100	2869 a	0.41 a	2.27 a	24.9 a	26.9 a
N 40	2807 a	0.41 a	2.66 b	23.8 a	27.2 a
Dobla 100	3216 a	0.49 a	2.36 ab	31.7 a	24.2 b
Dobla 40	2986 a	0.48 a	2.66 a	27.7 a	26.2 b
Tina 100	2602 b	0.33 b	2.20 b	18.2 b	27.7 a
Tina 40	2629 b	0.36 b	2.66 a	20.0 b	29.3 a
1989					
N 100	2060 a	0.37 a	1.04 a	23.4 a	27.3 a
N 40	1960 a	0.38 a	0.91 a	20.7 a	29.6 a
Dobla 100	2186 a	0.44 a	1.17 a	26.7 a	26.7 a
Dobla 40	1914 a	0.44 a	1.06 ab	23.0 a	29.7 a
Tina 100	1916 a	0.31 b	0.92 ab	21.2 a	27.7 a
Tina 40	1978 a	0.33 b	0.77 b	18.5 a	29.5 a

HI, harvest index; NEP, number ears per plant; NGE, number of grains per ear; GW, grain weight

In this marginal cropping system, barley cultivars with high efficiency in radiation capture can be used to improve yields. However, when the water supply is restricted, production is less dependent on intercepting radiation; water–nitrogen relationships are much more important in this Mediterranean system.

References

Cantero-Martinez, C. (1989) *Caracterización agronómica y respuesta a la fertilización nitrogenada de genotipos de cebada y triticale en condiciones de secano en la Segarra. Lérida*. Ph.D. Thesis. UPC, Lérida

Soil Survey (1987) *Keys to soil taxonomy*. 3rd Edn. SMSS Technical Monograph 6. Ithaca, New York

Villar, J.M. (1989) *Evapotranspiración y productividad del agua en cebada y triticale en condiciones de secano en la Segarra (Lleida)*. Ph.D. Thesis. UPC, Lleida

EFFECTS OF ELEVATED CO_2 LEVELS ON RESOURCE CAPTURE AND USE IN GROUNDNUT AND SORGHUM CROPS

S.C. CLIFFORD, I.M. STRONACH, S.N. AZAM-ALI and N.M.J. CROUT
Tropical Crops Research Unit, Nottingham University, Sutton Bonington Campus, Loughborough, Leicestershire, LE12 5RD, UK

In 1991 groundnut (*Arachis hypogaea* L., a C_3 legume) and sorghum (*Sorghum bicolor* L. Moench., a C_4 cereal), were grown as undisturbed stands in controlled-environment glasshouses in either 350 ppmv (control) or 700 ppmv (elevated) atmospheric CO_2 concentration under non-limiting conditions.

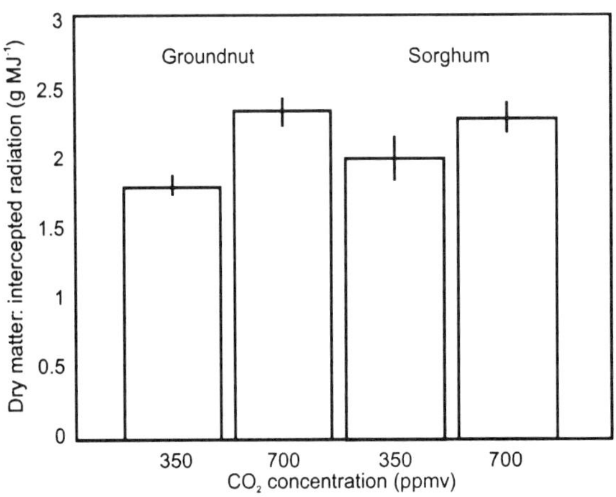

Figure 1 Dry matter:intercepted radiation ratio for groundnut and sorghum at 350 ppmv and 700 ppmv CO_2 concentrations. Error bars show one standard error about the mean

For groundnut elevated CO_2 accelerated canopy closure and allowed plants to support 25% more leaves compared with control levels. The mean leaf size and thickness remained unchanged. The dry matter:intercepted radiation ratio increased from 1.80 (±0.09) g MJ^{-1} in the control to 2.33 (±0.11) g MJ^{-1} in elevated CO_2, as shown in *Figure 1*. A similar large increase was seen in the dry matter:water use ratio, from 2.98 (±0.18) g kg^{-1} to 4.44 (±0.23) g kg^{-1} as can be seen in *Figure 2*. The average pod yield was increased by ~ 31% as demonstrated in *Figure 3*.

Preliminary data for sorghum indicated that elevated CO_2 had a relatively small effect on both dry matter:intercepted radiation ratio and dry matter:water use ratio (*Figures 1* and *2*), with a yield increase of no more than 6% (*Figure 3*).

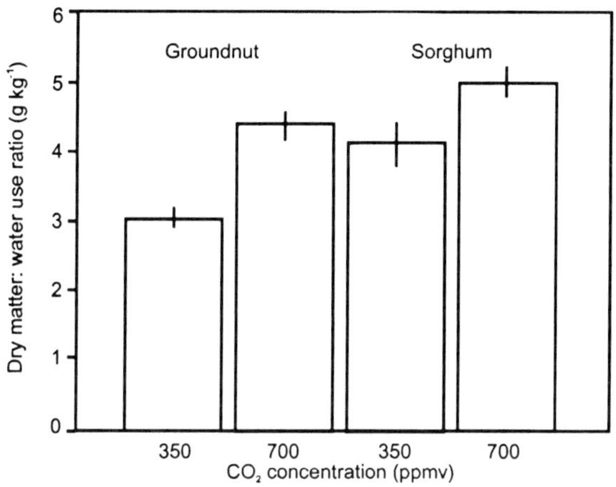

Figure 2 Dry matter:water use ratio for groundnut and sorghum at 350 ppmv and 700 ppmv CO_2 concentrations. Error bars show one standard error about the mean

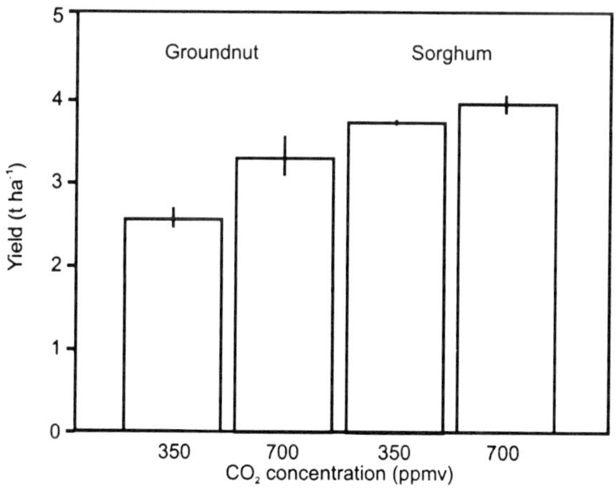

Figure 3 Pod or seed yield at final harvest for groundnut and sorghum at 350 ppmv and 700 ppmv CO_2 concentrations. Error bars show one standard error about the mean

Neither crop showed any change in proportion of dry matter partitioned to the reproductive elements, while measurements of net photosynthesis showed increased rates in groundnut at elevated atmospheric CO_2. The implication was that the increased yield occurring in elevated atmospheric CO_2 was driven by changes in the rate of net photosynthesis and dry matter:intercepted radiation ratio rather than by changes in partitioning. The increased dry matter:water use ratios observed for elevated CO_2 concentrations would be of great significance in the semi-arid tropics, which are particularly susceptible to drought and to small changes in climate.

A STATISTICAL TECHNIQUE FOR ANALYSING SPATIAL TRENDS IN RESOURCE CAPTURE

J.E. CORLETT[1], S. LANGTON*[2] and C.R. BLACK
*Department of Physiology and Environmental Science, University of Nottingham, Sutton Bonington, Loughborough, Leics, LE12 5RD, UK; *Central Science Laboratory, Tangley Place, Worplesdon, Guildford, Surrey, GU3 3LQ, UK*

In an alley cropping experiment in India (ICRISAT, Patancheru), growth, development and micrometeorological variables were measured row-by-row for pearl millet (*Pennisetum glaucum* (L.) R.Br.) grown between hedgerows of leucaena (*Leucaena leucocephala* (Lam.) de Wit). The treatments were: LM5, five rows of millet (inter-row spacing 47 cm) between hedgerows of leucaena (alley width 2.8 m); LM6, six rows of millet between leucaena hedgerows (alley width 3.3 m); LM6P, as LM6 but with vertical polythene barriers, 50 cm deep, separating the millet and leucaena root systems. The leucaena hedges were lopped to 70 cm before sowing the millet in both the 1986 and 1987 rainy seasons, and also 30 days after sowing in 1986. The polythene barrier effectively removed below-ground competition in 1986, but in 1987 excavation showed that leucaena roots had grown under the barrier and up into the millet rooting zone.

The treatment means for any variable could be compared using standard analysis of variance (ANOVA), but row-to-row differences posed problems because row position cannot be randomized. Rowell and Walters (1976) proposed the 'analysis of contrasts' to analyse trends in data collected over time, and this was applied in a modified form to compare spatial trends in microclimate and productivity across the alleys of all three treatments.

Linear and quadratic regression coefficients were calculated for each replicate and treatment and these coefficients were then analysed in a standard ANOVA. *Figure 1* shows a typical data set and *Table 1* a summary of the ANOVA. The treatment and year × treatment variances have been split into two contrasts: LM6P *versus* LM6 and LM5, and LM6 *versus* LM5.

The analysis shows that the treatments differed in their linear trend, with greater asymmetry across the alley in LM6P than in the other two treatments. The linear trends

[1]Present address: Horticulture Research International, Wellesbourne, Warwick CV35 9EF, UK
[2]S.L. provided statistical advice while on ODA contract at Rothamsted Experimental Station, Harpenden, Herts AL5 2JQ, UK

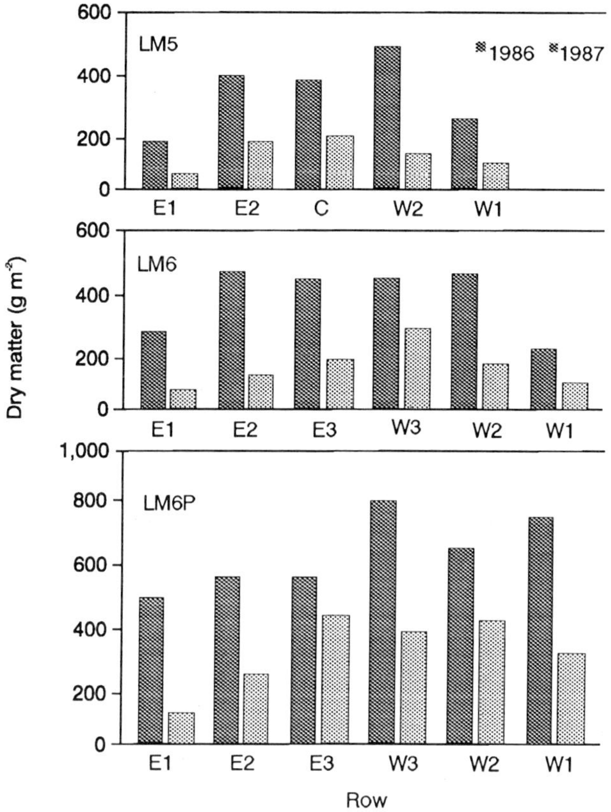

Figure 1 Row-by-row above-ground dry matter for millet at final harvest. Row E1, first millet row of the east of the leucaena hedge; Row C, central millet row in the alley

Table 1 SUMMARY OF ANALYSIS OF VARIANCE

Variate		*Linear coefficients*		*Quadratic coefficients*	
Source	*d.f.*	*Var. ratio*	*Significance*	*Var. ratio*	*Significance*
Treatment	2	6.9	*	9.6	*
LM6P *v* LM6 & LM5	1	12.8	*	5.8	
LM6 *v* LM5	1	1.0		13.4	*
Residual	6				
Year	1	53.2	***	16.6	**
Year × treatment	2	17.8	**	1.8	
Year × LM6P *v* LM6 and LM5	1	33.1	***	3.6	
Year × LM6 *v* LM5	1	2.5		0	
Residual	6				
Total	17				

*P = <0.05; **P = <0.01; *** P = 0.001

were more pronounced in 1987 than in 1986 and there was a significant year × treatment interaction. The treatments also differed significantly in their quadratic trends, with LM5 showing a more pronounced trend than LM6. There was a significant difference between the years, with treatments showing stronger quadratic trends in 1987 than 1986.

These results agree with other measurements (Corlett, 1989) that showed that below-ground competition was symmetrical, giving negative quadratic trends, and that any asymmetry was due to above-ground competition which tended to give a skewed quadratic effect. The asymmetry was due to the hedge orientation (NNW) and to the westerly wind that bent the hedges towards the east over the adjacent millet rows and shaded them from sun and rain (Monteith, Ong and Corlett, 1991).

This type of analysis is useful whenever it is desirable to compare trends in either space or time. Polynomials need not be used; for example, Yates (1982) described analysis of data with sigmoidal trends. Applications include: comparing horizontal profiles of plant, soil or meteorological variables (e.g. effects of different wind breaks); or comparing vertical profiles of variables such as soil moisture, root length density or horizontal wind speed.

References

Corlett, J.E., (1989) *Leucaena/millet alley cropping in India: microclimate and productivity*. Ph.D. Thesis, Nottingham University

Monteith, J.L., Ong, C.K. and Corlett, J.E., (1991) Microclimate interactions in agroforestry systems. *Forest Ecology and Management*, **45**, 31–44

Rowell, J.G. and Walters, D.E. (1976) Analysing data with repeated measurements on each experimental unit. *Journal of Agricultural Science (Camb)*, **87**, 423–432

Yates, F. (1982) Regression models for repeated measurements: Reader reaction. *Biometrics*, **38**, 850–853

VARIATION IN DARK RESPIRATION DURING THE GROWTH OF FIELD BEAN (*VICIA FABA* L.) UNDER DIFFERENT SOIL WATER REGIMES AND DENSITIES

L.C. COSTA[1], M.D. DENNETT, and J.I.L. MORISON*

*Departments of Agricultural Botany and *Meteorology, University of Reading, Reading, UK*

Introduction

Together with the analysis of resource capture by crops, it is important to understand how plants utilize those resources in their growth and maintenance processes. Since respiration is no longer considered energetically wasteful, but an essential plant process, the consideration of respiration in analyses of plant productivity could increase our ability to understand the link between resources captured by plants, and the efficiency of their utilization. Such information is very important, since the final productivity of a crop depends on

[1]Permanent address: Universidade Federal de Vicosa, DEA, 36570 – Vicosa-MG-Brasil

the amounts of resources that it is able to capture and on the efficiency of utilization of these resources.

The link between respiration and crop productivity in legume species is still not well known. The components of respiration, growth and maintenance, are expected to vary during leaf ontogeny and with environmental factors such as temperature and light. Under water stress the rate of respiration could either increase or decrease.

The most important factors in the control of respiration in plants are: substrate available and the amount of glucose, energy demand and the electron transport pathway.

Materials and methods

A field experiment was carried out in the summer of 1991 to examine the behaviour of the dark respiration of leaves inserted at different nodes of *Vicia faba* (cv. Gobo). The influence of leaf age, water stress, density and environmental conditions were followed. The plants were grown at different levels of soil water availability and at different densities. Dark respiration of leaves inserted at different nodes (4, 7, 10 and 20) on the plants, was followed during the growth of the crop, to quantify the dark respiration in different phases. Dark respiration was measured using a portable infrared gas exchange system (ADC). The treatments considered here are the factorial combination of high (49 plants m^{-2}) and low density (24 plants m^{-2}) with two water treatments W1 (fully irrigated) and W3 (crop growing only on the water available in the soil at planting).

Results

Specific respiration rates in leaves 4 and 7 were greater in the lower density plots than in the higher density plots ($P < 0.05$). The differences were more clear, for the well irrigated treatment, later on in the season, due the increase in Leaf Area Index. However, there were no statistically significant differences for leaves 10 and 20, which indicate that at the higher density the size of the substrate pool available for respiration, for the lower leaves, was smaller. Under water stress the results indicate that the substrate was still less available for the lower leaves. *Figures 1* and *2* show those results.

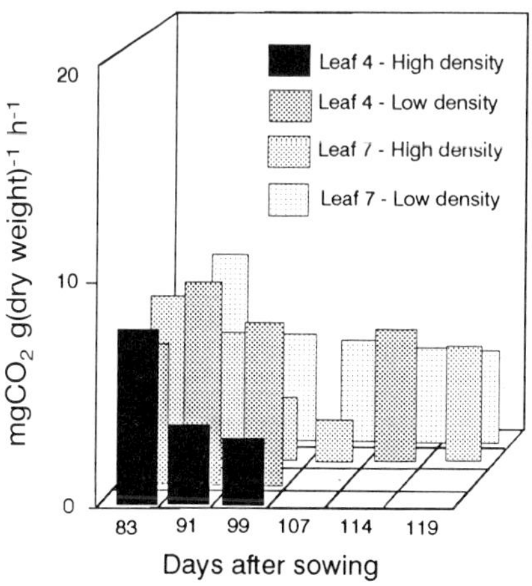

Figure 1 Specific respiration for leaves 4 and 7, in the well irrigated treatment, for both high and low densities

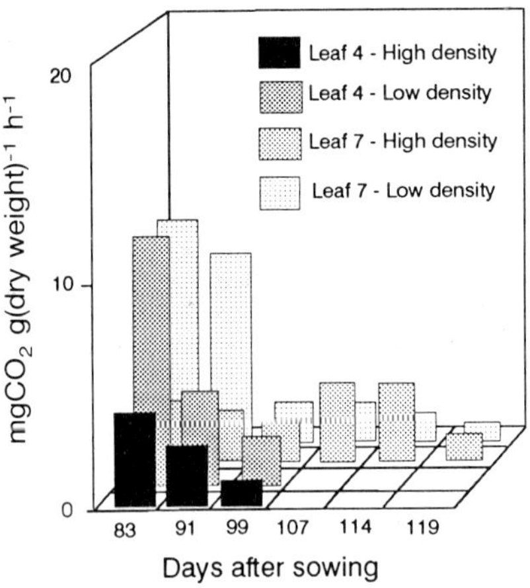

Figure 2 Specific respiration for leaves 4 and 7, in the dry treatment, for both high and low densities

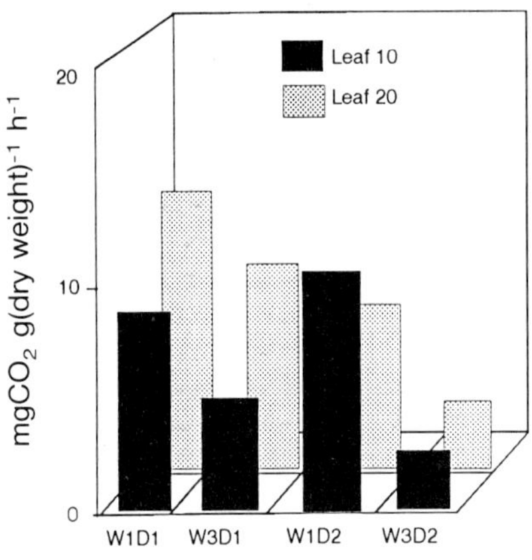

Figure 3 Specific respiration for leaves 10 and 20 at the time of full expansion. The treatments are: W1D1 – well irrigated, high density; W3D1 – dry, high density; W1D2 – well irrigated, low density; W3D2 – dry, low density

The rate of respiration of leaves 10 and 20 were compared at the same physiological age; the time of full expansion. The results presented in *Figure 3* indicate an increase in respiration for the later leaf, a fact that could be explained by two main reasons: the greater amount of light available at the time of growth of leaf 20 and by the presence of pods that are known to be sinks for substrate.

Conclusion

The results indicate that under field conditions the control of respiration is mainly due to the amount of substrate available, which depends on the amount of light. Water stress decreased the rate of respiration by decreasing the growth rate of the plant and consequently the growth component of respiration. Experiments are under way to find the variation of maintenance and growth respiration in different leaves during the growth of the plant and its link with productivity. The influence of the presence of sinks and the amount of substrate supply available for respiration will be quantified.

THE STATE OF PHYTOCHROME IN *POPULUS* CANOPIES, AND THE MEDIATED ACTION OF PROXIMITY PERCEPTION ON CANOPY GROWTH

I.R. GILBERT, G.E. SEAVERS, P.G. JARVIS* and H. SMITH
Department of Botany, University of Leicester, Leicester, UK;
**Department of Ecology & Resource Management, University of Edinburgh, Edinburgh, UK*

The signal transducing photoreceptor phytochrome can perceive variations in the relative fluxes or red and far-red light in actinic radiation. This property has been proposed as

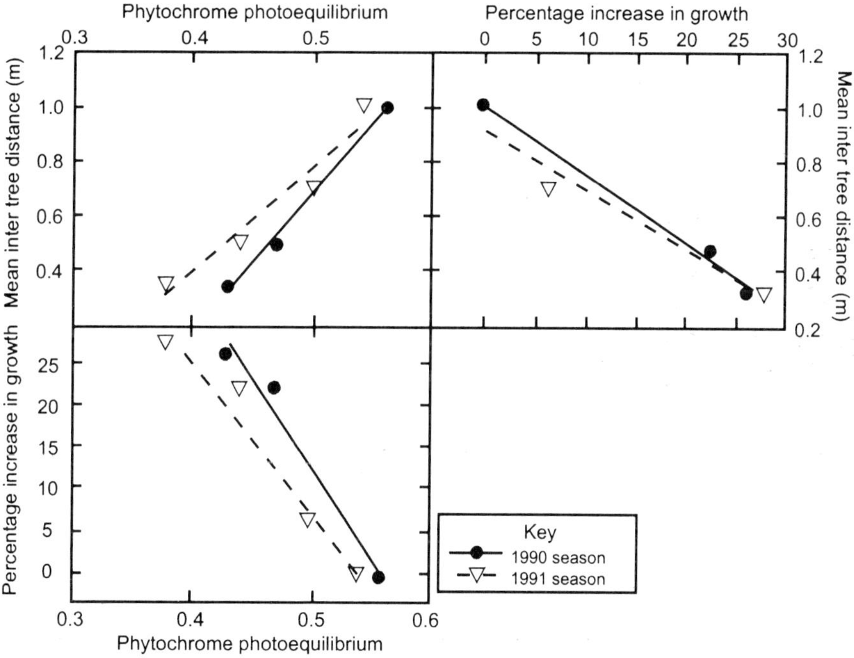

Figure 1 The linear relationships between the phytochrome photoequilibrium calculated from horizontally-propagated radiation, the inter-tree spacing, and the percentage increase in growth of the *Populus* 'Beaupré' main stems. Data were averaged from the start of growth until the photoequilibrium values began to rise, after the peak rate of elongation had passed, and leaf senescence started. Data from 1990 and 1991 are presented together, with the regression lines for each set of data points. The percentage increase in growth was related to trees grown at the intertree spacing of 1 m

conveying ecological value in the perception of neighbouring plants and the induction of shade avoidance responses. The work reported here has evaluated this potential by quantitatively determining the effect of neighbour proximity on the elongation growth of young tree canopies of *Populus trichocarpa* × *deltoides* 'Beaupré', and by relating these parameters to the long-term vectoral light quality inside each canopy. Radiation was measured in 10° horizontal wedges thoughout the growing season in four tree canopies of different densities. The light environment, in which the plants grew, was not artificially modified, but was found to be modified in quality and quantity by the plants themselves.

Linear relationships were found between increase in elongation rate, plant density and the phytochrome photoequilibrium calculated to be established by light propagated approximately horizontally within the canopies as shown in *Figure 1*. The photoequilibrium, calculated from horizontally propagated radiation measured in the vertical middle of each canopy, linearly decreased in each canopy until the peak in elongation growth was passed (*Figure 2*). The lower leaves then started to senesce; and the photoequilibrium started to increase. The depression in the photoequilibrium was greater in the denser canopies, than the less dense, and was found to be similar for two consecutive field seasons.

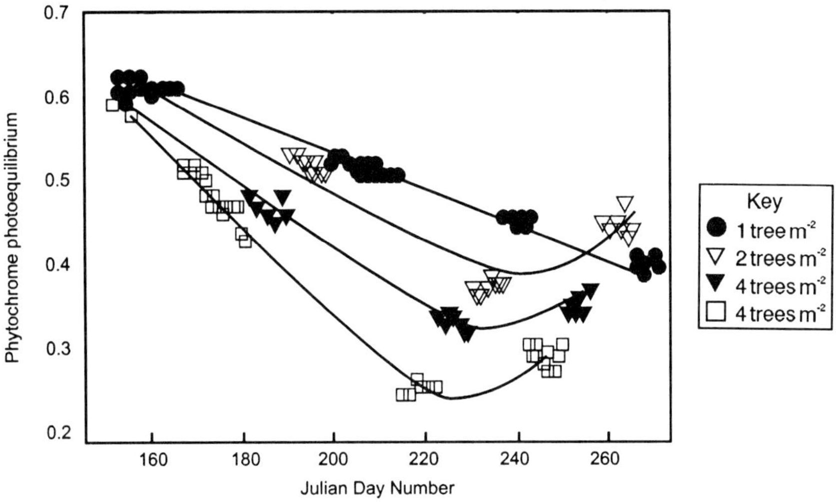

Figure 2 The calculated phytochrome photoequilibria from horizontally-propagated radiation for four canopy densities in the 1991 growth season. Data points show the actual measured values. The straight lines show the interpolated photoequilibrium values for each canopy density. All values are plotted against the Julian calendar

The findings support the proposed role of phytochrome in proximity perception in the natural environment, and provide a quantitative basis for investigating the competitive interactions between plants growing in dense stands. The competitive interactions between *Populus* 'Beaupré' trees in developing canopies can be fully accounted for on the basis of phytochrome mediated perception signals from neighbour plants, and the resultant shade avoidance. The conclusion holds valid when the canopy is young and mutual shading cannot occur, and even when the canopy has closed and mutual shading does occur.

OZONE AFFECTS YIELD OF WATERMELON PLANTS GROWN AT THE EBRO DELTA (SPAIN)

B.S. GIMENO, V. BERMEJO, A. TARRUELL* and J.Mª. SALLERAS†
*Ciemat, Instituto De Medio Ambiente, Av. Complutense 22, 28040 Madrid; *Instituto De Salud Carlos III, Majadahonda, 28220 Madrid; †Migjorn SA, Carretera del canal s/n, San Jaume d'Enveija, Tarragona, Spain*

In 1987 suspected effects of ozone on crops were described for the first time in Spain. These symptoms were detected earlier on the Migjorn farm, located at the Ebro Delta, during the summer of 1976. Since then, they have been observed every growing season.

During 1988, 1989 and 1990, a study using open-top chambers was performed at the Ebro Delta to assess the effects of ambient levels of ozone on visual injury, yield and quality of watermelon plants. An open top chamber experimental field was established at Migjorn farm, three treatments were defined (ambient air (AA), non-filtered air (NF) and charcoal filtered air (CF)), and five replications per treatment were used.

Four watermelon plants (*Citrullus lanatus* L cv.Toro) per plot were transplanted from greenhouse. Standard cultural practices were used, except for the trickle irrigation system.

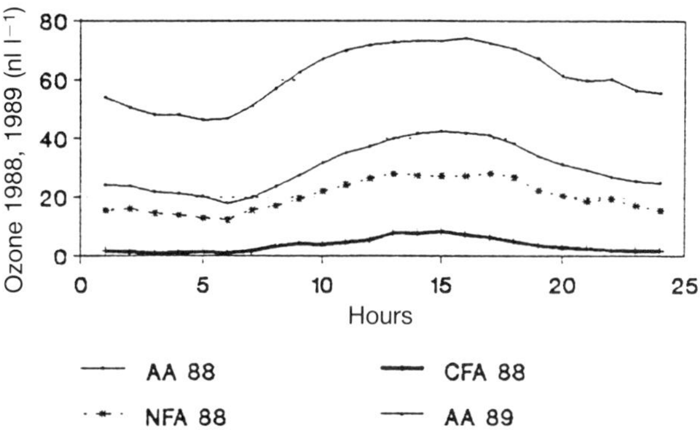

Figure 1 Ozone hourly mean during the watermelon growing season of 1988 for ambient air (AA 88), non-filtered air (NFA 88) and charcoal filtered air (CFA 88). 1989 ambient air data (AA 89) are presented as well

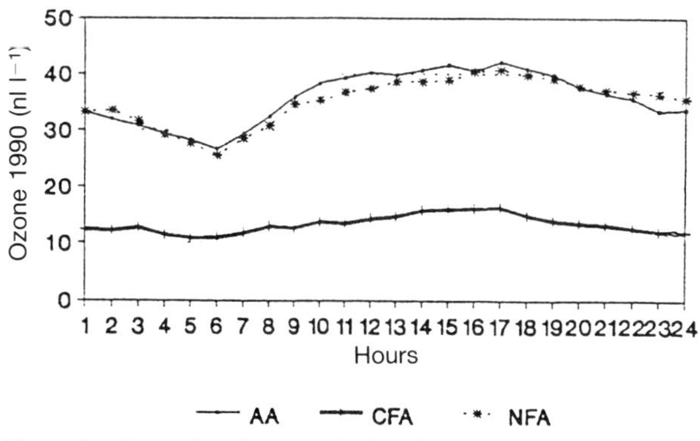

Figure 2 Ozone hourly mean during the watermelon growing season of 1990 for ambient air (AA), non-filtered air (NFA) and charcoal filtered air (CFA)

Fruits were collected on a weekly basis, following their full maturity. They were weighed, and their organoleptic characteristics (brix degrees) were determined by means of a manual refractometer.

SO_2, NO_x and O_3 levels were monitored in every plot, using a time sharing system. SO_2 and NO_x (NO and NO_2) levels were always around their detection limit. O_3 levels varied through the different years, being highest in 1989 and lowest in 1990 (*Figures 1* and *2*).

A chamber effect was detected: there was a delay in fruiting time inside the chambers compared with AA plots. The effect was most visible in the last harvest of every year, in which fruit number was much higher in chambers (CF and NF) than in ambient plots (although they were not viable commercially).

During the three sampling years, the number of fruits was significantly higher ($P< 0.05$) in filtered chambers than in non-filtered or ambient plots (*Figure 3*).

The final weight yield of commercial fruits was always higher but not statistically significant in the CF treatment than in NF or AA. Since a delay in fruiting time was detected in the chambers, and those newly formed fruits could not reach their full maturity, the fruit number per plot was multiplied by an average mature weight (6 kg). Following this procedure the number of fruit and the final yield were significantly higher ($P< 0.05$) in CF chambers than in NF (*Figure 4*).

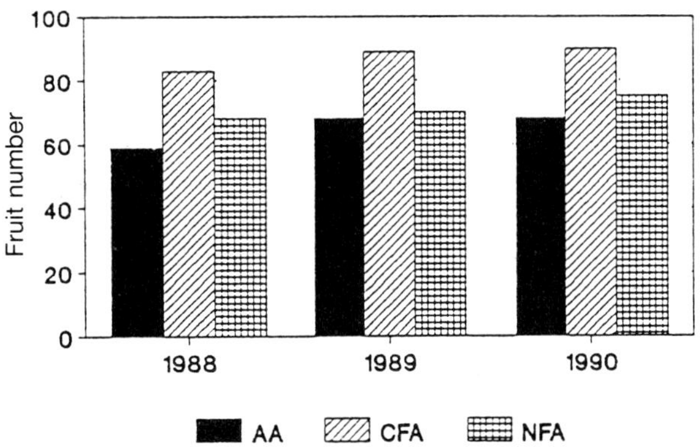

Figure 3 Watermelon yield of 1988, 1989 and 1990 expressed as fruit number for ambient air (AA), charcoal filtered air (CFA) and non-filtered air (NFA)

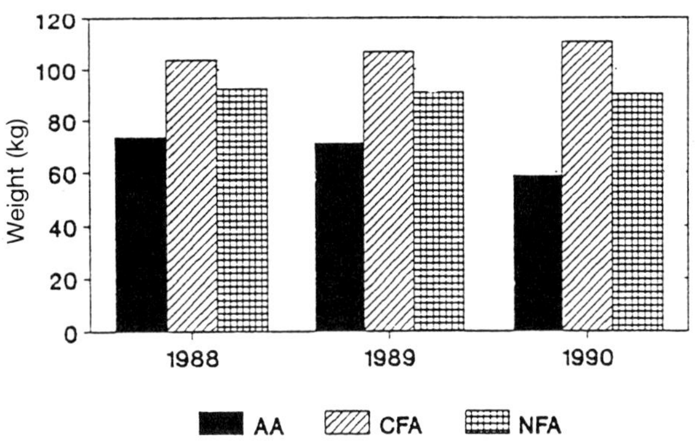

Figure 4 Watermelon yield of 1988, 1989 and 1990 and expressed as total fruit weight (kg) for ambient air (AA), charcoal filtered air (CFA) and non-filtered air (NFA)

ROOT SYSTEMS OF SOME TREE SPECIES AND IMPLICATIONS FOR RESOURCE CAPTURE IN AGROFORESTRY

P.A. HUXLEY
International Centre for Research in Agroforestry (ICRAF), P.O. Box 30677, Nairobi, Kenya

Two characteristics of the root systems of woody plants are especially important when they are grown in mixtures with crops and/or grasses. First, the fine roots of many tree species are much more 'clumped' than those of herbaceous plants, the roots of which tend to fill out the occupied space, except at the periphery (see Weaver, 1926; Weaver and Bruwer, 1927; Kutschera, 1960; Kutschera *et al.*, 1982). Clumping occurs because tree roots tend to exploit new or unused soil volume through the extension of 'long' roots that often remain unbranched for some time before forming a proliferation of fine roots. Second, some (but by no means all) tree roots systems can root relatively more deeply than most crop or grass species and many, either sooner or later, form extensive lateral systems; species, soil type and management can affect these characteristics, which are more prevalent and accentuated in drier than in wetter regions. In agroforestry, where woody and non-woody species are grown in close proximity, these characteristics of tree root systems will need to be taken into account when considering the relative efficiency of below-ground resource capture.

Root profiles obtained in ICRAF's Field Station, Machakos (Rao *et al.*, 1992), have identified *Cassia siamea*, for example, as a multi-purpose tree (MPT) species that has a particularly extensive root system (over 15 m in 4½ years, for unpruned trees) so that its roots can quickly traverse, and may even extend beyond, adjacent cropped plots. Again, a root profile (14 m long × 2 m deep) of *Grevillea robusta* grown as a hedgerow, co-existed with the immediately adjacent maize crop in the top 1.3 m but it occupied the lower profile alone (Huxley *et al.*, in preparation). Where tree roots both share and solely occupy soil volume, as in these two cases, then the uptake of water and soluble nutrients is likely to occur preferentially from the solely occupied volume of soil, either seasonally or even diurnally, as long as the resistance to water traversing the major roots remains less than that obtaining in the depletion zones around the crop roots in the shared parts.

Even where tree roots share macro-space with a crop, there may be less direct competition for resources than at first appears because of 'clumpiness'; especially of the tree root system.

Root numbers of both species were recorded concurrently in the *G. robusta*/maize profile, using a 5 × 5 cm grid unit (a 'micro-unit'); so that it was possible to examine the extent to which these small soil units were solely occupied by either species, or were shared. Relatively few were extensively shared and there were micro-units in shared macro-space that remained empty; i.e. they remained available for tree or crop roots in the future. In this situation of limited interference, competition for below-ground resources may not depend so much on direct contention at any one time for resources by individual roots (Tardieu *et al.*, 1922), but more on the number and rapidity with which micro-units of soil volume are occupied by the roots of each species 'pre-emptive' competition). Both root 'clumpiness' and early fine root extension growth could be important characteristics to know about when selecting MPTs for agroforestry. The extensive data sets are being examined more thoroughly using various mathematical techniques.

References

Kutschera, L. (1960) *Wurselatlas mittel europäischer Ackerrunkräuter and Kulturpflanzen.* Frankfurt am Main: DLG Verlag GMBI

Kutschera, L., Lichtenegger under Von Monika Sobotik, M. (1982) *Wurzelatlas mitteleuropäischer Grünlandpflazen. Band I: Monocotyledoneae*. Stuttgart: Gustav Fischer Verlag, 516 pp
Huxley, P.A., Pinney, A., Akunda, A., and Muraya, P. (in preparation). A tree/crop interface orientation experiment with a *Grevillea robusta* hedgerow and maize
Rao, M.R., Muraya, P. and Huxley, P.A. (1993) Observations of some tree root systems grown in agroforestry intercrop situations, and their graphical presentation. *Experimental Agriculture*, **29**, 183–194.
Tardieu, F., Bruckler, L. and Lafolie, F. (1992) Root clumping may affect root water potential and the resistance to soil-root water transport. *Plant and Soil*, **140**, 291–301
Weaver, J.E. (1926) *Root development of field crops*. New York: McGraw–Hill, 291 pp
Weaver, J.E. and Bruwer, W.E. (1927) *Root development of vegetable crops*, New York: McGraw Hill, 351 pp

DEVELOPMENT AND TESTING OF A NEW MODEL FOR DETERMINING THE ALLOCATION BETWEEN SHOOT AND ROOT. PART II: INFLUENCE OF WATER AND NITROGEN STRESS

T. KARVONEN, J. KLEEMOLA and J. PELTONEN
Department of Plant Production, University of Helsinki, Finland

Introduction

The effect of water and/or nitrogen stress on allocation between shoot and root is discussed. The basic idea of the new method is to maximize the daily growth of plant tissue; the method is presented more thoroughly by Kleemola and Karvonen in Part I. The influence of water and nitrogen stress is taken into account by a two step process. Firstly, the specific activity of roots is related to uptake of water and nitrogen under optimum conditions. Secondly, the influence of too dry soil or deficiency of mineral nitrogen in the rooted zone is taken into account. Soil water content and available mineral nitrogen in the rooted zone are determined by using simple methods for soil water and nitrogen balance. In addition to including water and nitrogen stress into the model, a few procedures presented in Part I were modified. The effect of nitrogen on growth was assessed by assuming that in addition to daily uptake of nitrogen almost all (99%) of plant nitrogen was available when determining the nitrogen stress on growth. Also the effect of temperature on growth and nitrogen depletion from shedding plant parts were taken into account.

Results

The model was tested against winter wheat data published by Groot, Willigen and Verberne (1991). The growing period 1982–83 was used for testing and the experimental farm was 'the Eest'.

Under both nitrogen treatments (0 kgN ha^{-1} and 160 kgN ha^{-1}) the model overestimated the measured data; the overestimation was smaller under high nitrogen treatment (*Figure 1 a, b*). The measurements indicate that the biomass of crop increased also during the latest part of the growing season, whereas the simulated shoot biomass decreased in the end due to low leaf area index (low photosynthesis) and high maintenance respiration. The parameters affecting the uptake of nitrogen (BmRootSat and NTakeMax) did not have much effect on

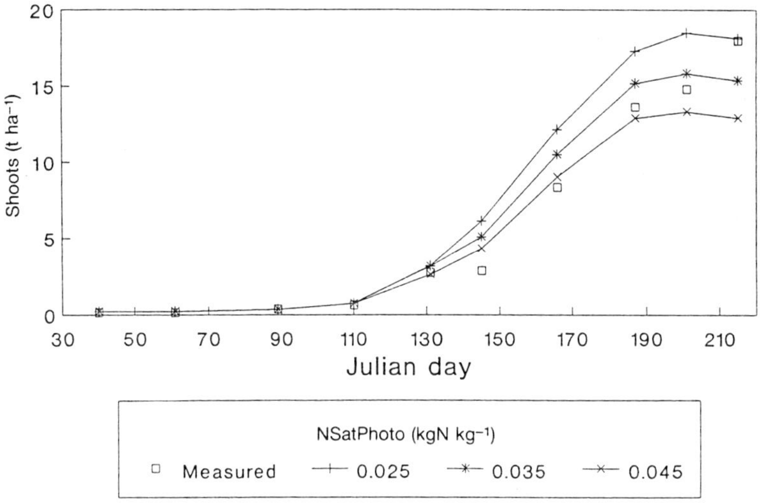

Figure 1a The measured and simulated biomass of shoots under low nitrogen treatment (0 kgN ha^{-1}); the effect of parameter NSatPhoto (see text) on simulated biomass of shoots

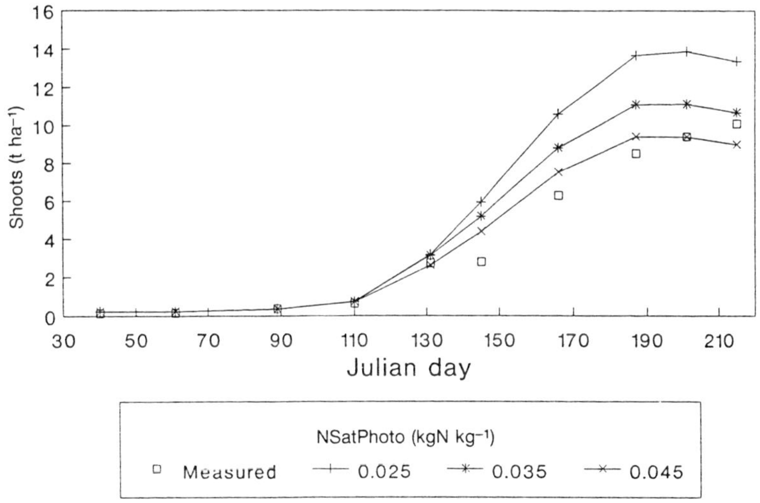

Figure 1b The measured and simulated biomass of shoots under high nitrogen treatment (160 kgN ha^{-1}); the effect of parameter NSatPhoto (see text) on simulated biomass of shoots

the calculated shoot biomass (results not shown) under low nitrogen treatment, whereas the effect was clear under high nitrogen treatment. The parameter that relates leaf nitrogen concentration and maximum photosynthesis of a single leaf (NSatPhoto) affected shoot biomass clearly under both treatments (*Figure 1 a, b*). The *Figures 2 a and b* present the effect of different values of BmRootSat and NTakeMax on grain dry weight. The overall effect is the same than that for shoot biomass under different treatments because the biomass of grains is computed using prescribed harvest index function: the influence

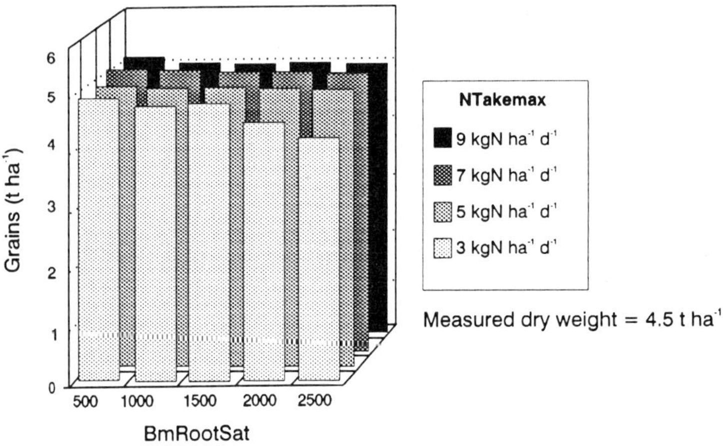

Figure 2a The effect of altering the parameters BmRootSat and NTakeMax (see text) on simulated biomass of grains under low nitrogen treatment (0 kgN ha^{-1})

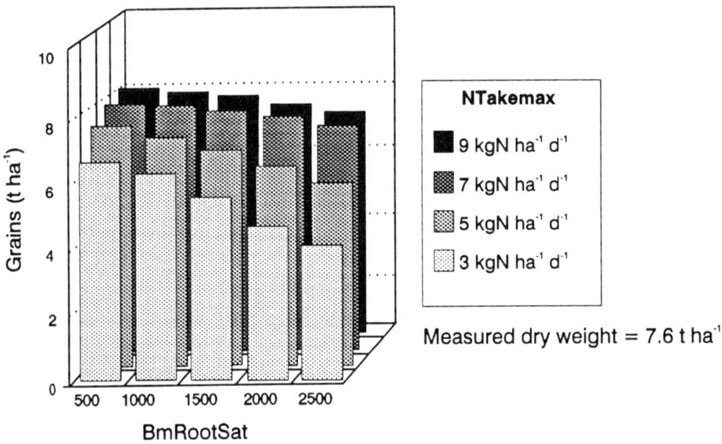

Figure 2b The effect of diverting the parameters BmRootSat and NTakeMax (see text) on simulated biomass of grains under high nitrogen treatment (160 kgN ha^{-1})

of change in parameters was greater under high nitrogen treatment. As the biomass of roots needed to saturate the maximum nitrogen uptake (BMRootSat) increases a plant has to invest more biomass to roots in order to maintain the balance between roots and shoots and to be able to satisfy the demand for nitrogen. Hence, the biomass of shoots (and grains) is decreased.

Discussion

The model seems to fail in simulation between days 130 and 145; after that the actual and computed growth rates are nearly the same except at the end of the growing season. This indicates that during that period there has been a stress which was not taken into account by the model. The fact that the model did not respond to changing the nitrogen uptake parameters under low nitrogen treatment but had influence under high nitrogen treatment was caused by the low content of the mineral N in the soil. There was no use for higher maximum uptake rate if there was not enough N available.

The present model is capable of determining a dynamic parameter for allocation between root and shoot; the allocation among shoot organs is still prescribed. This is a lack, because relations of dry masses of shoot organs may vary under stresses. Other procedures that also lack physiological basis are determination of the growth of plant tissue and nitrogen uptake. The temperature limit of growth is relatively crucial for the growth during the winter time, small change in a parameter value may affect big differences to cumulative biomass. The model for nitrogen uptake is rather simple, only the amount of dry mass of root affects the potential nitrogen uptake; at least a procedure for rooting depth would be necessary to add to the model.

One objective for the present research was to develop a simple model for determining the allocation between shoot and root. However, there are still many relatively poorly known or measurable parameters in the model (root biomass needed for saturation of nitrogen uptake, maximum uptake of nitrogen for closed canopy, amount of nitrogen and carbon reserves that are available for growth processes etc.). Moreover, the use of parameters determining maximum or optimum levels are criticized by some researchers (Willigen and Noordwijk, 1987).

References

Groot, J.R.R, Willigen, P. de and Verberne, E.L.J. (1991). Special issue, nitrogen turnover in the soil-crop system. *Fertilizer Research*, **27** (2–3), 141–383

Willigen, P. de and Noordwijk, M. van (1987) *Roots, Plant Production and Nutrient Use Efficiency*. Ph.D. Thesis. Agricultural University of Wageningen. 282 pp

DEVELOPMENT AND TESTING OF A NEW MODEL FOR DETERMINING THE ALLOCATION BETWEEN SHOOT AND ROOT. PART I: OPTIMUM GROWING CONDITIONS

J. KLEEMOLA and T. KARVONEN
Department of Plant Production, University of Helsinki, Finland

Introduction

Several crop growth models relate the partitioning between shoot and root to the development stage of a crop or directly to the temperature sum (DD). This method has one disadvantage: if a crop is grown under stress, the partitioning is different from that under optimum conditions. The models using the partitioning method mentioned above are not capable of taking into account the effect of stresses on partitioning, unless the partitioning parameters are changed.

Model description

This study presents a method for determining a dynamic allocation parameter of assimilates between shoot and root. The present method is based on a simple idea that a plant allocates assimilates supplied by the photosynthesis in such a way that the growth of plant tissue is maximized (β_{root} is optimized). As the growth is maximized, the ratio of demand and supply of carbon and nitrogen is maintained in balance. Hence, this assumption is a simplification of the hypothesis that proposes that there is a functional equilibrium between the size and activity of shoot and root (Thornley, 1972).

The partitioning model has four submodels which are used in determining the actual daily growth of the crop: 1. photosynthesis, 2. potential growth, 3. nitrogen uptake, and 4. nitrogen demand. The potential growth is assumed to be the maximum daily increment of biomass of plant tissue.

In the present model the net CO_2 assimilation of canopy added to the amount of available reserves is regarded as a carbon source. The computation of daily net assimilation of canopy is based on the light response curve of a single leaf. The maximum value of the curve (PhotoMax) is a function of leaf nitrogen content (van Keulen and Seligman, 1987). The instantaneous canopy assimilation and daily photosynthetic integral are derived from the photosynthesis of single leaf using methods presented by Goudriaan (1986), Spitters, Toussaint and Goudriaan (1986) and Spitters (1986). The daily net carbon dioxide assimilation is further converted to glucose according to Penning de Vries *et al.* (1989). The amount of available reserves (20%) is added to the net assimilation and the maintenance respiration is subtracted in order to determine the amount of assimilates available for the growth of plant tissue (Graf *et al*,. 1989). Potential growth (carbon demand) is a function of development stage and biomass. A relative growth rate parameter (μ) determines the potential growth: the potential growth is computed by multiplying the biomass of crop and μ. The value of μ is reduced during the growing season as the crop is altering.

Nitrogen uptake (nitrogen supply) is determined using a simple method which relates the biomass of roots and nitrogen uptake (*Figure 1*). The two parameters needed are the maximum uptake rate of N (NTakeMax) and the root biomass which saturates the daily uptake (BmRootSat). The computation of the nitrogen demand of the crop (a function of development stage, DVS) is based on the difference between optimum and actual nitrogen concentrations of the crop (van Keulen and Seligman, 1987). The DVS is the ratio of temperature sum accumulated since emergence to temperature sum needed for ripening.

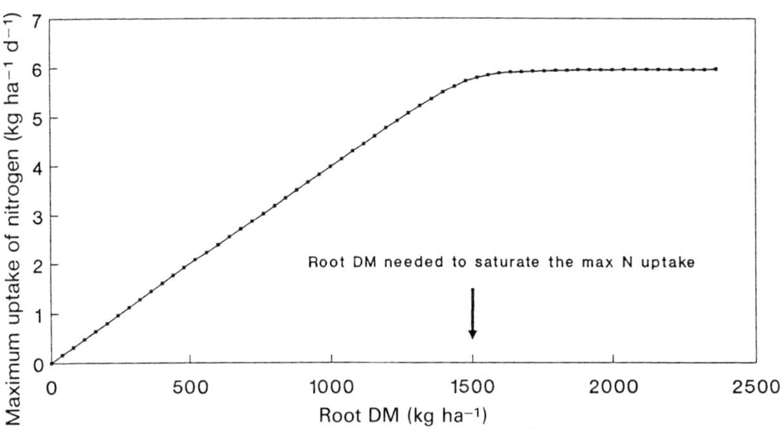

Figure 1 Relation between biomass of roots and nitrogen uptake

The supply:demand ratios of carbon and nitrogen are used in computing the stress factor which further determines the ratio between actual and potential growth.

Sensitivity analysis

In sensitivity analysis barley (*Hordeum vulgare*) was used as a test crop (i.e. parameters were for barley). The computation period was from emergence to anthesis (0.0–0.5) and the timestep of computations was one day. The parameters used in the computation of

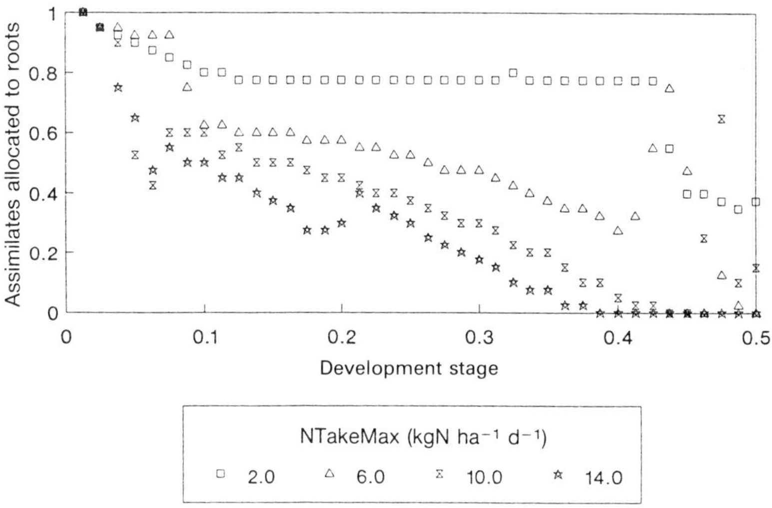

Figure 2 Allocation to roots using four values of parameter for maximum uptake of nitrogen

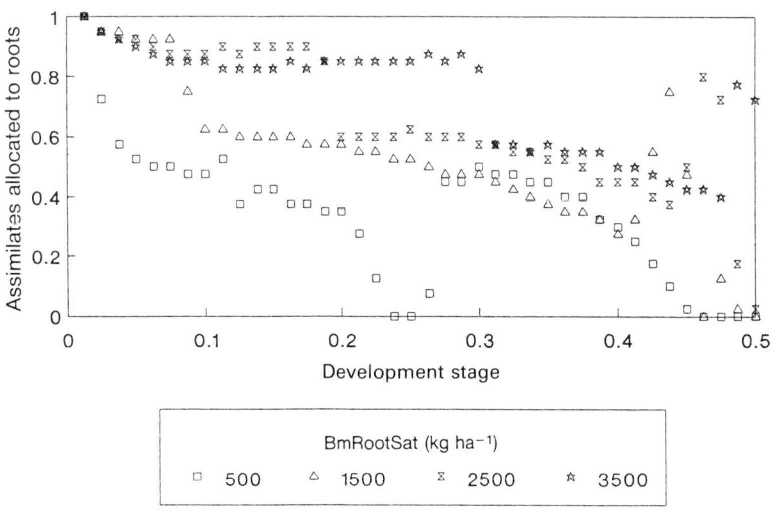

Figure 3 Allocation to roots using four values of parameter for biomass of roots needed to saturate the maximum uptake of nitrogen

daily assimilates, growth respiration and maintenance respiration were adopted from the literature (Penning de Vries *et al.*, 1989). The parameters for dry matter partitioning among shoots were computed using unpublished data from a glasshouse experiment established in the Department of Plant Production.

Figures 2 and *3* show the sensitivity of the model for the two parameters that determine the daily nitrogen uptake. In can be seen that the allocation to roots decreases as the maximum uptake of nitrogen increases and as the amount of root dry matter needed for maximum N uptake decreases. The influence of the parameters BmRootSat and NTakeMax on root:shoot ratio is given in *Figure 4*. Root:shoot ratio seemed to increase steadily as

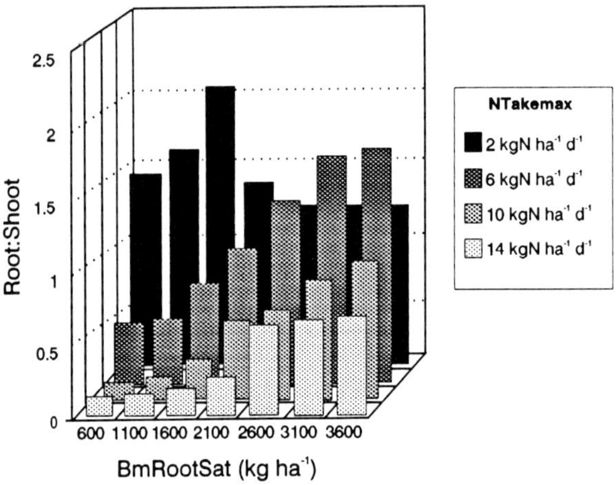

Figure 4 Root:shoot ratio using four values of parameter for maximum uptake of nitrogen and seven values of parameter for biomass of roots needed to saturate the maximum uptake of nitrogen

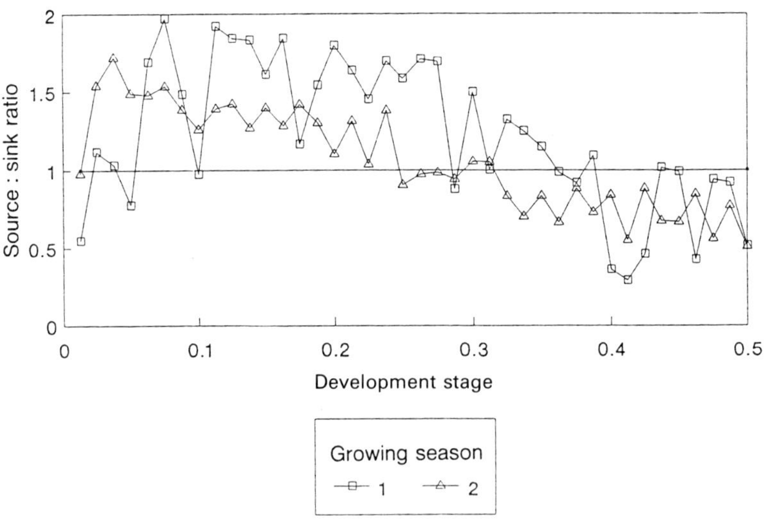

Figure 5 Source:sink ratio during two growing seasons with different conditions of solar radiation

the availability of nitrogen decreased. The value 2 kgN ha^{-1} d^{-1} of maximum uptake of nitrogen was an exception: the root:shoot ratio began to decrease as high values of the BmRootSat was used. This was due to the fact that when such parameter values were used, the nitrogen stress was severe and the crop never began to grow; the root:shoot ratios were mainly determined by initial values of biomasses of roots and shoots. *Figure 5* presents the ability of the model to simulate the source:sink ratio during the growing season. During the early part of the growing season the size of sink restricts the growth, whereas later on the photosynthesis becomes the limiting factor. The lower curve indicates a growing period when the solar radiation is relatively low; photosynthesis becomes the limiting factor earlier than under high radiation conditions. The increase in PhotoMax resulted in an increased root:shoot ratio (*Figure 6*). This kind of response has been noticed for example in experiments done under low and high CO_2 concentrations: high CO_2 increases maximum photosynthesis and also root:shoot ratio.

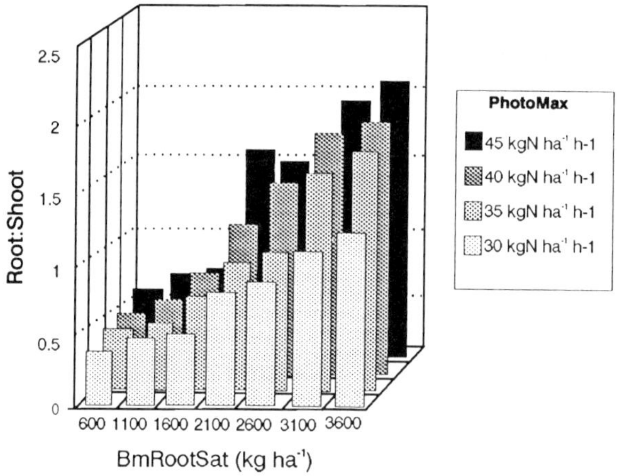

Figure 6 Root:shoot ratio using four values of parameter for photosynthetic maximum of a leaf and seven values of parameter for biomass of roots needed to saturate the maximum uptake of nitrogen

Discussion

In general, the model responded as expected to changes in the parameter values. For example, the parameter β_{root} was lower if nitrogen was more easily available. On the other hand, an increase in the shoot activity (maximum photosynthesis of a leaf) resulted in higher root:shoot ratio. However, there are also some cases that need more discussion. In certain cases the allocation parameter tended to vary a lot on successive days (*Figures 2* and *3*). Usually, if photosynthesis is the limiting factor, β_{root} is small whereas growth as a limiting factor results in bigger β_{root}. During days when e.g. photosynthesis is the limiting factor an optimum between zero and one was difficult to determine: in such cases the β_{root} was usually zero.

In *Figure 3* β_{root} rises sharply after development stage 0.21 (BmrootSat equals 500 kg). This may have been affected by the fact that the model maximizes the daily growth; the low values of β_{root} resulted in nitrogen stress later on during the growing period which increased the values of β_{root}.

Development needs:

- In the present model the partitioning within shoot is computed using prescribed functions of development stage; thus the model is not capable of taking into account the influence of stresses on partitioning among shoot organs.

- The parameters used in the computation of growth are difficult to determine due to the fact that the measurements done in the field usually include reserves, i.e. the weight of plant organ may rise without any growth of tissue.
- If one wants to do simulations over the whole growing period, nitrogen depletion from shedding plant parts has to be taken into account; the present model is not capable of simulating the translocation of nitrogen.
- A mechanistic nitrogen uptake model is needed.

References

Goudriaan, J. (1986) A simple and fast numerical method for the computation of daily totals of crop photosynthesis. *Agricultural and Forest Meteorology* **38**, 249–254

Graf, B., Rakotobe, O., Zahner, P. Delucchi, V. and Gutierrez, A.P. (1990) A simulation model for the dynamics of rice growth and development: Part I – The carbon balance. *Agricultural Systems*, **32**: 341–365

Penning de Vries, F.W.T., Jansen, D.M., ten Berge, H.F.M. and Bakema, A. (1989). *Simulation of Ecophysiological Processes of Growth in Several Annual Crops*. Pudoc: Wageningen 271 p.

Spitters, C.J.T., Toussaint, H.A.J.M. and Goudriaan, J. (1986) Separating the diffuse and direct component of global radiation and its implications for modeling canopy photosynthesis. Part I: Components of incoming radiation. *Agricultural and Forest Meteorology*, **38**, 217–229

Spitters, C.J.T. (1986) Separating the diffuse and direct component of global radiation and its implications for modeling canopy photosynthesis. Part II: Calculation of canopy photosynthesis. *Agricultural and Forest Meteorology*, **38**, 231–242

Thornley, J.H.M. (1972) A balanced quantitative model for root:shoot ratios in vegetative plants. *Annals of Botany* **36**, 431–441

van Keulen, H. and Seligman, N.G. (1987) *Simulation of Water Use, Nitrogen Nutrition and Growth of a Spring Wheat Crop*, Pudoc, Wageningen 1987, 310 p.

MAPPING SIMULATION RESULTS IN VECTOR SPACE

K. KOBAYASHI
Division of Agrometeorology, National Institute of Agro-Environmental Sciences, Tsukuba, Japan

Crop growth models are a powerful tool in understanding as well as predicting the impact of global environment changes on crop production (Curry *et al.*, 1990; Muchow and Sinclair, 1991). The method described here is to help understand the contribution of the individual changes, e.g. warming and CO_2 increase, to the simulated crop yield change, and to assess the average and variation of the change if the simulation is performed repeatedly over a number of cases, e.g. years.

Suppose that a growth model simulates the impact of changes in m factors on crop yield y. Let yb be the simulated yield for the base case, e.g. current environment, $yall$ be the simulated yield with changes of all factors being accounted for, e.g. future environment, and y_j be the simulated yield with only the jth factor being changed from the base case. The changes are expressed on a relative basis against the base-case yield yb as:

$$call = (yall - yb)/yb, \text{ and } c_j = (y_j - yb)/yb,$$

where *call* is relative yield change due to the changes of all the m factors, and c_j is the relative change solely due to the jth factor. *call* can be written as:

$$call = \sum_{j=1}^{m} c_j + \left(call - \sum_{j=1}^{m} c_j \right)$$
$$= \sum_{j=1}^{m} c_j + e \qquad (1)$$

The first term of the right-hand side of Equation 1 is sum of contributions of the individual changes, whereas the second term is a residual. If e is negligible compared with $\Sigma\, c_j$, *call* can be approximated by the sum of the individual effects. Because of intrinsic non-linearity and non-additivity of the growth model, the residual could be of magnitude comparable to $\Sigma\, c_j$, in which case an estimation of the contribution of the individual changes would be inadequate. Thus e, the residual term, is a measure of the approximation error.

If the simulation is performed repeatedly over n cases, e.g. years or locations, the above discussion holds for each case, and the simulation results can be expressed as vectors as shown in Equation 2.

$$\begin{pmatrix} call_1 \\ call_2 \\ \vdots \\ call_n \end{pmatrix} = \begin{pmatrix} \sum_{j=1}^{m} c_{j1} \\ \sum_{j=1}^{m} c_{j2} \\ \vdots \\ \sum_{j=1}^{m} c_{jn} \end{pmatrix} + \begin{pmatrix} e_1 \\ e_2 \\ \vdots \\ e_n \end{pmatrix} \text{ or } \mathbf{call} = \sum_{j=1}^{m} \mathbf{c}_j + \mathbf{e}. \qquad (2)$$

The simulation results are thus mapped into an n–dimensional space spanned by the vectors in Equation 2. Since the vectors $\Sigma\, \mathbf{c}_j$ and $\mathbf{e}$ span a two-dimensional subspace, i.e. a plane, the relationship between them, and hence ***call*** can easily be viewed, and magnitude of the vectors is expressed by their norms. If the residual vector $\mathbf{e}$ is small enough compared with $\Sigma\, \mathbf{c}_j$, the whole change ***call*** can be approximated by the sum of individual changes $\Sigma\, \mathbf{c}_j$.

Each vector in Equation 2 can be expressed as the sum of average vectors of order one and deviation vectors of order n-1. The n-dimensional space is thus partitioned to a one-dimensional subspace for the average and an (n-1)-dimensional subspace for the deviation. The average vectors reside in a line, and can be expressed by scalars as:

$$\overline{call} = \sum_{j=1}^{m} \bar{c}_j + \bar{e}, \qquad (3)$$

where *call*, c_j, e are the averages across the n cases. The average yield change on the left side of Equation 3 can thus be expressed as the sum of the individual changes and a residual on the right side of Equation 3.

Viewing the deviation vectors is not so simple as above. However, there are few needs for viewing all the vectors at once. If the m factors are classified into two groups, the deviation vectors are expressed as:

$$\boldsymbol{cd\,all} = \sum_{j=1}^{m_1} \boldsymbol{cd}_j + \sum_{j=m_1+1}^{m} \boldsymbol{cd}_j + \boldsymbol{ed}$$

$$= \boldsymbol{cd}_{g1} + \boldsymbol{cd}_{g2} + \boldsymbol{ed}, \qquad (4)$$

where $\boldsymbol{cd}_{g_1}$ and $\boldsymbol{cd}_{g_2}$ are sum of the vectors in the respective groups, and m_1 is the number of factors which belong to the first group. The vectors in Equation 4 span a three-dimensional subspace, which can be viewed rather easily. Furthermore, providing ***ed*** is negligible, the whole deviation vector ***cd all*** can simply be expressed by the vector sum of the first two terms on the right-hand side of Equation 4.

In the above steps, the author has tried to view the vectors as directly as possible. There are various ways to summarize the multi-dimensional vectors. The vectors could be projected to a subspace of lower order. Even a simple comparison of the vector norms would give a rough idea of the magnitude of yield changes due to the individual factors. Nevertheless, direct viewing would facilitate correct use of the model summaries, and help find the most appropriate way to characterize the simulation results.

In this paper, a comparison between only two situations has been addressed. It is easy to extend the method for comparing a number of situations, e.g. past, current, and future environments.

References

Curry, R.B., Peart, R.M., Jones, J.W., Boote, K.J. and Allen, Jr, L.H. (1990). Simulation as a tool for analyzing crop response to climate change. *Transactions of the ASAE*, **33**, 981–990

Muchow, R.C. and Sinclair, T.R. (1991) Water deficit effects on maize yields modeled under current and 'greenhouse' climates. *Agronomy Journal*, **83**, 1052–1059

SENSITIVITY OF THE AFRC WINTER WHEAT MODEL TO CHANGES IN TEMPERATURE AND RADIATION

Z. KOCABAS, J. CRAIGON, R.A.C. MITCHELL*, J.N. PERRY*, G.J.S. ROSS* and M.H. UNSWORTH
*University of Nottingham, Sutton Bonington Campus, and *Rothamsted Experimental Station, Harpenden, UK*

The AFRC winter wheat model (Porter, 1984; Weir, *et al.* 1984) simulates the phenological development and canopy development of winter wheat, and it predicts the total dry matter and final grain yield. The main input variables to the model are daily temperature and radiation. The model was run using a range of simulated means and variances of daily temperature and radiation. To do this, double Fourier curves were fitted to daily temperature and radiation based upon 12 years' data from Rothamsted Experimental Station. These functions and the variances around the trends were used to simulate the driving variables of the model.

When AFRCWHEAT1 was run using simulated changes in daily temperature it was found that small systematic changes in temperature of 0.1°C resulted in stepped increases

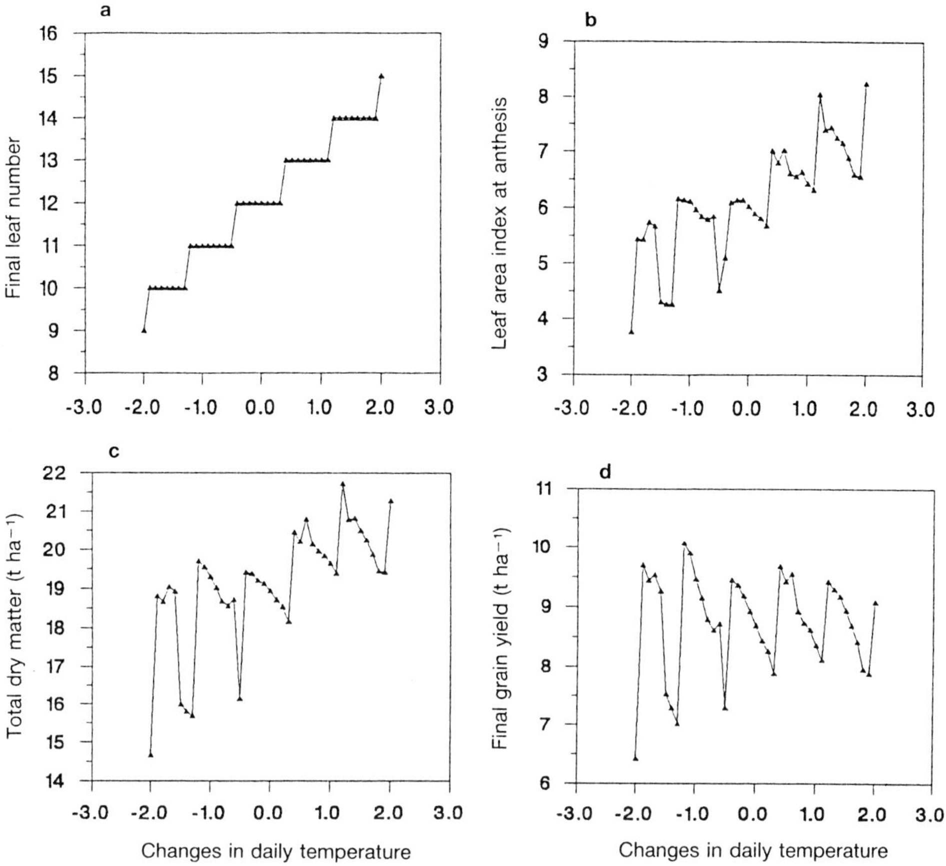

Figure 1 Predicted values of (**a**) final leaf number, (**b**) leaf area index at anthesis, (**c**) total dry matter (t ha^{-1}) and (**d**) final grain yield (t ha^{-1}) plotted against increments (°C) in daily mean temperature

in leaf number (*Figure 1a*). There were abrupt changes in predicted values of 'leaf area index at anthesis' (*Figure 1b*), and 'dry matter' (*Figure 1c*) and 'final grain yield' (*Figure 1d*) which coincided with the changes in integer leaf number (*Figure 1a*).

Grain yield was predicted by running the model for all 25 combinations of five changes to daily temperature and five changes to daily radiation, with 100 independent Monte Carlo replicates for each combination. *Figure 2a* shows that grain yield was insensitive to temperature differences before emergence ($P>0.05$). Increasing temperature between emergence and anthesis had a positive effect on grain yield, but a negative effect in the period from anthesis to end of grain filling ($P<0.01$). Increasing radiation after terminal spikelet positively affected yield ($P<0.01$) (*Figure 2b*). The effects for the whole growing season were similar to those for the period from anthesis to end of grain filling.

Increasing the variance of climatic variables did not affect the mean but increased the variability of yield. Changing the variance of radiation had a larger effect on yield variability than did the same proportional changes in temperature variation (*Figures 3a and 3b*). (Further information in Kocabas *et al.*, 1993)

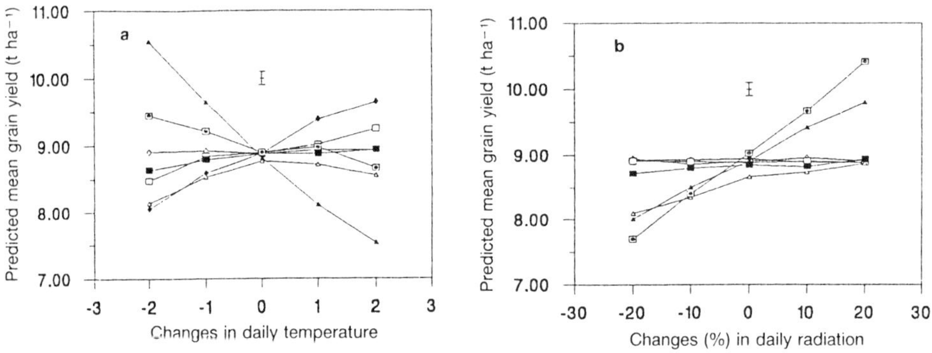

Figure 2 Mean predicted grain yield (t ha^{-1}), plotted against (**a**) increments (°C) in daily mean temperature and (**b**) increments (%) in daily radiation during the periods: sowing to emergence (–◇–); emergence to double ridge (–◆–); double ridge to terminal spikelet (–□–); terminal spikelet to beginning of ear growth (–■–); beginning of earth growth to anthesis (–△–); anthesis to end of grain filling (–▲–); and over the whole growing season (–⊡–). (Bar indicates the 95% confidence interval for the mean of grain yield)

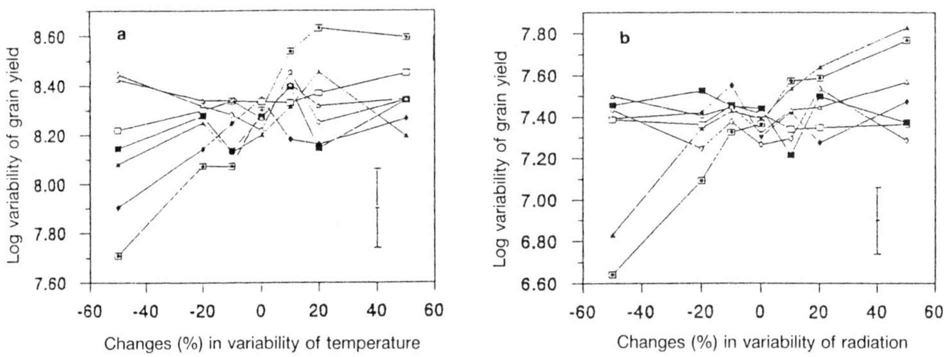

Figure 3 Log variability of final grain yield, plotted against increments (%) in variability of (**a**) daily mean temperature and (**b**) daily radiation during the periods: sowing to emergence; emergence to double ridge; double ridge to terminal spikelet; terminal spikelet to beginning of ear growth; beginning of ear growth to anthesis; anthesis to end of grain filling; and over the whole growing season (symbols are the same as in *Figure 2*). (Bar indicates the 95% confidence interval for the log variability of grain yield)

Conclusions

The AFRC winter wheat model is widely used to help researchers to understand the factors influencing the growth of wheat. Yet this study shows that it is very sensitive to changes in temperature, mainly because the number of leaves produced is recorded as an integer. The model must therefore be used with care when predicting the effects of small changes in temperature because for any one season a small systematic error in temperature measurements may cause a dramatic change in the model's predictions. The average of a number of Monte Carlo simulations confirmed that increasing temperature during canopy development has positive effects on grain yield. However, after anthesis the effects are negative as the duration of the grain filling period is shortened. Increasing daily radiation after terminal spikelet, particularly during the phase from anthesis to end of grain filling, leads to greater grain yield. Higher temperatures throughout the whole growing season result in lower grain yield, whereas increasing daily radiation increases grain yield. Increasing the variability in either driving variable led to greater variability of grain yield, but the model was most sensitive to the variability of radiation.

References

Kocabas, Z., Mitchell, R.A.C., Craigon, J. and Perry, J.N. (1993). Sensitivity analysis of the AFRCWHEAT1 crop model: the effect of changes in radiation and temperature. *Journal of Agricultural Science, Cambridge*, **120**, 149–158

Porter, J.R, (1984) A model of canopy development in winter wheat. *Journal of Agricultural Science, Cambridge,* **102**, 383–392

Weir, A.H., Bragg, P.L., Porter J.R. and Rayner, J.H. (1984) A winter wheat crop simulation model without water and nutrient limitations. *Journal of Agricultural Science, Cambridge,* **102**, 371–382

THE EFFECT OF ATMOSPHERIC CO_2 ENRICHMENT ON THE SALT MARSH SPECIES *ELYMUS ATHERICUS*. INTERACTIONS WITH UV-B RADIATION AND SOIL SALINITY

G.M. LENSSEN, J.W.M. VAN DE STAAIJ AND J. ROZEMA
Department of Ecology and Ecotoxicology, Vrije Universiteit, De Boelelaan 1087, 1081 HV Amsterdam, The Netherlands

The aim of this study was to examine the response of the C_3 salt marsh species *Elymus athericus* to elevated CO_2. Special attention was paid to the interaction of CO_2 with enhanced UV-B radiation and soil salinity. Plant growth of many species is depressed by high levels of UV-B radiation (Caldwell, Teramura and Teveni, 1989). The combined effect of CO_2 enrichment and soil salinity may be particularly important to the salt marsh vegetation. Growth of salt marsh species is limited by direct and indirect effects of flooding with sea water, and these plant species respond markedly to gradients of soil salinity.

Experiments were performed in glasshouses by growing plants in pots. In one greenhouse the CO_2 concentration was about 380 μl l^{-1} while in another the CO_2 level was raised to a level of 720 μl l^{-1}. In the first experiment half of the plants in each compartment received 1 W m^{-2} UV-B radiation while the other half received 2 W m^{-2}. In the second experiment plants were grown in soil containing either 0 or 300 mmol NaCl.

The CO_2 concentration of 720 μl l^{-1} increased growth of *E. athericus*, resulting in higher root, shoot and leaf dry weight. The relative growth stimulation by elevated CO_2 was higher under enhanced UV-B radiation (*Figure 1.*) but the interaction term of CO_2 and UV-B on total plant weight was not significant. In another experiment (van de Staaij *et al.*, 1992) the relative growth stimulation of *E. athericus* at elevated CO_2 was lower under high UV-B radiation. The relative growth response to CO_2 enrichment was also higher under saline conditions (*Figure 2.*). The interaction term of CO_2 and salt tended to be significant. In a previous experiment (Lenssen and Rozema, 1990), the effect of elevated CO_2 on growth of *E. pycnanthus* (=*E. athericus*) was equal under different salinity levels.

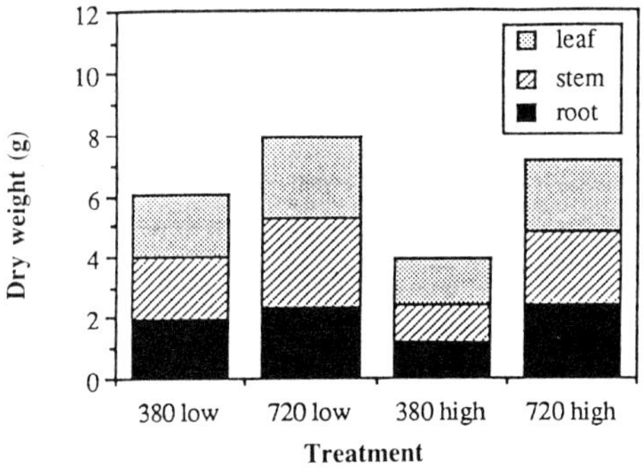

Figure 1 The effect of CO_2 (380 and 720 μl l^{-1}) and UV-B radiation (1 (low) and 2 (high) W m^{-2}) on dry weight of root, stem, leaf and total plant of *Elymus athericus*. Dry weights were determined 7 weeks after start of the experiment

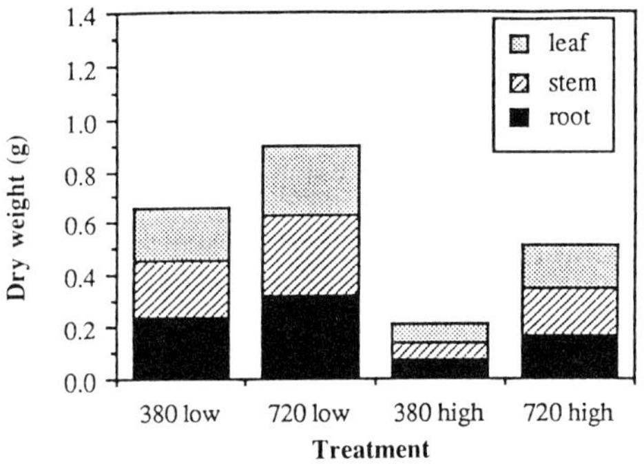

Figure 2 The effect of CO_2 (380 and 720 μl l^{-1}) at low (0) and high (300 mM NaCl) soil salinity on dry weight of root, stem, leaf and total plant of *Elymus athericus*. Dry weights were determined 5 weeks after start of the experiment. (Data of Lenssen *et al.*, 1992)

The increased growth of *E. athericus* at elevated CO_2 was a result of a higher net photosynthesis. This effect was not changed by UV-B or salt treatment (data not shown). Net leaf photosynthesis (measured at the end of the experiment) was not affected by the UV-B treatment (data not shown). In another experiment (van de Staaij, Rozema and Stroetenga, 1990) UV-B radiation reduced net photosynthesis of the salt marsh species *Aster tripolium*. The production of UV-B absorbing compounds in leaves of *E. athericus* was stimulated by UV-B radiation (data not shown). These compounds probably protected the photosynthetic apparatus against UV-B radiation. Since the accumulation of these compounds took some time, it cannot be excluded that during the first period of the experiment high UV-B had a negative effect on net photosynthesis. This negative effect could be responsible for the observed growth reduction in enhanced UV-B radiation. UV-B effects on cell division (Dickson and Caldwell, 1978) might also have contributed to the lower dry weights. High salinity caused a marked growth reduction which could be related to a reduced leaf expansion rate under saline conditions. Although net leaf photosynthesis was slightly lower under high salinity this seemed not to be the primary cause of the reduced growth (data not shown). At the moment the authors are investigating whether CO_2 and salinity might cause interactive growth effects by affecting factors (turgor, cell wall elasticity) which control leaf expansion.

The observed results do not clearly show any statistical interaction between CO_2 and UV-B and CO_2 and salt. These combined effects have also been studied for other C_3 species (Schwarz and Gale, 1984; Zeroni and Gale, 1989; Lenssen and Rozema, 1990; Rozema *et al.*, 1990a; Rozema, Lenssen and Van de Staaij, 1990; Teramura, Sullivan and Ziska, 1990; Rozema *et al.*, 1991; Rozema, 1992). They showed that the relative growth stimulation by elevated CO_2 was lower, equal or higher under enhanced UV-B radiation and equal or higher under saline conditions. Weak or absence of any interactions can be caused by the fact that CO_2, UV-B and salt affect different processes within the plant. CO_2 primarily affects photosynthesis, salt seems to affect leaf elongation, while UV-B might affect cell division.

References

Caldwell, M.M., Teramura, A.H. and Teveni, M.M. (1989) The changing solar ultraviolet climate and the ecological consequences for higher plants. *Tree*, **4**, 363–367

Dickson, J.G. and Caldwell, M.M. (1978) Leaf development of *Rumex patientia* L. (Polygonaceae) exposed to UV radiation. *American Journal of Botany*, **65**, 857–863

Lenssen, G.M., Lamers, J., Stroetenga, M. and Rozema J. (1992) Interactive effects of atmospheric CO_2 enrichment, salinity and flooding on growth of C_3 (*Elymus athericus*) and C_4 (*Spartina anglica*) salt marsh species. *Vegetatio* (in press)

Lenssen, G.M. and Rozema, J. (1990) The effect of atmospheric CO_2-enrichment and salinity on growth, photosynthesis and water relations of salt marsh species. In *The Greenhouse Effect and Primary Productivity in European Agro-ecosystems*, pp 64–67 Edited by J. Goudriaan, H. van Keulen and H.H. van Laar. Wageningen: Pudoc

Rozema, J. (1992) Plant responses to atmospheric carbon dioxide enrichment: interactions with some soil and atmospheric conditions. *Vegetatio* (in press)

Rozema, J., Dorel, F., Janissen, R., Lenssen, G., Broekman, R., Arp, W. and Drake, B.G. (1991) Effect of elevated atmospheric CO_2 on growth, photosynthesis and water relations of salt marsh grass species. *Aquatic Botany*, **39**, 45–55

Rozema, J., Lenssen, G.M., Broekman, R.A. and Arp, W.P. (1990) Effects of atmospheric carbon dioxide enrichment on salt-marsh plants. In *Expected effects of climatic change on marine coastal ecosystems*, pp 49–54 Edited by J.J. Beukema, W.J. Wolff, and J.J.W.M. Brouns. Dordrecht: Kluwer

Rozema, J., Lenssen, G.M. and van de Staaij, J.W.M. (1990) The combined effect of increased atmospheric CO_2 and UV-B radiation on some agricultural and salt marsh species. In *The Greenhouse Effect and Primary Productivity in European Agro-ecosystems* pp 68–71 Edited by J. Goudriaan, H. van Keulen and H.H. van Laar, Wageningen: Pudoc

Schwarz, M. and Gale, J. (1984) Growth response to salinity at high levels of carbon dioxide. *Journal of Experimental Botany*, **35**, 193–196

Teramura, A.H., Sullivan, J.H. and Ziska, L.W. (1990) Interaction of elevated ultraviolet-B radiation and CO_2 on productivity and photosynthetic characteristics in wheat, rice and soybean. *Plant Physiology*, **94**, 470–475

van de Staaij, J.W.M., Lenssen, G.M., Stroetenga, M. and Rozema, J. (1992) The combined effect of elevated CO_2 levels and UV-B radiation on growth characteristics of *Elymus athericus*. *Vegetatio* (in press)

van de Staaij, J.W.M., Rozema, J. and Stroetenga, M. (1990) Expected changes in Dutch coastal vegetation resulting from enhanced levels of solar UV-B. In *Expected Effects of Climatic Change on Marine Coastal Ecosystems pp 211–217* Edited by J.J. Beukema, W.J. Wolff, and J.J.W.M. Brouns. Dordrecht: Kluwer

Zeroni, M. and Gale, J. (1989). Response of 'Sonia' roses to salinity at three levels of ambient CO_2. *Journal of Horticultural Science*, **64**, 503–511

HEAT BALANCE MEASUREMENTS OF TRANSPIRATION IN PERENNIAL PIGEONPEA–GROUNDNUT AGROFORESTRY SYSTEMS

F.M. MARSHALL[1,3], C.R. BLACK and C.K. ONG[2,3]
Department of Physiology and Enviromental Science, University of Nottingham, Sutton Bonington, Loughborough, Leics, LE12 5RD, UK

Complementarity of resource utilization is crucial to the success of intercropping and agroforestry systems. Increased capture and/or more efficient use of limiting resources, such as light, water and nutrients, are key factors.

Field techniques for measuring light interception and conversion efficiency in mixed communities are well established, but techniques for partitioning water use between the components of mixed communities have only recently become available. A major advance has been the heat balance approach for measuring sap flux through the stems of individual plants. Several studies indicate that this approach can measure transpiration with an accuracy of ±10% or better in relatively large plants (Baker and van Bavel, 1987; Steinberg, van Bavel and McFarland, 1989). The technique is sensitive to small changes in sap flux and has a fast response time.

A heat balance system similar to that described by Ishida, Campbell and Calissendorff (1991) was used in a field trial at ICRISAT, India in which the water use, light interception and productivity of groundnut–perennial pigeonpea agroforestry systems were examined. Line-planted (5.4 m pigeonpea alleys) and dispersed arrangements (1.8 × 1.2 m pigeonpea spacing) were compared. The pigeonpea population was 0.5 plants m^{-2} in both arrangements, while groundnut was sown at 10 cm intervals in rows 30 cm apart. Sole pigeonpea and groundnut were also grown for comparison. The heat balance was used to measure transpiration by the dispersed and line planted pigeonpea between February 1990 and January 1991, when the pigeonpea were 7–18 months old.

Typical diurnal trends for transpiration by individual trees are shown in *Figure 1*. During the dry season (*Figure 1a*), shortly after the plants had been cut for fodder, leaf area was small in both treatments and there was no significant difference in transpiration. However, in the rainy season (*Figure 1b*), transpiration was five times higher at midday in the dispersed planting than in the line arrangement. Because of their wider spacing, the dispersed plants developed a fuller canopy and exploited the soil profile more thoroughly for water. Transpiration followed the diurnal changes in vapour pressure deficit closely, though irradiance, soil moisture availability and leaf area were also important in determining water use. Such detailed studies of transpiration in relation to microclimatic conditions contribute significantly to our understanding of the success or failure of specific agroforestry systems, and may provide invaluable information for modelling crop growth.

The heat balance approach permits daily transpiration to be followed much more accurately than was previously possible using porometry or soil water balance approaches. The daily values can also be used to calculate cumulative transpiration (*Figure 2*), which again emphasizes the much greater water use of the dispersed pigeonpea. The cumulative data allow the proportion of annual rainfall used by the tree component to

[1]Present address: Henry Doubleday Research Association, Ryton-on-Dunsmore, Coventry, CV8 3LG, UK
[2]International Crop Research Institute for the Semi-arid Tropics, Patancheru PO, Andhra Pradesh 502324, India
[3]Present address: International Centre for Agroforestry Research, ICRAF House, PO Box 30677, Gigiri, Nairobi, Kenya

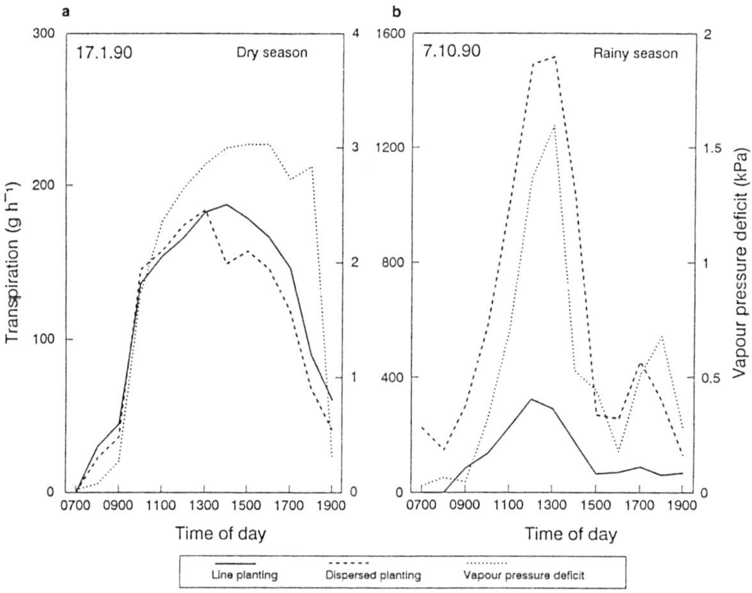

Figure 1 Typical diurnal time courses of transpiration and leaf to air vapour pressure difference in perennial pigeonpea during (**a**) the dry and (**b**) the rainy seasons; $n = 3$

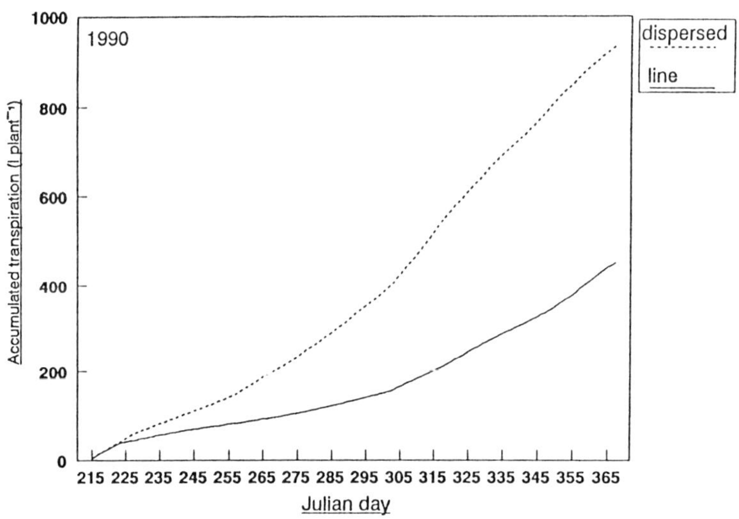

Figure 2 Relation between accumulated transpiration and time in perennial pigeon pea; $n = 3$.

be calculated. However, although water use and dry matter production differed greatly between treatments, the water use ratio (WUR) was similar (*Table 1*).

In agroforestry it is important not only to increase the proportion of rainfall used by the tree and crop components during the rainy season, but also to utilize off-season rainfall. Transpiration by pigeonpea accounted for 60% of the annual rainfall in the dispersed

Table 1 DRY MATTER PRODUCTION, TRANSPIRATION AND WATER USE RATIO IN PIGEONPEA

	Dry matter (t ha^{-1})	*Transpiration* (mm)	*WUR* (kg ha^{-1} mm^{-1})
Dispersed	4.52	580	8.88
Line-planted	2.09	256	8.18

planting and 30% in the line-planted arrangement. These results can be used in conjunction with other standard techniques to compare the overall water budget of these systems.

Recent developments suggest that the heat balance can be miniaturized for field use with small annual crops, thereby enabling transpiration to be measured continuously and non-intrusively in each component of mixing cropping systems. The authors are currently modifying this approach for use on both the tree and crop components of *Grevillea robusta*-maize and cowpea systems in Kenya.

Acknowledgements

F.M. Marshall was supported by the Natural Environment Research Council.

References

Baker, J.M. and van Bavel, C.H.M. (1987) Measurements of mass flow in the stems of herbaceous plants. *Plant, Cell and Environment*, **10**, 777–82

Ishida, T., Campbell, G.S. and Calissendorff, C. (1991) Improved heat balance method for determining sap flow rate. *Agricultural and Forest Meteorology*, **56**, 36–48

Steinberg, S., van Bavel, C.H.M. and McFarland, M.J. (1989) A gauge to measure mass flow rate of sap in stems and trunks of woody trees. *Journal of American Society of Horticultural Science*, **114**, 466–472

SOYBEAN PUBESCENCE – DOES IT REALLY MAKE A DIFFERENCE?

M.J. MORRISON and H. VOLDENG
Plant Research Centre, Agriculture Canada, Ottawa, Canada

Near-isogenic soybean lines, differing in pubescence colour (grey or brown) and pubescence density (normal and dense) were developed to examine the effects of trichome colour and number on the growth and yield of soybean. Three isogenic series were planted in four-row plots at Ottawa (45° 24' N, 75° 43' W) in a randomized block design. On cloudless days during the growing season, the percentage of incident radiation (450–800 nm, in 50 nm increments) reflected from the canopy was measured using a ‘Cropscan’. PAR (photosynthetic active radiation) sensors above and below the canopy were used to measure the amount of intercepted light. Leaf temperature, transpiration and photosynthesis were measured with a LICOR 6200 portable photosynthesis measuring system. Leaves were sampled and the number of trichomes per square centimetre counted. At maturity the seed was harvested and the yield determined.

In 1991, mean maximum daily temperatures for the months of June and July were 1–2°C warmer than the 75 year average. One hundred millimetres less precipitation fell in June and July than average. Plots were sampled during and after the drought (20 July and 6 August). Densely pubescent soybeans had two to four times as many trichomes as the normal lines (*Table 1*). Grey-dense (GD) soybeans had significantly more trichomes than brown-dense (BD). Grey soybeans required more time to mature than brown types. BD plants required significantly more days to mature than brown-normal (BN) ones. There were no significant differences in light interception measured on 20 July among soybean types. However, on 6 August, brown soybean canopies intercepted significantly more PAR radiation than grey canopies and BN soybeans intercepted significantly more PAR than BD canopies. Generally, brown soybeans reflected significantly more short wave-radiation and less near-IR radiation than grey soybeans (*Figure 1*). When measured on 6 August, GD canopies reflected significantly more radiation (450–600 nm) than GN canopies. On 20 July, the BN soybeans had a significantly lower leaf temperature and higher transpiration and photosynthetic rates than all other types (*Table 1*). On 6 August, there were no differences in leaf temperature and photosynthetic rates; the brown soybeans had higher transpiration rates than the grey. Grey pubescent soybean had significantly higher yields than brown pubescent types. There were no significant yield differences between pubescence density types within each colour type.

In summary, pubescence density did not influence the agronomic and radiation interception characteristics studied as much as pubescence colour. Soybeans with grey pubescence were higher yielding, reflected more near IR radiation and had lower transpiration rates than brown pubescent soybeans. Brown pubescent soybeans matured earlier, reflected more short wave radiation and appeared to function at higher rates under drought conditions than grey. Pubescence colour is linked to genes for maturity and therefore may not be responsible for the observed delay. Further work is being conducted to sever this linkage. Increased trichome number is not a detriment to soybean production and may improve the adaptability of soybean in areas prone to high radiation, high temperature and low moisture.

Table 1 EFFECT OF LEAF PUBESCENCE COLOUR AND DENSITY ON AGRONOMIC AND RADIATION INTERCEPTION CHARACTERISTICS MEASURED AT TWO DATES

	Grey		*Brown*	
	Normal	*Dense*	*Normal*	*Dense*
Leaf hairs (cm^{-2})	30.4 c	120.5 a	42.5 c	93.6 b
Yield ($g\ m^{-2}$)	162.3 a	191.4 a	115.5 b	98.7 b
Maturity (days)	99.7 a	102.5 a	90.5 c	95.0 b
20 July				
Light interception (%)	35.2 a	39.1 a	43.6 a	40.0 a
Leaf temperature (°C)	27.8 a	28.0 a	27.4 b	28.1 a
Transpiration ($\mu mol\ m^{-2}\ s^{-1}$)	7.44 b	7.34 b	9.03 a	8.13 b
Photosynthesis ($\mu mol\ m^{-2}\ s^{-1}$)	15.15 b	14.69 b	18.20 a	15.74 ab
6 August				
Light interception (%)	27.9 c	29.6 c	48.8 a	38.5 b
Leaf temperature (°C)	27.4 a	27.3 a	27.4 a	27.3 a
Transpiration ($\mu mol\ m^{-2}\ s^{-1}$)	8.26 b	8.17 b	9.77 a	10.36 a
Photosynthesis ($\mu mol\ m^{-2}\ s^{-1}$)	20.43 a	21.11 a	18.12 a	21.17 a

a–c means among pubescence types, followed by the same letter are not significantly different at the LSD = 0.05 level of probability.

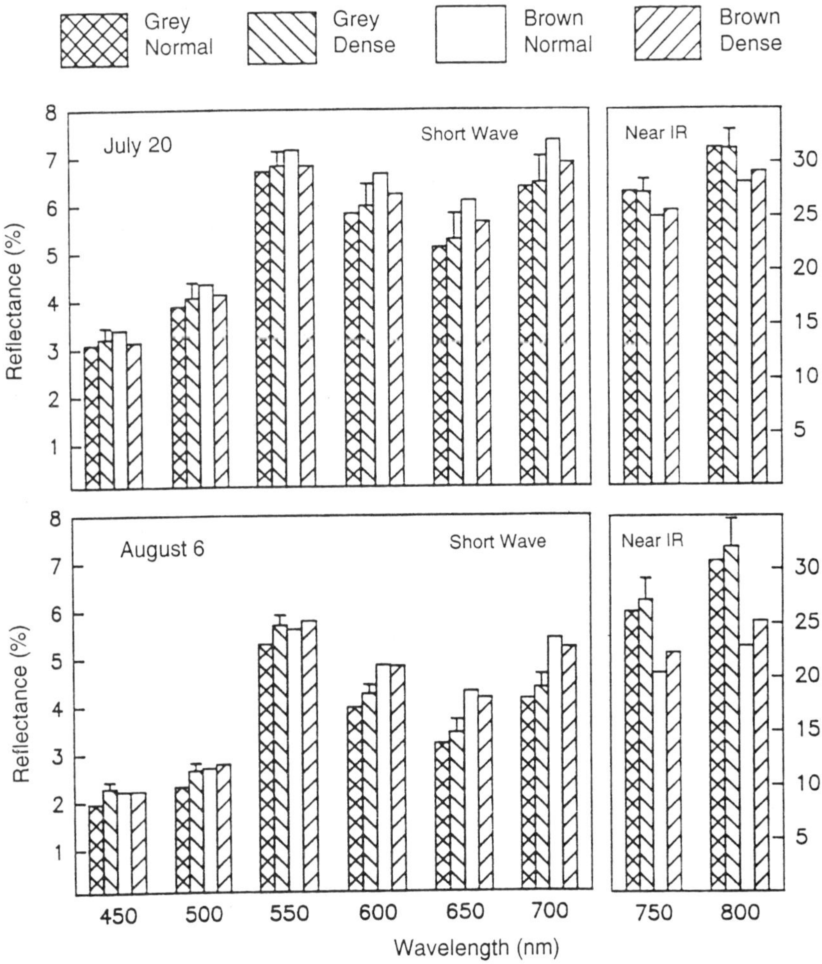

Figure 1 Effect of soybean pubescence density and colour on light reflected from the canopy (percentage) at different wavelengths. Error bars represent LSD at P=0.05 among pubescence types

EFFECTS OF OZONE ON PHOTOSYNTHESIS AND CARBON TRANSPORT OF SPRING WHEAT EXPOSED IN OPEN TOP CHAMBERS

L. MORTENSEN and K.C. ENGVILD*
*National Environmental Research Institute, Department of Emissions and Air Pollution, Frederiksborgvej 399, DK-4000 Roskilde, Denmark; *Risø National Laboratory, Environmental Science and Technology Department, DK-4000 Roskilde, Denmark*

In order to determine if altered carbon allocation patterns in ozone treated plants (Cooley and Manning, 1987) could be explained by reduced photosynthesis or by a diminished translocation speed of the assimilated carbon, spring wheat (*Triticum aestivum* L. cv. Ralle) was exposed in open top chambers to either carbon filtered air, non-filtered air, or non-filtered air with 30 ppb ozone added 8 h d^{-1}. Photosynthesis of the flag leaf was measured before, during and after anthesis with a portable infrared gas analysis system (ADC model LCA2) linked to a Parkinson leaf cuvette (ADC model PLC(N)).

At the same time $^{14}CO_2$ was administered for 5 min to sections of the flag leaf or the second leaf of some of the plants, which were harvested after 20, 40, 60 or 240 min, mounted on cardboard and covered with plastic film, quickfrozen in solid CO_2 and lyophilized. Autoradiograms were prepared on X-ray film and radioactivity was determined in plant sections by scintillation counting of samples decolorized in hypochlorite.

As shown in *Table 1*, ozone addition decreased net assimilation rate after anthesis, but not before. *Figure 1* shows no differences in the translocation speed of radioactive assimilate between control plants and ozone treated plants.

Table 1 EFFECTS OF OZONE ON NET ASSIMILATION RATE (mmol m^{-2} s^{-1}) OF THE FLAG LEAF MEASURED BEFORE AND AFTER ANTHESIS

	Treatment		
Time of measurement:	*CF*	*NF*	*NFO*
Before anthesis	17.3	16.6	17.4
After anthesis	13.6	11.7	2.7

CF, charcoal-filtered air; NF, non-filtered ambient air; NFO, NF plus 30 ppb O_3 8 hours daily (9am-5pm, GMT+1)

It can be concluded that ozone has no effect on translocation velocity of assimilates. This agrees with results of Spence, Rykiel and Sharpe (1990). However, the variability in the experiment is so large that minor indirect effects of the reduced photosynthesis or effects on phloem loading cannot be excluded.

Acknowledgement

This study was financed by a CEC subcontract to University of Nottingham, the National Environmental Research Institute, and the Risø National Laboratory. Photosynthetic equipment was kindly lent from University of Copenhagen, Institute of Plant Ecology.

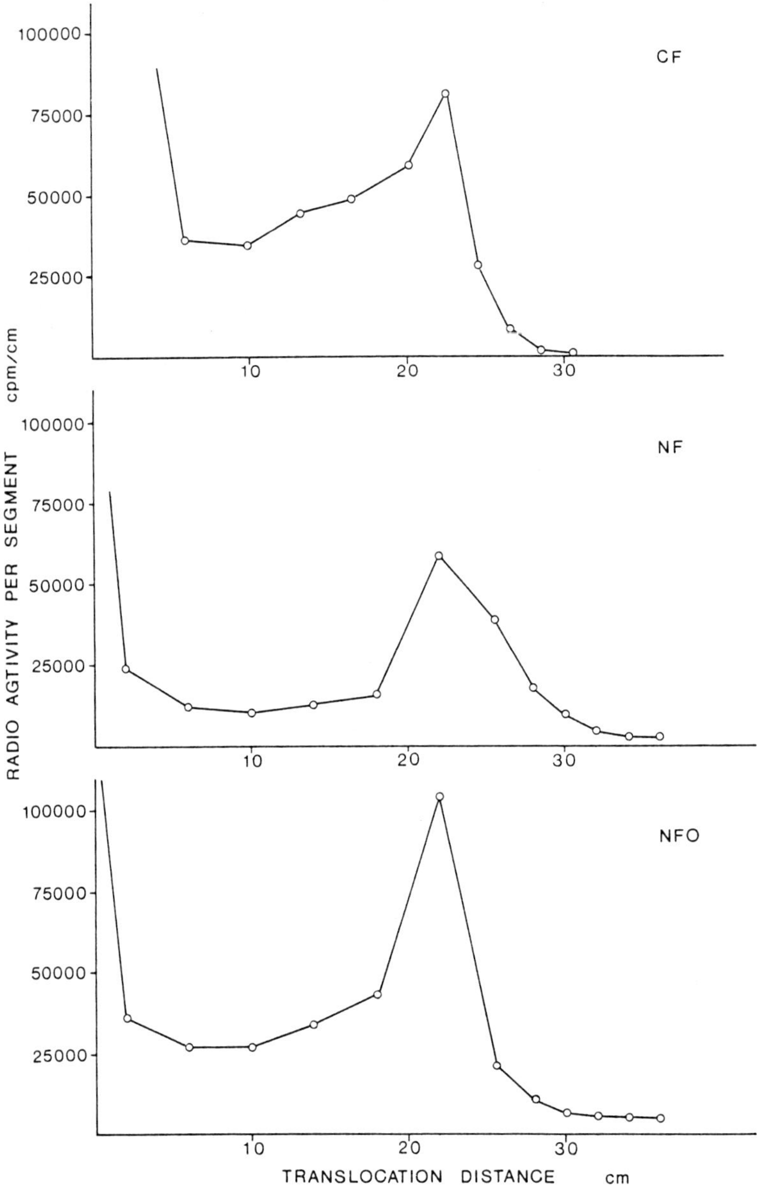

Figure 1 The amount of tranported ^{14}C and translocation distance in spring wheat, 60 min after exposure to $^{14}CO_2$ for 5 min

References

Cooley, D.J. and W.J. Manning (1987) The impact of assimilate partitioning in plants. *Environmental Pollution*, **47**, 95–113

Spence, R.D., Rykiel, E.J. and Sharpe, P.H.J. (1990) Ozone alters carbon allocation in loblolly pine: assessment with carbon-V labelling. *Environmental Pollution*, **64**, 93–106

IS PLANT GROWTH IN COMPACTED SOIL REGULATED BY A ROOT-SOURCED CHEMICAL SIGNAL?

B.J. MULHOLLAND, J.A. ROBERTS, I.B. TAYLOR and C.R. BLACK.
University of Nottingham, Department of Physiology and Environmental Science, Sutton Bonington, UK

Soil compaction is a serious problem in agriculture, causing several factors to interact to reduce plant growth; these include waterlogging, poor aeration, nutrient imbalances and mechanical impedance. Masle and Passioura (1987) postulated a role for a root-sourced chemical signal which regulated shoot physiology and produced an integrated plant response when the roots encountered compacted soil. Hormonal signals, particularly abscisic acid (ABA), have been implicated in root:shoot signalling and plant responses

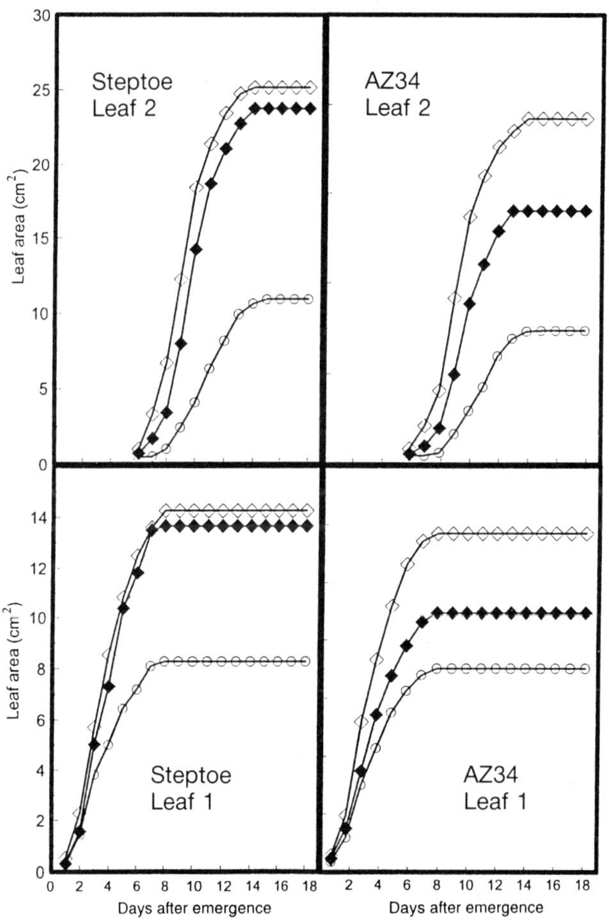

Figure 1 Leaf expansion in leaves 1 and 2 of cv. Steptoe and the ABA-deficient mutant, AZ34, at bulk densities of 1.4 (◇), 1.6 (◆) and 1.7 g cm^{-3} (○)

to drought (Zhang and Davies, 1990) and waterlogging (Jackson, 1991), while Tardieu *et al.* (1991, 1992) detected enhanced ABA concentrations in the xylem sap and reduced stomatal conductance in maize plants grown on compacted field plots. It therefore appears probable that a root-sourced hormonal signal may be involved in plant responses to soil compaction.

Normal and ABA deficient mutants of barley (cv. Steptoe and AZ34 respectively) were grown in cylindrical PVC tubes 150 mm tall by 75 mm in diameter under controlled environment conditions of 18/12°C day/night with a 14-h photoperiod. Sandy loam soil was packed to bulk densities of 1.1, 1.4, 1.5, 1.6 and 1.7 g cm^{-3}, to produce a realistic range of soil strength. Plants were harvested 6, 12 and 18 days after emergence (DAE). Before harvest, stomatal conductance was measured on the youngest expanded leaves, before determining the ABA content and water, solute and turgor potentials of the same leaves. Total leaf area and shoot fresh and dry weight were measured. Xylem exudate was collected by attaching tight-fitting silicone rubber tubing to the decapitated root systems and subsequently assayed for ABA, pH and inorganic cations and anions. Daily non-destructive measurements of leaf length and breadth were made for all leaves of all plants, and used to calculate leaf area. Data are presented here for the 1.4, 1.6 and 1.7 g cm^{-3} treatments.

AZ34 exhibited slower leaf expansion and smaller final leaf areas than Steptoe at all bulk densities. This effect was particularly apparent at 1.6 g cm^{-3} (*Figure 1*). Although both genotypes showed no difference in the rate of leaf emergence between treatments, leaf expansion rates (cm^2 d^{-1}) for leaves 1 and 2 in the 1.7 g cm^{-3} treatment were 26 and 60% lower in Steptoe and 48 and 67% lower in AZ34 than in the 1.4 g cm^{-3} treatment.

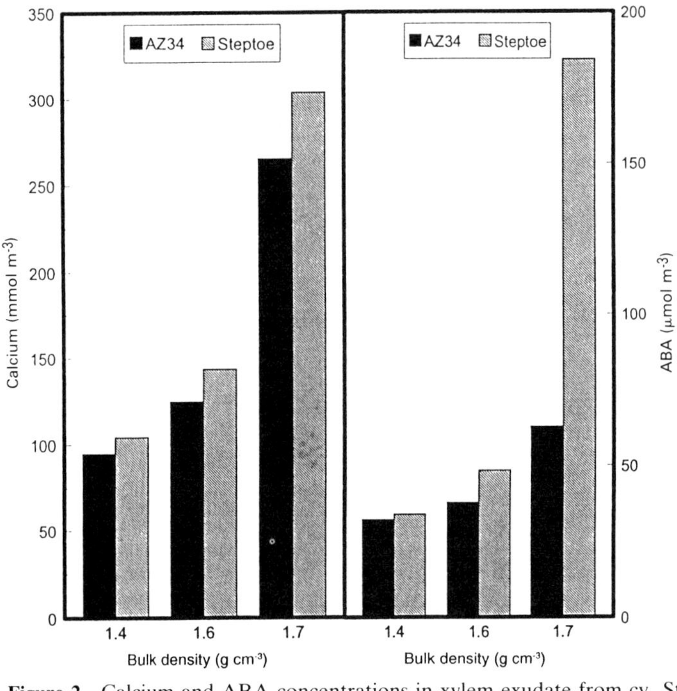

Figure 2 Calcium and ABA concentrations in xylem exudate from cv. Steptoe and the ABA-deficient mutant, AZ34, 12 days after emergence (DAE) at bulk densities of 1.4, 1.6 and 1.7 g cm^{-3}. Values are the means of 20 sap aliquots (1.4 and 1.6 g cm^{-3}) or 10 sap aliquots (1.7 g cm^{-3}) for both cultivars

Because the duration of leaf expansion was unaffected by bulk density, the final leaf area was consequently reduced by 40 and 55% in leaves 1 and 2 of Steptoe, and by 40 and 62% in leaves 1 and 2 of AZ34 at 1.7 g cm^{-3}. Total dry matter accumulation was also reduced by 54% in Steptoe and 45% in AZ34.

Water, solute and turgor potentials differed little between treatments, but stomatal conductance was reduced by 20% at the highest bulk density in both genotypes. Steptoe showed a five-fold increase in the ABA content and a three-fold increase in the calcium concentration of xylem exudate as bulk density increased from 1.4 to 1.7 g cm^{-3} (*Figure 2*). This was in contrast with AZ34, in which there was only a two-fold increase in xylem ABA content but a similar three-fold increase in calcium concentration. Magnesium, sodium and pH levels showed no significant trends.

Thus, in barley a threshold soil bulk density exists between 1.6 and 1.7 g cm^{-3}, above which growth is severely reduced. A marked increase in ABA xylem sap concentration is observed at 1.7 g cm^{-3} in the normal wild-type cultivar, but the increase is much smaller in the ABA-deficient mutant, AZ34, although the response of xylem sap calcium is similar in both genotypes. AZ34 exhibits slower leaf expansion and smaller final leaf areas, particularly at 1.6 g cm^{-3} than cv Steptoe. It is possible that the reduced ability of AZ34 to produce ABA in response to increasing soil bulk density may adversely affect its ability to grow under such conditions as compared with the wild-type Steptoe. The authors propose therefore a positive role for ABA in maintaining leaf expansion in plants growing in compacted soil.

Acknowledgement

B.J. Mulholland is supported by a NERC postgraduate scholarship.

References

Jackson, M.B. (1991) Regulation of water relationships in flooded plants by ABA from leaves, roots and xylem sap. In *Abscisic Acid Physiology and Biochemistry* pp 217–226 Edited by W.J. Davies and H.G. Jones. Bios Scientific Publishers, UK

Masle, A. and Passioura, J.B. (1987) Impact of soil strength on the growth of young wheat plants. *Australian Journal of Plant Physiology*, **14**, 643–656

Tardieu, F., Katerji, N., Bethenod, O., Zhang, J. and Davies, W.J. (1991) Maize stomatal conductance in the field: its relationship with soil and plant water potentials, mechanical constraints and root messages. *Plant, Cell and Environment*, **14**, 121–126

Tardieu, F., Zhang, J., Katerji, N., Bethenod, O., Palmer, S. and Davies, W.J. (1992) Xylem ABA controls the stomatal conductance of field-grown maize subjected to soil compaction or soil drying. *Plant, Cell and Environment*, **15**, 193–197

Zhang, J. and Davies W.J. (1990) Changes in the concentration of ABA in the xylem sap as a function of changing soil water status can account for changes in leaf conductance and growth. *Plant, Cell and Environment*, **13**, 271–285

TREE–CROP COMPETITION FOR RESOURCES ABOVE AND BELOW THE GROUND

Chin K. ONG and M.R. RAO
International Center for Research in Agroforestry, Nairobi, Kenya

When trees and crops are grown close together, they inevitably compete. This competition is more acute when prunings from the trees are removed for use as fodder and fuelwood, rather than applied to the crops as mulch. One of the most important challenges of agroforestry research is to develop systems that enhance the positive interactions of trees and crops and keep competition to a minimum.

Studies on tree–crop competition are in progress at ICRAF's research station in Machakos, Kenya. At an altitude of 1600 m, the climate is transitional between dry subhumid and semi-arid, with bimodal rainfall averaging 760 mm a year. Soils are variable, but predominately haplic lixisols, and most of the land is sloping.

Above-ground interactions between trees and crops may include interception of light and rainfall, as well as modifications of temperature, windspeed and humidity. Studies at Machakos of *Cassia siamea* interplanted with maize showed that the tree canopy was intercepting about 20% of the seasonal rainfall, but this was not reflected in lower maize yields. Presumably, the negative redistribution of rainfall was offset by the positive mulch effect.

Below-ground competition for water and nutrients is poorly understood, but root excavations have indicated an extensive and dense spread of tree roots that could result in serious competition with crops. At Machakos, roots of 5-year-old cassia trees extended up to 14 m out from the stems.

An experiment began at Machakos in November 1989 to assess the importance of above- and below-ground competition between single rows of *Leucaena leucocephala* and adjacent maize. The leucaena was either managed as hedgerows by regular pruning to 0.5 m above ground or allowed to grow freely as an upperstorey tree. All prunings were removed from the site.

In a separate set of treatments, galvanized iron sheets were buried between trees and crops to a depth of 1 m to act as root barriers. The objective was to rule out below-ground competition. All tree and crop rows were arranged on contours irrespective of orientation, since previous evidence showed that orientation *per se* had only a minor influence on tree–crop interactions at Machakos.

Results were obtained from three successive maize crops. Grain yields (dry weight) were measured by individual rows for 10 rows on each side of the leucaena. Results from the first season are not presented because the leucaena was newly transplanted and there were no apparent differences between the treatments and control plots of sole maize.

The first important results were obtained furing the first rains of 1991 (*Figure 1*). Maize yield declined linearly from rows closer to the upperstorey trees. Maize yields next to hedgerows followed a similar pattern except that rows just adjacent to hedges yielded 10 to 15 higher than the next rows out. In both treatments, the influence of the leucaena extended to about 5 m. Beyond this distance, maize yields were close to the level achieved on sole maize plots, which averaged 3.28 t ha^{-1}. Total maize yields from plots with upperstorey trees were only 54% of sole maize yields. The trend was the same when the leucaena was pruned, but yield loss in the four maize rows closest to the hedgerows was relatively small, resulting in an overall yield reduction of only 20%.

The root barriers had only a minor influence on tree–crop interactions. Maize yields on plots with upperstorey trees were 12% higher with the galvanized sheets than without. The difference on plots with hedgerows was only 7%.

These observations suggest that even in a fairly good season with 333 mm of rain the negative effect of competition was substantial for both pruned hedgerows and upperstorey

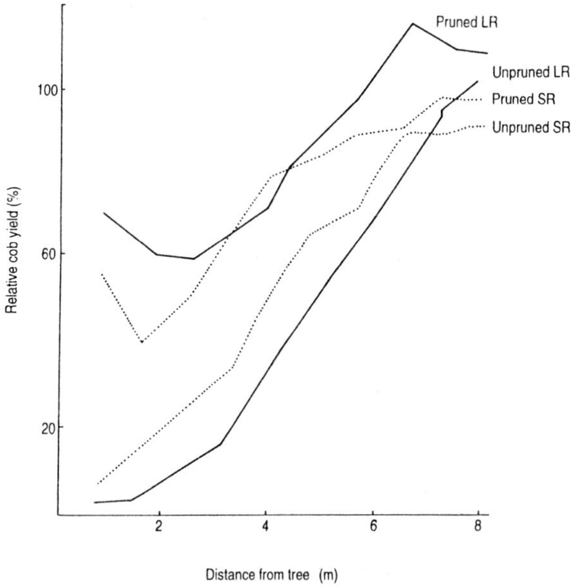

Figure 1 Effect of distance and pruning on competition between *Leucaena leucocephala* and maize at Machakos, Kenya: maize grain yields as percentage of yield under sole cropping. LR refers to long rains for 20 March – 31 June and SR for short rains for 20 October – 31 January

trees. Either below-ground competition was of minor importance or the root barriers were ineffective.

During the second growing season of 1991, total rainfall was only 214 mm and there was a long dry spell following crop emergence. Maize grain yields on sole plots averaged only 1.5 t ha^{-1}, about 40% of the average for the previous season. Again, the effects of the leucaena rows extended to a distance of 5 m, and, again, the root barriers had a negligible effect.

Despite the difference in rainfall, the negative influence or leucaena was similar during the two growing seasons. In the second season, overall reductions in crop yields were 46 to 55% on plots with upperstorey trees and 20 to 18% on plots with hedgerows.

In an attempt to discover the causes of these reductions in maize yield, rain gauges and tube solarimeters were installed along the tree–crop interfaces to measure the redistribution of rainfall and the shading from tree canopies. Neutron access tubes were installed to a depth of 2.0 m to measure the soil moisture profile, and tree transpiration rates were monitored using heat-balance equipment.

For the pruned hedgerows, interception of rainfall and shading were negligible, indicating that competition was primarily below ground. However, the root barriers were ineffective in preventing root competition.

The upperstorey trees had a major influence on rainfall redistribution for the three nearest crop rows. Their influence was greatest during small showers (less than 10 mm per day) when the nearest crop row received only 55% of the rainfall that fell in the open. When rainfall was heavier (more than 30 mm per day) the nearest row received 75% of the rain that fell in the open.

These results indicate that the rain shadow created by upperstorey canopies can account for substantial losses in crop yield, especially in areas with a high proportion of small showers. In addition, shading can extend 3.8 m away from the trees, creating a 25% reduction in solar radiation. However, in the semi-arid environment of Machakos, such a reduction in solar radiation might not be sufficient to explain the 35 to 70% loss in crop yield.

Measurements of transpiration rates showed negligible differences between upperstorey trees and pruned hedgerows during both rainy and dry periods, even though their canopies differed substantially in size. An explanation could be that the canopy of the hedgerows consisted of young, actively transpiring leaves, whereas the upperstorey canopy consisted of a large proportion of older leaves and reproductive structures.

Soil moisture profiles were essentially similar on plots and hedgerows and with upperstorey trees, implying that crops and trees under the two systems may use different proportions of the total water available. The partition of water use by the two components is being studied in 1992.

In summary, current evidence indicates that in the seasonally dry tropics competition between leucaena and maize for below ground growth resources extends well beyond the tree canopy. Separation of below-ground interactions by vertical root barriers was not effective because, over time, the tree roots arched back towards the upper soil layers to compete with adjacent crops. Previous experience with sunken-pit planting at Machakos confirmed this phenomenon of root redirection.

LIGHT AND NITROGEN CAPTURE BY A GRASS SPECIES (*FESTUCA ARUNDINACEA*) INTERCROPPED OR NOT WITH A LEGUME SPECIES (*TRIFOLIUM REPENS*)

H. SINOQUET, P. CRUZ, R. TOURNEBIZE and B. MOULIA
INRA, Station Agropédoclimatique de la Zone Caraïbe, BP 1232, 9718S Pointe-à-Pitre Cedex, Guadeloupe

Introduction

Intercropping generally uses environmental resources such as light, water and nitrogen in a more efficient way. That is especially the case in extensive conditions where the potential production is not reached. The higher efficiency of mixed crops may be the consequence of the exploitation of different aerial and underground spaces by the components, resulting in complementary resource capture.

Resource capture in pure stands may be approached at the whole canopy level. In mixed stands the share of resources used by each species should be distinguished, so global concepts, as gap frequency for the light or available soil reserve of water, become ineffective.

This work applied a diagnostic method (Cruz, Moulia and Sinoquet, 1989) for separating light and nitrogen capture by the components of a well mixed tall fescue and white clover canopy, in order to analyse the effect of legume presence on the growth and mineral nutrition of the grass.

Materials and methods

Plots of pure and mixed tall fescue were grown at two levels of nitrogen supply (0 and 150 units per hectare) at Lusignan, France. Two blocks with three replicates of each treatment were assigned for growth measurements and a third one for solar radiation measurements.

Aerial biomass, leaf area index and nitrogen content were measured once a week during the summer regrowth in 1986.

Light absorption in pure and mixed tall fescue was estimated by a short-wave radiation model for well-mixed canopies (Sinoquet *et al.*, 1990). Nitrogen nutrition level of the tall fescue in both stands was expressed by the relation between nitrogen uptake and aerial biomass production (Lemaire and Salette, 1984; Cruz and Lemaire, 1986).

Results and discussion

LAI of tall fescue in mixed stands was higher than or similar to that of pure stands in N0 and N150 treatments respectively (*Figure 1*). Nevertheless total solar energy harvested by the mixed grass was 30% lower at the two levels of nitrogen supply (*Figure 2*). Shortage in radiation capture by the mixed grass was the consequence of competition for light by the companion legume species. However lower absorption efficiencies were counterbalanced by higher radiation use efficiencies (*Figure 3*), resulting in similar total harvested biomass of both mixed and pure grasses.

Lastly, higher radiation-use efficiencies of mixed grasses may be related to better nitrogen nutrition (*Figure 4*): at the end of regrowth, nitrogen uptake by mixed grasses was 32% (N0) and 12% (N150) greater than that of pure stands.

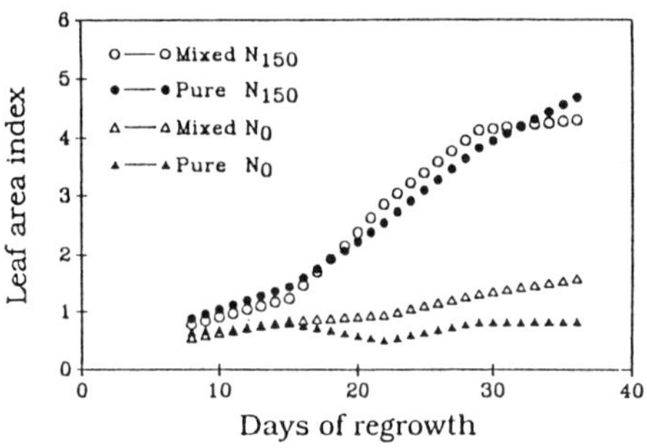

Figure 1 Leaf area indices of tall fescue in pure and mixed stands during regrowth

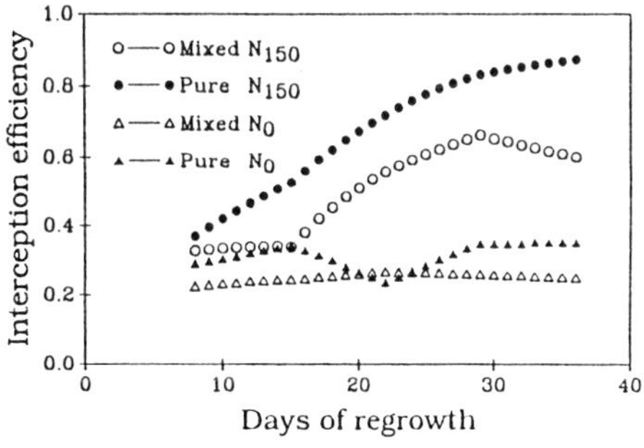

Figure 2 PAR interception efficiencies of tall fescue in pure and mixed stands during regrowth

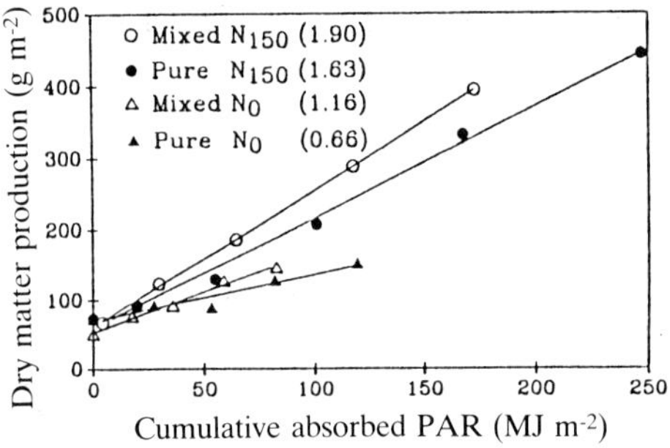

Figure 3 Aerial dry matter production of tall fescue in pure and mixed stands as a function of cumulative absorbed PAR

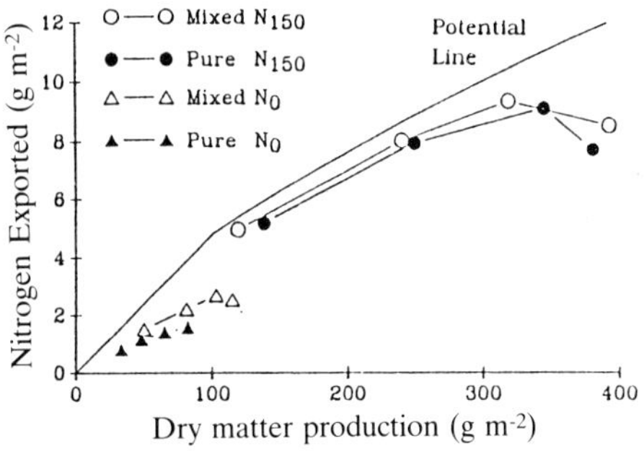

Figure 4 Nitrogen export of tall fescue in pure and mixed stands as a function of aerial dry matter production

Conclusion and Perspectives

From an agronomical viewpoint, intercropping appeared to be advantageous because of additional legume production and better nitrogen content of mixed grass, especially in case of low fertilizer supply. Presently this approach is applied to other mixed canopies: annual species, grass–shrubs mixtures, tropical herbage mixtures. The next stage will be the inclusion of water capture, by using an energy balance model for each species.

References

Cruz, P. and Lemaire, G. (1986) Analyse des relations de compétition dans une association de luzerne (*Medicago sativa L.*) et de dactyle (*Dactylis glomerata L.*) I. Effets sur la nutrition azotée des deux espèces. *Agronomie*, **6**, 735–742

Cruz, P., Moulia, B. and Sinquet H. (1989) Extension d'une méthode d'analyse de la croissance d'un couvert végétal aux cultures associées: bases théoriques. *Proceedings of the 25th CFCS Meeting*, pp 268–280. Ed. INRA Antilles-Guyane

Lemaire, G. and Salette, J. (1984) Relation entre dynamique de croissance et dynamique de prélèvement d'azote pour un peuplement de graminées fourragères. I. Etude de l'effet du milieu. *Agronomie*, **4**, 423–430

Sinoquet, H., Moulia, B., Gastal, F., Bonhomme, R. and Varlet-Grancher, C. (1990) Modeling the radiative balance of the components of a well-mixed canopy: application to a white clover – tall fescue mixture. *Acta Oecologica*, **11**, 469–486

INCREASING RESOURCES AVAILABLE TO CONIFER SEEDLINGS

D.L. SPITTLEHOUSE
B.C. Ministry of Forests, Victoria, B.C., V8W 3E7, Canada

There are major biological constraints to survival and growth of white spruce seedlings in the interior of British Columbia. Problems are caused by low soil temperature, excessive soil moisture, poor soil aeration, low light due to a dense vegetation cover, and vegetation and snow press. The ability of a mounding treatment to improve the microclimate and, consequently, the survival and growth of white spruce seedlings, was investigated at a site 75 km ESE of Prince George, B.C. (53° 40' N, 123° 40' W). Soils are gleyed, grey luvisols, 20% coarse fragments, on a 15% NE-facing slope, 1100 m elevation. The competing vegetation consists of a 1–2 m tall, dense cover of ladyfern, indian hellebore, thimbleberry, devil's club and alder.

Mounds consist of 0.2 m of soil over an inverted humus layer, a layer of 'undisturbed' humus and undisturbed mineral soil. The mounds created an area clear of vegetation about 0.6 m in diameter, thus tripling the amount of sunlight reaching the seedlings and the soil surface. Mounding improved drainage and soil aeration. The reduction in shading, removal of the insulating organic layer and a drier soil improved soil warming, e.g., doubled the growing degrees (5°C base) at 0.1 m. Seedlings planted in mounds quickly produced new roots. After 5 years, height, biomass and nutrient content of seedlings on mounds were two to three times those of the untreated seedlings. Growth form was much improved on the mounds.

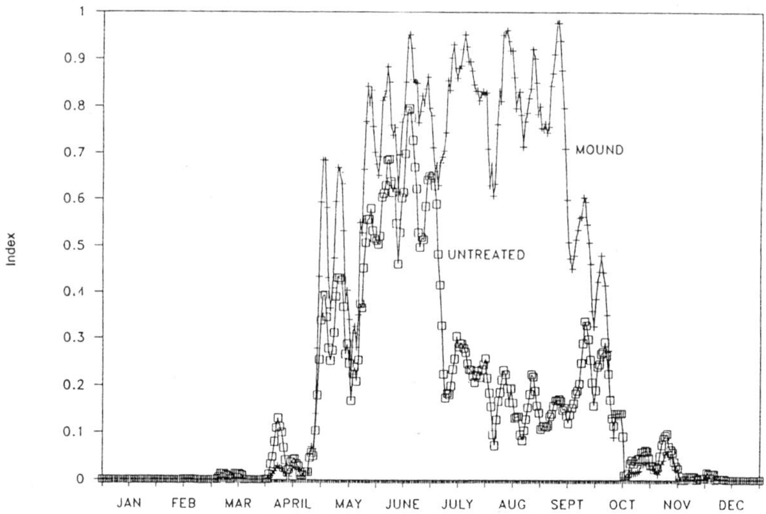

Figure 1. Five day running average of the photosynthetic index for spruce seedlings on a mound (+) and in an untreated area (□). Air and soil temperatures regulate the index for October through to May. Light and vapour pressure deficit are occasionally limiting from June through September. Soil moisture is not limiting. The lower index for the untreated area is due to a low soil temperature (May and June), and low light caused by the growth of the competing vegetation (July to September)

Effect of the microenvironment on photosynthesis

An index was developed to assess the photosynthetic potential of the microenvironment. It is calculated for individual weather variables, and these individual indices are then combined to give the photosynthetic index. The index varies from 0 (no potential for photosynthesis for the day) to a value of 1 for optimum conditions.

The index is calculated daily from photosynthetically active radiation (L), maximum vapour pressure deficit (V), average (A) and minimum (N) air temperature at 0.25 m, average soil temperature at 0.1 m (S), and average 0.1–0.3 m soil water potential (M). The combined index is calculated as the lesser of the individual L, V, A and M indices multiplied by the S and the N indices. Optimum conditions (index = 1) are L >30 mol m^{-2} d^{-1}, V <1.6 kPa, 14 <A < 21°C, N >0°C, S >10°C, and M >-0.2 MPa. The functions used are based on data collected in this project (Draper *et al.*, 1988; Spittlehouse and Stathers, 1990) and in the literature (e.g., DeLucia and Smith, 1987). The photosynthetic index for the mounds and an untreated area are shown in *Figure 1*.

References

DeLucia E.H. and Smith, W.K. (1987) Air and soil temperature limitations on photosynthesis in Engelmann spruce during summer. *Canadian Journal of Forest Research*, **17**, 527–533

Draper, D.A, Spittlehouse, D.L., Binder, W.D. and Letchford, T. (1988) Field measurement of photosynthetically active radiation. In *Proceedings, Combined Meeting of the Western Forest Nursery Associations*, pp 212–218 Edited by T.D. Landis, General Tech Report RM-167, U.S.D.A. Forest Service, Rocky Mountain Forest Range Experiment Station, Ft Collins, CO

Spittlehouse, D.L. and Stathers, R.J. (1990) *Seedling Microclimate*. Land Management Report 65, B.C. Ministry of Forests, Victoria, B.C.

MODELLING LIGHT CAPTURE BY SUNFLOWER CANOPIES

M.D. STEVEN, C.S.T. DAUGHTRY* and F. BARET†

*Department of Geography, University of Nottingham, Nottingham NG7 2RD, UK; *USDA Remote Sensing Lab, Beltsville, Md 20705, USA; †INRA Bioclimatologie, B.P.91, F84140 Montfavet, France*

Introduction

In the classical analysis of radiation interception by canopies of black leaves (Monteith and Unsworth, 1990), extinction coefficients can be determined from the ratio of the shadow area cast by leaves to the actual area of those leaves. With the assumption that the leaves are randomly distributed in space and that the leaf orientation is random in azimuth, the leaves can be reassembled to form simple shapes for which the shadow area is easily determined. For example, leaves at a fixed inclination, can be modelled by a vertical cone. However, many real canopies exhibit preferred orientations, and in some such as sunflower, leaves may move diurnally to maximize interception of light from the sun.

Modelling

In this paper, the classical analysis is extended to cones and other axially symmetric shapes whose axes are not vertical. To compute the interception by a tilted cone, the following steps are required:

1. calculate solar angles relative to the base of the tilted cone and compute the shadow area on this plane;
2. project onto the plane perpendicular to the solar beam and then back to the horizontal plane.

This analysis preserves the advantages of symmetry but allows a range of leaf angle distributions with non-random azimuths.

Results

Computations were made for conical leaf angle distributions with representative leaf inclinations. The cone axes were tilted in various fixed directions or made to track the solar azimuth with a fixed degree of tilt. The computed diurnal variations of solar beam transmission exhibit interesting asymmetric features (*Figure 1*). The asymmetries are greater for small leaf inclinations. Conical distributions are a useful tool for exploring the nature of these asymmetries, but the analysis can equally be applied to more realistic distributions of leaf angle such as the ellipsoidal distribution proposed by Campbell (1986).

Previous published work (Shell and Lang, 1976) on the sun-tracking action of sunflower canopies indicates that the orientation of leaves may lag 1–2 h behind the actual solar position. As shown in *Figure 2*, the computed effect of this lag on the light budget is less than 2%, indicating that there is little to gain from a more efficient sun-tracking strategy. A separate program was written to account for the interception of diffuse solar irradiance, approximated by 48 point sources. When diffuse transmission is included the transmittances

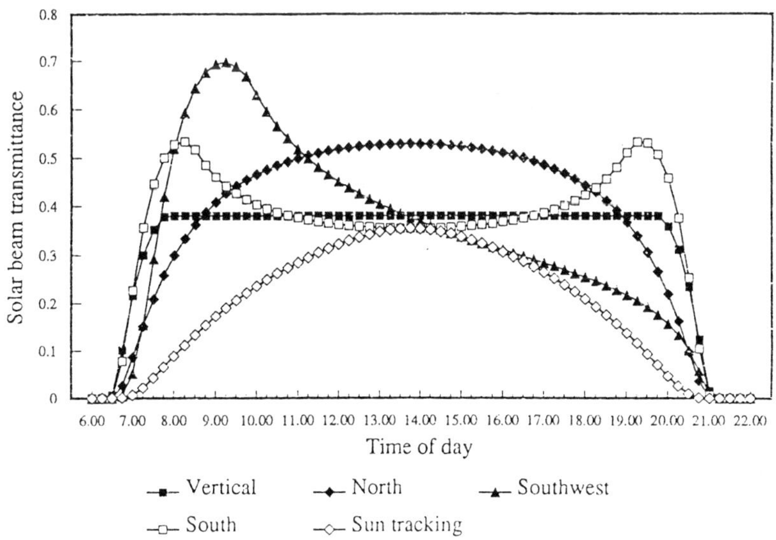

Figure 1 Diurnal variation of simulated transmittance by conical leaf angle distributions with leaf inclination 15°, tilted at 30°. LAI = 1

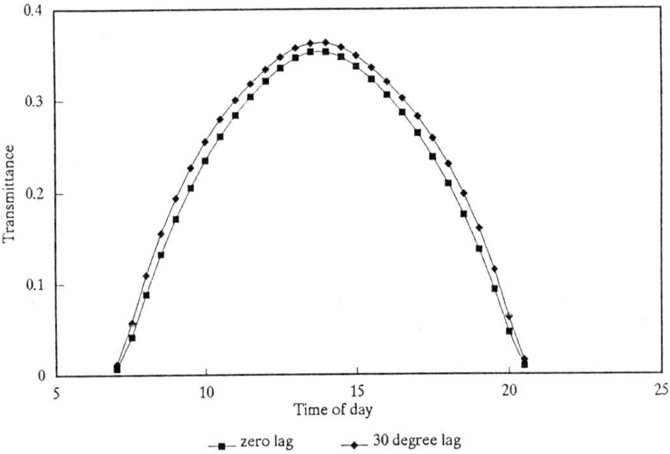

Figure 2 Modelling of PAR transmittance with an efficient sun tracking strategy compared with tracking with a 30° lag in azimuth; for LAI = 1, 15° conical distribution and 30° tilt

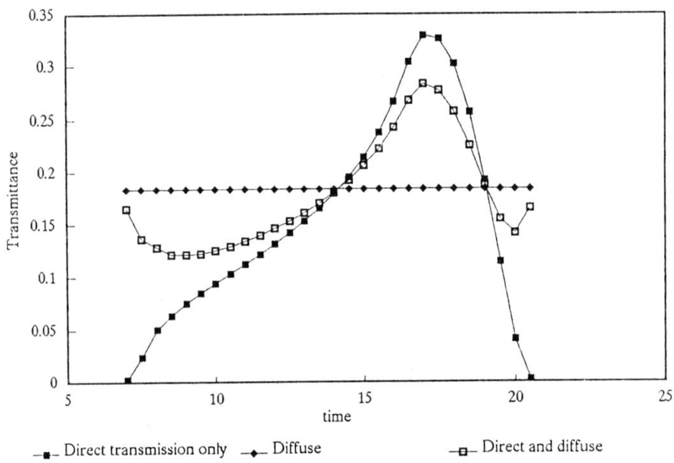

Figure 3 Comparison of direct beam, diffuse and combined PAR transmittance for a 30° conical distribution tilted 30° to the southeast with LAI = 2

rise near sunrise and sunset (*Figure 3*). These features are compatible with the pattern seen in measurements on sunflowers.

Conclusions

Light capture by many asymmetric canopies can be modelled using simple symmetric leaf angle distributions tilted from the vertical axis. Data from sunflower canopies are at least qualitatively consistent with a sun-tracking mechanism that lags somewhat behind the actual solar position. Work is continuing to fit the model parameters to canopy data and compute the effects on the light budgets of canopies. A similar modelling approach could be applied to studies of the light budget of symmetric vegetation canopies growing on slopes.

References

Campbell, G.S. (1986) Extinction coefficients for radiation in plant canopies calculated using an ellipsoidal inclination angle distribution. *Agricultural and Forest Meteorology*, **36**, 317–321

Monteith, J.L. and Unsworth, M.H. (1990) *Principles of Environmental Physics*, 2nd Edn. London: Edward Arnold

Shell, G.S.G. and Lang, A.R.G. (1976) Movements of sunflower leaves over a 24-h period. *Agricultural Meteorology*, **16**, 161–170

EFFECT OF GROWTH TEMPERATURE ON THE PHOTOSYNTHETIC PERFORMANCE OF IMMATURE MAIZE CANOPIES

C.M. STIRLING, V.H. RODRIGO* and P.J. THORNE†
*Department of Biology, University of Essex, UK; *Rubber Research Institute of Sri Lanka, Agalawatta, Sri Lanka; †Natural Resource Institute, Chatham Maritime, Kent, UK*

The susceptibility of photosynthesis to chilling temperatures poses a major limitation to early growth of maize in temperate climates (Baker, Long and Ort, 1988; Stirling *et al.*, 1991). This study evaluates the effects of low temperature on the parameters of the light response curve for CO_2 assimilation and the relative importance of these parameters to canopy carbon gain in immature stands. To impose a range of temperatures during early canopy development, caryopses of maize (*Zea mays* cv. LG11) were sown on two dates, 1 May and 28 June 1991. During the period of measurement, the early-sown crop experienced temperatures below 10°C on 33 occasions compared with only one occasion in the late-sown treatment.

Light response curves for CO_2 assimilation were fitted using the quadratic equation:

$$A = 1/2\theta\{\phi Q + A_{sat} - [(\phi Q + A_{sat})^2 - 4\phi Q A_{sat}\theta]^{0.5}\} \quad (1)$$

where Q is the incident photon flux density, A is the rate of CO_2 assimilation, ϕ, θ and A_{sat} are the initial slope, convexity and upper asymptote of the light response curve.

Low temperatures were associated with a marked reduction in CO_2 assimilation rates at all light levels in the early-sown treatment (*Table 1*). While estimated values for ϕ and A_{sat} remained depressed until 28 June, approximately 3 weeks after chilling conditions had ceased, θ had recovered to a value similar to that observed at warmer temperatures in the late-sown treatment (*Table 1*).

A mathematical model was developed to assess the relative sensitivity of canopy carbon gain to fractional reductions in the parameters, ϕ, θ and A_{sat} of the light response curve. *Figure 1* shows diurnal changes in the rate of CO_2 assimilation for an immature maize canopy at the 5th leaf stage and with a leaf area index of 0.3. The canopy is divided into three leaf inclination classes of 15°, 45° and 75°, comprising 17%, 56% and 27% of the total leaf area respectively (Ross, 1981). Simulations are for clear conditions on 21 June

Table 1 MODELLED PARAMETERS OF THE PHOTOSYNTHETIC LIGHT-RESPONSE CURVE FITTED TO DATA FROM THE EARLY- (ES) AND LATE-SOWN (LS) TREATMENTS USING EQUATION 1. A_{sat} ϕ AND θ ARE THE UPPER ASYMPTOTE, INITIAL SLOPE AND CONVEXITY OF THE LIGHT RESPONSE CURVE RESPECTIVELY, AND RSD REPRESENTS THE RESIDUAL STANDARD DEVIATION FOR FITTED CURVES

Date	*Treatment*	A_{sat}	ϕ	θ	*RSD*
5 June	ES	8.28	0.008	0.749	0.757
14 June	ES	17.35	0.016	0.754	0.770
19 June	ES	19.07	0.024	0.542	0.566
28 June	ES	19.51	0.018	0.955	0.973
3 August	ES	37.01	0.029	0.988	1.000
19 July	LS	25.39	0.046	0.970	0.928
22 July	LS	33.50	0.031	0.937	0.968

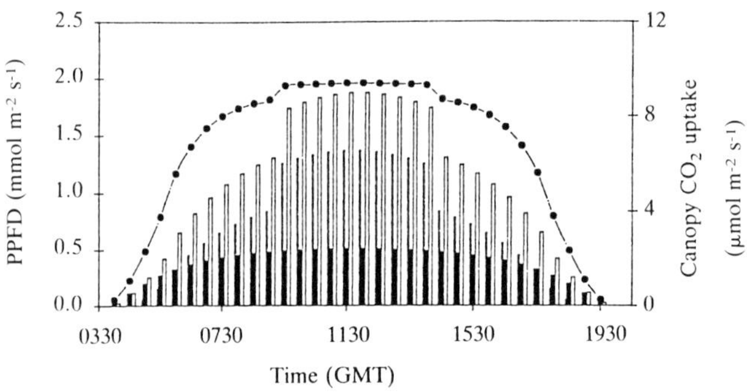

Figure 1 Simulated hourly incident total (□), direct (▧) and diffuse (■) radiation and canopy CO_2 uptake per unit ground area (●)

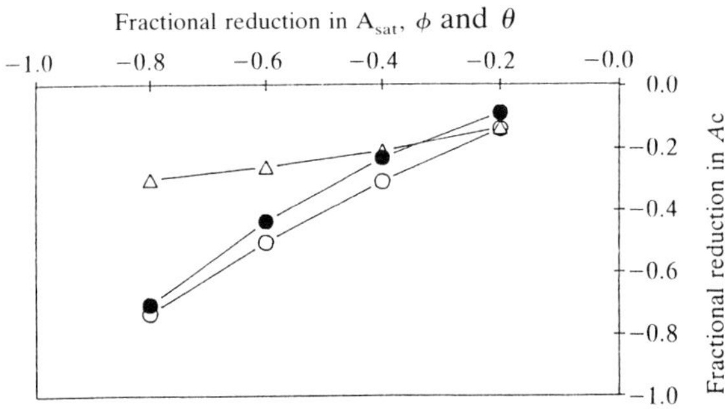

Figure 2 Sensitivity of daily canopy CO_2 uptake per unit ground area (A_c) to fractional reductions in the initial slope (●), convexity (△) and upper asymptote (○) of the light response for CO_2 assimilation

at a latitude of 52°N and for unstressed leaves where A_{sat} = 33.5 μmol m^{-2} s^{-1}, ϕ = 0.031 and θ = 0.937 (see *Table 1*).

Simulations suggest that under clear conditions, when chilling temperatures are most prevalent, canopy carbon gain of immature stands is highly sensitive to reductions in both

ϕ and A_{sat}, and least sensitive to changes in the convexity of the light response curve (*Figure 2*).

References

Baker, N.R., Long, S.P. and Ort, D.R. (1988) Photosynthesis and temperature with particular reference to effects on quantum yield. *Plants and Temperature*, pp 347–375 Edited by S.P. Long, and F.I. Woodward. Cambridge: Company of Biologists

Ross, J. (1981) *The Radiation Regime and Architecture of Plant Stands*. The Hague: Dr. J.W. Publishers

Stirling, C.M., Nie, G.Y., Aguilera, C., Nugawela, A., Long, S.P. and Baker, N.R. (1991) Photosynthetic productivity of an immature maize crop: changes in quantum yield of CO_2 assimilation, conversion efficiency and phylakoid proteins. *Plant Cell and Environment*, **14**, 947–954

CONCLUDING REMARKS

J. ELSTON
Universities of Leeds and Reading, UK

I want to start by thanking, on all our behalf if I may, the organizers and all their assistants for the admirable organization of this meeting. They and all those who have presented papers and posters have helped make this a pleasant occasion and a most stimulating and useful meeting.

This cannot be a summary or a conclusion; at the moment resource capture is not summarizable and is certainly not concluded. We have instead a patchwork. Some work on mechanisms and consequences seems reasonably complete while other parts of resource capture are not understood. Resource capture is a concrete part of the ecological process of competition. Monteith (pp 1–15) sets out clearly the nature of the resources for both external and internal competition. We have to consider the nature and consequence of the processes and the possibilities of successful manipulation that follow understanding. The processes occur in complex systems of environments and plants which are not fully open to precise experiment.

Resource use by natural vegetation

The processes of competition and selection *should* tend to optimize resource capture in natural vegetation. Any ineffective use of one resource by one genotype implies an opportunity for another genotype to exploit that environmental resource more fully and gain a competitive advantage. Gradually that resource will be used more and more effectively as evolutionary change proceeds. However plants require many environmental resources and respond to other environmental factors. Maximizing the exploitation of one, such as light, may affect deleteriously the use of another, such as water. Adaptation to the variations in resources, in space and time, may imply that precise requirements are not advantageous.

Certainly plant succession leads to a relatively stable natural vegetation in which the equilibrium between plants, animals and environment implies a relatively effective use of environmental resources. Ecologists have thought about the 'strategies' that have evolved in plants to allow a fuller exploitation of particular environments. Grime (1974), in an admirable paper, organized this approach with his C-S-R model. In his scheme, environments are polarized for high and low levels of stress (here, factors which restrict production) and for disturbance (here, activities which damage). The resulting 2 × 2 table defines three sorts of extreme habitats: low stress and low disturbance occupied by 'competitors', high stress, low disturbance occupied by 'stress' tolerators' and low stress and high disturbance occupied by 'ruderals'. These three extremes are abbreviated to C, S and R respectively. There is no type for the fourth cell, high stress and high disturbance, which is unoccupied. Grime is well aware of the range of difficulties of this model (Grime, Hodgson and Hunt, 1988) from the use of the word 'strategy' to the use of stereotypes. Both Grime and Harper (see, for example, Harper, 1977) have worked to make the ecological schemes about competition more rigorous. How far can we now extend their analyses, using the ideas and information recorded here? How far can we use the C-S-R scheme to explain what happens in fields, forests and glasshouses?

We now understand adequately and can model precisely the interception of radiation by complete canopies (Goudriaan, pp 99–110). We are beginning to extend the models to mixed canopies and to isolated plants. Campbell and van Evert (pp 35–52) have invited us to consider the costs of producing new leaf by showing the benefit, in intercepted irradiance, of new leaf. The approach is attractive even if the details are arguable. Should a ruderal have a lower breakeven point than a stress tolerator? Does the experimental detail exist to make the calculations now? Jarvis (pp 351–374) has given an excellent discussion of measurements in a coniferous forest, an extreme stress tolerator.

The production and partition of new plant mass

Whole-plant physiologists have used the idea of internal competition, especially for assimilate and for N, to analyse the changes in plants which follow developmental changes and the production of new mass, though the concept has been difficult to quantify. We have some ideas about the mechanisms for allocation, driven by the plant growth regulators. Long (pp 17–34) and Jones and Lynn (pp 111–123) have told us about the production of photosynthate by single leaves and discussed the partition of that assimilate. Does it matter that our understanding of partition is so poor? After all plant breeders have manipulated partition, as shown by the harvest index, most successfully. Hamblin (pp 211–232) has stressed, correctly and emphatically, the importance of development but we can often model that precisely using accumulated temperatures (which is one of the ways in which plants count time), or daylength or change in daylength. How much of a scientific base do we need, to be sure that our descriptions are reasonable? Will expolinear descriptions of organ growth be sufficient? Surely not.

We have a danger of slack phrasing which holds the possibility of slack thinking. We should not use teleonomy because plants don't reason nor do they appear to reason. Instead they are made up of systems with inputs, outputs and mechanism and feedback controls. Let us therefore not think of a plant as a rational economic being which has to choose how to distribute new mass between several possible sinks. It grows in a more-or-less uncertain environment and so those allocations will be tempered by the amount of risk and will have a time frame determined by the environment and the developmental pattern of the plant. The plant has some investment appraisal system – in Darwinian life it is in the genetic background of the fittest that survive. The new mass is distributed in some particular way and the opportunities of other possible distributions are foregone. Choices are made, but not by plants but between plants.

Within a plant, some resources may be in surplus and some in deficit. Systems tend to work to bring balance but there may be occasions when there are economies of waste for some products and economies of stress for others. Plants and crops don't have problems of optimization – we have problems of understanding how they operate and are adapted to particular environments. A flexibility of responses to environmental variation may be important and may explain some difficulties in the analysis of experiments.

It is clear that we have neglected the study of roots and root systems – Ong and Black (pp 255–278) and Scott, Jaggard and Sylvester-Bradley (pp 279–302) have clearly shown the importance of water and nitrogen uptake in different cropping systems but Robinson (pp 53–76) and Gregory (pp 77–97) have shown how poorly placed we are. Clearly we should distinguish between nutrients such as, K and Ca which move by mass flow in solution and, for example, P where more complex uptake mechanisms exist. Are there any general sets of measurements or of techniques that we should recommend for the measurement and analysis of root systems?

Natural and artificial selection

The production of new variation or new combinations of variation competition, relative reproductive success and survival provide the mechanisms that move plants towards full resource use. Plants are complex systems but even they cannot use all environmental resources fully. However natural selection must be a powerful force towards the maximum possible use of resources in climax vegetation. There is clear experimental evidence to show how plants are adapted to exploit differences in soil chemistry and that canopies intercept photosynthetically active radiation effectively.

Selection in cultivated plants by breeders, though impressive, may not necessarily produce systems as effective. First, physiological ideas and information and appropriate assays for selection may not be available. Secondly, the process of selection may be arbitrary, may be on a relatively narrow range of variation and for a relatively short time. Breeders may select for inappropriate opportunities and agronomists, while trying to provide more desirable environments, may not know what is required of them.

We are used to thinking of leaves as compromises; attempts to optimize the fluxes of CO_2 and water to the benefit of a species in one or more sort of environments. We should think of roots in a similar way, as much of an evolutionary compromise as leaves are and as much adapted to their environments. Clearly the root systems of ruderals grow into their resource zones as leaves grow up into canopies.

Models and modelling

We have seen models used in all sorts of ways. Azam-Ali, Crout and Bradley (pp 125–148) used one to organize measurements of efficiencies. Kropff (pp 233–253) and Day (pp 303–314) used models, amongst other things, to aid management decisions. Jarvis (pp 351–374) has used measurements and models to look at coniferous forests as sinks for CO_2 while Parsons (pp 315–349) used models to examine and improve grassland management.

How realistic do models have to be? That depends on their function and on the scale of their approximation. We treat closed canopies as uniform green jellies, rather than structured masses, successfully. The effects of pests (van Emden and Hadley, pp 149–165), diseases (Madeira and Clark, pp 167–188) and pollution (Unsworth and Colls, pp 189–209) seem almost ungeneralizable. How should we simplify now for immediate advantage, in understanding or predicting?

Should we be helping modellers and measurers to a common experience by recommending protocols and appropriate techniques? No measurement on a canopy without one on a root system?

What attention should we give to spatial variation? Both canopies and root systems are spatially complex and soils are heterogeneous. Can we ignore spatial variation in canopies? Surely we have to account for it in root systems.

As I age, I tend to confuse conventional ideas with the truth. However the progress of experimental science is sometimes closer to a zigzag course than to a series of successive approximating in which each approximation is an improvement. Thomas (pp 375–385) offers an unconventional view, resource rejection rather than resource use and allocation. Is it closer to the truth?

References

Grime, J.P. (1974) Vegetation classification by reference to strategies. *Nature (London)* **250**, 26–31

Grime, J.P., Hodgson, J.G. and Hunt, R. (1988) *Comparative Plant Ecology*. London: Unwin Hyman, 742 pp

Harper, J.L. (1977) *Population Biology of Plants*. London: Academic Press, 892 pp

LIST OF PARTICIPANTS

Abdulwakil, I-H.	University of Nottingham School of Agriculture, Sutton Bonington, Loughborough, Leics. LE12 5RD, UK
Abreu, F.	Instituto Superior de Agronomia, Dept Ciencias do Ambiente, Agrometeorologia, Tapada da Ajuda, 1399 Lisboa Codex, Portugal
Abreu, J.	Instituto Superior de Agronomia, Dept Ciencias do Ambiente, Agrometeorologia, Tapada da Ajuda, 1399 Lisboa Codex, Portugal
Acevedo, E.	CIMMYT, Apartado Postal 6-641, 06600 Mexico DF, Mexico
Adjei, P.	University of Nottingham School of Agriculture, Sutton Bonington, Loughborough, Leics. LE12 5RD, UK
Ahmed, A.	University of Nottingham School of Agriculture, Sutton Bonington, Loughborough, Leics. LE12 5RD, UK
Addo-quaye, A.	Plant Environmental Laboratory, University of Reading, Cutbush Lane, Shinfield, Reading RG2 9AD, UK
Aikman, D.	Horticulture Research International, Worthing Road, Littlehampton, West Sussex BN17 6LP, UK
Alcalde, J.	Crop Physiology Section, University of Reading, 239 Beech Lane, Lower Earley, Reading, Berks. RG6 2UP, UK
Allen, R.	119 Haydn Road, Nottingham, UK
Atherton, J.	University of Nottingham School of Agriculture, Sutton Bonington, Loughborough, Leics. LE12 5RD, UK
Atkinson, L.	MAFF, Room G35 Nobel House, 17 Smith Square, London SW10P 3JR, UK
Azam-Ali, S.	University of Nottingham School of Agriculture, Sutton Bonington, Loughborough, Leics. LE12 5RD, UK

Baggs, E.	University of Nottingham School of Agriculture, Sutton Bonington, Loughborough, Leics. LE12 5RD, UK
Benjamin, L.	Horticulture Research International, Wellesbourne, Warwick CV35 9EF, UK
Bermejo, V.	CIEMAT, Instituto de Medio Ambiente, IMA 3B, Avenida Complutense, 22, 28040 Madrid, Spain
Black, C.	University of Nottingham School of Agriculture, Sutton Bonington, Loughborough, Leics. LE12 5RD, UK
Blair, A.	ADAS, Boxworth Experimental Farm, Boxworth, Cambridge CB3 8NN, UK
Bradley, R.	University of Nottingham School of Agriculture, Sutton Bonington, Loughborough, Leics. LE12 5RD, UK
Burgess, P.	Silsoe College, Silsoe, Bedford MK45 4DT, UK
Campbell, G.	Department of Crop and Soil Sciences, Washington State University, Pullman, WA 99164 6420, USA
Cannell, M.	Institute of Terrestrial Ecology, Edinburgh Research Station, Bush Estate, Penicuik, Midlothian EH26 0QB, UK
Cantero-Martinez, C.	UPC – IRTA, Alcalde Rovira Roure, 177, 25006 – Lleida, Spain
Chigumira, F.	University of Nottingham School of Agriculture, Sutton Bonington, Loughborough, Leics. LE12 5RD, UK
Clark, J.	University of Nottingham School of Agriculture, Sutton Bonington, Loughborough, Leics. LE12 5RD, UK
Claydon, K.	University of Nottingham School of Agriculture, Sutton Bonington, Loughborough, Leics. LE12 5RD, UK
Clifford, S.	University of Nottingham School of Agriculture, Sutton Bonington, Loughborough, Leics. LE12 5RD, UK
Cockshull, K.	Horticulture Research International, Worthing Road, Littlehampton, West Sussex BN17 6LP, UK
Collins, R.	Welsh Plant Breeding Station, Plas Gogerddan, Aberystwyth, Dyfed SY23 3EB, UK
Colls, J.	University of Nottingham School of Agriculture, Sutton Bonington, Loughborough, Leics. LE12 5RD, UK
Cook, J.	c/o Crop Science Dept, Harper Adams Agricultural College, Newport, Shropshire TF10 8NB, UK
Corlett, J.	Horticulture Research International, Wellesbourne, Warwick CV35 9EF, UK
Corrie, J	University of Nottingham School of Agriculture, Sutton Bonington, Loughborough, Leics. LE12 5RD, UK

Costa, L.	Department of Agricultural Botany, Plant Science Laboratories, University of Reading, Whiteknights, PO Box 221, Reading RG6 2AS, UK
Cozar, S.E.	CIEMAT, Instituto de Medio Ambiente, IMA 3B, Avenida Complutense, 22, 28040 Madrid, Spain
Craigon, J.	University of Nottingham School of Agriculture, Sutton Bonington, Loughborough, Leics. LE12 5RD, UK
Crout, N.	University of Nottingham School of Agriculture, Sutton Bonington, Loughborough, Leics. LE12 5RD, UK
Daughtry, C.	USDA/ARS, Remote Sensing Research Laboratory, 10300 Baltimore Avenue, Building 001 Room 334, Barc-West, Beltsville, Maryland 20705-2350, USA
Day, B.	Silsoe Research Institute, Wrest Park, Silsoe, Bedford, UK
de Costa, J.	Department of Agricultural Botany, Plant Science Laboratories, University of Reading, Whiteknights, PO Box 221, Reading RG6 2AS, UK
de Lucas, C.	Department of Agricultural Botany, University of Reading, Earley Gate, Whiteknights Road, Reading RG6 2AS, UK
de Ruiter, J.	DSRI Crop Research, Private Bag, Palmerston North, New Zealand
de Temmerman, L.	Institute for Chemical Research, Leuvensesteenweg 17, B-3080 Tervuren, Belgium
del Pozo, A.	Department of Agricultural Botany, Plant Science Laboratories, University of Reading, Whiteknights, PO Box 221, Reading RG6 2AS, UK
Dunham, R.	AFRC Institute of Arable Crops Research, Broom's Barn Experimental Station, Higham, Bury St Edmunds, Suffolk IP28 6NP, UK
Eastwood, J.	Department of Geography, University of Nottingham, UK
Edmeades, G.	Plant Environmental Laboratory, University of Reading, Cutbush Lane, Shinfield, Reading, UK
Elston, J.	Pound Cottage, Horsington, Near Templecomb, Somerset BA8 0EF, UK
Fara, M.	University of Nottingham School of Agriculture, Sutton Bonington, Loughborough, Leics. LE12 5RD, UK
Firman, D.	Cambridge University Farm, Huntingdon Road, Cambridge CB3 0LH, UK
Fothergill, M.	AFRC – IGER, Plas Gogerddan, Aberystwyth, Dyfed SY23 3EB, UK
Foulkes, J.	University of Nottingham School of Agriculture, Sutton Bonington, Loughborough, Leics. LE12 5RD, UK

Frost, D.	Department of Agriculture, University of Newcastle, King George VI Building, Newcastle-upon-Tyne NE1 7RU, UK
Fuhrer, F.	Swiss Federal Research Station for Agricultural Chemistry and Environmental Hygiene, CH-3097 Liebfeld-Bern, Switzerland
Gharres, S.	University of Nottingham School of Agriculture, Sutton Bonington, Loughborough, Leics. LE12 5RD, UK
Gilbert, I.	Department of Botany, University of Leicester, University Road, Leicester LE1 7RH, UK
Giorio, P.	ISPAIM – CNR, Naples, Italy
Goded, M. U.	Instuto Pirenaicoe Ecologia (CSIC), Avda Regimiento Galicia s/n, Apartado 64, 22700 – JACA (Huesca), Spain
Goudriann, J.	Agricultural University, Wageningen, The Netherlands
Gregory, P.	CSIRO, Dryland Crops & Soils Research Unit, Private Bag, P O Wembley, Western Australia 6014, Australia
Grema, A.	Department Agricultural Water Management, Silsoe College, Bedford MK45 4DT, UK
Grime, V.	University College of North Wales, c/o Department of Soil Science, University of Reading, London Road, Reading RG2 0DH, UK
Grime, J. P.	Unit of Comparative Ecology, Sheffield, UK
Grindlay, D.	University of Nottingham School of Agriculture, Sutton Bonington, Loughborough, Leics. LE12 5RD, UK
Gubbins, M.	University of Nottingham School of Agriculture, Sutton Bonington, Loughborough, Leics. LE12 5RD, UK
Habash, D.	1 Barr Close, Wivenhoe, Colchester, Essex CO7 9RG, UK
Hamblin, J.	ICARDA, PO Box 5466, Aleppo, Syria
Hand, D.	Horticulture Research International, Worthing Road, Littlehampton, West Sussex BN17 6LP, UK
Harbour, A.	Neoplants Ltd, 197 Kirkham Road, Freckleton, Lancs, UK
Hardwick, T.	CAB International, Wallingford, Oxon OX10 8DE, UK
Hargreaves, N.	Delta-T Devices, 128 Low Road, Burwell, Cambridge CB5 0EJ, UK
Haydock, P.	Crop & Environment Research Centre, Harper Adams Agricultural College, Newport, Shropshire TF10 8NB, UK
Hogge, M.	Cambridge University Farm, Huntingdon Road, Cambridge CB3 0LH, UK
Huxley, P.	ICRAF, PO Box 30677, Nairobi, Kenya

Jaggard, K.	AFRC Institute of Arable Crops Research, Broom's Barn Experimental Station, Higham, Bury St Edmunds, Suffolk IP28 6NP, UK
Jarvis, P.	University of Edinburgh, Edinburgh, UK
Johnson, J.	Department of Forestry, University of Florida, Institute of Food & Agricultural Sciences, 118 Newins-Ziegler Hall, 0303 IFAS, Gainesville, Florida 32611-0303, USA
Jones, H.	Horticulture Research International, Wellesbourne, Warwick CV35 9EF, UK
Kantar, F.	University of Nottingham School of Agriculture, Sutton Bonington, Loughborough, Leics. LE12 5RD, UK
Karvonen, K.	Department of Crop Production, University of Helsinki, 00710 Helsinki, Finland
Kawimbe, C.	University of Nottingham School of Agriculture, Sutton Bonington, Loughborough, Leics. LE12 5RD, UK
Kerkham, G.	University of Nottingham School of Agriculture, Sutton Bonington, Loughborough, Leics. LE12 5RD, UK
Kerr, P.	University of Nottingham School of Agriculture, Sutton Bonington, Loughborough, Leics. LE12 5RD, UK
Kerstiens, G.	Division of Biological Sciences, University of Lancaster, Lancaster LA1 4YQ, UK
Kettlewell, P.	Harper Adams Agricultural College, Newport, Shropshire TF10 8NB, UK
Kleemola, J.	Department of Plant Productivity, Helsinki University, Viikki, C-House, 00710 Helsinki, Finland
Kocabas, Z.	University of Nottingham School of Agriculture, Sutton Bonington, Loughborough, Leics. LE12 5RD, UK
Kobayashi, K.	National Institute of Agro-Environmental Science, 3-1-1 Kannondai, Tsukuba, Ibavaki 305, Japan
Kropff, M.	IRRI, PO Box 933, 1099 Manila, The Philippines
Lenssen, G.	Department of Ecology & Ecotoxicology, De Boelelaan 1087, 1081 HV Amsterdam, The Netherlands
Leuning, R.	CSIRO, Centre Environental Mechanics, GPO Box 821, Canberra, ACT 2601, Australia
Leverenz, J.	Department of Plant Physiology, University of Umeå, S-90187 Umeå, Sweden
Lewis, J.	c/o Crop Science Department, Harpers Adams Agricultural College, Newport, Shropshire TF10 8NB, UK
Livingstone, N.	Department of Biology, University of Victoria, Victoria BC, Canada
Long, S.	University of Essex, Department of Biology, Colchester CO4 3SQ, UK

Mabika, V.	University of Nottingham School of Agriculture, Sutton Bonington, Loughborough, Leics. LE12 5RD, UK
McCartney, A.	Rothamsted Experimental Station, Harpenden, Herts AL5 2JQ, UK
McWilliam, S.	University of Nottingham School of Agriculture, Sutton Bonington, Loughborough, Leics. LE12 5RD, UK
Madeira, A. C. de	Agrometeorologia, Instituto Superior de Agronomia, Tapada da Ajuda, 1300 Lisboa, Portugal
Marshall, F.	University of Nottingham School of Agriculture, Sutton Bonington, Loughborough, Leics. LE12 5RD, UK
Maryah, U.	Silsoe College, Bedford MK45 4DT, UK
Mburu, M.	Department of Soil Science, University of Reading, London Road, Reading RG1 5AG, UK
Mebrat, T.	University of Nottingham School of Agriculture, Sutton Bonington, Loughborough, Leics. LE12 5RD, UK
Meinke, H.	QDPI/CSIRO, PO Box 102, 203 Tor Street, Toowoomba, Queensland 4350, Australia
Mgeni, I.	University of Nottingham School of Agriculture, Sutton Bonington, Loughborough, Leics. LE12 5RD, UK
Monaghan, J.	c/o Crop Science Department, Harper Adams Agricultural College, Newport, Shropshire TF10 8NB, UK
Monteith, J.	21 St Albans Road, Edinburgh EH9 2LT, UK
Morrison, M.	Agriculture Canada, Plant Research Centre, Central Experimental Farm, Building 75, Ottawa, Ontario K1A 0C6, Canada
Mortensen, E.	National Environmental Research Institute, Department of Emissions and Air Pollution, Frederiksborgvej 399, DK Roskilde, Denmark
Mtikani, L.	University of Nottingham School of Agriculture, Sutton Bonington, Loughborough, Leics. LE12 5RD, UK
Mueller, P.	Imperial College of Science, Technology & Medicine, Department of Biology, Silwood Park, Sunninghill, Ascot, Berkshire SL5 7PY, UK
Mukelabai, S.	University of Nottingham School of Agriculture, Sutton Bonington, Loughborough, Leics. LE12 5RD, UK
Mulholland, B.	University of Nottingham School of Agriculture, Sutton Bonington, Loughborough, Leics. LE12 5RD, UK
Newman, R.	ADC, Pindar Road, Hoddesdon, Herts EN11 0AQ, UK
Ng'etich, W.	Silsoe College, Silsoe, Bedford MK45 4DT, UK

Nilsson, H-E.	Department of Plant and Forest Protection, Laboratory of Phytopathometry, Swedish University of Agricultural Science, PO Box 7044, S-75007 Uppsala, Sweden
Norton, G.	University of Nottingham School of Agriculture, Sutton Bonington, Loughborough, Leics. LE12 5RD, UK
Nyakanda, C.	University of Nottingham School of Agriculture, Sutton Bonington, Loughborough, Leics. LE12 5RD, UK
Nyalemegbe, K.	Department of Agricultural Botany, Plant Science Laboratories, University of Reading, Whiteknights, PO Box 221, Reading RG6 2AS, UK
O'Donovan, G.	AFRC – IGER, Plas Gogerddan, Aberystwyth, Dyfed SY23 3EB, UK
Ögren, E.	Department of Plant Physiology, University of Umeå, S-90187 Umeå, Sweden
Ong, C.	ICRAF, ICRAF House, PO Box 30677, Nairobi, Gigiri, Kenya
Ozpinar, H.	University of Nottingham School of Agriculture, Sutton Bonington, Loughborough, Leics. LE12 5RD, UK
Parsons, A.	AFRC – IGER, North Wyke Research Station, Okehampton, Devon EX20 2SB, UK
Phiri, C.	University of Nottingham School of Agriculture, Sutton Bonington, Loughborough, Leics. LE12 5RD, UK
Pihl, G.	IVL, PO Pox 47086, S-40258 Göteborg, Sweden
Place, A.	Cranfield Institute of Technology, Silsoe Campus, Silsoe, Beds, UK
Pleijel, H.	IVL, PO Box 47086, S-40258 Göteborg, Sweden
Potter, E.	Delta-T Devices, 128 Low Road, Burwell, Cambridge CB5 0EJ, UK
Powell, K.	Biological Sciences Department, University of Durham, South Road, Durham DH1 3LE, UK
Pudney, K.	University of Nottingham School of Agriculture, Sutton Bonington, Loughborough, Leics. LE12 5RD, UK
Rhodes, I.	AFRC – IGER, Plas Gogerddan, Dyfed SY23 3EB, UK
Riddell-Black, D.	Medmenham, PO Box 16, 16 Henley Road, Marlow, Bucks SL7 2HD, UK
Roberts, J. A.	University of Nottingham School of Agriculture, Sutton Bonington, Loughborough, Leics. LE12 5RD, UK
Robinson, D.	Cellular and Environmental Physiology Department, Scottish Crop Research Institute, Invergowrie, Dundee DD2 5DA, UK

Rosenfeld, A.	University of Nottingham School of Agriculture, Sutton Bonington, Loughborough, Leics. LE12 5RD, UK
Santiveri, P.	UPC-IRTA, Alcalde Rovira Roure 177, 25006 Lleida, Spain
Sattin, M.	Centro Studio Diserbanti CNR, Via Gradenigo 6, 35131 Padova, Italy
Scott, K.	University of Nottingham School of Agriculture, Sutton Bonington, Loughborough, Leics. LE12 5RD, UK
Shibles, R.	Department of Agronomy, IOWA State University, Ames, IA 50011-1010, USA
Spittlehouse, D.	British Columbia Ministry of Forests, Research Branch, 31 Bastion Square, Victoria BC, V8W 3E7, Canada
Stafford, J.	University of Nottingham School of Agriculture, Sutton Bonington, Loughborough, Leics. LE12 5RD, UK
Steven, M.	Department of Geography, University of Nottingham, UK
Stirling, C.	Department of Biology, University of Essex, Wivenhoe Park, Colchester CO4 3SQ, UK
Stronach, I.	University of Nottingham School of Agriculture, Sutton Bonington, Loughborough, Leics. LE12 5RD, UK
Sylvester-Bradley, R.	Soil Science Department, ADAS, Block C Government Buildings, Brooklands Avenue, Cambridge CB3 7NZ, UK
Taylor, E.	Iowa State University, 2104 Agronomy Hall, Ames IA 50011, USA
Taylor, I.	University of Nottingham School of Agriculture, Sutton Bonington, Loughborough, Leics. LE12 5RD, UK
Terry, A.	University of Sheffield, Sheffield, UK
Thomas, H.	Cell Biology Department, AFRC Institute of Grassland & Environmental Research, Plas Gogerddan, Aberystwyth, Dyfed, SY23 3EB, UK
Thompson, J.	University of Nottingham School of Agriculture, Sutton Bonington, Loughborough, Leics. LE12 5RD, UK
Thornton, B.	Macaulay Land Use Research Institute, Craigiebuckler, Aberdeen AB9 2QJ, UK
Todd, J.	University of Nottingham School of Agriculture, Sutton Bonington, Loughborough, Leics. LE12 5RD, UK
Torquebiau, E.	ICRAF, PO Box 30677, Nairobi, Kenya
Tournebize, R.	INRA des Antilles-Guyane, Station Agropedo-Climatique, BP 1232, F-97184 Pointe a Pitre Cedex, Guadeloupe (French Antilles)

Ullah, J.	University of Nottingham School of Agriculture, Sutton Bonington, Loughborough, Leics. LE12 5RD, UK
Unsworth, M.	University of Nottingham School of Agriculture, Sutton Bonington, Loughborough, Leics. LE12 5RD, UK
Upadhyay, A.	Plant Environment Laboratory, University of Reading, Cutbush Lane, Shinfield, Reading RG2 9AD, UK
Ussi, M.	University of Nottingham School of Agriculture, Sutton Bonington, Loughborough, Leics. LE12 5RD, UK
van Emden, H. F.	University of Reading, Reading, UK
van Oosten, J-J.	Faculté Sciences Agronomiques, Unité d'Enseignement et de Recherche de Biologie Vegétale, 2a Avenue de la Faculté, 5030 Gembloux, Belgium
Visser, A.	Department of Ecology & Ecotoxicology, De Boelelaan 1087, 1081 HV Amsterdam, The Netherlands
Wright, W.	University of Nottingham School of Agriculture, Sutton Bonington, Loughborough, Leics. LE12 5RD, UK
Zhang, P.	University of Nottingham School of Agriculture, Sutton Bonington, Loughborough, Leics. LE12 5RD, UK

INDEX

G

H

I

L

M

N

O

W

Y

Z